Comprehensive Physical Science

THIRD EDITION of Physics with Chemistry

J. V. Portch, M.Sc.,
Senior Curriculum Officer
Gwent Education Committee

and **O. J. Simpson,** B.Sc., F.R.I.C., F.C.S.
Head of Chemistry Department
Wanstead High School, London, E.11

Edward Arnold

© J. V. Portch and O. J. Simpson 1976
First published as *Physics with Chemistry* 1967
by Edward Arnold (Publishers) Ltd.,
25 Hill Street, London W1X 8LL

Second Edition 1970
Reprinted 1971, 1973, 1974

Third Edition 1976

ISBN: 0 7131 0020 6

All Rights Reserved. No part of this publication may be reproduced, stored in a retrieval system, or transmitted in any form or by any means, electronic, mechanical, photocopying, recording or otherwise, without the prior permission of Edward Arnold (Publishers) Ltd.

Other books by O. J. Simpson:
Basic Chemistry (with C. F. Dingle and R. A. Southcott)
Basic Chemical Calculations 3rd edition
Multiple Choice Chemistry
Practice Questions in Chemistry

Unit Studies in Science
The Nature of Substances
The Chemistry of Air and Water
Chemistry of Non-metals
Chemistry of Metals
The Chemistry of Carbon Compounds

Photoset in Malta by St Paul's Press Ltd
and printed in Singapore

Foreword to the First Edition

The appearance of a new book requires some explanation. The Nuffield Foundation, now turning its attention to work at Advanced Level of the G.C.E., recommends a merger of the common topics of physics and chemistry in one subject, physical science. At the lower ordinary level stage this joint subject is already examined by most Boards, but there are few books which cover both the physics and chemistry in their pages. This book sets out to fill this gap.

In many schools the physics and the chemistry of physical science are taught separately by members of the two departments, each of whom may possibly know little of what is being taught by the other at any particular time. It is of course essential when, say, 'electrolysis' is being studied, that the purely physical aspects such as, for example, the amount of copper deposited by a given current in a given time, should not be divorced from the chemical theory concerning the mechanism of the process. Similar overlapping will be found with the 'gas laws' and other topics.

The Regional Examining Boards set up to administer the Certificate of Secondary Education examinations have mostly included physical science amongst their subjects; in chemistry they have stressed the importance of studying at this level a fair cross-section of organic chemistry, particularly that of substances with a considerable industrial application. The older Boards also have brought this aspect into their revised syllabuses. In physics a number of Boards include a choice of special topics for study in addition to a basic syllabus on G.C.E. lines. It is hoped that this book will be found adequate in providing all the basic material necessary for the C.S.E. examination also and, indeed, the greater part of that needed for the separate subjects, physics and chemistry.

We would like to think that this book will be helpful in schools not only in this country but also in Africa and Asia where facilities for studying the separate sciences are often not available. One of the authors has considerable experience in marking examination papers from abroad and has borne in mind the peculiar difficulties that may arise due to climatic conditions and also to the widely-differing backgrounds of the students. It is intended that this book shall cover the last $2-2\frac{1}{2}$ years of work for the physical science papers of the various examining Boards. No two syllabuses are identical and the teacher may omit certain parts of the book which are not required. It is assumed that pupils will already have spent 2–3 years studying chemistry and physics or general science and topics that are normally covered in these years have had to be omitted from the book (except in revision outline) in order to keep down its size. In chemistry these include the simple processes of solution, crystallization, etc.; symbols; oxygen and elementary combustion; hydrogen and water (excepting hardness); carbon dioxide; chalk; acids and bases; simple experiments bearing on the position of metals in the electrochemical series.

In physics, the subject matter is determined almost entirely by the existing examination syllabuses, but in places the approach and experimental methods have been influenced by the work of the American Physical Sciences Study Committee and the A.S.E. publication 'Physics for Grammar Schools'.

When teaching it is often desirable to pose questions and to leave the students to find the answers. However, in order to make it possible for students to work without a teacher, to make good work missed through absence, or to revise for examinations, a textbook must supply the answers.

The stock system of nomenclature has been introduced and SI units are used throughout.

Almost all the questions and problems have been taken from examination papers set by the different Boards; in the case of the C.S.E. examination some of the problems are from specimen papers and an occasional one from a physics or chemistry paper has been included to give greater variety. Permis-

---

Experiments marked with an asterisk, *, are meant to be carried out by the teacher and not by the pupil as extra care in performing these may be needed.

sion to reproduce these problems has been readily given by the Boards listed and our thanks are due to them.

University of Cambridge Local	*(C)*
London University	*(L)*
Joint Matriculation	*(F)*
Oxford and Cambridge	*(OC)*
Oxford Local	*(O)*
Southern Universities'	*(S)*
East Midland Regional	*(EM)*
Metropolitan Regional	*(MR)*
Middlesex Regional	*(MX)*
North Regional	*(NR)*
North-Western Secondary School	*(NW)*
Southern Regional	*(SR)*
South-East Regional Examinations	*(SE)*
Welsh Regional Examination	*(WR)*
West Midland Regional Examination	*(WM)*

Foreword to the Third Edition

The considerable success which followed the publication of the second edition has led us to bring the book 'up-to-date' by inclusion of new topics which have appeared in recently revised syllabuses. The accounts of the Periodic Table and of atomic structure have been extended and sections devoted to carbon and to the movement of particles introduced. The energy and rates of reaction details have been expanded while deletions include reference to chemical equivalents, normality and some of the more theoretical aspects of various chemical laws. A number of the new style of 'semi-structured' questions have been added where appropriate.

J.V.P. O.J.S.

Contents

	List of Elements	
1	Atomic Structure and the Classification of the Elements	1
2	Measuring Length, Mass and Density	3
3	Time, Velocity and Acceleration	10
4	Forces and Motion	17
5	Chlorine, the Green Gas, and its Compounds; The Halogens	21
6	How Elements Combine	28
7	Molecules and Molecular Properties of Matter	34
8	Pressure in Liquids and Gases	39
9	Chemical Arithmetic	45
10	Equations and What They Tell Us	49
11	Archimedes' Principle and Floating Bodies	53
12	Statics	57
13	Sulphur, the Yellow Solid, and the Sulphides	65
14	The Oxides and Acids of Sulphur	67
15	Machines, Work, Energy and Power	74
16	Expansion of Solids and Liquids. Temperature	85
17	Expansion of Gases	90
18	How Heat gets from One Place to Another	96
19	Heat and Mechanical Energy	102
20	Measuring Quantities of Heat Energy	106
21	Change of State and Latent Heat	109
22	The Air We Breathe	115
23	Some Simple Chemical Laws	123
24	Ammonia and the Oxides of Nitrogen	127
25	Nitric Acid	135
26	Carbon, Carbon Monoxide and Fuels; Sources of Energy	141
27	More Laws of Chemistry; Volume Relationships.	153
28	Light Rays	159
29	Reflection	163
30	Refraction	170
31	Lenses; Optical Instruments and the Eye	175
32	The Spectrum and Colour	183
33	Waves	187
34	Sound	192
35	Solution	200
36	Electric Currents	208
37	The Heating Effect of Electric Currents	216
38	Using Electrical Energy in Chemical Changes	221
39	Chemical Cells	234
40	Standard Solutions; Volumetric Analysis and pH	238
41	The Periodic Table; Order among the Elements	244
42	The Winning of Metals	250
43	Compounds of Metals	258
44	Ferromagnetism	271

45	Electromagnetism and the Motor Effect	276
46	Induced Currents	282
47	Static Electricity	289
48	States of Matter; Structure; Reaction Rates; Equilibrium	295
49	The Electron; Radioactivity	304
50	Small-scale Preparations and Identifications in Chemistry	315
51	Simple Organic Chemistry; Long Chains and Giant Molecules	325

Miscellaneous Questions	335
Answers to Numerical Questions	339
Useful Data	341
Some Common Chemical Formulae and Approximate Atomic Masses	342
International Atomic Masses	343
Logarithms	344
Antilogarithms	346
Index	348

Quantities, Units and Abbreviations

QUANTITY	UNITS	ABBREVIATION
Length	metre	m
	millimetre	mm
	centimetre	cm
	kilometre	km
Mass	kilogram	kg
	gram	g
Time	second	s
	hour	h
Area	square metre	m^2
	square centimetre	cm^2
Volume	cubic metre	m^3
	cubic centimetre	cm^3
	cubic decimetre (litre)	dm^3
Density	kilogram per cubic metre	kg/m^3
	gram per cubic centimetre	g/cm^3
Velocity or Speed	metre per second	m/s
	centimetre per second	cm/s
Acceleration	metre per second squared	m/s^2
	centimetre per second squared	cm/s^2
Force or Weight	newton	N
Work or Energy	joule	J
	kilowatt hour	kW h
Power	watt	W
	kilowatt	kW
	megawatt	MW
Pressure	newton per metre squared	N/m^2
Moment of a force	newton metre	Nm
Temperature	degree Celsius	°C
	Kelvin	K
Specific heat capacity	joule per kilogram degree Celsius	J/kg°C
Specific latent heat	joule per kilogram	J/kg
Frequency	hertz	Hz
Electric current	ampere	A
	milliampere	mA
Charge	coulomb	C
Potential difference	volt	V
Resistance	ohm	Ω
Resistivity	ohm metre	Ωm

List of the Elements

Atomic Number	Name	Symbol	Atomic Mass	Number of Neutrons	Number of Protons
1	HYDROGEN	H	1·008	—	1
2	HELIUM	He	4·00	2	2
3	LITHIUM	Li	6·94	4	3
4	BERYLLIUM	Be	9·01	5	4
5	BORON	B	10·81	6	5
6	CARBON	C	12·01	6	6
7	NITROGEN	N	14·01	7	7
8	OXYGEN	O	16·00	8	8
9	FLUORINE	F	19·00	10	9
10	NEON	Ne	20·18	10	10
11	SODIUM	Na	22·99	12	11
12	MAGNESIUM	Mg	24·31	12	12
13	ALUMINIUM	Al	26·98	14	13
14	SILICON	Si	28·09	14	14
15	PHOSPHORUS	P	30·99	16	15
16	SULPHUR	S	32·06	16	16
17	CHLORINE	Cl	35·45	18	17
18	ARGON	Ar	39·95	22	18
19	POTASSIUM	K	39·10	20	19
20	CALCIUM	Ca	40·08	20	20
21	SCANDIUM	Sc	44·96	24	21
22	TITANIUM	Ti	47·90	26	22
23	VANADIUM	V	50·94	28	23
24	CHROMIUM	Cr	52·00	28	24
25	MANGANESE	Mn	54·94	30	25
26	IRON	Fe	55·85	30	26
27	COBALT	Co	58·93	32	27
28	NICKEL	Ni	58·71	30	28
29	COPPER	Cu	63·54	34	29
30	ZINC	Zn	65·37	34	30
31	GALLIUM	Gn	69·72	38	31
32	GERMANIUM	Ge	72·59	42	32
33	ARSENIC	As	74·92	42	33
34	SELENIUM	Se	78·96	46	34
35	BROMINE	Br	79·91	44	35
36	KRYPTON	Kr	83·80	48	36
37	RUBIDIUM	Rb	85·47	48	37
38	STRONTIUM	Sr	87·62	50	38
39	YTTRIUM	Y	88·91	50	39
40	ZIRCONIUM	Zr	91·22	50	40
41	NIOBIUM	Nb	92·91	52	41
42	MOLYBDENUM	Mo	95·94	54	42
43	TECHNETIUM*	Tc	(97)	(54)	43
44	RUTHENIUM	Ru	101·1	58	44
45	RHODIUM	Rh	102·9	58	45
46	PALLADIUM	Pd	106·4	60	46
47	SILVER	Ag	107·9	60	47
48	CADMIUM	Cd	112·4	66	48
49	INDIUM	In	114·8	64	49
50	TIN	Sn	118·7	70	50
51	ANTIMONY	Sb	121·8	70	51
52	TELLURIUM	Te	127·6	78	52
53	IODINE	I	126·9	74	53
54	XENON	Xe	131·3	78	54
55	CAESIUM	Cs	132·9	78	55
56	BARIUM	Ba	137·3	82	56
57	LANTHANUM	La	138·9	82	57
58	CERIUM	Ce	140·1	82	58
59	PRASEODYMIUM	Pr	140·9	82	59
60	NEODYMIUM	Nd	144·2	82	60
61	PROMETHIUM*	Pm	(145)	(84)	61
62	SAMARIUM	Sm	150·4	90	62
63	EUROPIUM	Eu	152·0	90	63
64	GADOLINIUM	Gd	157·3	94	64
65	TERBIUM	Tb	158·9	94	65
66	DYSPROSIUM	Dy	162·5	98	66
67	HOLMIUM	Ho	164·9	98	67
68	ERBIUM	Er	167·3	98	68
69	THULIUM	Tm	168·9	100	69
70	YTTERBIUM	Yb	173·0	104	70
71	LUTETIUM	Lu	175·0	104	71
72	HAFNIUM	Hf	178·5	108	72
73	TANTALUM	Ta	180·9	108	73
74	TUNGSTEN	W	183·9	110	74
75	RHENIUM	Re	186·2	110	75
76	OSMIUM	Os	190·2	116	76
77	IRIDIUM	Ir	192·2	116	77
78	PLATINUM	Pt	195·1	116	78
79	GOLD	Au	197·0	118	79
80	MERCURY	Hg	200·6	122	80
81	THALLIUM	Tl	204·4	124	81
82	LEAD	Pb	207·2	126	82
83	BISMUTH	Bi	209·0	126	83
84	POLONIUM*	Po	(209)	(125)	84
85	ASTATINE*	At	(210)	(125)	85
86	RADON*	Rn	(222)	(136)	86
87	FRANCIUM*	Fr	(223)	(136)	87
88	RADIUM*	Ra	(226)	(138)	88
89	ACTINIUM*	Ac	(227)	(138)	89
90	THORIUM*	Th	232·0	142	90
91	PROTACTINIUM*	Pa	(231)	(140)	91
92	URANIUM*	U	238	146	92
93	NEPTUNIUM*	Np	(237)	(144)	93
94	PLUTONIUM*	Pu	(244)	(150)	94
95	AMERICIUM*	Am	(243)	(148)	95
96	CURIUM*	Cm	(247)	(151)	96
97	BERKELIUM*	Bk	(247)	(150)	97
98	CALIFORNIUM*	Cf	(251)	(153)	98
99	EINSTEINIUM*	Es	(254)	(155)	99
100	FERMIUM*	Fm	(253)	(153)	100
101	MENDELEVIUM*	Mv	(256)	(155)	101
102	NOBELIUM*	No	(253)	(151)	102
103	LAWRENCIUM*	Lw	(257)	(154)	103
104	KURCHATOVIUM*	Ku			104

NOTES: 1. Elements marked * are unstable.
2. The number of neutrons in the nucleus is calculated for the most commonly found isotope.

1 Atomic Structure and the Classification of the Elements

When you were studying science in junior classes you must have become aware of the fact that some elements resemble each other very closely, but that others are quite different. Some experiments were designed to show that the metallic elements could be arranged in such an order that there was a gradual change in properties in passing up or down the series. Such a list is called the Electrochemical or Electromotive Series; it is an 'activity' or 'reactivity' series. Near the top of the table come sodium and potassium and calcium, three very active metals which oxidize easily in air and attack water vigorously: these are followed by magnesium—slower in its action on water—and zinc—not attacked by water, but which burns when heated in steam. Lower down still come iron (oxidizes in steam but does not burn) then lead and near the bottom copper and mercury which are not affected at all by steam. We shall frequently refer to this series when studying the properties of various metals and their compounds.

No two elements are exactly the same and various attempts have been made to draw up a table that would show which elements do resemble each other. A table of this kind is called a Periodic Table and part of a modern version of one is shown in Table 1.2. In such an arrangement the elements are numbered consecutively from hydrogen upwards; this is the lightest of the elements and is adopted for many purposes as a standard of reference; the heaviest element known at present is number 104, called kurchatovium.

Every atom contains smaller particles inside it; these, called protons, neutrons, and electrons are the 'building bricks' of which all matter is made. For simplicity an atom can be compared to a miniature solar system with the planets revolving round the sun. The protons and neutrons in an atom are concentrated at the centre in the nucleus and the electrons move around in various circular or elliptical orbits at varying distances from the nucleus.

Table 1.1 Electrochemical or Activity Series of the Metals

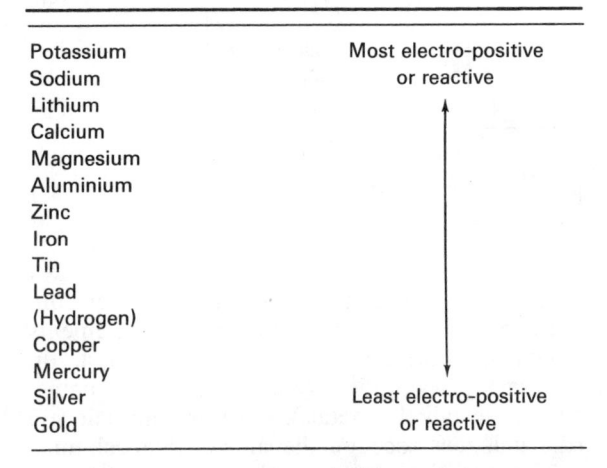

Potassium	Most electro-positive
Sodium	or reactive
Lithium	↑
Calcium	
Magnesium	
Aluminium	
Zinc	
Iron	
Tin	
Lead	
(Hydrogen)	
Copper	
Mercury	
Silver	Least electro-positive
Gold	or reactive

Each proton or neutron weighs almost as much as an atom of hydrogen whereas it takes about 1840 electrons to equal the mass of one proton. The proton itself is so incredibly minute that it would take six hundred thousand million million million (6×10^{23}) to weigh just one gram. Each proton and electron carries an electric charge and this is considered to be the unit charge of electricity. Protons are charged positively and electrons negatively and every atom has exactly the same number of protons as electrons so that it is always electrically neutral.

The hydrogen atom has 1 proton (and therefore 1 electron) and the next heaviest, helium, has 2 protons and 2 neutrons (and 2 electrons). Each succeeding element has one more proton (and one more electron); the number of neutrons may not alter so regularly as these are neutral particles. Table 1.2 shows the number of protons (electrons) and neutrons present in the first twenty elements. The number of protons present in the atom is called the Atomic Number of the element. More about these particles will be found in Chapters 6 and 49.

Table 1.2 Part of Periodic Table

Group / Period	I	II	III	IV	V	VI	VII	0
1	Hydrogen H 1p **1**							Helium He 2p **2** 2n
2	Lithium Li 3p **3** 4n	Beryllium Be 4p **4** 5n	Boron B 5p **5** 5n	Carbon C 6p **6** 6n	Nitrogen N 7p **7** 7n	Oxygen O 8p **8** 8n	Fluorine F 9p **9** 10n	Neon Ne 10p **10** 10n
3	Sodium Na 11p **11** 12n	Magnesium Mg 12p **12** 12n	Aluminium Al 13p **13** 14n	Silicon Si 14p **14** 14n	Phosphorus P 15p **15** 16n	Sulphur S 16p **16** 16n	Chlorine Cl 17p **17** 18n	Argon Ar 18p **18** 22n
4	Potassium K 19p **19** 20n	Calcium Ca 20p **20** 20n					Bromine Br 35p **35** 45n	Krypton Kr 36p **36** 48n
5	Rubidium Rb 37p **37** 49p						Iodine I 53p **53** 74n	Xenon Xe 54p **54** 77n

Although this table is only a small extract from the complete one it can be seen that those elements which resemble each other, helium, neon, argon (the rare gases in the air), lithium, sodium, potassium (the 'alkali' metals), fluorine and chlorine (the 'halogens') occur in the same vertical columns.

This may all seem rather advanced chemistry for the beginning of a course and a study of the Periodic Table does belong to more advanced work, but a knowledge of its construction and of the way in which elements are built up and join together to form compounds is very important. This will be considered further in a later chapter when the subject of valency or combining power is studied. The Periodic Table will be studied in more detail in Chapter 41.

Exercise 1

1 (a) Which of the following numbers indicates the total number of atoms present in one molecule of calcium hydrogencarbonate, $Ca(HCO_3)_2$? 6, 9, 11, 12, 32.

(b) Which of the following elements does not have any neutrons in its atom? Sodium, hydrogen, oxygen, sulphur, carbon. (MR)

2 Calculate the total number of atoms present in the following formulae:
(a) $4Al_2(SO_4)_3$
(b) $3Ca_3(PO_4)_2$
(c) $5Pb(NO_3)_2$
(d) $3NaHSO_4$
(e) $2K_3Fe(CN)_6$
(f) $6C_{12}H_{22}O_{11}$
(g) $4Ba(HCO_3)_2$
(h) $3(NH_4)_2CO_3$
(i) $5Na_2HPO_4$

3 Study Table 1.2, copy and complete the following:
(a) The elements in the same period, from left to right, gradually change from _____ to _____.
(b) With the exception of _____ the number of neutrons in an atom is never _____ than the number of protons.
(c) As we go along a period from left to right the number of _____ increases regularly.

4 Elements may be represented by atomic symbols in this way: ^1_1H; $^{12}_6C$; $^{28}_{14}Si$; $^{40}_{20}Ca$. What do you think the small numbers stand for?

5 Give the name and symbol in each case of an element which is (a) metallic, and less dense than water, (b) normally a liquid, and metallic, (c) normally gaseous, and coloured, (d) capable of existing in more than one well defined form. (S)

6 (i) Give the name of a gaseous element in each

case which is:
(a) inflammable;
(b) poisonous;
(c) able to relight a glowing splint;
(d) unreactive, yet present in many compounds.
(ii) Give the name of an element in each case which is:
(e) metallic, and dissolves in cold water giving a white precipitate;
(f) a greenish gas;
(g) non-metallic and glows in the dark;
(h) used in one of its forms for making 'lead' pencils. (S)

7 A, B, C and D are four metals:
(i) A and D react slowly with water to give hydrogen;
(ii) the oxides of A, B and D are unaffected by heat and hydrogen;
(iii) D rapidly forms its oxide on exposure to the atmosphere;
(iv) A, B and D react with dilute sulphuric acid to yield hydrogen.
Arrange the four metals in the order in which they would appear in the activity series. (SE)

2 Measuring Length, Mass and Density

One of the most important parts of a scientist's work is the organization and presentation in the simplest possible way of the results of his experiments. These results may often at first sight seem very complicated and not capable of being fitted into an orderly scheme. However the scientist often knows from experience techniques that are most likely to sort them out. In physics particularly this frequently involves plotting graphs. In order to start to understand how this can be done and what information can be gained from graphs it is convenient to plot the results of very simple experiments.

Experiment 2.1. Take a number of cylindrical tins or jars of various diameters, and measure as carefully as you can both their diameters and circumferences. A piece of cotton may be used to measure the circumferences. Tabulate your results as in Table 2.1 which shows a typical set of values.

Plot a graph of circumference against diameter. Since we should generally say that the circumference depends on the diameter, D is the independent variable and should be plotted on the horizontal axis. C is therefore the dependent variable. The typical values quoted are plotted in Fig. 2.1.

Table 2.1

Diameter in cm (D)	Circumference in cm (C)
2·1	6·6
4·0	12·6
5·5	17·3
6·3	19·8
8·2	25·7

We shall find that the results of many experiments may be plotted to give such a straight line graph passing through the origin of the axes. It is important to be able to interpret such graphs when they arise. Fig. 2.2 shows graphs of the equations $y = \frac{1}{2}x$, $y = x$, $y = 2x$ plotted on the same axes. You can show for yourself, that any equation of the form: $y = $ some number (a constant) $\times x$, will give a straight line graph passing through the

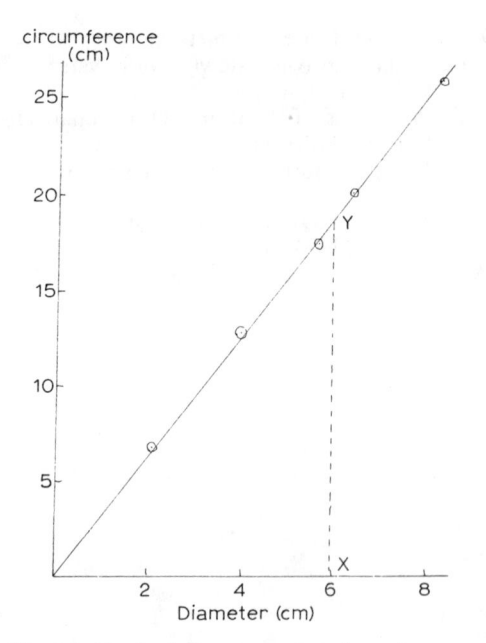

Fig. 2.1. Graph of diameter against circumference

origin. Thus the graph in Fig. 2.1 shows that:

Circumference = a constant × diameter.

The gradients of the three graphs in Fig. 2.2 calculated from AB/OA, AC/OA and AD/OA are (i) $\frac{1}{2}$, (ii) 1 and (iii) 2. Thus the constant in the equation is the gradient of the graph.

The gradient of the graph in Fig. 2.1 is:

$$\frac{XY}{OX} = \frac{18 \cdot 6}{6} = 3 \cdot 1$$

(XY and OX are measured in terms of the units on the axes).

Thus this graph shows that:
Circumference = 3·1 × diameter.
The constant here is clearly π.

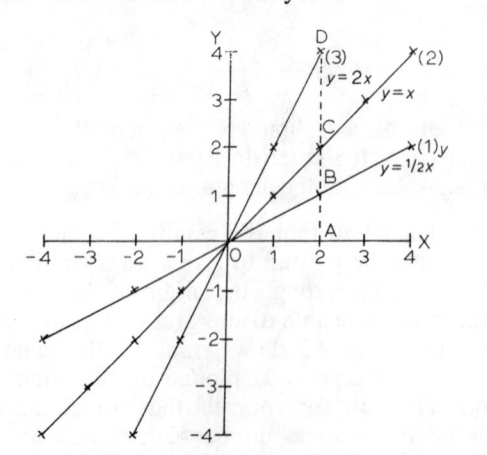

Fig. 2.2. Straight line graphs

In the very simple example chosen here the graph serves only to average the results obtained. In most cases the relationship between the two variables is not so obvious and the graph makes this relationship much clearer.

Equation (1) may be rewritten:

Circumference is proportional to diameter or $C \alpha D$.

If the graph in Fig. 2.1 was an exercise in algebra it might be concluded that the points off the line arose from errors in calculation or bad plotting. When a graph of experimental results is plotted however some scatter of points about the line is to be expected. The line in Fig. 2.1 is drawn so that the points are evenly distributed on both sides of it. The scatter of points arises on account of 'random errors' which will be present in any experiment. This does not mean that an experimenter can be careless in making measurements. Indeed he should always be as accurate as his apparatus permits.

More Accurate Measurement of Lengths and Distances

It is possible to measure lengths of about a metre, with a ruler, to the nearest millimetre. This represents an accuracy of one part in a thousand. With rather complicated optical apparatus scientists today are able to measure lengths of this size to one part in 10 million. This kind of accuracy is only rarely required, but it is often desirable to improve on what one can do with a ruler and we shall now mention some ways in which this may be done.

Many accurate measuring instruments are based on a device known as a vernier (Fig. 2.3).

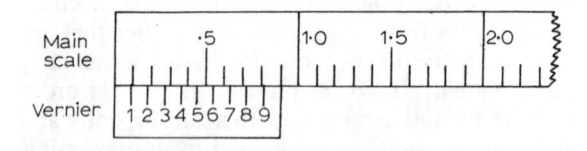

Fig. 2.3. The Vernier Scale

The vernier is a small scale which moves along the main scale and its object is most often to enable measurements to be made to $\frac{1}{10}$th part of the smallest division on the main scale. It is equal in length to nine small divisions on the main scale and is divided up into ten equal divisions. Thus if the smallest divisions on the main scale in Fig. 2.4 are millimetres then the vernier scale divisions are 0·9 mm or 0·09 cm.

Fig. 2.4 shows a pair of calipers incorporating

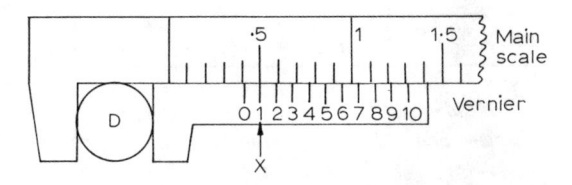

Fig. 2.4. Vernier calipers

a vernier scale being used to measure the diameter of a disc D. The value to the nearest millimeter is given by the position of the zero of the vernier scale. In this case it is 0.4 cm. The second figure after the decimal point is given by the line on the vernier scale which coincides exactly with a line on the main scale. This occurs at X in the diagram. Thus the full reading is 0·41 cm.

From zero on the main scale to X is 0·5 cm. From zero on the vernier to X is $1 \times \cdot 09 = 0\cdot 09$ cm. Thus from 0 on the main scale to 0 on the vernier is: $0\cdot 5 - 0\cdot 09 = 0\cdot 41$ cm.

The micrometer screw gauge (Fig. 2.5) is a device for measuring the size of objects even more accurately than can be done with a vernier. Commonly the micrometer will measure to $\frac{1}{100}$ mm. The instrument shown in Fig. 2.5(a) is based on a screw

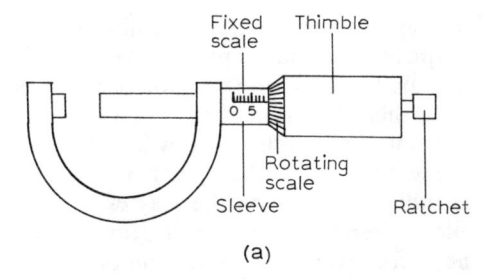

Fig. 2.5a. Micrometer screw gauge

whose 'pitch' is 0·5 mm. This means that when the thimble is turned through one complete revolution it will move forward or back by 0·5 mm (one small division on the fixed scale). The circular scale is divided into 50 divisions so that each of them represents a forward or back movement of $\frac{1}{50}$th of 0·5 mm or $\frac{1}{100}$ mm. The reading on the micrometer shown (Fig. 2.5(b)) is between 3·5 and 4·0

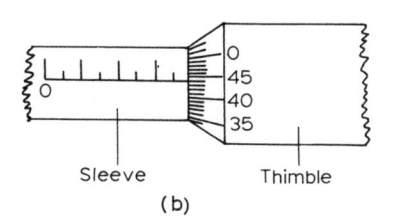

Fig. 2.5b. Fixed and rotating scales of micrometer screw gauge

mm on the main scale. The circular scale tells us that the correct value is 3.94 mm. Before or after making a reading, a micrometer should be checked for zero error by closing the jaws right up. No correction of the value is necessary if the zero on the circular scale coincides with the datum line on the sleeve.

Fig. 2.6 shows three micrometers all with jaws fully closed. Fig. 2.6(a) shows no zero error. All readings taken with instrument in Fig. 2.6(b) should be reduced by 0·02 mm and those taken with the micrometer shown in Fig. 2.6(c) should have 0.03 mm added to them.

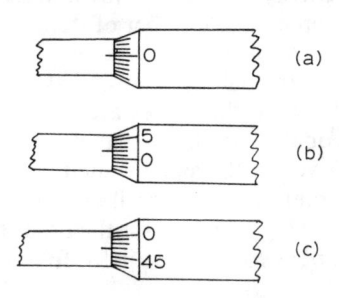

Fig. 2.6. Micrometer zero errors

Mass, Weight and their Measurement

In everyday life we use the word 'weight' in what the scientist regards as a rather careless way. We see, for example, on a jar of jam, 'Net weight 454 g'. The scientist would argue that the gram is not the unit of weight but of mass. For everyday purposes the difference between mass and weight is, perhaps, not an important one. However for scientific purposes we need to be rather more careful in our use of the two terms. In fact it is easier for us to understand the difference between the two terms now that we have seen on television men walking on the moon, than it was before the days of space travel. Astronauts, and all the things they take with them, 'weigh' much less on the moon than they do no earth. You will have seen on television broadcasts from the moon how things, when they are dropped, fall much more slowly than we are used to. This is because the pull of the moon's gravity on an object on the surface of the moon is much less than would be the pull of the earth's gravity on the same object if it were on the surface of the earth. This pull of gravity on an object is what we mean by weight. Weight, then, is a force and should be measured in the units of force. The unit of force, which we shall define later in this book, is the newton.

5

Although we say that an object weighs less on the moon than on earth, we do not, of course, mean that the amount of matter in the object changes on the journey from earth to moon. The 'mass' of an object, which does not depend on where the object is in the universe, is a measure of the number and type of atoms of which it is made up. The unit of mass is the kilogram. The standard kilogram is a piece of platinum kept in Paris.

When we use instruments to 'weigh' things, it is not always clear whether it is mass or weight which we are in fact measuring. When we use the ordinary chemical beam balance, we are comparing the force of gravity on the unknown mass in one pan with the force on a number of standard masses in the other. In general we alter these standard masses until the balance shows that the two forces are equal. When the forces are equal, the masses will be equal. It follows that we are measuring mass and we get the same value for our unknown mass whether we measure it on earth or on the moon. A spring balance, on the other hand, measures weight as you may see in the following experiment.

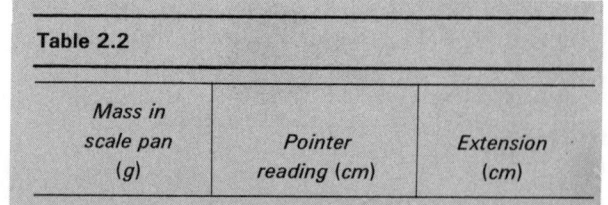

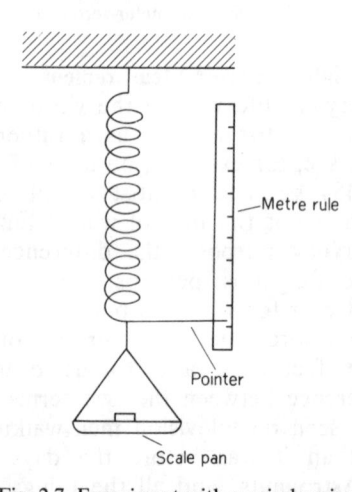

Table 2.2

Mass in scale pan (g)	Pointer reading (cm)	Extension (cm)

mass in the scale pan. Plot a graph of mass in the scale pan against the extension. Fig. 2.8 shows the sort of graph you will probably obtain. It shows that the extension is proportional to the mass in the scale pan.

$$E \propto W$$

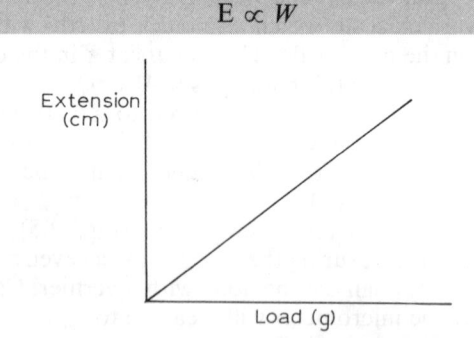

Fig. 2.8. Load–extension graph for a spiral spring

This graph may be used as a calibration graph for the spring. If a body of unknown mass is placed on the scale pan and the extension of the spring is observed, then the mass of the object may be obtained from the graph. You will note, though, that the extension of the spring really depends on the pull of gravity on the masses in the scale pan and a completely different graph would be obtained, for example, on the moon. A spring balance calibrated in grams is only useful if gravity stays the same. One calibrated in newtons is always valuable.

A practical form of spring balance is shown in Fig. 2.9.

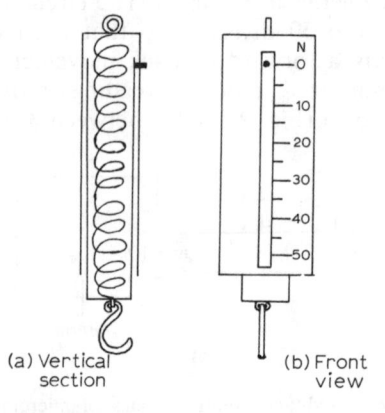

Fig. 2.7. Experiment with a spiral spring

Experiment 2.2. Fix one end of a spiral spring in a clamp and from the other suspend a scale pan (Fig. 2.7). Attach also a piece of wire to the bottom end of the spring to act as a pointer. Clamp a metre ruler in a vertical position as in the diagram so that it may be used to measure vertical displacements of the pointer. First note the position of the pointer on the ruler and then add masses, say 10 g at a time, noting the pointer reading each time. Take care not to overload the spring so that it gets a permanent extension. Tabulate your results as in Table 2.2.

The extension in each case is the difference between the pointer reading and that with no

Fig. 2.9. A spring balance

Density

It is common to refer to materials such as lead as heavy, and materials such as cork as light. Of course it is possible to have a piece of lead which is 'light' if it is small enough and a piece of cork which is 'heavy' if it is big enough. We really mean that if we had equal *volumes* of lead and cork, the lead would be very much heavier than the cork. Lead is said to have a high 'density' and cork a low density. The density of a material is defined as the mass of unit volume.

If the mass is measured in kilograms then the unit volume will be one cubic metre ($1 m^3$), and the density will be in kilograms per cubic metre (kg/m^3). This is the strictly correct unit for density, however in simple laboratory experiments it is often more convenient to measure mass in grams and volumes in cubic centimetres.

The density of lead is $11 \cdot 4$ g/cm^3 and of cork $0 \cdot 24$ g/cm^3.

Since, as we shall see in Chapter 15, the volumes of objects change as the temperature changes, it is really necessary to state the temperature with a value of density. These volume changes are not large for solids and, for most work, their effect on density may be neglected. For liquids and gases the effect of temperature on density is much more significant. The density of water at 4°C is $1 \cdot 00$ g/cm^3. This, of course, is not a coincidence. The standard kilogram was chosen to have a mass very close to that of 1000 cm^3 of water at 4°C.

In practice one does not measure the density of a material by finding the mass of one cubic centimetre of it. As an alternative to the definition given previously, density may be defined as $\frac{Mass}{Volume}$.

In simple determinations of density, it is sufficient to measure the mass and volume of an object.

Measurement of Density

Experiment 2.3. Regular solids. The volumes of regular solids, e.g. cubes, cuboids, spheres, cylinders, cones, etc. may be found by calculation from the appropriate dimensions. These may be measured with the aid of calipers or a micrometer screw gauge, or simply with a ruler. If the solid is of metal accurately machined, and the dimensions are carefully measured with a micrometer it may be appropriate to find the mass with a chemical balance. Otherwise the mass may be found with a spring or level balance. Typical results of two such experiments with the necessary calculations are shown below.

(a) For a cylinder of brass. (Micrometer and chemical balance)

Diameter = $2 \cdot 54$ cm
Height = $2 \cdot 56$ cm
Mass = $109 \cdot 1$ g

$$\text{Volume of cylinder} = \pi \times \left(\frac{2 \cdot 54}{2}\right)^2 \times 2 \cdot 56$$
$$= 12 \cdot 9 \text{ cm}^3$$

$$\text{Density of brass} = \frac{\text{Mass}}{\text{Volume}}$$
$$= \frac{109 \cdot 1}{12 \cdot 9}$$
$$= 8 \cdot 42 \text{ g/cm}^3$$

(b) For a cuboid of wood. (Ruler and spring balance)

Length = $0 \cdot 3$ m
Width = $0 \cdot 1$ m
Height = $0 \cdot 1$ m
Mass = $2 \cdot 1$ kg

$$\text{Volume of block} = 0 \cdot 3 \times 0 \cdot 1 \times 0 \cdot 1$$
$$= 0 \cdot 003 \text{ m}^3$$

Hence density of wood
$$= \frac{\text{Mass}}{\text{Volume}}$$
$$= \frac{2 \cdot 1}{0 \cdot 003}$$
$$= 700 \text{ kg/m}^3$$

Experiment 2.4. Liquids. Weigh an empty measuring cylinder on a lever balance. Now pour some liquid into it and weigh again. Note the volume of the liquid as indicated by the measuring cylinder. The following are typical results obtained with methylated spirits.

Mass of measuring cylinder = 68 g
Mass of measuring cylinder
+ methylated spirits = 124 g
Mass of meths. = 56 g
Volume of meths. = 69 cm^3

$$\text{Density of meths.} = \frac{\text{Mass}}{\text{Volume}}$$
$$= \frac{56}{69}$$
$$= 0 \cdot 81 \text{ g/cm}^3$$

Experiment 2.5. Irregular solids. The volumes of irregular solids are most easily found by 'displacement of water'. If the solid is small enough to go into a measuring cylinder, first put some water into the cylinder and note the reading. Now gently lower the solid into the water and note the new reading. If the solid is completely immersed, this second reading gives the volume of the water and the solid together. Thus the volume of the solid

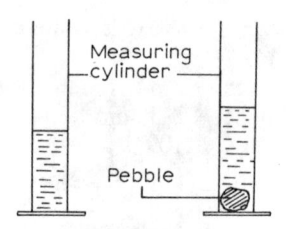

Fig. 2.10. To find the volume of an irregular solid

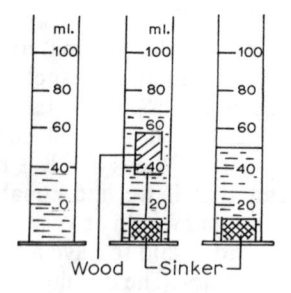

Fig. 2.12. To find the volume of a floating object using a sinker

may be found by subtraction. Finally dry the solid and weigh it on a lever balance. This experiment is illustrated in Fig. 2.10 and typical results are shown for a pebble.

First measuring cylinder
reading = 42 cm³
Second measuring cylinder
reading = 65 cm³
Volume of pebble = 23 cm³
Mass of pebble = 60 g

$$\text{Density} = \frac{\text{Mass}}{\text{Volume}} = \frac{60}{23} = 2\cdot 6 \text{ g/cm}^3$$

If the irregular solid is too large to go in a measuring cylinder, its volume may be found with the aid of a displacement or 'Eureka' can (Fig. 2.11).

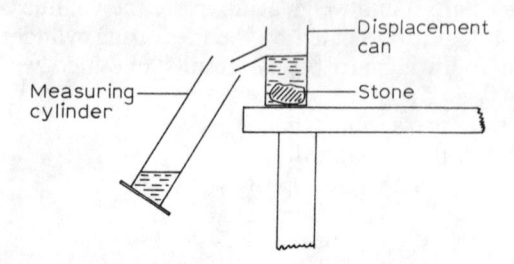

Fig. 2.11. The use of a displacement or eureka can

Place this can on the bench and pour in water until some comes out of the spout. When it has finished dripping, place a measuring cylinder under the spout and gently lower the solid into the can. The volume of water collected is equal to the volume of the solid. Find the mass of the solid and calculate the density as before.

Solids which float in water

The methods described in Experiment 2.5 may be used with solids which float in water.

However it is necessary to attach to the floating object a piece of metal which will make it sink. A separate experiment must therefore be carried out to find the volume of the sinker. This will be subtracted from the volume of the sinker and the object under test. The results of such an experiment are shown in Fig. 2.12.

This shows that the volume of wood + sinker = 30 cm³ and the volume of sinker = 10 cm³ ∴ volume of wood = 20 cm³.

Relative Density

All the experiments to determine density described so far involve finding a mass and a volume. Mass may be found with high accuracy quite easily. Measurement of volume, on the other hand, may not be carried out so accurately. It is not possible to measure density by carrying out only weighings, but it is possible to compare the densities of two liquids in this way. All we must do is find the masses of equal volumes of the liquids.

$$\frac{\text{Density of liquid } A}{\text{Density of liquid } B} = \frac{\text{Mass of 1 cm}^3 \text{ of } A}{\text{Mass of 1 cm}^3 \text{ of } B}$$
$$= \frac{\text{Mass of any volume of } A}{\text{Mass of THE SAME volume of } B}$$

It is convenient to deal with a quantity which we call Relative Density. This is defined as the ratio of the density of a substance to the density of water. It is the ratio of the left-hand side of the equations above where liquid B is water. The relative density of a substance may thus also be defined as:

$$\frac{\text{Mass of substance}}{\text{Mass of an equal volume of water}}$$

We can measure relative density by finding just two masses.

Density is measured in g/cm³ or kg/m³. Relative Density, being the ratio of two densities or of two masses, has itself no dimensions and it does not matter in what units the two densities or two masses are measured as long as it is the same system for both.

Experiment 2.6. The Relative Density of a Liquid. A density bottle (Fig. 2.13) is a vessel designed so that when carefully filled at a certain temperature it will always contain exactly the same volume of liquid. It has a narrow neck into which fits a ground glass stopper with a vertical narrow bore hole in it. Find, with a chemical balance, the mass of the empty density bottle together with its stopper. Fill it with the liquid whose relative density is to be found and put in the stopper.

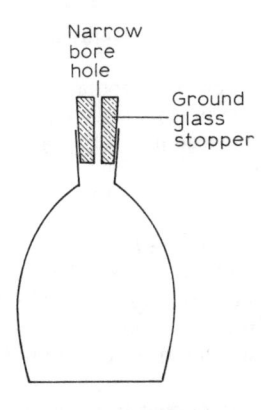

Fig. 2.13. A density bottle

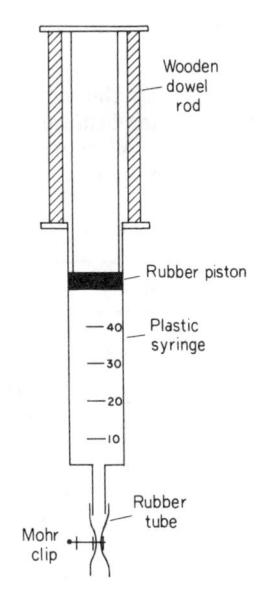

Fig. 2.14. Density of air experiment

The surplus liquid will escape through the hole in it. Carefully dry the outside of the bottle and find its mass full of liquid. Pour away the liquid, wash the density bottle and fill it in the same way with water. Finally find the mass of the bottle full of water. The calculation for a typical experiment with glycerine is shown below.

$$\text{Mass of empty density bottle} = 25.2 \text{ g}$$
$$\text{Mass of density bottle} + \text{glycerine} = 60.4 \text{ g}$$
$$\text{Mass of density bottle} + \text{water} = 53.4 \text{ g}$$
$$\therefore \text{Mass of glycerine} = 35.2 \text{ g}$$
$$\text{and Mass of equal volume of water} = 28.2 \text{ g}$$

Relative Density =
$$\frac{\text{Mass of glycerine}}{\text{Mass of an equal volume of water}}$$
$$= \frac{35.2}{28.2} = 1.25$$

Density of Air

Gases generally have very low densities. Any experiment to measure the density of air must therefore involve finding a rather small mass unless a large volume is used. A top pan balance sensitive to 0.01 g or a balance of similar sensitivity is needed for the experiment suggested below.

Experiment 2.7. The Density of Air. A 50 cm³ plastic syringe of the type shown in Fig. 2.14 is required and the outlet should be made airtight with a short piece of rubber tube fitted with a screw clip. If you attempt to pull back the piston of the syringe with the outlet sealed, you will find that it is very hard to do so and the piston will return to its original position when released. However, you can fix the piston in a withdrawn position by pushing in two pieces of wood as shown in the diagram. You should obtain two pieces of wood of the right length to hold the piston at the 50 cm³ mark. Start with the clip on the rubber tubing open and the piston right in. Close the clip securely, pull back the piston to the 50 cm³ mark and fix it in this position. What does the space inside the syringe now contain? Now find the mass of the syringe together with all of its fittings, on a top pan balance. Next open the clip on the rubber tube and listen carefully as you do so. You should be able to hear the air going into the syringe. Finally find the mass of syringe plus fittings again. It should have increased by the mass of air which has entered it. In fact the change in mass should be the mass of 50 cm³ of air. A typical set of results might be as follows:

$$\text{Mass of syringe} + \text{fittings} + \text{air} = 98.76 \text{ g}$$
$$\text{Mass of syringe} + \text{fittings} = 98.70 \text{ g}$$
$$\text{Thus, mass of 50 cm}^3 \text{ of air} = 0.06 \text{ g}$$
$$\text{and mass of 1 cm}^3 \text{ of air} = 0.0012 \text{ g}$$

This is the density of air for the conditions of the experiment. We shall see in later chapters that the densities of gases are very much affected by factors such as the temperature.

Exercise 2

1 Give instructions, with the aid of a labelled diagram, how to make a spring balance to weigh in kg up to 10 kg, given: a spring; a support to hang it on; a single 1 kg mass; and a piece of cardboard to make a scale. (*WM*)

2 Complete the following table:

Substance	Glass	Water
Mass	20 g	250 kg
Volume		
Density	2·5 g/cm^3	1000 kg/m^3

(*S*)

3 Milk of good quality has a density of 1·03 g/cm^3. How would you expect its density to change if:
(*a*) the cream was removed from it,
(*b*) Water was added to it,
(*c*) It was placed inside a milk-cooler? (*EM*)

4 Describe, giving all practical details, how you would (*a*) find the density of a given piece of cork of irregular shape, (*b*) verify the accuracy of a 25 cm^3 pipette.

Calculate the relative density of an alloy which has the following composition: 40% by weight of a metal *A* of relative density 8 and 60% by weight of a metal *B* of relative density 3.

Assume that no final expansion or contraction occurred as a result of the alloying process. (*J*)

5 80 cm^3 of water are mixed with 140 cm^3 of liquid of relative density 0·83. What is the density of the mixture if there is no change in total volume on mixing? (*J*)

6 Define *density* and *relative density*.

Describe how you would find the relative density of methylated spirit, using a density bottle.

250 cm^3 of water are mixed with 1000 cm^3 of methylated spirit. Calculate the density of the mixture, assuming that there is no change of volume as a result of the mixing.

(The relative density of methylated spirit is 0·80.) (*L*)

7 A measuring cylinder contains 60 cm^3 of water. An iron block 6 cm in length is completely immersed in the water, and the reading on the cylinder is found to increase to 81 cm^3.
(*a*) Calculate the average cross sectional area of the block.
(*b*) If the density of iron is 7·9 g/cm^3 calculate the mass of the block. (*C*)

8 Describe how a density bottle may be used in an experiment to measure the density of a liquid. Why is the density bottle method considered to be particularly accurate?

100 cm^3 of water of density 1·0 g/cm^3 is mixed with 60 cm^3 of a liquid of density 1·8 g/cm^3. What is the density of the mixture? How would the mixing be carried out if the second liquid was concentrated sulphuric acid? (*C*)

3
Time, Velocity and Acceleration

An event which is repeated regularly may be used to measure time. Our units of time are based on the revolution of the earth on its axis. We say that the time for one revolution of the earth is 24 hours and our basic unit of time, the second, is $\frac{1}{3600}$ hours.

Other regularly repeated events are the oscillations of a pendulum or balance wheel in a clock or watch. Electric clocks depend on the regular alternations of the current supplied from the power station (see Chapter 46). This will only be regular if the engineers in the power station ensure that the generators turn always at the same speed. We shall now investigate under what conditions the swings of a simple pendulum are regular.

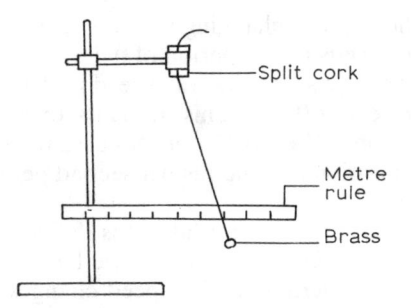

Fig. 3.1. Simple pendulum experiments

Experiment 3.1. A simple pendulum consists of a small mass suspended on the end of a light piece of string or thread. There are really only three things which can vary with such a pendulum. They are:

(1) The size of the swings
(2) The mass suspended
and (3) The length of the string

We shall consider how the time of swing of the pendulum is affected by varying each of these in turn.

(1) Set up a simple pendulum as shown in Fig. 3.1 with a 20 gram mass suspended from a thread which at its upper end passes through a split cork mounted in a clamp. Fix a metre rule horizontally just above the brass mass so that the size of the swings may be measured. Pull the mass aside through a small measured distance and use a stop watch or clock to time swings. In order to obtain an accurate value of the time for one swing (from one side to the other and back again) it is necessary to time, say, 10 swings and divide the value by 10. Repeat the experiment with the same mass suspended and the same length of thread but with different initial displacements. Tabulate your results as follows:

Table 3.1

Initial displacement (cm)	Time for 10 swings (s)	Time for 1 swing (s)

Your results will probably show that the time of swing is only very slightly affected by the size of the swings. This is a very important result since if a pendulum is used in a clock, the swings may get smaller and it is essential that the clock should not go fast or slow as this happens.

(2) Repeat the experiment without changing the length of the pendulum and keeping the initial displacement constant in each case, but using a range of different masses suspended. You will probably find little or no variation in the time of swing.

(3) Finally repeat the experiment with a fixed mass suspended and a fixed, small, initial displacement, varying this time the length of the pendulum. Table 3.2 shows a typical set of values for such an experiment:

Table 3.2

Length (l, cm)	Time for 10 swings (s)	Time for 1 swing (T s)	T^2 (s^2)
25	10·1	1·01	1·02
50	14·3	1·43	2·04
75	17·5	1·75	3·06
100	20·2	2·02	4·08
125	22·6	2·26	5·10
150	24·8	2·48	6·15

You will see from these results that the time of swing increases but T is *not* simply proportional to l. If it were, the time of a 50 cm pendulum would be twice that of one of 25 cm length. It does seem however from results 1 and 4 in table 3.2 that the time of swing is doubled when the length is increased four times. This suggests that the relationship between length and time might be: $T^2 \propto l$.

The final column of the table shows values of T^2 and it confirms this relationship. If you plot a graph of T^2 against l it will be a straight line passing through the origin. You could find out from your graph how long the pendulum would have to be to have a time of swing of 1 second. Such a pendulum could be used for timing purposes and as long as the swings were never very large it would be quite reliable even though the size of the swings gradually got less.

The bob of the pendulum in this experiment is an example of something oscillating. There are many other examples in Physics. The time for one swing is known as the *period* of the oscillation. The number of swings in one second is known as the *frequency* (n) of the oscillation. You may be able to see that:

$$T = \frac{1}{n}$$

The distance between the rest position of the pendulum bob and the extreme point of its swing is known as the *amplitude* of the oscillation.

Speed and Acceleration

Figure 3.2 represents a ball rolling on a smooth table by the side of a cm scale. In order to get a picture of this kind we could take a number of photographs of the ball in quick succession, all on the same piece of film. Suppose that, in obtaining Fig. 3.2, the photographs were taken at regular intervals of say $\frac{1}{10}$ s. It is now an easy matter to calculate the speed of the ball. It takes $\frac{1}{10}$ s to move from position 1 to position 2, a distance of 10 cm. We could thus say that its speed is 100 cm/s. We would equally well say that it takes 0·5 s to move from 1 to 6, a distance of 50 cm. This gives the same value for the speed of 1 m/s. We are dealing here with a case of constant speed and we shall get the same answer whatever distance and time interval we consider.

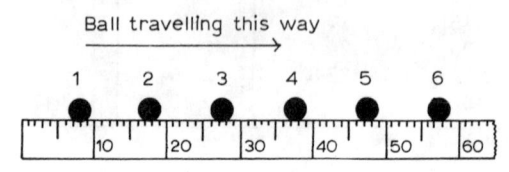

Fig. 3.2. Constant velocity

Figure 3.3 represents a photograph obtained in a similar way with photographs taken at $\frac{1}{10}$ s intervals. Here again the distance travelled in 0·5 s is 50 cm. However in 0·1 s between 1 and 2 only 2 cm are covered and the average speed during this time is 20 cm/s while in 0·1 s between 5 and 6, 18 cm are covered and the average speed is 180 cm/s. In this case the speed is not constant, but gradually increasing throughout. The ball is undergoing an acceleration. This might happen if it were rolling down a sloping surface. In cases where the speed is getting less we say there is a retardation.

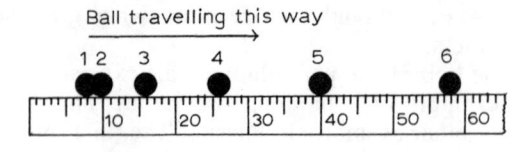

Fig. 3.3. Constant acceleration

The speeds during each of the five $\frac{1}{10}$ s periods of the motion shown in Fig. 3.3 are as follows:

1 to 2	2 cm in 0·1 s	20 cm/s
2 to 3	6 cm in 0·1 s	60 cm/s
3 to 4	10 cm in 0·1 s	100 cm/s
4 to 5	14 cm in 0·1 s	140 cm/s
5 to 6	18 cm in 0·1 s	180 cm/s

As the speed is changing continuously these are average speeds over a period of 0·1 s in each case. However they show that the speed is changing by 40 cm/s every 0·1 s. This amounts to 400 cm/s every second. We say that in this case the ball has an acceleration of 4 metres per second per second or 4 m/s².

Acceleration (or retardation) is defined as *rate of change of velocity*. In Fig. 3.3 we have a case of constant acceleration. The speed changes by the same amount for each 0·1 s interval of time. If an acceleration is constant, it is the change in velocity in unit time. We have been using the words speed and velocity as though they meant exactly the same thing. In fact speed has size only, while velocity has direction also. Thus one may speak of a speed of 50 km/h but strictly one should refer to a velocity of say 50 km/h in a north westerly direction.

You may have seen in advertisements for motor cars statements like: '0 to 90 km per hour in 12 seconds'. If the car had constant acceleration in going from rest to 90 km/h, (which is unlikely) then the acceleration was $\frac{90}{12} = 7\cdot 5$ kilometres per hour per second.

We have calculated the acceleration here as:

$$\frac{\text{Change in velocity}}{\text{time}}$$

Thus if the car starts with a velocity u and finally, after a time t, has a velocity v, the change in velocity
$$= v - u$$
and the acceleration $a = \dfrac{v - u}{t}$
$$\therefore at = v - u$$
$$\text{or } v = u + at \quad\quad (1)$$

This equation may be used to solve simple problems on constant accelerations and retardations, e.g. a car travelling at 60 km/h comes to rest with constant retardation in $2\frac{1}{2}$ s. What is the retardation?

We could of course deal with this in mixed units (km/h and s) as in the example above, but it is more satisfactory to start by changing the speed to m/s.

$$60 \text{ km/h} = 1 \text{ km/minute} = \frac{1000}{60} = 16\tfrac{2}{3} \text{ m/s}$$

Thus $v = 0$ m/s $u = 16\tfrac{2}{3}$ m/s and $t = 2\tfrac{1}{2}$ s, and from equation (1)
$$0 = 16\tfrac{2}{3} + 2\tfrac{1}{2} a$$
$$\text{or } a = -\frac{16\tfrac{2}{3}}{2\tfrac{1}{2}} = -6\cdot 7 \text{ m/s}^2$$

The minus sign tells us that we have a retardation and not an acceleration.

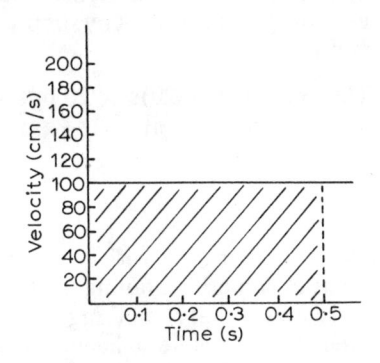

Fig. 3.4. Velocity time-graph (constant velocity)

Velocity-time Graphs

Figs. 3.4 and 3.5 are graphs of velocity against time for the motions shown in Figs. 3.2 and 3.3 respectively. In Fig. 3.2 the speed is constant at 100 cm/s for 0.5 s. The area shaded under the graph in Fig. 3.4 is velocity × time = distance travelled in

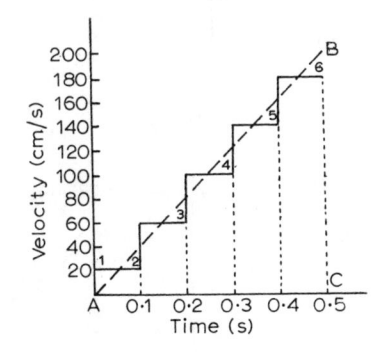

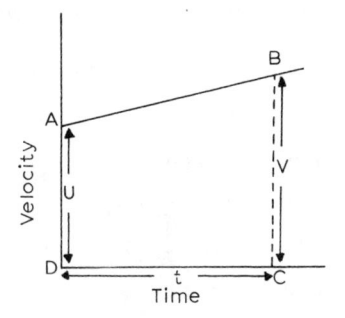

Fig. 3.5. Velocity-time graph (constant acceleration)

this time ($100 \times 0.5 = 50$ cm). In Fig. 3.5, the area under the 'stepped' graph is:

$(20 \times 0.1) + (60 \times 0.1) + (100 \times 0.1) + (140 \times 0.1) + (180 \times 0.1)$
$\quad = 50$ cm

which is again the distance travelled in the 0.5 s. However in a case of constant acceleration the velocity will increase continuously and not in sudden jumps. Thus the dotted line AB properly indicates the way in which the velocity changes. The area under this line (the area of triangle ABC) is:

$$\tfrac{1}{2} \times AC \times BC$$
$$= \tfrac{1}{2} \times 200 \times 0.5$$
$$= 50 \text{ cm}$$

which is the distance travelled.

To summarize, the velocity time graph of something travelling with constant acceleration (or retardation) will be a straight line and the area under this line will be the distance travelled.

Fig. 3.6. Velocity-time graph (constant acceleration)

Fig. 3.6 shows a velocity time graph of something having an initial velocity u which increases with constant acceleration to a velocity v, in a time t. The distance travelled (s) in this time will be given by the area of the quadrilateral $ABCD$ which is:

$\tfrac{1}{2}(AD + BC) \times DC$

or $\tfrac{1}{2}(v + u) \times t$

($\tfrac{1}{2}(v + u)$ is average velocity during the time)

$$\therefore s = \tfrac{1}{2}(v + u)t \qquad (2)$$

Substituting for v from equation (1) above we have

$$s = \tfrac{1}{2}(u + at + u)t$$
$$\therefore s = ut + \tfrac{1}{2}at^2 \qquad (3)$$

Equations (1) and (3) together with one other obtained from them enable us to solve almost all problems dealing with constant accelerations and retardations.

From equation (1)

$$\frac{v - u}{a} = t$$

Substituting this value of t in equation (2) we have:

$$s = \frac{(v + u)(v - u)}{2} \cdot \frac{1}{a}$$

$$s = \frac{v^2 - u^2}{2a}$$

and $2as = v^2 - u^2$

or $v^2 = u^2 + 2as \qquad (4)$

In order to understand the use of these equations

and velocity-time graphs we will solve a simple problem with them.

A car accelerates from rest to 20 m/s in 8 s with uniform acceleration, travels at this speed for 20 s and finally, when the brakes are applied, comes to rest again with constant retardation in a further 3 s. How far has the car moved?

Considering first the acceleration we have:

$$v = 20 \text{ m/s} \quad u = 0 \quad t = 8 \text{ s}$$

Thus from equation (1) we find

$$20 = 0 + 8a$$
$$\text{or } a = 2\tfrac{1}{2} \text{ m/s}^2$$

and from equation (3)

$$s = (0 \times 8) + (\tfrac{1}{2} \times 2\tfrac{1}{2} \times 64)$$
$$= 80 \text{ m}$$

During the period of constant speed (20 s at 20 m/s) the distance travelled is $20 \times 20 = 400$ m. During the slowing down:

$$u = 20 \text{ m/s} \quad v = 0 \text{ and } t = 3 \text{ s},$$

and from equation (1)

$$0 = 20 + 3a$$
$$\text{or } a = \frac{-20}{3} \text{ m/s}^2$$

$$\therefore s = \left(20 \times 3\right) - \left(\tfrac{1}{2} \times \tfrac{20}{3} \times 9\right)$$
$$= 30 \text{ m}$$

Thus the total distance is:

$$80 + 400 + 30 = 510 \text{ metres}.$$

We can readily check this value by means of a velocity–time graph (Fig. 3.7).

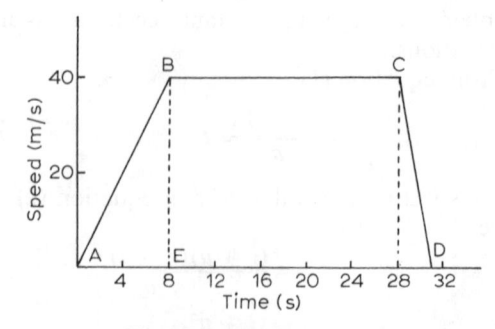

Fig. 3.7. Velocity-time graph

The line *AB* represents the acceleration, *BC* the constant speed and *CD* the retardation. The distance travelled will be given by the area of the trapezium *ABCD*, which is:

$$\tfrac{1}{2}(AD + BC) \times BE$$

This must be calculated in terms of the units on the axes of the graph, and not by simply measuring lengths on it.

$$\text{Thus } s = \tfrac{1}{2}(31 + 20)\text{s} \times 20 \text{ m/s}$$
$$= 25 \cdot 5 \times 20$$
$$= 510 \text{ m}$$

Free Fall

One form of acceleration with which we are all familiar is that of something falling freely. Ancient scientists were aware that something very light like a feather fell slowly, while a heavier object fell much more quickly. They therefore supposed that the acceleration depended on the weight of the falling object. The following experiments show that they were really wrong in assuming this.

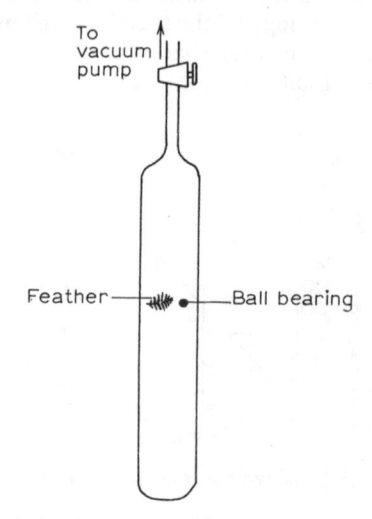

Fig. 3.8. Free fall in a vacuum

Experiment 3.2. Place a small feather and a ball bearing in a thick walled glass tube (Fig. 3.8). Invert the tube a few times and observe that the feather falls much more slowly than the ball bearing. Now connect the tube by a length of rubber pressure tubing to a vacuum pump and remove the air. Close the tap and disconnect the tube from the pump. Invert the tube a few more times. You will see that the feather falls much more quickly than before and in fact there may be little difference between the rates of fall of the two objects.

Experiment 3.3. Release small and large metal weights simultaneously from the same height above the ground (having previously made sure that they will cause no damage on impact). You will find that they strike the ground at the same instant.

These two experiments suggest that the slow fall of light objects is due to air resistance, and that, in the absence of air, all objects would take the same time to fall through the same height.

Acceleration due to Gravity

The next step towards understanding free fall is to find out if a freely falling object has a constant acceleration and, if it has, to measure this acceleration. It is not possible to time a falling object with an ordinary stopwatch, unless the height of fall is very big. If you drop something from the top floor of a three storey building it will only take perhaps two seconds to hit the ground, and your reaction time in operating a stop-watch will be such that an accurate measurement of the time of fall is very hard to obtain. A number of methods for direct timing of a falling object over much smaller heights than this are possible with apparatus available in many school laboratories. One of these possibilities is suggested in the following experiment:

Experiment 3.4. C in Fig. 3.9 represents an electrically operated device for measuring time intervals to $\frac{1}{100}$ or perhaps $\frac{1}{1000}$ second. It may be an electric stop-clock. It may be an electronic counter with a built in device for supplying 100 or 1000 regular pulses of electricity per second which will be registered. In this case each count will represent a time of 0·001 s. The timing device will operate as long as the circuit *ABCDES* is complete. *S* is a switch which can change from position 1 to position 2 in a very short time. When it is in position 1 it completes the circuit *AMRF* so that a current flows through the electromagnet *M*. When it is in position 2 it completes the timer circuit. *E* is a 'trap door' arranged so that when an object falls on it it opens and breaks the timer circuit.

Put the switch in position 1 and use the rheostat to adjust the current in the electromagnet until it JUST supports a steel ball-bearing. Close the trap door and situate it a measured distance (h) vertically below the electromagnet. Attach the ball-bearing to the magnet and operate the switch. The ball will be released and at the same time the timer circuit will be completed. The timer will go on functioning until its circuit is broken again by the impact of the ball on the trap door. In this way the time of fall (t) is found. Make two or three more timings for the same height before repeating the experiment for a range of values of h.

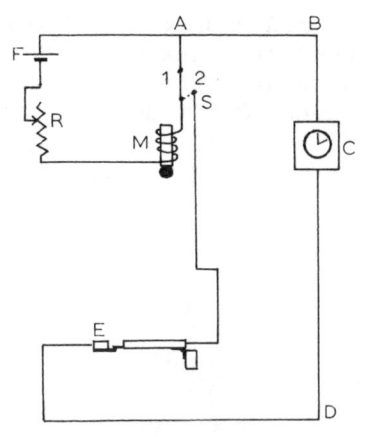

Fig. 3.9. Experiment to measure the acceleration due to gravity

If the acceleration of the falling ball-bearing is uniform then from our second equation of motion ($s = ut + \frac{1}{2}at^2$):

$$h = \tfrac{1}{2}at^2$$

($u = o$, since the ball starts from rest and $s = h$)

So if a is constant and we plot h against t^2 our equation reduces to the form $y = mx$ and we should obtain a straight line graph, the slope of which is ($\frac{1}{2} \times$ acceleration due to gravity). The following table (3.3) shows some typical results from such an experiment.

Table 3.3

h (m)	t (s)	t^2 (s^2)
0·20	0·20	0·040
0·40	0·29	0·084
0·60	0·35	0·122
0·80	0·41	0·168
1·00	0·46	0·212

A graph of these figures is shown in Fig. 3.10.

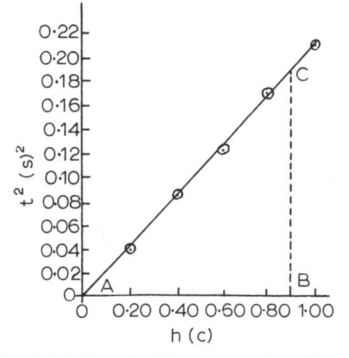

Fig. 3.10. Graph of height against (time)^2

We have plotted h on the x axis and not t^2 as was indicated above. This is because h is the quantity which we varied (h is the independent variable) and t depends on it (t is the dependent variable).

Now a (the acceleration due to gravity)

$$= \frac{2h}{t^2}$$

from the graph it is $2 \times \dfrac{AB}{BC}$

$$= \frac{2 \times 0.90 \,\text{m}}{0.19 \;\;\text{s}^2}$$

$$= 9.52 \text{ m/s}^2$$

More accurate experiments show that this quantity (usually given the letter g) is around 9.80 m/s^2. It is not exactly the same at all places on the earth's surface. This value of g applies to an object projected upwards (in which case it become a retardation) as well as to something falling.

We will take an example to show how the acceleration due to gravity may be used.

A ball is thrown vertically upwards with a speed of 20 m/s. How high will it go and how long will it take to reach the ground again?

The ball will start to fall when its upward speed has fallen to zero. Thus for the upward journey:

$v = 0$ $u = 20$ m/s and $a = -9.8$ m/s.

$$v = u + at$$
$$0 = 20 + (-9.8 \times t)$$
$$t = \frac{20}{9.8} \text{ s} = 2.04 \text{ s}$$

also $s = ut + \tfrac{1}{2}at^2$

so $s = (20 \times 2.04) - (\tfrac{1}{2} \times 9.8 \times 2.04^2)$

$$= 20.4 \text{ m}$$

Thus it takes 2·04 s to reach its maximum height of 20·4 m. On the downward journey

$s = 20.4$ m $u = 0$ and $a = +9.8$ m/s

hence $20.4 = (0 \times t) + (\tfrac{1}{2} \times 9.8 \times t^2)$

$$20.4 = 4.9 t^2$$

$$t = 2.04 \text{ s}.$$

Thus the total time the ball is in the air is 4·08 s.

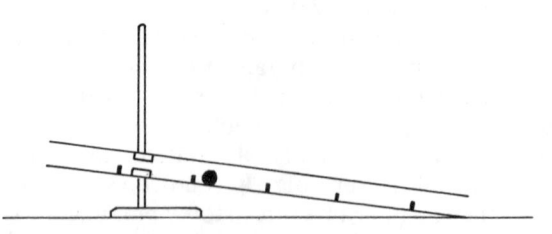

Fig. 3.11. Acceleration down an inclined plane

Experiment 3.5. Movement on an inclined plane

An example of an accelerating body which can be studied rather more easily than the case of free fall is that of a ball rolling down a smooth inclined plane. The apparatus may consist of a long piece of straight wide bore glass tube, or a length of wood with a groove cut along it in which a ball-bearing may roll. Mark off equal lengths of say 20 cm along the track (Fig. 3.11) and clamp it at a convenient shallow slope. Release the ball-bearing from rest at the top of the track and, using a stop-watch, find the time the ball takes to reach the first mark. Repeat the experiment several times taking the time to reach the second, third and subsequent marks. In this way you can obtain a set of values for distance (s) and time (t), which may be analysed as in the free fall experiment. Once again you should obtain a straight line graph of s against t^2 showing that we are dealing with a case of constant acceleration. However the value of the acceleration which you obtain from the graph will be much less than g.

Exercise 3

1 Describe an experiment to illustrate motion with uniform acceleration. Explain how you would establish that the acceleration is uniform.

A body moves from rest with uniform acceleration and travels 90 m in the fifth second. Calculate the velocity of the body 10 s after starting. (J)

2 Define *average velocity* and *acceleration*.

A lift ascends from the bottom of a mine shaft with an acceleration of 2 m/s² for 4 s, continues with uniform velocity for a further 6 s, and is then brought to rest at the surface with a retardation of 1 m/s².

Draw the velocity time graph for the motion, and by using it or otherwise deduce (*a*) the depth of the shaft, and (*b*) the average velocity of the lift during the motion. (L)

3 Define *velocity* and *acceleration*.

A body, starting from rest, moves for 4 s with a uniform acceleration of 3 m/s² and then continues with the velocity so acquired for another 3 s. It is then brought to rest in 5 s more by a uniform retardation. Find the velocity at each of the times of 1 s, 2, 3, 4, 7 and 12 s from the start and hence draw a velocity-time graph. Find EITHER from the graph OR by calculation, the distance travelled by the body from the beginning of the third second to the end of the seventh second. (*F*)

4 In a car test it was stated that, from a standing start, a certain car covered 600 m in 30 s. Calculate the acceleration, assuming this to have been uniform, and also the final velocity reached at the end of 30 s. (*O*)

5 What is meant by *the acceleration due to gravity*? Describe an experiment to determine its value at a given place.

A stone is projected vertically upwards from the ground. If the greatest height reached is 49 m, find the velocity with which it is projected, and the total time during which it is in the air. (*L*)

6 A simple pendulum is made to oscillate in air. The bob is heavy and made of iron. A strong magnet is placed near to, and directly beneath, the mean (lowest) position for the pendulum and the pendulum is observed to decrease in *period*. The bob is observed to travel faster through its mean (lowest) position with the magnet in place than without it.

(*a*) What is meant by the period of the pendulum?

(*b*) Name a form of energy, measured at the mean (lowest) position of the oscillation, which changes in value when the magnet is placed in position. Has this energy increased or decreased in value? (*L*)

4
Forces and Motion

When an astronaut steps out of his space capsule orbiting the earth at about 30 000 km/h, he remains close to it. This is not because he is being dragged along by it or because he has his own private propulsion system. Inside the space ship he was travelling at this high speed and when he steps out he will continue to do so unless some force slows him down. In fact the satellite itself has no propulsion system once it is in orbit. Those rockets it has are only for guidance and slowing it down for re-entry into the earth's atmosphere. All of this is rather strange to us because in everyday life we become aware that to keep things moving we have to keep pushing them. We cannot, unfortunately, switch off the car engine once the car is moving at the required speed. If we do it will quickly come to rest. The essential difference between the behaviour of the satellite and that of the motor car is that the latter has to contend with friction and air resistance. These are both forces trying to stop the vehicle from moving. The satellite which is outside the earth's atmosphere and therefore in a near vacuum, experiences no friction or air resistance. Even with things on the earth's surface we can greatly reduce friction and if we do we find that moving things do tend to go on moving much further. You may for example have pushed a stone along the surface of a frozen pond and been surprised at how far it travelled. In the laboratory there are a number of ways in which friction may be reduced even further. Most of these depend on the 'hovercraft' effect where an object is supported on a cushion of air. Fig. 4.1 shows one simple method of achieving this. A piece of square section steel tube has small holes drilled at regular intervals along one of its faces. One end of the tube is sealed up and the other is connected by a hose to a vacuum cleaner arranged to act as a blower. Air entering the tube escapes through the small holes and a light 'saddle' shaped vehicle will ride on these jets very smoothly. Once set in motion by a gentle

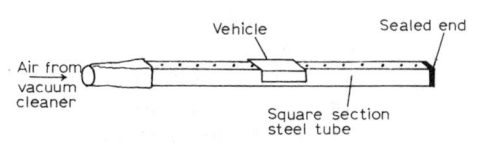

Fig. 4.1. Air track apparatus

push the vehicle will continue moving with a very nearly constant speed.

The conclusions we might come to from these considerations were summarised by Newton in the first of his 'Laws of Motion'. It states:

Things which are at rest will remain at rest and things which are moving will continue moving at constant velocity unless they are acted upon by a resultant force.

This law means that to produce an acceleration (or retardation) there must be a force. If there is no acceleration or retardation there is no *net* force. If a motor car is travelling at constant velocity along the road there is, of course, a number of forces acting on it. But Newton's first law tells us that forces in a forward direction due to the engine will be just equal and opposite to frictional and air resistance forces so that the resultant force is zero.

Force, Mass and Acceleration

Newton's second law of motion deals with the way in which the acceleration which an object has depends on the force producing the acceleration and the mass of the object. A most convenient set of apparatus for investigating these relationships has been developed by the Physical Sciences Study Committee in the United States. It is shown in Fig. 4.2. The accelerated object is a wooden trolley

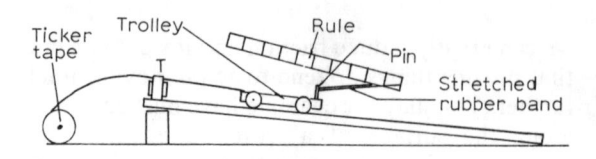

Fig. 4.2. Trolley and ticker tape apparatus

mounted on ball-bearing wheels which moves on a flat, smooth surface so that the friction is very low. For accurate work, what friction there is may be compensated for by allowing the trolley to move down a shallow slope. The gradient should be adjusted until, when the trolley is given a gentle push down the slope, it will continue to move with constant speed. The trolley, as it moves, pulls behind it a narrow strip of paper which passes through the device T. This is a vibrator, somewhat similar in action to an electric bell (see chapter 43) which at regular small intervals of time pushes the tape against a small disc of carbon paper and so makes a series of black dots on the paper tape. If the vibrator operates on a mains A.C. electricity supply it will make 100 dots per second and the time interval between two dots is $\frac{1}{100}$ s. Thus the distance between two dots on the tape is the distance travelled in $\frac{1}{100}$ s. If the trolley has a constant speed the dots will be evenly spaced. If it is accelerating the distance between dots will gradually increase (see Fig. 4.3).

(a) Ticker tape for constant speed

Trolley going this way →

(b) Ticker tape showing acceleration

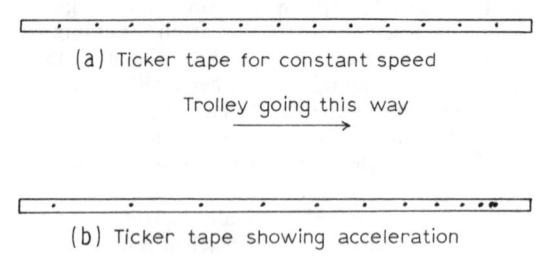

Fig. 4.3.

A convenient method of applying a constant force to the trolley is by means of a stretched rubber band (Fig. 4.3). One end of the band is attached to the trolley while the other is looped around a rule which is held in the hand. If the same mark on the rule is always level with the pin on the trolley then the force exerted by the rubber band on the trolley is always the same. By introducing a second and then a third similar rubber band, stretched by the same amount, it is possible to double and treble the force on the trolley.

Experiment 4.1. Applying a constant force by means of one rubber band, obtain a tape record of the motion of the trolley down the incline. Since the time interval between dots is $\frac{1}{100}$ s, 10 dots corresponds to 0·1 s. By measuring the length of tape corresponding to each set of 10 intervals find the mean speed in cm/s for each 0·1 s time interval of the motion. Plot a velocity-time graph as in Fig. 3.5 (in the previous chapter). Your graph should be a straight line showing that in this case a constant force has produced a constant acceleration.

Experiment 4.2. Repeat Experiment 4.1 using 2, 3 and 4 similar rubber bands, i.e. with 2, 3 and 4 times the accelerating force. Calculate the acceleration in all four cases as follows.

If the acceleration is uniform we may use the expression $v^2 = u^2 + 2as$. Measure the lengths of two ten-dot lengths of tape, one in the early part of the motion (l_1) and one near the end (l_2) (Fig. 4.4). The average speeds during these two intervals of 0·1 s will be $\frac{l_1}{0·1}$ cm/s and $\frac{l_2}{0·1}$ cm/s. These are the values of u and v respectively. Measure the distance (s) between the fifth dots of each set. Sub-

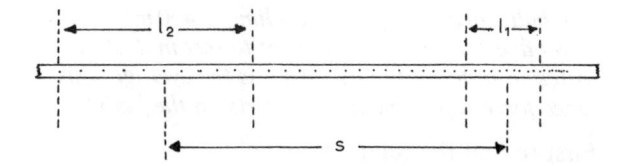

Fig. 4.4. Finding acceleration from tape record

stitute the values of u, v and s in $v^2 = u^2 + 2as$ and calculate the acceleration.

You should observe that your accelerations are in the ratio $1:2:3:4$. From this we can conclude that:

acceleration (a) $\propto$ accelerating force (f)

Experiment 4.3. Find the mass of the trolley. Now measure the acceleration of it by the method described above when it is loaded with various known masses and accelerated always by one rubber band stretched by a fixed amount. Draw a table of results as follows:

Mass of trolley + load (m g)	Acceleration (cm/s^2)	$\frac{1}{m} g^{-1}$

You will see, no doubt as you would expect, that as the mass of the trolley is increased, the acceleration produced by a certain accelerating force gets less. You will see that if the mass of the trolley is doubled the acceleration is halved. This suggests a relation of the form

$$\text{Acceleration} \propto \frac{1}{\text{mass of trolley}}$$

To verify this plot a graph of a against $\frac{1}{m}$. It should be a straight line passing through the origin indicating that $a \propto \frac{1}{m}$.

To summarize the conclusions of these experiments:

(1) A constant force acting on a mass produces a uniform acceleration.

(2) The acceleration is proportional to the accelerating force. $a \propto f$.

(3) The acceleration produced by a certain force is inversely proportional to the mass of the body.

$$a \propto \frac{1}{m}$$

Conclusions (2) and (3) may be combined to

$$a \propto \frac{f}{m}$$

or $f \propto ma$.

This is really Newton's second law of motion although it is not stated in quite the same way as Newton stated it.

Units of Force

The relationship between force, mass and acceleration quoted above may be written:

$$f = kma$$

where k is a constant.

This relationship provides us with the opportunity to define our unit of force, the newton.

1 newton is that force which when applied to a mass of 1 kg gives it an acceleration of 1 m/s².

Referring again to our equation:

$$f = kma$$

We now know that if f is 1 newton and m is 1 kg, then a will be 1 m/s².

so $1 = k \times 1 \times 1$
and $k = 1$

It follows then that if we use the 'right' units, newtons for force and kilograms for mass and metres per second² for acceleration, our equation becomes

$$f = ma.$$

We have seen that if a 1 kg mass (or any other mass for that matter) is allowed to fall freely under the influence of this downward force, it will fall with a uniform acceleration of 9·8 m/s². Thus using the relation $f = ma$,

$$f \text{ newtons} = 1 \times 9\cdot 8$$

where f is the gravitational force which causes the 1 kg mass to have an acceleration of 9·8 m/s². The gravitational force is thus 9·8 N. By similar reasoning the weight of any body whose mass is m kg, must be $m \times g$ N, where g is the appropriate gravitational acceleration. g will of course only be 9·8 m/s² near the surface of the earth. To simplify the arithmetic it is convenient and usually accurate enough to call g, 10 m/s². The weight of a 1 kg mass will thus be 10 N. This practice is frequently adopted in later chapters of this book.

Mass

We are now in a position to be a little more

precise about the meaning of mass. In chapter 2 we suggested that the mass of a body is a measure of the number and type of atoms of which it is made up. We can now see that it is related to its 'inertia', that is to the difficulty with which its state of rest or uniform velocity may be changed. Having chosen a certain piece of material to be our standard of mass, we may say if a certain force gives this an acceleration a, and the same force gives another piece of material an acceleration $2a$, then the mass of our second piece of material is half that of our standard.

Use of the Formula f = ma

Finally in this chapter we will take a simple example to illustrate the application of this important law at which we have arrived.

A bullet of mass 1 g and travelling at 400 m/s strikes a wooden target and is brought to rest in it. It penetrates to a depth of 2 cm. What is the average resistance force which the target exerts on the bullet?

First to find the retardation:

$$u = 400 \text{ m/s} \quad v = 0 \quad s = 0\cdot02 \text{ m}$$
$$v^2 = u^2 + 2as$$
$$\text{so } 0^2 = 400^2 + 2 \times a \times 0\cdot02$$
$$\text{and } a = \frac{160\,000}{0\cdot04}$$
$$t = 4\,000\,000 \text{ m/s}^2$$

Now from $f = ma$ since $m = 0\cdot001$ kg
$$f = 0\cdot001 \times 4\,000\,000$$
$$= 40\,000 \text{ N}.$$

Exercise 4

1 Explain the meaning of the terms *energy* and *force*. State *one* unit in which each can be measured.

A car of total mass 1000 kg is travelling at 60 km/h along a horizontal road. Find the constant retarding force required to reduce it to rest while the car travels 50 m.

2 A horizontal force of 50 N acts on a mass of 10 kg. Calculate the horizontal velocity of the body after 10 s if it starts from rest.

3 Define *acceleration*, and *force*.

A motor car moves from rest with constant acceleration down a hill 300 m long. After 30 s it has travelled 100 m. Find its acceleration, the time it takes to reach the foot of the hill and its velocity when it arrives there.

A car of mass 1200 kg travelling at 50 km/h is brought to rest in 50 m. Find the retarding force.

5
Chlorine, the Green Gas, and its Compounds; The Halogens

One of the chemicals you have already used in the laboratory is hydrochloric acid. The three common acids, hydrochloric, nitric and sulphuric, are called 'mineral' acids because the raw materials from which they are made are naturally-occurring minerals, salt, nitre and sulphur.

Experiment 5.1. Preparation of Hydrogen Chloride.
(i) Take two small test-tubes and in one place a little powdered salt. In the other put a small lump of rock-salt, impure sodium chloride as mined (see p. 264). To each add sufficient concentrated sulphuric acid just to cover the solid. Notice a rapid evolution of gas in the first case and a slower rate of action in the second. The gas given off is hydrogen chloride.

$$NaCl(s) + H_2SO_4(aq) \rightarrow NaHSO_4(aq) + HCl(g)$$

Breathe across the mouth of the tubes; the gas becomes misty.
(ii) When making the gas on a larger scale it is best regulated by using rock-salt with concentrated acid and heating as necessary.

Set up the apparatus of Fig. 5.1 and collect a few jars of the dry gas. At the temperature used in the laboratory sodium hydrogensulphate is formed in the flask and NOT sodium sulphate. Be careful not to let the gas escape into the air in much quantity as it is decidedly unpleasant. Test jars of the gas as below.
(i) Put in a piece of dry blue litmus paper. Any change?
(ii) Repeat with damp litmus paper. Any difference?
(iii) Remove the cover from a jar and notice that the gas becomes visible. This is because it dissolves in the moisture present in the air to form minute droplets of hydrochloric acid which stay suspended in the air. It is this action which makes breathing the gas unpleasant.
(iv) See if the gas will burn.
(v) Plunge a lighted splint into a jar of the gas; is the light extinguished?
(vi) Invert a jar of ammonia over the hydrogen chloride jar and remove the covers. A white solid, ammonium chloride, forms and gradually falls to the bottom of the jar.

$$NH_3(g) + HCl(g) \rightarrow NH_4Cl(s)$$

This is a test for identifying either ammonia or hydrogen chloride if one of them is known.

In industry most hydrogen chloride is now prepared by the controlled burning of hydrogen in an atmosphere of chlorine.

Hydrogen chloride is a gas which, when completely dry or when liquefied, does not behave as an acid. However, it very readily dissolves in water to give a solution having all the properties we associate with acids: it is hydrochloric acid. Because of its great solubility in water (450 volumes in one volume of water at room temperature) precautions must be taken when a solution is being made to prevent water sucking back into the preparation flask. These may take the form of:
(a) having the delivery tube *just* touching the surface of the water;

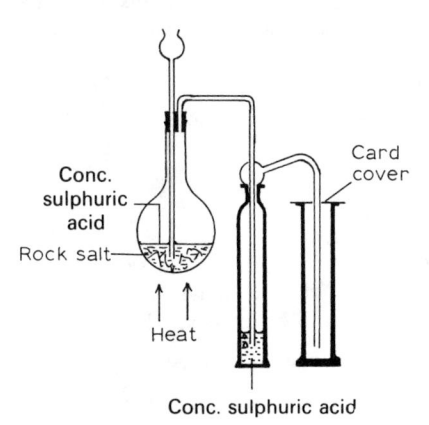

Fig. 5.1

(b) fixing a funnel to the end of the delivery tube and having this *just* touching the water (Fig. 5.2);
(c) fitting a Bunsen-valve to the end of the delivery tube. A slit about 5 mm. long is cut in a piece of rubber tubing closed at one end. The other end fits the delivery tube. Gas under pressure can escape through the slit, but water cannot enter from the outside.

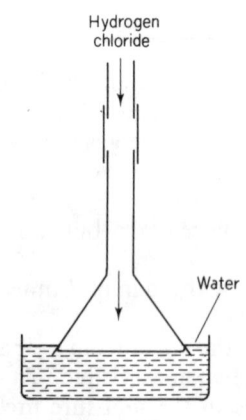

Fig. 5.2

The great solubility of the gas can be demonstrated by means of the 'Fountain experiment' described in connection with ammonia (p. 128).

Experiment 5.2. Testing Solutions of Hydrogen Chloride.
Two solutions of the gas will be required:
 (i) in methylbenzene; bubble the gas through the liquid for some time in the fume cupboard.
 (ii) in water; bench strength dilute hydrochloric acid will be very suitable.
Test separate portions of each solution as follows:
(a) Put a drop of it on neutral litmus or Universal indicator paper.
(b) Add a piece of marble or a measure of anhydrous sodium carbonate.
(c) Put a piece of magnesium or zinc into the solution.
(d) Insert two carbon rods connected via a small lamp to a 6-volt d.c. electricity supply into the solution, about 2 cm apart, and switch on. Is any action or smell noticeable? Does the lamp glow? Insert pieces of red litmus paper close to each electrode; is there any change?
When reaction (c) in the aqueous solution ceases, evaporate some of the liquid to dryness by gentle heating; a white solid is left. What is this?

Write equations for the reactions (b) and (c) in the solution in water.

The above experiment shows you that the solution of hydrogen chloride in the organic liquid does not conduct an electric current, has no action on metals such as magnesium, zinc and iron and has very little action on chalk or sodium carbonate. It only behaves as an acid in the presence of water. Hydrochloric acid is readily electrolyzed, and, like all acids, yields hydrogen at the negative electrode (cathode).

You will find out later that the solvent employed in making a solution of a substance often plays a large part in determining the type of properties which the solution has.

The solution of the gas in water is hydrochloric acid. It reacts with marble to liberate carbon dioxide.

$$CaCO_3(s) + 2HCl(aq) \rightarrow CaCl_2(aq) + H_2O(l) + CO_2(g)$$

It is decomposed by many metals with evolution of hydrogen.

$$Mg(s) + 2HCl(aq) \rightarrow MgCl_2(aq) + H_2(g)$$

Hydrogen chloride does not burn nor does it easily support combustion although it will combine with heated sodium or iron (see p. 265) to form chlorides. It is denser than air (1·25:1) and has a pungent smell. The salts formed from it or from its solution are called chlorides.

Experiment 5.3. Preparation of Chlorides from Hydrogen Chloride.
Various methods are available for the preparation of chlorides. Their synthesis using chlorine is described on pp 25 and 266. Formation of sodium chloride or iron(II) chloride from the metal and hydrogen chloride has been referred to above. Here we shall consider methods using hydrochloric acid.
 (i) Neutralization of the acid by an alkali such as sodium hydroxide produces a salt and water. This preparation you will already have carried out.

$$NaOH(aq) + HCl(aq) \rightarrow NaCl(aq) + H_2O(l)$$

(ii) Add sufficient calcium carbonate or zinc oxide to warm bench hydrochloric acid in a beaker until no more will dissolve. Warm the suspension and filter to remove excess solid.

$$CaCO_3(s) + 2HCl(aq) \rightarrow CaCl_2(aq) + H_2O(l) + CO_2(g)$$

$$ZnO + 2HCl \rightarrow ZnCl_2 + H_2O$$

Evaporate about half the water away and then put a drop of the solution on a watch glass and cool it to see if crystallization occurs. When it does, leave the bulk of the liquid in a covered dish to crystallize. Calcium or zinc chloride will be formed.

(iii) To a solution of lead(II) nitrate add hydrochloric acid. White lead(II) chloride, only sparingly soluble in water, is precipitated.

$$Pb(NO_3)_2(aq) + 2HCl(aq) \rightarrow PbCl_2(s) + 2HNO_3(aq)$$

If the solution is heated the chloride is dissolved, being reprecipitated on cooling.

Experiment 5.4. Recognition of Chlorides.
(i) To a solution of (a) hydrochloric acid, (b) solium chloride add a little silver nitrate solution. A white precipitate of silver chloride is formed (silver and lead (II) chlorides are the only common chlorides insoluble in cold water).

$$AgNO_3(aq) + NaCl(aq) \rightarrow AgCl(s) + NaNO_3(aq)$$

Divide the suspension into three parts;
(a) Leave—the colour darkens.
(b) Add dilute nitric acid and shake—no change.
(c) Add dilute ammonia solution and shake—the precipitate dissolves.

Here we have the standard method of testing for the presence of a chloride. If, to a solution of the suspected substance is added dilute nitric acid followed by silver nitrate solution and a white precipitate soluble in a dilute ammonia solution is obtained, then a chloride is present. All soluble metallic chlorides behave in this way.

(ii) Mix a little solid chloride with an equal amount of granular manganese(IV) oxide and cover with concentrated sulphuric acid. Heat; chlorine is evolved and it can be recognised by its bleaching action on moist litmus paper held at the mouth of the tube and by its greenish colour.

*Experiment 5.5. Preparation of Chlorine.
One element of great importance in industry is chlorine; in the laboratory it is made by heating hydrochloric acid with some oxidizing agent which will remove the hydrogen and leave the chlorine. A very suitable one is manganese(IV) oxide and this is warmed with concentrated hydrochloric acid in the apparatus shown in Fig. 5.3. The gas evolved always contains some acid fumes and these are removed by passing the chlorine through

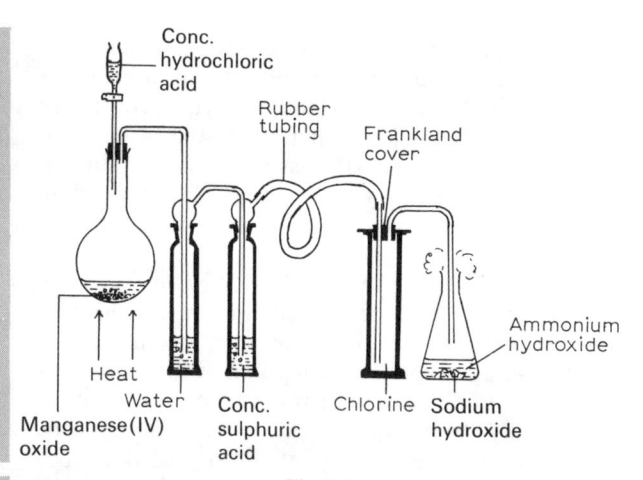

Fig. 5.3.

a Dreschel bottle containing water. If the gas is required dry it is then passed through concentrated sulphuric acid in a second bottle. If the preparation is carried out in a fume cupboard the gas can be collected in jars in the usual way by displacement of air upwards. This is perhaps the best way as chlorine is extremely poisonous, but it can be prepared quite safely on the open bench by absorbing any excess gas in a trap containing some stick sodium hydroxide and 1:1 ammonia solution. The manganese(IV) oxide used in this experiment does *not* behave as a catalyst as it does in the preparation of oxygen from a mixture of it with potassium chlorate(V), but is changed into manganese(II) chloride.

The reaction can be represented by three partial equations which can then be added together to give the final result:

$$MnO_2 = M\cancel{nO} + \cancel{O}$$
$$\cancel{MnO} + 2HCl = MnCl_2 + H_2O$$
$$2HCl + \cancel{O} = Cl_2 + H_2O$$

Adding, $MnO_2 + 4HCl \rightarrow MnCl_2 + Cl_2 + 2H_2O$

If the gas is not required dry it can be collected over brine or hot water.

Properties of Chlorine
Chlorine is a pale-green poisonous gas which acts painfully on the lining of the nose and mouth and on the lungs. If it is accidentally inhaled a speedy 'first-aid' remedy is to take a good sniff at a bottle of ammonia solution (dilute) or to gargle with milk. It is quite a heavy gas, being about $2\frac{1}{2}$ times as dense as air or $35\frac{1}{2}$ times as heavy as the same volume of hydrogen. The atom contains 17 protons along with either 18 or 20 neutrons,

three atoms out of every four having 18 neutrons. The molecule of chlorine, like that of other gases such as hydrogen, nitrogen and oxygen, consists of two atoms joined together. One dm^3 of water at room temperature will dissolve about $2\frac{1}{2}$ dm^3 of chlorine to form a yellow green solution smelling strongly of the gas.

Experiment 5.6. The Action of Chlorine on Water.
Invert a test-tube containing water, through which chlorine has been bubbling, in a dish of the same solution and leave exposed to the light (preferably bright sunshine) for a time. The solution gradually loses its colour and a small amount of a colourless gas collects at the top of the tube. If there is sufficient it can be tested with a glowing splint which will be re-ignited, identifying the gas as oxygen.

Now dip a piece of blue litmus in the solution and notice that it is first turned red and then colourless or bleached.

Again using partial equations the first reaction is explained by:

$$H_2O + Cl_2 \equiv HCl + HClO$$
chloric (I) acid

The chloric(I) acid is fairly easily decomposed and each molecule of it is capable of oxidizing certain coloured dyestuffs such as litmus:

$$HClO + dyestuff \rightarrow HCl + \text{Oxidized dyestuff}$$

If this oxidized dyestuff is a colourless compound we say the dye has been bleached. In the absence of any such substance the available oxygen atoms join together to form molecules. The final result of the action of chlorine and water is to form hydrogen chloride and oxygen.

$$2H_2O(l) + 2Cl_2(g) \rightarrow 4HCl(aq) + O_2(g)$$

Notice from this experiment that water *must* be present before any bleaching by chlorine can occur; in its absence no chloric(I) acid can be formed. Household bleaches such as Domestos, Parazone and Milton contain chloric(I) acid.

The following experiment will show the necessity for water to be present before bleaching will occur:

Experiment 5.7. Chlorine as a Bleaching Agent.
Take a piece of some dyed cloth (choose a strong colour such as Turkey Red) and with a rod dipped in water mark a cross on it. Place the piece in a jar of chlorine prepared and dried as in Experiment 5.5. In a very short time the cross will be visible in a lighter colour as it becomes bleached, but the rest of the material is unchanged.

The above experiments show one particular property of chlorine, the strong affinity or attraction it has for hydrogen, even when the latter is combined with some other element such as oxygen in water. A few other experiments will illustrate this even more clearly:

Experiment 5.8. The Attraction of Chlorine for Hydrogen.
(i) Invert a jar of chlorine over one of hydrogen for a few minutes and then apply a light to each jar. There is a loud report as the two gases combine to form hydrogen chloride.

$$H_2 + Cl_2 \rightarrow 2HCl$$

If the mixed gases are left exposed to strong sunlight they will slowly join together without any light being applied.

(ii) Connect a silica or metal jet, bent as shown in Fig. 5.4, to a hydrogen generator and collect a test-tube of gas. Hold this near a bunsen flame and note if it burns *quietly*. Only if it does so (why?), light the gas at the jet and lower it into a jar of chlorine. White fumes of hydrogen chloride are formed and the pale blue hydrogen flame becomes leaden in colour.

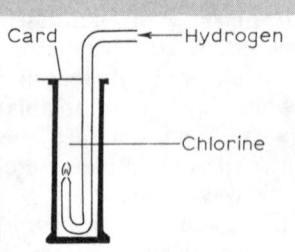

Fig. 5.4. Burning hydrogen in chlorine

(iii) Heat some *pure* turpentine—a hydrocarbon—in a test-tube until it boils (being careful that it does not take fire) and then pour it quickly into a jar of chlorine. Immediately blackish-grey fumes appear.

$$C_{10}H_{16}(l) + 8Cl_2(g) \rightarrow 10C(s) + 16HCl(g)$$

(iv) Wax is also a hydrocarbon. Light a taper or candle and lower it into a jar of chlorine. Notice once more the fumes of hydrogen chloride which this time again become dark grey as black particles of carbon are released from the wax and mix with the gas. The flame becomes reddish as the other spectrum colours which make up its normal appearance are unable to penetrate through the mass of carbon particles to reach the observer's eye. The reason for this may become clear after you have studied Chapter 33.

Because chlorine is so very active (it is the second most active element known) it will combine with many other elements, both metallic and non-metallic. Here is another method of making chlorides. In this case the chloride is made by direct union, i.e. it is synthesized. In the following experiments it is not necessary to dry the chlorine first unless this is specifically stated.

Experiment 5.9. Synthesis of Chlorides.
(i) Place a small piece of yellow phosphorus (dry) in a deflagrating spoon and warm just a little before inserting it into a jar of chlorine. It immediately catches fire and forms white fumes just as it did when placed in oxygen. But these fumes are phosphorus pentachloride, not phosphorus pentoxide.

$$2P(s) + 5Cl_2(g) \rightarrow 2PCl_5(s)$$

(ii) Repeat the above experiment using sodium. The metal burns with a brilliant yellow flame and the fumes are of sodium chloride.

$$2Na(s) + Cl_2(g) \rightarrow 2NaCl(s)$$

If dry hydrogen chloride is passed over heated sodium the metal takes fire and again the chloride is formed.

$$Na(s) + 2HCl(g) \rightarrow 2NaCl(s) + H_2(g)$$

(iii) Light a strip of magnesium and plunge it into chlorine. It burns brilliantly forming white magnesium chloride.

$$Mg(s) + Cl_2(g) \rightarrow MgCl_2(s)$$

(iv) Pass a stream of dry chlorine through a long combustion tube containing iron wire (steel wool) which is being heated (Fig. 5.5). The metal changes into black crystals of iron(III) chloride (ferric chloride) which sublime and are deposited in the flask at the end of the tube. (If the gas is not dry a moist brown deposit or solution will be obtained.)

$$2Fe(s) + 3Cl_2(g) \rightarrow 2FeCl_3(s)$$

If this experiment is repeated using dry hydrogen chloride instead of chlorine the product is iron(II) chloride (ferrous chloride).

$$Fe(s) + 2HCl(g) \rightarrow FeCl_2(s)$$

(v) Obtain a sheet of Dutch metal (an alloy of zinc and copper which looks like gold leaf) and roll it into a ball. Drop it into a jar of chlorine and quickly put back the cover plate. In a very short time fumes of zinc and copper chlorides form, possibly accompanied by a flash of flame.

$$Zn/Cu + 2Cl_2 \rightarrow ZnCl_2 + CuCl_2$$

Bleaching Powder

Chlorine was first prepared by Scheele in the eighteenth century and was soon used as a bleaching agent. The poisonous nature of the gas and the difficulty of transporting it from one place to another, however, led to attempts to find a much less obnoxious substance. It was discovered that slaked lime would absorb the chlorine to form a new substance (called bleaching powder) and that the chlorine could be released from this powder when and where required by the addition of a dilute acid. In the laboratory we can prepare this ourselves.

Experiment 5.10. Preparation of Bleaching Powder.
Add some calcium hydroxide (slaked lime) to a jar of chlorine and shake; continue to add powder until the colour of the gas has completely gone. The new substance formed, bleaching powder or 'chloride of lime', has only a faint smell of chlorine.

$$Ca(OH)_2(aq) + Cl_2(g) \rightarrow CaOCl_2(s) + H_2O(l)$$

Place a very small amount of it in a few cm³ of dilute acid and notice the evolution of chlorine.

$$CaOCl_2(s) + 2HCl(aq) \rightarrow CaCl_2(aq) + H_2O(l) + Cl_2(g)$$

Add some water to the rest of the powder and make a suspension and arrange three other dishes alongside it containing respectively dilute hydrochloric or nitric acid, a solution of photographers' 'hypo' (sodium thiosulphate(VI)) and water. Now take a piece of coloured cloth and soak it in the bleaching powder suspension. Transfer it with tongs to the acid and notice that it is bleached in this dish. Next, dip it into the two remaining dishes

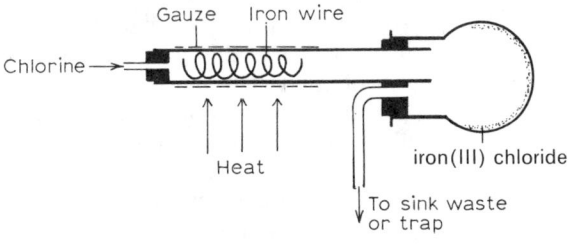

Fig. 5.5.

and dry. The thiosulphate(VI) acts as an 'anti-chlor' —i.e. it removes excess chlorine which might rot the material. Finally the water washes the different chemicals out of the cloth. The action of the thiosulphate(VI) can be represented by the equation:

$$Na_2S_2O_3(s) + Cl_2(g) + H_2O(l) \rightarrow Na_2SO_4(aq) + 2HCl(aq) + S(s)$$

Now that it is possible to transport liquefied chlorine in cylinders quite easily, this element is once more being used for bleaching.

Industrial Manufacture and Uses of Chlorine

Most chlorine is now manufactured by the electrolysis (i.e. splitting-up by means of an electric current) of a strong solution of common salt (see p. 259). The main substance produced is sodium hydroxide and chlorine is a very important by-product.

The amount of chlorine produced annually in the United Kingdom is about 700 000 Mg (One megagram (Mg) is approximately equal to one ton.) Its main uses include:

(a) Preparation of bleaching agents (Parazone, Milton, Domestos) and for direct use as a bleach.
(b) Manufacture of hydrogen chloride by the burning of hydrogen in chlorine and absorption of the gas in water to form hydrochloric acid.
(c) Sterilization of drinking water and swimming baths. A very minute proportion of chlorine is sufficient to kill bacteria which cause cholera, typhoid fever and other diseases.
(d) Manufacture of organic chemicals, many of which are used for dry cleaning (tetrachloromethane or carbon tetrachloride, CCl_4, Westrosol, C_2HCl_3) or as anaesthetics (trichloromethane or chloroform, $CHCl_3$). Some, such as polyvinyl chloride (PVC), are amongst the newer plastics and others, such as Dettol, TCP, and DDT, are used as disinfectants, germicides, weed killers, etc. Modern aerosols use a propellant called freon; chlorine is used in its manufacture as it is in that of neoprene (synthetic rubber) and of arcton (used as a refrigerant).

The Halogens

Chlorine is only one of an important group of elements know as the *halogens* (Gk. salt producers). All the halogens have names derived from Greek or Latin and are listed in the next column.

Fluorine is the most reactive of the elements

Element	Symbol	Appearance	Meaning of Name	Names of Salts
Fluorine	F	Gas	*'Fluere'* to flow	Fluorides
Chlorine	Cl	Greenish gas	*'Chloros'* green	Chlorides
Bromine	Br	Red-brown pungent liquid	*'Bromos'* a stench	Bromides
Iodine	I	Silvery solid forming a violet vapour	*'Ioedes'* like a violet	Iodides
Astatine	At	Solid		Astatides

and, like the other halogens, is never found in the native state. Its great reactivity has led to difficulties in its manufacture. The only suitable method available to obtain it from one of its compounds, such as potassium fluoride, is by electrolysis, but the problem is to prevent the gas from attacking the electrodes, as almost all metals react with fluorine. A special steel is now employed in modern cells along with a carbon anode that does not contain any graphite. Fluorine attacks water much more vigorously than does chlorine to give a variety of products. Bromine and iodine are much less reactive and often require to be heated first.

It has been mentioned before that compounds which are easily formed are usually difficult to split up again and vice versa. What would you think then about compounds of fluorine? Stable or unstable? Obviously, very stable. Those of chlorine, bromine and iodine are progressively less stable as will be shown in Experiment 5.11 later.

We have stated that chlorine is made industrially by the electrolysis of sodium chloride as brine. The main source of bromine is the bromides present in sea water (such as magnesium bromide) and of iodine is sodium iodate(V), $NaIO_3$, found among deposits of sodium nitrate, $NaNO_3$ (*caliche*), in northern Chile. Some iodine is also obtained from sea-water or sea-weed; some forms of sea-weeds and sponges have the ability to absorb iodine from any present in the sea, in the form of iodides, iodates or organic compounds.

Experiment 5.11. A Comparison of the Activity of the Halogens. (i) Pass a current of chlorine into test-tubes containing very dilute solutions of (*a*) potassium bromide, (*b*) potassium iodide (alternatively, chlorine water may be added).

Bromine is displaced from (*a*) and iodine from (*b*).

If some tetrachloromethane (better known as carbon tetrachloride) is added to the tubes and these are then shaken, the halogens liberated will be absorbed in this liquid to give either a reddish-brown solution (bromine) or a pink one (iodine). These solutions do not mix with the water and will form a layer below it in the tube.

If chlorine or chlorine water is not available, aqueous sodium chlorate(I) (made from chlorine and sodium hydroxide)* may be used. In the case of the bromide a drop or two of dilute sulphuric acid should be added: this is not necessary in the case of the less stable (and therefore more reactive) iodide.

(ii) Add a little bromine water to a dilute solution of potassium iodide containing some tetrachloromethane and shake. Again notice the pink colour in the bottom layer (brown in the water). Obviously iodine has again been displaced.

This experiment shows that chlorine combines more readily with potassium than does either bromine or iodine, since it displaces both these elements from their potassium salts:

Bromides and iodides are salts of hydrobromic and hydriodic acids, which correspond to hydrochloric acid.

$$Cl_2 + 2KBr \rightarrow Br_2 + 2KCl$$
$$Cl_2 + 2KI \rightarrow I_2 + 2KCl$$

*Household bleaches such as Domestos or Parazone are very suitable.

The reaction occurring in (ii) is expressed by

$$Br_2 + 2KI \rightarrow I_2 + 2KBr$$

This result shows that bromine is intermediate in activity between chlorine and iodine. Little is known about astatine, but it is the least reactive of all the halogens.

The uses of chlorine have been fully considered. Fluorine gas has few uses, but some of its more complicated compounds, such as the refrigerant, *freon 12* (CCl_2F_2), and the 'non-stick' coating, *Teflon* or PTFE (polytetrafluoroethene), used on cooking pans and in bearings, are well known and depend for their use on their stability and lack of reactivity.

Fluorine compounds are present in the enamel of our teeth, and traces of these compounds are essential for our health; some Water Boards 'fluorinate' drinking water by adding a little sodium fluoride with this end in view.

Bromine is used mainly in the manufacture of silver bromide for photography and in drug manufacture (potassium bromide is a sedative). Iodine is an element essential to health. A deficiency of it leads to an abnormal swelling of the thyroid gland in the neck. This condition is known as 'goitre' and can be controlled by adding an iodide to the diet in regulated quantities (usually in drinking water or mixed with common salt).

Exercise 5

1 Explain what happens when a solution of hydrogen chloride is made by passing the gas through an inverted funnel just touching the surface of some water in a beaker. Draw a diagram.

2 Draw a clearly labelled diagram of an apparatus you could use for preparing hydrogen chloride gas from common salt. Say in labelling whether heat is needed throughout the experiment or not, and if you use any acid in the preparation say whether the acid must be concentrated or dilute.

Dilute hydrochloric acid reacts with each of the three substances zinc, marble (calcium carbonate) and sodium hydroxide solution. Consider each in turn and for each say what products are formed, and where any gas is evolved say how you would attempt to test for the gas. (*EM*)

3 Describe how you would prepare specimens of (*a*) copper(II) chloride from copper(II) carbonate and (*b*) iron(III) chloride from iron(II) chloride. Give equations.

4 Write the equations for the action of (*a*) nitric acid and (*b*) sulphuric acid on bleaching powder. Which of the two acids is the more suitable for obtaining a large amount of chlorine from the bleaching powder? Explain why you think this is so.

5 Starting with common salt explain in detail how you would prepare from it (*a*) another chloride soluble in water and (*b*) a chloride insoluble in water. Give all equations.

6 Draw a labelled diagram of the apparatus you would use to produce from hydrochloric acid some jars of pure dry chlorine. Give the equation for the reaction.

What reactions occur between chlorine and (*a*) calcium hydroxide, (*b*) a cold solution of sodium hydroxide, (*c*) red hot iron? (*L*)

7 Describe, with a labelled diagram, the preparation of several gas jars of reasonably pure, dry chlorine. State TWO everyday uses of the gas.

Describe and explain what is seen to happen on introducing into separate jars of chlorine (*a*) a burning candle, (*b*) some dry white phosphorus, (*c*) a piece of damp blue litmus paper. (*S*)

8 Describe a laboratory method for the preparation and collection of hydrogen chloride.

How would you show that this gas is extremely soluble in water?

What is the action of the solution so formed when warmed with (*a*) magnesium, (*b*) sodium sulphite,

(c) manganese(IV) oxide (manganese dioxide)? (F)

9 (a) Name a substance that will liberate chlorine from concentrated hydrochloric acid, and give the equation. What type of reaction is this?

(b) Draw a labelled diagram showing how you would obtain a few gas jars of dry chlorine, free from hydrogen chloride.

(c) Describe what you observe when the following substances are lowered into jars of chlorine, and say what is formed: (i) phorphorus, (ii) red-hot iron filings or wire, (iii) EITHER a burning wax candle OR a filter paper soaked in warm turpentine. (C)

10 Draw a labelled diagram of the apparatus you could use to prepare jars of pure dry chlorine from hydrochloric acid. Write the equation for the reaction.

Explain how the bleaching action of chlorine differs from that of sulphur dioxide. (L)

11 Outline a method of preparing and collecting hydrogen chloride. (A sketch of the apparatus is not required.) Write the equation representing the reaction.

What do you *observe* when hydrochloric acid, of a concentration suitable to give a reaction, is added to each of the following substances: sodium carbonate, iron filings, lead(IV) oxide (lead dioxide)? In each case state the concentration of the acid used, whether heat is required, and write the equation representing the reaction.

'Hydrochloric acid is a monobasic acid.' What does this statement mean? (J)

12 By the use of the raw materials air, water, rock salt and sulphur, describe how (a) an acid, and (b) an alkali, can be manufactured on an industrial scale.

Your answer should indicate the names and formulae of the acid and alkali, the chemicals used, the raw materials from which they have been obtained and the conditions for the reactions together with the relevant chemical equations.

Name the substance obtained by reacting together your named acid and alkali and give one large scale use of this substance. (L)

13 Dry chlorine was passed over heated iron and the product collected in a dry flask.

(a) (i) What is the name of the reaction product?
(ii) What is the colour of the product?

(b) (i) How is dry chlorine prepared in the laboratory? Draw a labelled diagram to illustrate your answer. No description is required.
(ii) What are the products of the reaction between chlorine and sodium hydroxide solution?

(c) Iron powder reacts with dry hydrogen chloride. Write a balanced equation for the reaction.

(d) Draw a labelled diagram for the preparation of dry hydrogen chloride.

(e) Briefly describe an experiment to show the very soluble nature of hydrogen chloride.

(f) In an experiment a mixture of chlorine and hydrogen in a sealed contained was left standing for several days. The colour of chlorine gradually faded and tests showed that hydrogen chloride was formed. How would you increase the rate of reaction between chlorine and hydrogen?

(g) What volume of dry chlorine at room temperature would be required to react completely with 1.40 g of iron? Assume that the molar volume of chlorine at r.t.p. is 24 dm^3.

6
How Elements Combine

In the first chapter of the book you were introduced to the fundamental particles of chemistry, the proton, neutron and electron, which together make up the atoms of the different elements. In order to study the ways in which chemicals react with each other it is essential to know how these particles are arranged in the atom. It would be helpful here to read Chapter 1 once more.

It is very difficult at this elementary stage to give a satisfactory picture of the real nature of the electron, but an explanation of the behaviour of elements can be given if we consider it to be an extremely small particle carrying a single electric charge of negative sign. As explained earlier, the number of electrons varies and they can be regarded as existing in various layers around the nucleus. As the number of electrons increases the number of layers must also increase in order to accommodate the extra electrons. There is only one layer or shell in the hydrogen and helium atoms and this holds the one or two electrons present. In all the other atoms there can be up to eight electrons in the outermost shell; when this number is reached a new shell is formed. It is these electrons in this shell which determine the behaviour and chemical properties of the atoms; they are referred to as valency electrons.

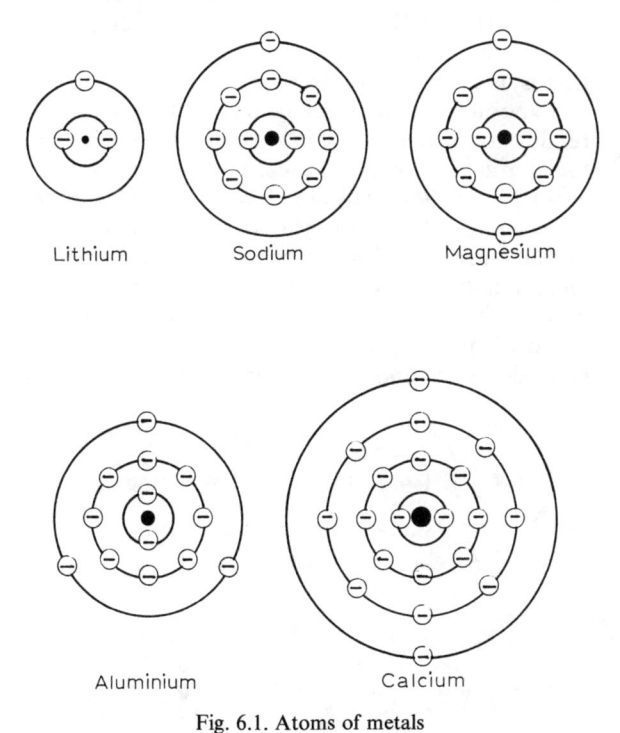

Fig. 6.1. Atoms of metals

The electronic structures of some simple metals and non-metals are shown in figures 6.1 and 6.2.

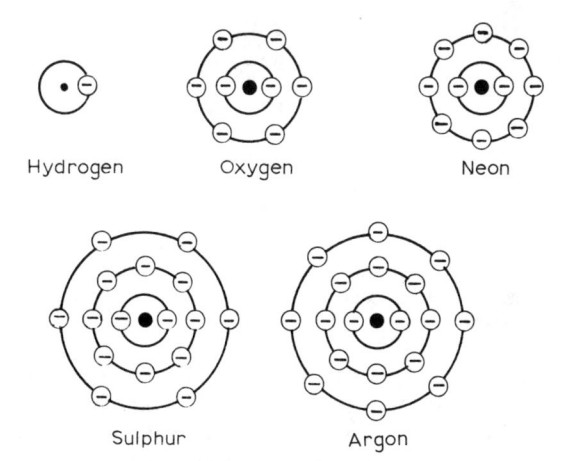

Fig. 6.2. Atoms of non-metals

The shells are also referred to as *energy levels*, the amount of energy (e.g. as heat) which has to be supplied in order to break off an electron from an atom depending on its distance from the nucleus, to which the protons attract it by virtue of their opposite charge.

The total number of protons and neutrons in an atom is known as the *mass number* of the element and can be shown by putting a small figure at the top left-hand corner of the symbol, e.g. $^{12}_{6}C$. The figure at the bottom is the *atomic number*. Thus $^{12}_{6}C$ means that the atom of carbon has six protons and a total of twelve protons and neutrons; therefore it must also have six neutrons. As each proton is balanced by an electron there must also be six electrons. The figures for elements with atomic numbers 1–20 are shown here.

1	Hydrogen	$^{1}_{1}H$
2	Helium	$^{4}_{2}He$
3	Lithium	$^{7}_{3}Li$
4	Beryllium	$^{9}_{4}Be$
5	Boron	$^{11}_{5}B$
6	Carbon	$^{12}_{6}C$
7	Nitrogen	$^{14}_{7}N$
8	Oxygen	$^{16}_{8}O$
9	Fluorine	$^{19}_{9}F$
10	Neon	$^{20}_{10}Ne$
11	Sodium	$^{23}_{11}Na$
12	Magnesium	$^{24}_{12}Mg$
13	Aluminium	$^{27}_{13}Al$
14	Silicon	$^{28}_{14}Si$
15	Phosphorus	$^{31}_{15}P$
16	Sulphur	$^{32}_{16}S$
17	Chlorine	$^{35}_{17}Cl$
18	Argon	$^{40}_{18}Ar$
19	Potassium	$^{39}_{19}K$
20	Calcium	$^{40}_{20}Ca$

From these two figures the number of neutrons can easily be calculated.

When metals and non-metals combine together they can gain or lose electrons in order to achieve a structure having eight electrons in the outermost shell. The metal atoms *lose* the one, two or three electrons they have in this layer and so leave some protons in the nucleus with no corresponding electrons to maintain neutrality. These atoms become *ions* with one, two or three positive charges. It is easier for the atoms of non-metals to *capture* electrons in order to complete the outermost group and so they then have excess electrons and become ions with negative charges.

Sodium atom
Na

Chlorine atom
Cl

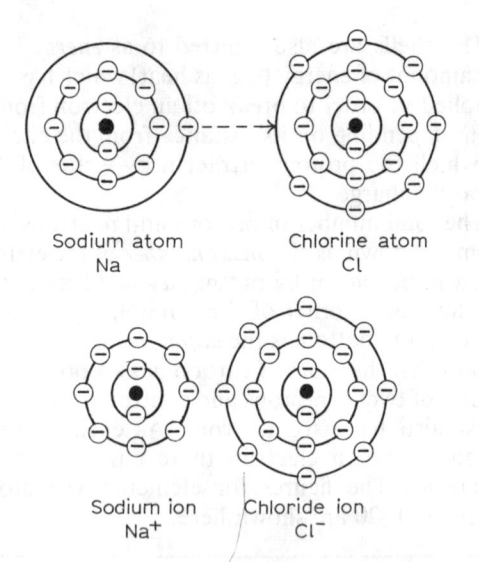

Sodium ion
Na$^+$

Chloride ion
Cl$^-$

Fig. 6.3. Formation of sodium chloride from sodium and chlorine

(Note that metal ions are smaller than the corresponding atoms)

Thus we have H$^+$, Li$^+$, Mg^{2+}, Ca^{2+}, Al^{3+}, O^{2-}, S^{2-}, Cl$^-$ and Br$^-$, the plus and minus signs showing the charge of each ion. (Notice that hydrogen and metals form positively charged ions and that non-metals form negatively charged ones.) It should be fairly simple to understand how this loss or gain takes place by studying Fig. 6.3 which represents the formation of sodium chloride from its elements.

$$Na + Cl \rightarrow Na^+Cl^- \text{ or } NaCl$$

It should be apparent that an atom of magnesium would join with two atoms of chlorine (it has two electrons to lose) to form magnesium chloride with the formula MgCl$_2$. This is shown in Fig. 6.4.

Groups of atoms such as NH$_4$ (ammonium), OH (hydroxide), NO$_3$ (nitrate), CO$_3$ (carbonate), SO$_4$ (sulphate) and HCO$_3$ (hydrogencarbonate) also form ions; the first one is positively charged and the others negatively, e.g. NH$_4^+$, OH$^-$, NO$_3^-$, CO$_3^{2-}$, SO$_4^{2-}$, and HCO$_3^-$.

The charge on each ion is its valency or combining power and indicates the number of monovalent ions of opposite charge with which it will combine.

You should notice here that the compounds formed are neutral electrically so the positive and negative charges must always balance. These compounds formed by the transference of electrons are called *electrovalent* or *ionic* ones. Their formation is as example of *electrovalency*.

Some elements such as carbon or nitrogen do not readily form ions, but join with other elements by the sharing, not transference, of electrons. As one atom of carbon will join with four atoms of hydrogen or chlorine in this way it has a valency of four; this type of combination is called *covalency* and a number of elements including hydrogen and chlorine are able to react in both ways. Examples of covalent compounds are ammonia, hydrogen chloride, water and most organic compounds.

Oxygen

Nitrogen

Water

Ammonia

Methane

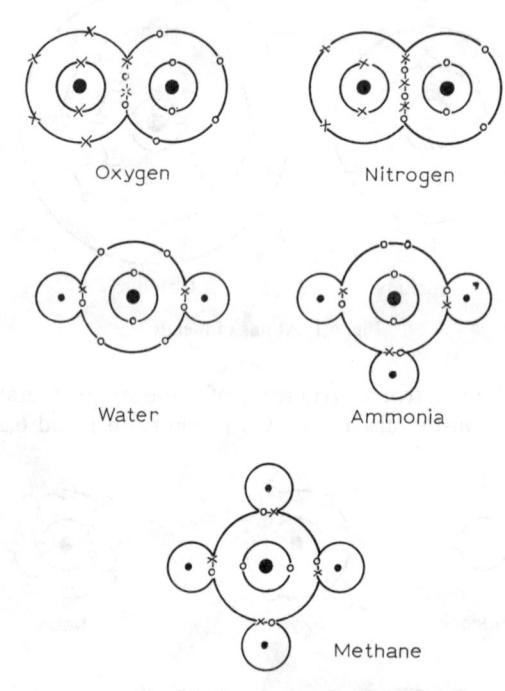

Fig. 6.5. Covalent molecules

(the electrons are represented by X and O for convenience only in showing to which atoms they originally belonged; they are, of course, all the same)

Let us see how this happens when nitrogen and hydrogen join to form ammonia (Fig. 6.5). Nitrogen with five outer electrons needs another three to complete this shell while each hydrogen atom

Chlorine atom

Magnesium atom

Chlorine atom

$$Mg + Cl_2 \longrightarrow Mg^{2+}(Cl^-)_2$$

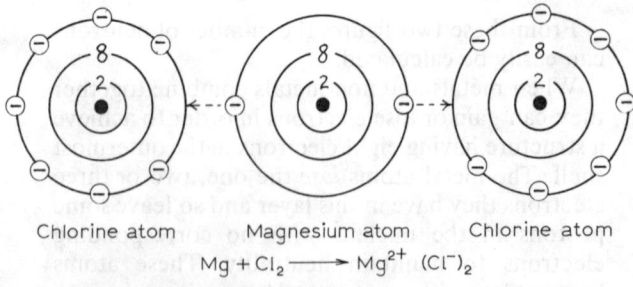

Fig. 6.4. Transference of two electrons from an atom of magnesium in forming magnesium chloride

(only the valency electrons are shown, the complete orbits being indicated by the number of electrons in that shell) needs one more electron. This balance is achieved when the three hydrogen electrons and three from the nitrogen are shared by the two atoms, each retaining a half-share in each electron. The diagram should make this clear. Such a compound is a *covalent* one and does not contain any ions. The nitrogen atom by losing a half-share in three electrons and gaining a half-share in three others still remains electrically neutral: its structure can be shown by joining the atoms with single lines (called covalent bonds) representing each pair of electrons shared. The structural formulae and the way in which the atoms are arranged in space in these compounds are depicted in Fig. 6.6 (remember molecules are three-dimensional).

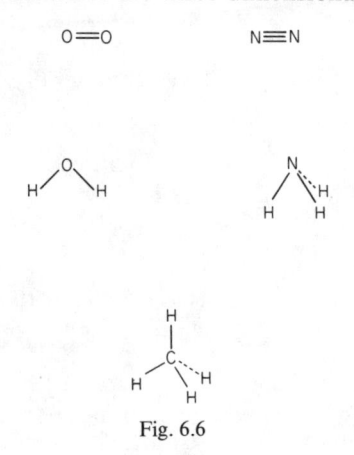

Fig. 6.6

It has already been stated that the smallest part of an element which can exist alone is called a molecule and that such a molecule may contain one atom as in the rare gases helium, neon, argon, etc. or more. In the gases hydrogen, oxygen, nitrogen, fluorine and chlorine and in the other halogens, bromine and iodine there are two atoms in the molecule, i.e. they are diatomic. The two atoms are joined together by covalent bonds and some of these are also shown in Fig. 6.5.

Electrovalent substances when fused or dissolved in water will conduct an electric current and this property will be considered in the chapter on electrolysis.

The valency of an element is the number of atoms of hydrogen or chlorine with which one atom of the element will combine.

Chlorine is included as a standard as well as hydrogen because many metals do not combine with hydrogen. They may, however, displace this gas from a dilute acid and the number of atoms of hydrogen displaced by one atom of the metal is again a measure of the valency of the metal, e.g.:

$$Zn + H_2SO_4 \rightarrow ZnSO_4 + H_2$$

Here one atom of zinc displaces two atoms of hydrogen showing that it has a valency of two. This equation could be written as an **ionic equation** (more about ionic equations will be found in Chapter 10):

$$Zn + 2H^+ \rightarrow Zn^{2+} + H_2$$

which brings out the relation between the valency and the number of electrons lost (in this case) or gained. It is possible for some elements to have a variable valency, e.g. iron can form both iron(II) (ferrous) ions as in $FeCl_2$ and iron(III) (ferric) ions as in $FeCl_3$, but as a general rule one valency is more common than the other. The table given on page 32 shows the usual valency or valencies of some common elements and ions. The great majority of metals have a valency of two; learn the exceptions to this.

The term 'oxidation number' is sometimes used for valency or combining power; thus iron is said to have oxidation numbers of $+2$ and $+3$ and chlorine an oxidation number of -1.

The elements and ions are said to be either *monovalent*, *bivalent*, *tervalent*, *tetravalent* and *pentavalent* (Greek) or *univalent*, *divalent*, *trivalent*, *quadravalent* and *quinquevalent* (Latin); one should try not to mix these terms although it must be admitted that the prefixes most commonly used are *mono-*, *di-*, *tri-*, *tetra-* and *penta-*!

It is time now to gather together the various definitions of valency that we have had:

1. The valency of an element is the number of electrons it gains or loses in forming an ion; this is electrovalency.
2. The valency of a radical or group is the charge on the ion; this is also electrovalency.
3. The valency of an element is the number of pairs of electrons it shares with others in forming a molecule; this is covalency.
4. The valency of an element is the number of atoms of hydrogen or chlorine which one atom of the element will combine with or displace; this may be either kind.

Naming Compounds

Where an element only exerts one particular valency and so forms only one series of compounds there is no difficulty about naming the compound, e.g.

    NaCl   sodium chloride
    $AgNO_3$ silver nitrate
    $MgSO_4$ magnesium sulphate

Some metals, however, as mentioned above

have a variable valency and so may form more than one series of compounds, e.g.

$$Cu_2O \text{ and } CuO; FeCl_2 \text{ and } FeCl_3;$$
$$SnCl_2 \text{ and } SnCl_4.$$

Until very recently these were distinguished by using the suffix '-*ous*' for the compound in which the metal has the lower valency and '-*ic*' for the other one e.g.

Cu_2O	cuprous oxide
CuO	cupric oxide
$FeCl_2$	ferrous chloride
$FeCl_3$	ferric chloride
$SnCl_2$	stannous chloride
$SnCl_4$	stannic chloride

You will still meet these names in books and so should be familiar with them. However, a method is now employed which indicates the actual valency of the metal, e.g.

Cu_2O	copper(I) oxide
CuO	copper(II) oxide
$FeCl_2$	iron(II) chloride
$FeCl_3$	iron(III) chloride
$SnCl_2$	tin(II) chloride
$SnCl_4$	tin(IV) chloride

This method is not used with non-metals of variable valency; PCl_3 is phosphorus trichloride and PCl_5 is phosphorus pentachloride, NO is nitrogen oxide, NO_2 is nitrogen dioxide and N_2O_4 is dinitrogen tetroxide.

Table 6.1 Usual Valencies of Common Elements and Ions

I	II	III	IV	V
H^+	Cu^{2+}	Al^{3+}	C	N
Li^+	Ca^{2+}	Fe^{3+}	Si	(in N_2O_5)
Na^+	Mg^{2+}	Cr^{3+}	S	P
K^+	Fe^{2+}	N	(in SO_2)	(in PCl_5)
Ag^+	Zn^{2+}	P		
	and	(in PCl_3)		
NH_4^+	most			
	metals	PO_4^{3-}		
OH^-	O^{2-}			
Cl^-	S^{2-}			
Br^-	(in			
NO_3^-	sulphides)			
HCO_3^-	CO_3^{2-}			
HSO_4^-	SO_3^{2-}			
	SO_4^{2-}			

Experiment 6.1. Using valencies to work out formulae.

Cut a number of pieces of card 3 cm square and mark on them the symbols for the monovalent elements or ions, e.g.

| Na | H | NO_3 | NH_4 |

On 6 cm × 3 cm cards write the elements or groups having a valency of two.

| Ca | O | SO_4 | Fe (ous) |

Do likewise for the tervalent ones (9 cm × 3 cm)

| P | Fe (ic) | PO_4 |

and for tetravalent carbon and silicon (12 cm × 3 cm).

Let us now use these cards to work out some formulae.

Hydrogen Chloride

One hydrogen and one chlorine card fit together thus:

| H | Cl |

to form a rectangle 6 cm wide; this gives HCl as the formula.

Sodium Sulphate.

Here the cards are arranged as shown to make a rectangle again 6 cm wide and so the formula is Na_2SO_4.

Na	Na
SO_4	

Aluminium Oxide

The only way the cards can be fitted together to make a suitable rectangle is like this; the formula is Al_2O_3.

Al		Al	
O	O		O

Iron(III) Chloride
This must be FeCl$_3$:

	Fe	
Cl	Cl	Cl

Calcium Phosphate
Here we must have another 3:2 arrangement, Ca$_3$(PO$_4$)$_2$:

Ca	Ca	Ca
PO$_4$		PO$_4$

Tetrachloromethane (Carbon Tetrachloride)
This is CCl$_4$:

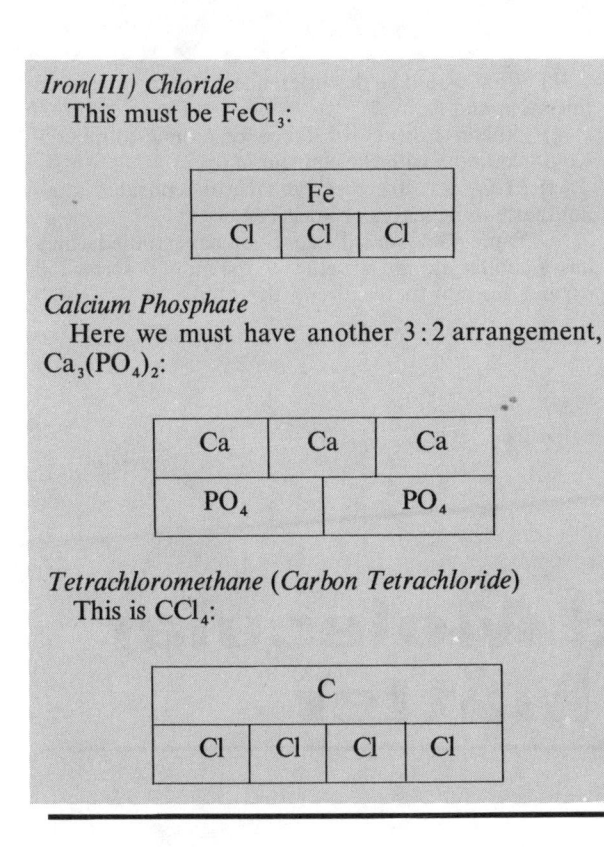

	C		
Cl	Cl	Cl	Cl

Using your knowledge of valency you should now be able to work out the formulae of inorganic compounds in the following way.

Write down the elements or ions which make up the compound and add the valencies at the top. Next, transpose the valencies and write them at the bottom, omitting the figure when the valency is one. Look at these examples. *Aluminium Oxide*: Start with Al^{3}O^2 and transpose to Al$_2$O$_3$ which is the formula. *Sodium carbonate*: First we have Na^{1}CO$_3{}^2$ and then Na$_2$CO$_3$ (as there is no need to write the suffix 1). *Copper(II) sulphate:* As before we write Cu^{2}SO$_4{}^2$ and then Cu$_2$(SO$_4$)$_2$ which becomes CuSO$_4$.

Write formulae for aluminium phosphate, potassium carbonate, magnesium hydrogen carbonate, iron(III) sulphate, silicon hydride, chromium(III) sulphide, zinc phosphate(V), iron(II) bromide, iron(III) hydroxide.

If an overhead projector is available these squares and rectangles can be cut out of coloured plastic and marked with the symbols. They can then be used on the projector for all the class to see.

Exercise 6

1 (*a*) For each of the three particles mentioned say what charge they carry and what are their approximate masses compared to one another: (i) proton, (ii) eletron, (iii) neutron.
(*b*) What do you understand by the term atomic number? Illustrate your answer by using as an example the element sodium.
(*c*) Name three electrovalent compounds and three covalent ones.
(*d*) How do electrovalent compounds differ from covalent ones in (i) melting point and (ii) the effect of an electric current? (*EM*)

2 Explain briefly what Dalton said in his theory about the structure of matter and the way in which elements combined with one another. What, approximately, was the date of Dalton's theory?
What do we now believe about the structure of atoms? Illustrate your answer by referring to the elements hydrogen, sodium and chlorine. Say briefly what the following contributed towards the modern theory of atomic structure: (*a*) Bohr and (*b*) J. J. Thomson. (*EM*)
(You will probably have to do some reading before you can answer this.)

3 Explain the difference between (*a*) electrovalency and (*b*) covalency. Draw diagrams to show the structures of (*c*) the hydrogen molecule, (*d*) sodium chloride and (*e*) the chlorine molecule, saying in each case whether the links are electrovalent or covalent. (*EM*)

4 (*a*) Which of the following elements has an outer shell of eight electrons? Calcium, neon, helium, oxygen, sulphur.
(*b*) Which one of the following compounds does not conduct an electric current? Sulphuric acid, copper(II) sulphate, potassium hydroxide, tetrachloromethane, sodium chloride.
(*c*) Copy and complete the sentence ... Atoms always try to get the same number of electrons in their outer shells as an ..... (*MR*)

5 (*a*) Draw diagrams showing the electrons in models of a sodium atom (2.8.1.) and a chlorine atom (2.8.7.).
   (i) Draw a line on the diagram to show the change of position of any electron when these two atoms combine.
   (ii) Why does this change take place?
   (iii) What are atoms now called?
   (iv) Write down the symbols for the atoms after the change.
   (v) What force holds the two atoms together?
   (vi) How would you actually join these atoms together in the laboratory?

(*b*) The formula for methane is
$$H-\underset{\underset{H}{|}}{\overset{\overset{H}{|}}{C}}-H$$

33

(i) How many covalent bonds are there in methane?
(ii) What is taking place at each line in the diagram?
(iii) Why does this take place?
(iv) Name one other covalent compound.
(c) Name two ways in which electrovalent compounds differ from covalent ones. (MR)

6 (a) The atomic number of an element containing 12 neutrons is 11. Draw a diagram to show the possible structure of an atom of this element.

(b) What would be the difference in structure between this atom and its ion?
(c) Explain how a substance of atomic number 9 would combine with the element in (a).
(d) From its structure would you consider atom number 9 to be active or inactive?
(e) Name a known substance you have studied which has a similar atomic structure to the atom described in (a) and describe its reaction with water. (SR)

7
Molecules and Molecular Properties of Matter

A large part of Science is concerned with studying the basic 'building blocks' of which matter is made. In different circumstances it is convenient to study different building blocks. For example, much of chemistry is concerned with the study of the interaction of atoms. In various Chapters of this book we shall be concerned with particles which are smaller than atoms, such as protons and neutrons. In Biology, the basic unit of matter is sometimes the 'cell'. Many of the physical properties of materials are best studied by considering the molecule as the building block. Molecules are the smallest particles of a substance which can have a separate existence. They will normally consist of groups of atoms, sometimes one or two, sometimes a large number. Perhaps it would be a good idea to start by getting an idea of the size of one sort of molecule. We can do this by an experiment based on the fact that certain oils spread on clean water to form a film which may be only one molecule thick. If we know the volume of oil we put on the water and the area of the film we may, from a simple calculation, find its thickness.

Experiment 7.1. Thoroughly clean a large dish, put some clean water in it and when the water has become still, sprinkle the surface with lycopodium powder. Now pick up some olive oil in a narrow bore capillary tube and measure the length of oil

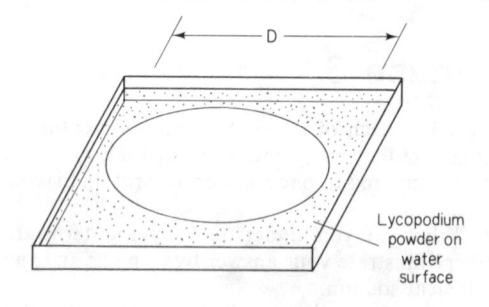

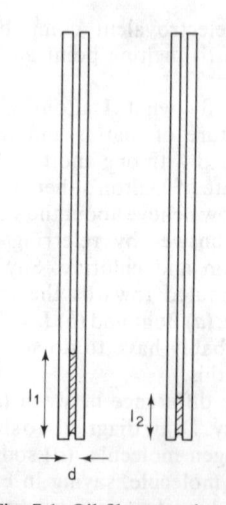

Fig. 7.1. Oil film experiment

in the tube. Release a drop of oil into the centre of the water surface and again measure the length of oil in the tube. The drop of oil will spread on the water surface to form a large circle free of lycopodium powder, whose diameter you can measure. In order to find the volume of the oil drop you will need to find the diameter of the capillary tube. This you can do with sufficient accuracy by comparing it with a finely ruled scale with the aid of a magnifying glass. Alternatively, you can take one drop from a teat pipette and then find the number of drops needed to fill 1 cm³ of a 10 cm³ graduated cylinder and so calculate the volume of one drop. The following is a typical set of results for such an experiment (note that we have given π a value of 3 in the calculation).

Initial length of oil in the
  capillary (l_1)   = 1.3 cm
Final length of oil in
  capillary (l_2)   = 1.1 cm
Diameter of capillary (d)   = 0.02 cm
Thus volume of oil drop =

$\pi \left(\dfrac{d}{2}\right)^2 (l_1 - l_2)$ $= 3 \times (0.01)^2 \times 0.2$
                              $= 6 \times 10^{-5}$ cm³
Diameter of oil film (D)   = 20 cm

Thus area of film = $\pi \left(\dfrac{D}{2}\right)^2 = 3 \times 10^2$ cm

And thickness of oil =
$$\dfrac{\text{volume}}{\text{area}} = \dfrac{6 \times 10^{-5}}{3 \times 10^2}$$
$$= 2 \times 10^{-7} \text{ cm}$$

We must not regard this as an accurate experiment, but it does give us an idea of the sort of size we must think of for molecules. Olive oil is quite a long molecule and we believe that in a film of this kind, the molecules will all be standing upright at the water surface.

Molecules in Motion

Although we believe that molecules in solids, liquids and gases are in continual motion, it is not easy to show by simple experiments the movement of molecules in a solid. We shall describe some experiments however which demonstrate the movement of molecules in liquids and gases.

Experiment 7.2. Diffusion in Liquids.
  (a) Fill a beaker with water and place in it a large crystal of copper(II) sulphate. Leave it for a day in a place where it will not be disturbed.
  (b) Fill a small tube with potassium manganate (VII) solution and stand it upright in a beaker. Carefully fill the beaker with water so that the tube is covered and leave for a day.
  In both cases you will find that after a day the colour has spread throughout the water.

Experiment 7.3. Diffusion in Gases.
  Warning: This Experiment should only be carried out by the teacher.
  Place a white screen behind a gas jar and place in it two drops of bromine; cover the jar. You will see that the brown bromine vapour gradually fills the gas jar. Take care, bromine is a very dangerous substance and all contact with liquid and vapour must be avoided.

These experiments suggest that the molecules in solids and liquids are moving. Furthermore, the more rapid diffusion (spreading out) in air may indicate that the movement is faster in gases. We shall describe one more experiment to support these ideas.

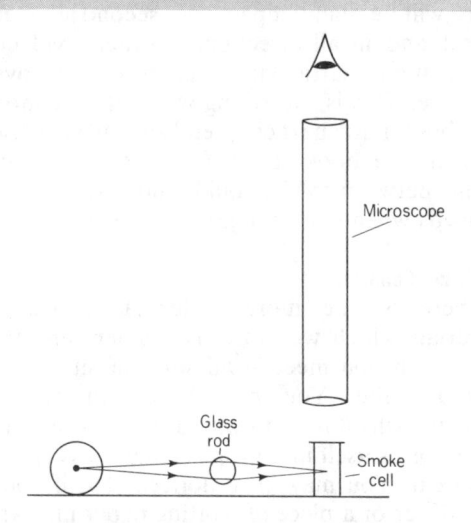

Fig. 7.2. Observation of Brownian movement

Experiment 7.4. Brownian Motion
  Brownian motion was discovered in the early 19th century by Robert Brown, who was studying pollen grains suspended in water under a microscope. We shall study the effect in a gas rather than in a liquid and a convenient apparatus is shown in Fig. 7.2. The vessel in which the Brownian motion takes place is a small transparent cylindrical cell which may be sealed with a thin glass cover slip. This is illuminated by a low voltage lamp which has a long straight filament. The light from this lamp is focused on to the cell by a glass rod, which is fixed parallel to the filament of the lamp and which acts as a lens. The cell is viewed from above by means of a microscope. Focus the microscope on the

middle of the cell and now introduce some smoke into the cell and put the cover slip on. Smoke from a piece of smouldering string is suitable and it may be collected in a teat pipette for injection into the cell. You should be able to see smoke particles through the microscope as points of light. Select individual particles and study them for as long as you can. You will find that they are always moving and constantly changing direction in a random way.

The Brownian motion may be explained by assuming that the air molecules are in constant movement and are all the time colliding with the smoke particles. These are many times heavier than the molecules, but they are still light enough to be knocked off course by the collisions. This theory implies, of course, that any object, large or small, which is situated in air or in any gas, is being bombarded in this way. However, for a large object, there will be many impacts per second all over the object and in all directions, so there will be no effect in any particular direction, only an overall pressure. This is something we shall have more to say about in the next chapter. We shall also discuss later in the book the differences, in molecular terms, between solids, liquids and gases, and what happens when one changes to another.

Surface Tension

There is one more molecular property of materials which we must discuss here and this is one which you meet in all sorts of situations in everyday life. You may have wondered, for example, why it is that water dripping slowly from a tap, forms itself into droplets that are very nearly spherical. You may have noticed that if you dip the corner of a piece of blotting paper into water, it will soak up through the paper to a considerable height above the water surface. You may have carried out the well known trick of floating a needle or razor blade on water. These are all effects of surface tension and the next two experiments may help you to understand them a little better.

Experiment 7.5. Make loops of copper wire as shown in Figs. 7.3(a) and (b). In Fig. 7.3(a) a straight piece of wire is just resting across the bent piece. In Fig. 7.3(b) a piece of cotton is tied across the loop so that it is slightly slack. Immerse each of the loops in turn in a dish containing a fairly strong solution of liquid detergent and lift them slowly out, keeping the loops horizontal. You will find that in case (a) you start to obtain a soap film on the loop but as you lift it clear of the solution, the loose wire will move to the right (Fig. 7.3(c)) and the film will disappear. With the second loop you should be able to obtain a good soap film. Now puncture the film on one side of the cotton only. You should see that the film shrinks and the cotton is stretched into the arc of a circle. (Fig. 7.3(d)).

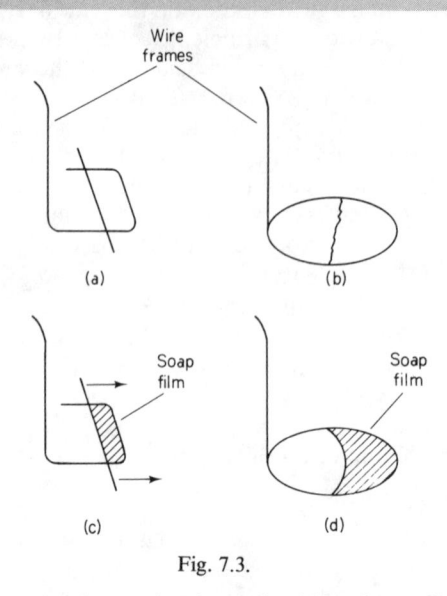

Fig. 7.3.

This experiment leads to the idea that a liquid surface has a skin which is in tension, like a piece of stretched rubber. This will shrink if it is given the chance, and a liquid surface will always tend to take up the smallest area possible. This will account also for the near spherical shape of a free falling droplet, for a sphere has the smallest surface area for a given volume. In fact, there is no skin on a liquid surface and the effect arises because of forces of attraction between molecules in the liquid. A detailed description of the theory of surface tension would be too complicated to give here, but you can get some idea by considering Fig. 7.4. A molecule in the body of a liquid is surrounded on all sides by other molecules all of which exert attractive forces on it. There is thus no net force in any direction. A molecule in the surface, on the other hand, has liquid molecules below it and air molecules above. These air molecules will be far

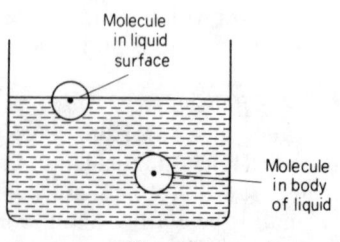

Fig. 7.4.

more spread out, in view of the much lower density of air. There will thus be a much stronger downward force on the surface molecule than the upward force. It is this downward force which gives a liquid surface its special properties.

Capillary Rise

Experiment 7.6. Obtain several pieces of narrow bore glass tube of different diameters, clean them and stand them upright in a dish of water (Fig. 7.5). Look carefully at the liquid levels in each tube. You will find that water has risen in all of them and that it has risen most in the narrowest tube.

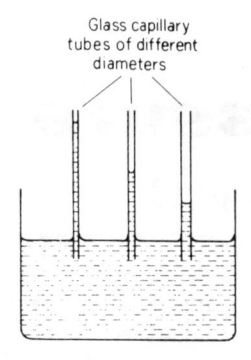

Fig. 7.5. Capillary rise

This is another surface tension effect, and one of considerable practical importance. It means that water will soak up through any porous material. Water from the ground will, for example, soak up through the brickwork of a house, unless a damp proof course is included just above ground level. If you repeat Experiment 7.5 using mercury instead of water you will find a capillary depression instead of a rise.

Exercise 7

1 Explain the following:
 (i) If a long clean glass tube of narrow bore is placed vertically in a beaker of clean water, the water rises a little way up the tube, and the water surface in the tube is curved.
 (ii) If a layer of distilled water is placed carefully on top of a quantity of saturated copper(II) sulphate solution in a sealed container and allowed to stand for a week *both* layers are then coloured blue. (*L*)

2 Describe two experiments which support the idea that the molecules of a gas are in continuous movement.

3 In an oil film experiment, a drop of oil of volume 10^{-4} cm^3 is placed on the surface of some water. If the length of the oil molecule is 3×10^{-7} cm, what is the largest area of oil film you could expect to find?

8
Pressure in Liquids and Gases

The idea of pressure is one that we all come across quite frequently in everyday life. The weather forecasters refer to regions of high and low pressure. The motorist has to pump up his car tyres to some pressure value specified by the manufacturer. The housewife may use a pressure cooker. We learn that most civil aircraft are pressurized. In addition to these everyday examples many people are concerned with pressure in the course of their work. The submarine builder and the deep-sea diver have to contend with the great pressures experienced at depth in the sea. The civil engineer building a dam and engineers in the gas and water supply industry are all very concerned with pressure. In order to understand some of the problems arising in these examples we must first be clear about what we mean by the term pressure.

Consider the ways in which it is possible to prevent heavy objects from sinking into soft surfaces (e.g. snow or soft mud). If a man wishes to avoid sinking into deep snow he will wear skis or snow-shoes on his feet. Vehicles intended for moving over muddy land, like agricultural tractors, will either be tracklaying, like a tank, or will have very thick tyres indeed. In all these cases the aim is to reduce the pressure on the soft surface. Of course it is not possible to reduce the weight of the man or the tractor. It is possible to distribute the weight over a larger area and this reduces the pressure. Pressure in these cases is the weight acting on unit area. Weight, as we mentioned in the last chapter, is a force, and in general pressure is defined as: *Force per unit area*.

Generally the force will be measured in newtons and the area in square metres, so pressures will be in newtons per square metre (N/m^2).

It will be of interest to calculate the pressure in one of the examples quoted above. A man wearing ordinary shoes may have 250 cm² (0.025 m²) in contact with the ground. If he has a mass of 80 kg his weight will be $80 \times g$ N. On the surface of the earth this will be about 80×9.8 N. Thus the average pressure under his feet is:

$$\frac{80 \times 9.8}{0.025} = 31\,000 \text{ N/m}^2 \quad \text{or} \quad 31 \text{ kN/m}^2$$

Of course the pressure depends on the force of gravity. On the moon, for example, where our man will experience only one fifth of the gravitational force that he does on earth, the pressure that he exerts on the ground will be correspondingly reduced by one fifth.

It should be noted that with the stiletto type of ladies high-heeled shoes, the wearer's mass is sometimes acting over an area of perhaps 1 cm². The pressure under the heel in this case is 250 times the value calculated previously. It is scarcely surprising that these shoes make dents in floor coverings.

Pressure in Liquids

We have mentioned the problems of submarine designers and deep-sea divers arising from pressure. Just as the mass of a man produces a pressure on the ground under his feet, so the mass of liquid above a surface immersed in that liquid will produce a pressure on it. The further down in a liquid a surface is, the greater the mass above and hence the greater the pressure.

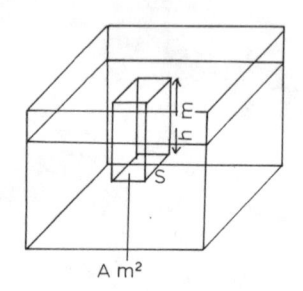

Fig. 8.1. To measure pressure at a depth in a liquid

Fig. 8.1 shows a surface S, of area A m² which is h m below the surface of a liquid. If the density of the liquid is d kg/m³ the volume of liquid in the cuboid above S is:

$$h \times A \text{ m}^3$$

and its weight

$$h \times A \times d \times g \text{ N}$$

Thus the pressure on S is:

$$\frac{\text{Force}}{\text{Area}} = \frac{h \times A \times d \times g}{A} \text{ N/m}^2$$
$$= h \times d \times g \text{ N/m}^2.$$

Thus, in a given liquid, the pressure below its surface is proportional to the depth.

Experiment 8.1. A simple experiment to show this increase of pressure with depth uses the apparatus shown in Fig. 8.2. It consists of a tall metal cylinder with three (or more) outlet tubes in the side fitted with taps or clips. Fill the cylinder with water and open all the outlet taps. Maintain the water level in the cylinder with water from the mains. You will observe that water from outlet A drops almost vertically down from the tube, that from B travels some distance out from the cylinder while the water from C travels a considerable distance out showing that the pressure at this level is higher than at B or A.

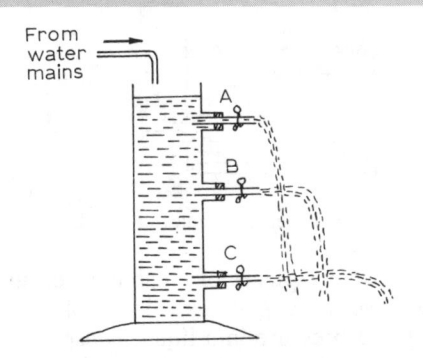

Fig. 8.2. Experiment to show pressure in a liquid

This experiment shows also another important effect of liquid pressure. In the cases of pressure arising from solid objects which we described earlier in this chapter, the pressure was always acting vertically downwards. In the experiment illustrated in Fig. 8.2, the pressure is clearly acting in a horizontal direction. The pressure at a level in a liquid is still due to the mass of liquid above that level. However, we shall see that it acts in all directions at the same time.

Measuring Pressure

Since the pressure in a liquid is proportional to the depth below the surface, it is possible, and often convenient, to measure pressures in terms of a height of liquid. A pressure of 10 m of water for example is the pressure 10 m down in water.

Since the pressure at a depth h in a liquid of density d is $h \times d \times g$, a pressure of 10 m of water is approximately:

$$10 \times 1000 \times 9 \cdot 8 \text{ N/m}^2$$
$$= 98 \text{ kN/m}^2$$

(1000 kg/m³ is the density of water).

Similarly a pressure of 76 cm of mercury is:

$$0 \cdot 76 \times 13\,600 \times 9 \cdot 8 \text{ N/m}^2$$
$$= 101 \text{ kN/m}^2$$

(13 600 kg/m³ is the density of mercury).

A device which is often used to measure liquid (and gas) pressures is the manometer (Fig. 8.3).

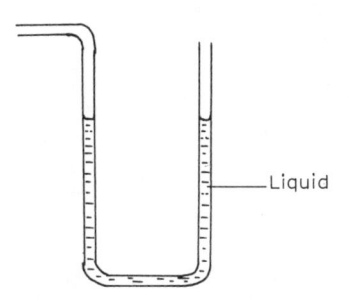

Fig. 8.3. A manometer

This consists simply of a U-tube containing a liquid; water or oil perhaps for low pressures and mercury for higher ones. Fig. 8.3 shows the liquid surfaces in the two limbs of the tube at the same level. This implies that the pressures on these two surfaces are the same. A difference in levels will indicate a difference between the pressures on the two surfaces. The size of the pressure difference may be measured by the difference in levels (h in Fig. 8.4). We shall use a water manometer in the following experiment.

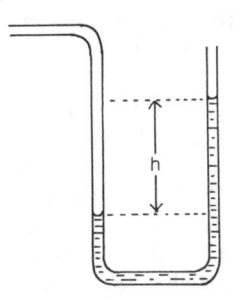

Fig. 8.4. Difference in levels shows greater pressure on liquid in left hand limb

Experiment 8.2. Across the top of a thistle funnel with a very short stem, fix a thin piece of stretched rubber. A piece of a toy balloon will serve quite well. Connect the stem of the funnel by a piece of rubber tube to one side of a water manometer. Now immerse the funnel in a trough of water as in Fig.

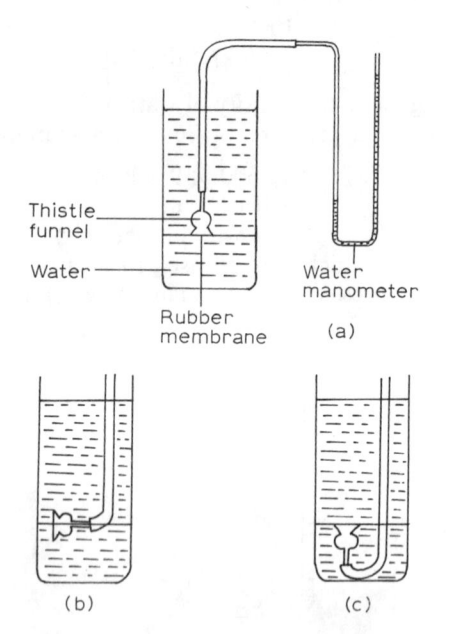

Fig. 8.5. The pressure at a point in a liquid is equal in all directions

8.5(a) with the rubber membrane level with a mark on the side of the trough. Note the difference in levels in the manometer. Repeat with the funnel in positions indicated in Figs. 8.5(b) and 8.5(c) noting the manometer readings in each case. You should observe that the manometer shows the same pressure with each arrangement. Our experiment shows that the pressure at a certain depth in a liquid is equal in the three directions chosen and further experiment shows that it is equal in all directions. The experiment shows also that the pressure is everywhere the same at a particular level in a liquid.

A consequence of this result is the well-known effect that 'a liquid finds its own level'. This may be illustrated by means of an apparatus, known as Pascal's Vases, shown in Fig. 8.6. Water is poured into the funnel shaped vessel on the left in the diagram and it will be seen that the level in all the assorted shaped tubes is the same. This must be so since, as we have seen in the previous experiment, the pressure must be the same at all points along the level AB. Hence points C, D, E and F must be equal distances below the surfaces.

Applications of Liquid Pressure

A very simple application of the principle we have just discussed is the water level gauge sometimes used with boilers. A pair of right angle metal tubes are welded or screwed into the side of the boiler as in Fig. 8.7 and a piece of glass tube is set in between them. The water level in the glass tube will be the same as that in the boiler.

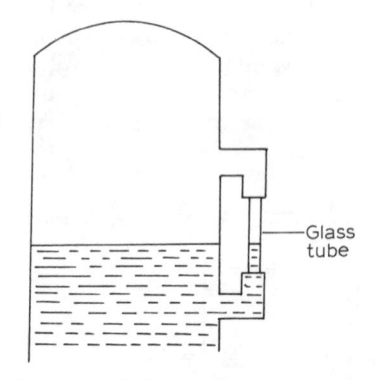

Fig. 8.7. Water level indicator on a boiler

A great many pieces of machinery of all types involve 'hydraulic systems'. These are systems whereby the pressure in a liquid is used to transmit a force from one place to another. In many cases the force may be made much bigger in the process. The simplest of such devices, in principle, is the hydraulic press shown in Fig. 8.8. A force is applied at the piston A which makes a good fit in its cylinder. The pressure in the liquid immediately below A is thus increased and the increased pressure is transmitted through the liquid to the piston B. This exerts a force on the object D which is to be compressed. This force will be considerably larger than the force applied at A. In order to

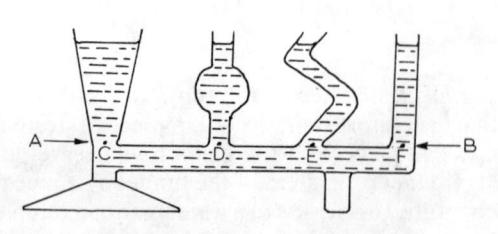

Fig. 8.6. 'A liquid finds its own level'

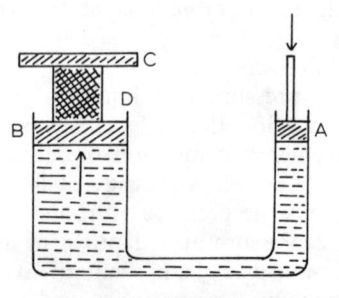

Fig. 8.8. Hydraulic press

see that this is so let us take some typical values. Suppose the force at A is 100 N and the cylinder at A of radius 1 cm. The pressure in the liquid under A will be:

$$\frac{\text{Force}}{\text{Area}} = \frac{100}{\pi \times 1^2} \text{N/cm}^2$$

and the pressure under B will have the same value.

Suppose the cylinder at B has a radius of 6 cm. Then the force exerted by B = Pressure × Area

$$= \frac{100}{\pi \times 1^2} \times \pi \times 6^2 \text{ N}$$
$$= 3600 \text{ N}.$$

Other examples of hydraulic systems working on this principle are hydraulic jacks and car-lift platforms used in garages and the braking systems of most modern motor vehicles. In a motor-car when a force is applied to the brake pedal a piston is moved to produce a pressure in a brake fluid contained in a 'master cylinder'. This pressure is transmitted through the fluid contained in pipes to the brake drums. In each of these the pressure acts on a piston which pushes the brake shoe against the inside of the drum to produce the friction required for braking.

Dams

If you have seen a large dam, or a photograph of one, you may have observed that it is very much thicker at the bottom than at the top. Fig. 8.9 shows a vertical cross section through a dam. The greater thickness is at the bottom of course because it is there that the water pressure on the dam is greatest. Large dams are most often built for the hydroelectric generation of electricity. This is sometimes achieved by mounting turbines (T in Fig. 8.9) in holes through the dam itself. The high pressure on the deep-water side of the dam causes water to flow through these holes at great speed turning the turbines as it does so.

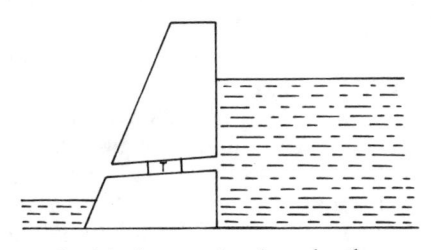

Fig. 8.9. Cross section through a dam

Domestic Water Supplies

If you try putting your thumb over the hot and cold taps at the kitchen sink (if the hot water is not too hot), you may find that it is not too difficult to stop the hot water flowing completely but much more difficult to stop the cold water. This will be due to the fact that the cold water is coming directly from the water supply company's main at high pressure while the hot is coming from a hot tank in the house at a much lower pressure. The hot water pressure is, in most domestic water systems, determined by the height (h) above the tap of the water level in a tank in the loft of the house (Fig. 8.10).

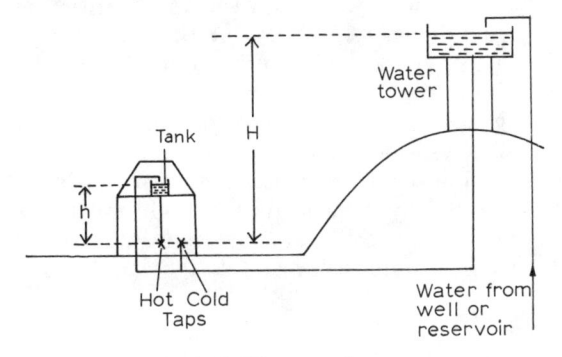

Fig. 8.10. Water supply system

The mains water in many cases comes in the first place from a spring or reservoir but in order to maintain a constant high pressure to the consumer it is pumped to the top of a water tower situated on high ground. In Fig. 8.10 the mains water pressure is determined by the height H.

Measurement of Gas Pressure

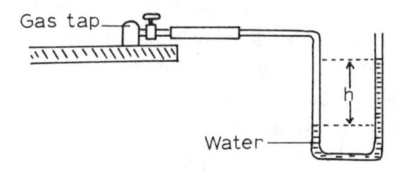

Fig. 8.11. Measuring gas pressure

Experiment 8.3. Connect a U tube containing water by means of a rubber tube to a gas tap. Turn the tap on and you will observe that the manometer shows a difference in levels corresponding to the excess of gas pressure over the atmospheric pressure. This may be said to be h m of water.

Atmospheric Pressure

We have been discussing the pressure on objects immersed in a liquid. We live all the time immersed (at a considerable depth) in a gas, namely air. There is, then, considerable pressure acting on us due to the atmosphere. The existence of atmospheric pressure may be shown by the following experiment.

Experiment 8.4. Heat a little water in the bottom of a fairly large metal can. (A 5 litre oil can will do very well.) When the water has been boiling for a few minutes, remove the heat and cork the can firmly with a rubber bung. Allow the can to cool. You will quite quickly see and hear the can collapsing (Fig. 8.12 (*b*)). When the water boils, steam fills the can and pushes out much of the air contained. However there is a steam pressure acting on the inside of the vessel to balance air pressure outside. When the can is corked and cooled, the steam condenses but no air is allowed back in to take its place. We thus still have atmospheric pressure on the outside but a very much reduced pressure inside and so the can caves in.

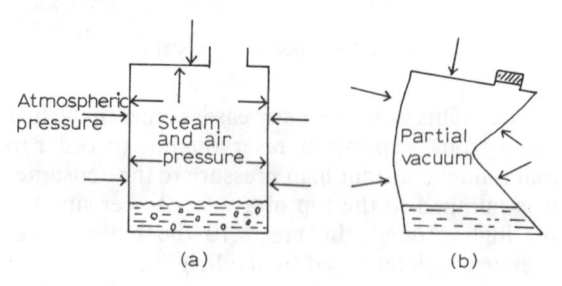

Fig. 8.12. Collapsing can experiment, to show atmospheric pressure

This tendency of vessels to cave in if the air is removed from inside can be a serious hazard; but it can be put to very good use. Several glass objects in everyday use have the air pumped out of them in manufacture. Examples are vacuum flasks (see Chapter 18), electric light bulbs and television picture tubes. If there is a flaw in the glass used in making these things there is a risk of them caving in violently or 'imploding' when the air is pumped out. Workers making them will generally wear protective eye glasses to protect them from flying glass.

The effect is widely used in industry to mould sheet plastics into complicated shapes, e.g. the inside of a domestic refrigerator with egg and milk compartments, etc. A flat sheet of plastic material is laid over a mould and the air between them is pumped out (Fig. 8.13). The atmospheric pressure on the other side of the plastic will force it against the mould.

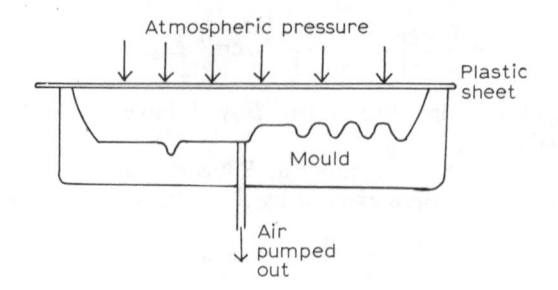

Fig. 8.13. Using atmospheric pressure to mould plastics

Experiment 8.4 shows not only the existence of atmospheric pressure but also that this pressure is quite large. We have seen that the pressure in a liquid is proportional to the depth and to the density of the liquid. Now the density of air is very small, about 1 kg/m³ under normal conditions. It follows then, if we assume that this expression for pressure applies to gases as well as to liquids, that the height of atmosphere over our heads must be rather large.

It is not easy to calculate the height of the atmosphere from a knowledge of atmospheric pressure, since the air density decreases as we go upwards. In fact until the development of satellites, there was considerable doubt about this. Even now, since the air simply becomes more and more rarefied the higher we go, it is not possible to specify an exact height. It is usually quoted as 300 to 500 km.

Measurement of Atmospheric Pressure

Experiment 8.5. Take a clean, dry length of thick-walled glass tube about a metre long and sealed at one end. Fill it completely with mercury. This is conveniently done by connecting a funnel by a piece of rubber tube to the open end. Now, having removed the funnel, place the thumb over the open end, invert the tube and remove the thumb when the open end is below the surface of some mercury contained in a dish (Fig. 8.14). The mercury will fall to a level about 76 cm above the surface in the dish. Now incline the tube to the vertical to take up positions (2) and (3) in Fig. 8.14. The vertical height of the mercury surface in the tube above the dish remains unchanged. If the tube is inclined further than in position (3) it will simply remain full to the top. We shall see now that this vertical

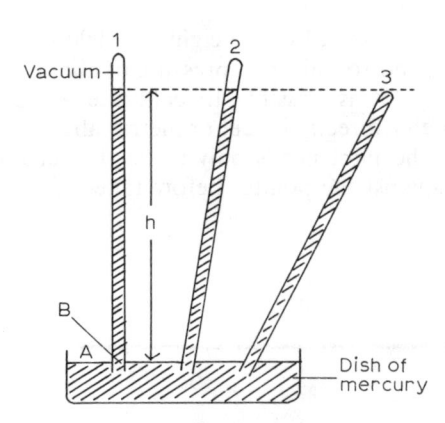

Fig. 8.14. Simple mercury barometer

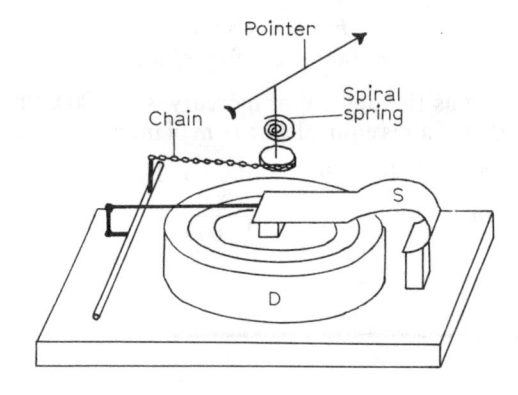

Fig. 8.15. An aneroid barometer

height (h in the diagram) gives the value of the atmospheric pressure in cm of mercury.

The pressure at A in the surface of mercury in the dish is atmospheric. Since the pressure at a certain level in a liquid will be everywhere the same, the pressure at B inside the tube will also be atmospheric. Now the pressure at B will be that due to h m of mercury + the pressure of any gas that may be present in the space above the column. If the apparatus has been carefully set up, only minute traces of gas should be present in this space and the pressure exerted negligible.

Thus, if the vertical height of the column is 0·76 m:

Pressure at B = Pressure at A
= Atmospheric pressure
= $0·76 \times 13\,600 \times 9·8$ N/m^2
= 101 000 N/m
= 101 kN/m^2

(13 600 kg/m^3 is the density of mercury.)

The apparatus we have used in Experiment 8.5 is a simple barometer. It was first investigated by Torricelli (1608–47) who concluded that the space above the mercury column was a vacuum and that atmospheric pressure supported this column.

Barometers and their Uses

It is clear that variations in atmospheric pressure will produce differences in the height of the column of our barometer. This simple apparatus may be modified to produce accurate instruments for use by scientists and meteorologists. It also provides the basis for some of the barometers used in the home. More often though, these instruments are based on a different principle. An aneroid barometer is illustrated in Fig. 8.15. The essential part is a flat cylindrical metal drum (D), the circular faces of which are corrugated to give it extra strength. This vessel has had some of the air pumped out of it so that there is a tendency for the atmospheric pressure on the outside to cause it to collapse. This is controlled by a steel spring S. However an increase in atmospheric pressure will cause the middle of the drum to move down. This small movement is transmitted by a system of levers and a chain to a pointer which moves around a scale. The scale may be calibrated in cm (of mercury) and on domestic barometers in usually marked Dry, Change, Rain etc.

A knowledge of atmospheric pressure is very important to anyone wishing to forecast the weather. As a rough guide, high pressure in summer means light winds, sunny and dry weather; in winter clear and cold. Low pressure usually indicates strong winds and generally stormy conditions. It is desirable to know not only the pressure at present but whether it is tending to go up or down, if one is to make a forecast. It is therefore common practice to tap a barometer lightly and observe the small movement of the pointer which usually results. In order to make more detailed forecasts of weather meteorologists are continuously observing atmospheric pressure, amongst other things, all over the world. Their values of the pressures are recorded on weather maps which show regions of high and low pressure and directions in which these regions are moving.

As we have pointed out, atmospheric pressure is caused by the weight of the air above us. It follows then that if we go up a high mountain or in an aircraft we should be able to observe a drop in pressure. For this reason, aneroid barometers may be used as height gauges or 'altimeters' in aircraft.

If we take the density of air to be 1 kg/m^3, then the pressure due to 1000 m of air will be:

$$h \times d \times g \, \text{N/m}^2$$
$$= 1000 \times 1 \times 9\cdot 8 \, \text{N/m}^2.$$

Now as the density of mercury is 13 600 kg/m³, 1000 m of air is equivalent to h m of mercury, where

$h \times 13\,600 \times 9\cdot 8 = 1000 \times 1 \times 9\cdot 8$

or $h = \dfrac{1000}{13\,600}$ m

$= 0\cdot 076$ m

or 7·6 cm.

Thus an increase in height of 1000 m should correspond to a drop in pressure of about 7 cm of mercury. It is possible to calibrate an aneroid barometer directly in feet or metres above ground level. The pilot needs only to set the zero of the scale against the pointer before take-off.

Exercise 8

1 A brick measuring 20 cm by 10 cm by 5 cm is placed flat on the floor. In what stable position of the brick would the greatest pressure on the ground be created?
If the brick has a mass of 8 kg what would be the pressure on the ground in this position? (*EM*)

2 The base of a rectangular vessel measures 10 cm by 18 cm. Water is poured in to a depth of 4 cm. What is the pressure on the base? What is the thrust on the base? (*J*)

3 Write down THREE facts about pressure in liquids.
The pressure of a gas supply is approximately 15 cm of water above atmospheric pressure. Describe an experiment you could perform to find the exact pressure of the gas supply. (*C*)

4 Distinguish carefully between *force* and *pressure*. Draw a large, labelled diagram of any named hydraulic machine.
The large piston of a hydraulic lift has a diameter of 25 cm but the inlet pipe has a diameter of 2 cm. What load could, in theory, be lifted by applying a force of 50 N at the inlet pipe? (*O*)

5 Describe two experiments to show that air exerts considerable pressure.
Explain how a simple mercury barometer is used to measure atmospheric pressure. (*L*)

6 Explain the effect on the height of the mercury column in a simple mercury barometer of (i) doubling the width of the tube, and (ii) inclining the tube to the vertical. Draw diagrams to illustrate your answers.
If the atmospheric pressure is 100 kN/m², what is the total pressure 4 m below the free surface of water in a tank?
(Take the density of water to be 1000 kg/m³.) (*S*)

7 Describe the apparatus you need in order to construct a simple mercury barometer. How is such a barometer made?
How would the barometric height be affected
(a) if there was a marked rise in temperature, the atmospheric pressure remaining the same,
(b) if the barometer was taken to the bottom of a deep quarry, the temperature there being the same as at the surface,
(c) if the barometer was slowly tilted to an angle of about 45° with the vertical?
On a day when the barometric height is 735 mm of mercury, what would the pressure of the atmosphere be in N/m²?
Briefly describe how the principle upon which an aneroid barometer works differs from that of a mercury barometer.
[Density of mercury = 13 600 kg/m³.]
(*C*)

9
Chemical Arithmetic

As protons and neutrons have mass, then each atom must also have mass. The lightest of all the atoms, hydrogen, has a mass of 0·000 000 000 000 000 000 000 167 g, i.e. there are 600 000 000 000 000 000 000 000 of them (600 thousand million million million) in one gram! As these masses are so minute they are not used in ordinary calculations. When, about the middle of the last century, scientists became interested in the relative mass of the various elements, they naturally compared them with the lightest of all elements, hydrogen, and a list of *relative atomic masses* (formerly called *atomic weights*) was drawn up with hydrogen = 1 as the standard of comparison. On this scale oxygen = 15·94.

Difficulties arose later, and as far more elements combine with oxygen than with hydrogen, it was decided to adopt oxygen = 16 as the new standard. By then it was known that hydrogen really consisted of two different varieties of the gas (called isotopes), one of atomic mass = 1 and the other having an atomic mass of 2. The average atomic mass of a sample of hydrogen on the oxygen = 16 scale is 1·008. A few years ago a new standard, carbon = 12, came into use. There are more compounds of carbon than of all the other elements.

The present definition of *Relative* Atomic Mass is as follows:

The Relative Atomic Mass of an element is the mass of one atom of it compared with the mass of one-twelfth of an atom of carbon.

However, as we shall not be using any atomic mass correct to more than one place of decimals, these three standards will all give the same results. A list of approximate relative atomic masses to be used in calculations is given on page 342.

The relative atomic mass of an element is strictly a number; if it is expressed in grams it is then the mass of a *mole* of atoms of the element; one mole of oxygen atoms would then be 16 g.

If we could count the number of atoms actually present in one gram of hydrogen we should find it to be the enormous number previously stated, six followed by twenty-three noughts, or 6×10^{23}.

If we were then to repeat this counting with 16 g of oxygen or 12 g of carbon we should get exactly the same figure. This number is so important that it has its own name, *Avogadro's number* or *constant*.

One mole of atoms of any element contains Avogadro's number (or constant) of atoms.

You have learned that many gaseous elements which occur naturally exist in units containing two atoms (can you name some?); these units are known as molecules. Other substances such as ammonia, NH_3, alcohol, C_2H_5OH, sugar, $C_{12}H_{22}O_{11}$, etc. also exist as molecules, the number of atoms—and hence the formula—in each being fixed.

The Relative Molecular Mass (M.M.) of an element or compound is the mass of one molecule of it compared with the mass of one twelfth of an atom of carbon.

However, in such compounds as sodium chloride, NaCl, lead(II) nitrate, $Pb(NO_3)_2$ and other salts and also in acids and bases in aqueous solution no discrete (i.e. separate) molecules exist. These compounds contain oppositely-charged ions, e.g. sodium and chloride ions (Na^+ and Cl^-) or lead(II) and nitrate ions (Pb^{2+} and NO_3^-) and are, as explained in Chapter 9, ionic or electrovalent compounds. Here then, we speak of a *mole of ions*.

The number of molecules of chlorine present in 71 g of the gas is 6×10^{23}, Avogadro's constant.

The number of molecules of water, H_2O, present in 18 g of the liquid is 6×10^{23}, Avogadro's constant.

The number of ions (Na^+ or Cl^-) present in 58·5 g of sodium chloride is again 6×10^{23}, Avogadro's constant, of each kind.

To sum up: any sample of matter which scientists examine contains millions and millions of atoms or molecules or ions; because these particles are too small to be handled individually chemists take as their unit that mass of an element or compound which contains Avogadro's constant of particles, whether these be atoms, ions or molecules; this unit is called the MOLE.

45

1 mole of carbon = 1 gram-atom of carbon and has a mass of 12 g.
1 mole of ammonia = 1 gram-molecule or 1 gram-formula of ammonia and has a mass of 17 g.
1 mole of sodium chloride has a mass of 58·5 g.

One-mole of a substance contains Avogadro's Constant of the particles in the formula of the substance. Its mass is the sum of the atomic masses in the formula.

EXAMPLE (a). *Find the mass of one mole of (i) sulphuric acid, (ii) calcium hydroxide, (iii) aluminium sulphate, (iv) potash alum.*

(i) Formula is H_2SO_4
Mass of 1 mole $= 2 \times 1 + 32 + 4 \times 16 = 98$ g
(ii) Formula is $Ca(OH)_2$
Mass of 1 mole $= 40 + 2(16 + 1) = 40 + 2 \times 17$
$= 74$ g
(iii) Formula is $Al_2(SO_4)_3$
Mass of 1 mole $= 2 \times 27 + 3(32 + 4 \times 16)$
$= 54 + 3 \times 96 = 342$ g

(iv) Formula is $K_2SO_4.Al_2(SO_4)_3.24H_2O$. This can be simplified to $K_2Al_2(SO_4)_3.24H_2O$ i.e. to $2[KAl(SO_4)_2.12H_2O]$.
Mass of 1 mole $= 2[39 + 27 + (2 \times 96) +$
$(12 \times 18)]$
$= 2(39 + 27 + 192 + 216)$
$= 2 \times 474$
$= 948$ g

You should get some practice in the use of the term 'mole' in different ways; try these simple problems (mentally if possible). Make use of the table of relative atomic masses (approximate) given at the end of the book.

A. What is the mass of
(i) 3 moles of hydrogen atoms
(ii) 3 moles of hydrogen molecules
(iii) 1/8 mole of oxygen atoms
(iv) 1/8 mole of ozone (O_3) molecules
(v) 1/6 mole of magnesium atoms
(vi) 5 moles of nitrogen atoms?

B. What is the mass of
(i) 0·5 mole sodium hydride (NaH) crystals
(ii) 1·5 moles of ethanol (C_2H_5OH)
(iii) 0·1 mole of calcium sulphate ($CaSO_4$)
(iv) 3 moles of water
(v) 0·6 mole of methanal (formaldehyde, HCHO)
(vi) 2 moles of hydrogen chloride (HCl)?

C. What fraction of a mole is present in 4 g of each of the following substances?
(i) sodium hydroxide(NaOH)? (ii) sulphur, (iii) magnesium oxide (MgO), (iv) lithium hydride (LiH), (v) methane (CH_4), (vi) calcium carbonate ($CaCO_3$).

Composition and Formulae

When research scientists prepare a hitherto unknown substance they have to analyze it both qualitatively and quantitatively in order to find out its composition. Qualitative analysis is concerned with applying tests to discover which elements are present and quantitative analysis is the means by which the amounts of each element in a given mass of the substance can be determined. The percentage composition of a substance has to be obtained before any formula can be assigned to it.

EXAMPLE (b). *What is the percentage composition of potassium nitrate?*

First write down the formula and calculate the mass of one mole
KNO_3
$39 + 14 + 3 \times 16 = 39 + 14 + 48 = 101$
Next express the mass of each element as a percentage of this.

$\%K = \dfrac{39}{101} \times 100 = 38·62$

$\%N = \dfrac{14}{101} \times 100 = 13·86$

$\%O = \dfrac{48}{101} \times 100 = 47·52$

Although the percentage of oxygen could be found from $100 - (38·62 + 13·86)$ it is best to work it out separately as this provides a check on the calculation.

EXAMPLE (c). *Find the percentage composition of copper(II) sulphate crystals, $CuSO_4.5H_2O$.*

Here the water of crystallization can be regarded as a single unit, as it comes off as water, and so the percentage of it should be worked out as such rather than as the percentage of hydrogen and the percentage of oxygen separately.
Mass of one mole
$= 63·5 + 32 + (4 \times 16) + (5 \times 18)$
$= 63·5 + 32 + 64 + 90 = 249·5$

$\%Cu = \dfrac{63·5}{249·5} \times 100 = 25·48$

$\%S = \dfrac{32}{249·5} \times 100 = 12·82$

$\%O = \dfrac{64}{249·5} \times 100 = 25·64$

$\%H_2O = \dfrac{90}{249·5} \times 100 = 36·06$

EXAMPLE (d). *Calculate the mass of each element present in 5 g of potassium chromate(VI) K_2CrO_4.*

Mass of one mole = 2 × 39 + 52 + 4 × 16
= 78 + 52 + 64 = 194

By proportion:

Mass of K = $\frac{78}{194}$ × 5 = 2·010 g

Mass of Cr = $\frac{52}{194}$ × 5 = 1·341 g

Mass of O = $\frac{64}{194}$ × 5 = 1·649 g

The method of working out the formula of a compound can easily be understood from one or two examples.

EXAMPLE (e). *Find the formula of a substance which has this percentage composition:*
C = 40%, H = 6·67%, O = 53·33%.

If we divide each percentage by the mass of a mole of that element we shall have figures which represent the numbers of moles of each element present in 100 grams.

Element	Moles
C	40 ÷ 12 = 3·33
H	6·67 ÷ 1 = 6·67
O	53·33 ÷ 16 = 3·33

If we divide these three results by the lowest value, 3·33, it will be seen that they are in the same proportion or ratio as 1:2:1, and therefore the formula could be CH$_2$O.

This, however, is not necessarily the *correct* formula for a molecule of the substance as the same percentage composition would result from a molecular formula of C$_2$H$_4$O$_2$, C$_3$H$_6$O$_3$, etc.

Thus, unless we are given any further data, we can only deduce the *simplest* or *empirical* formula. In order to determine the molecular formula we must know the mass of one mole or have some information which will enable us to work it out. If the vapour density of a gas or volatile substance is given the mass of a mole can be obtained from the relationship relative molecular mass = 2 × Vapour Density.* Problems of this kind will gain in clarity if set out in columns as shown below.

EXAMPLE (f). *Find the empirical formula of a substance with this percentage composition:*
C = 42·87%, H = 2·36%, N = 16·67%, O = 38·10%. *If the vapour density (V.D.) is 84, find the correct molecular formula.*

Element	%	A.M.	Moles	
C	42·87	12	3·57	3
H	2·36	1	2·36	2
N	16·67	14	1·19	1
O	38·10	16	2·37	2

The last column is obtained by dividing throughout by 1·19. The empirical formula is therefore C$_3$H$_2$NO$_2$.

This formula corresponds to a M.M. of 84. As the correct M.M. must be 168 (2 × V.D.), the molecular formula must be C$_6$H$_4$N$_2$O$_4$: dinitrobenzene is C$_6$H$_4$(NO$_2$)$_2$.

Sometimes when the first figures for the moles are divided by the lowest value, whole numbers do not result. This is the case when *more* than one atom of *each* element is present in the molecule. These figures, however, can usually be converted easily into whole numbers by some simple multiplication. If water of crystallization is present in a compound treat this as a unit, as explained when dealing with percentage composition.

EXAMPLE (g). *Find the simplest formula of a compound containing* 15·79% Al, 28·07% S *and* 56·14% O.

Element	%	A.M.	Moles	÷0·585	×2
Al	15·79	27	0·585	1	2
S	28·07	32	0·877	1·5	3
O	56·14	16	3·510	6	12

The simplest formula is therefore Al$_2$S$_3$O$_{12}$.

EXAMPLE (h). *A salt contains* 49·32% *of water of crystallization. The M.M. of the anhydrous salt is* 111. *How many moles of water of crystallization are present in one mole of the hydrated salt?*

Let us represent the anhydrous salt by *X* and set out as before.

	%	M.M.	Moles	
X	50·68	111	0·48	1
H$_2$O	49·32	18	2·74	6

Therefore one mole of the anhydrous salt is joined to six moles of water crystallization.

We shall next see how we can determine experimentally the simplest formula of a compound such as copper(II) oxide.

Experiment 9.1. To determine the empirical formula for copper(II) oxide.
Weigh a clean dry test-tube with a plug of glass wool or rocksil in it. Add a few copper turnings

---

*The vapour density of a gas or vapour is defined as the mass of a certain volume of the gas compared with the mass of the same volume of hydrogen under the same conditions and equals half the molar mass of the gas.

(about 1 g) and reweigh to find the exact mass of the copper added. (Call this x g.)

Add about 1 cm depth of concentrated nitric acid and stand the tube in the fume cupboard until all action has ceased and the copper has gone into solution. (Take out the wool when you add the acid and then replace it.) The solution will be one of copper(II) nitrate and on heating this decomposes to copper(II) oxide.

Heat the tube in the fume cupboard (gently at first) until all the liquid has gone and only a black powder is left. There should be no green or brown colour anywhere in the tube. Allow to cool and reweigh to find the mass of oxide formed. (Call this y g.)

In the compound formed we have x g of copper combined with $(y - x)$ g of oxygen. The number of moles of each present will be

$x/63 \cdot 5$ of copper and
$(y - x)/16$ of oxygen.

What is the ratio of these two numbers? If you have carried out the experiment accurately it should be close to 1:1, suggesting an empirical formula of CuO.

The method below is suitable for finding the number of molecules of water of crystallization in compounds which do not decompose on heating. Three such compounds are barium chloride, magnesium sulphate and sodium carbonate. When they are heated they leave the anhydrous compounds behind.

Experiment 9.2 To determine the Number of Molecules of Water of Crystallization in Barium Chloride Crystals.

Weigh a crucible alone and then containing about 5 g of powdered barium chloride crystals.

Place the crucible on a pipe-clay triangle on a tripod and heat gently. It is best to have a lid on the crucible to prevent any of the solid 'spitting' out; a tiny gap can be left for the escape of the steam. When no more steam comes off, allow the crucible to cool in a desiccator (why?) and then weigh again. Repeat the heating, cooling and weighing until there is no further change in mass.

If we write the formula for barium chloride crystals as $BaCl_2 \cdot xH_2O$ the equation for the action of heat is

$$BaCl_2 \cdot xH_2O \longrightarrow BaCl_2 + xH_2O$$

Suppose the masses are as follows:

Mass of crucible alone	$= a$ g
.. .. + crystals before heating	$= b$ g
.. .. .. after heating	$= c$ g
The mass of crystalline solid	$= (b - a)$ g
and mass of anhydrous solid	$= (c - a)$ g
The mass of water of crystallization must be	$(c - b)$ g

The M.M. of $BaCl_2 = 137.5 + (2 \times 35 \cdot 5) = 208 \cdot 5$
The M.M. of $H_2O = 2 + 16 = 18$
The number of moles of $BaCl_2 = (c - a)/208 \cdot 5$; call this m
the number of moles of water $= (c - b)/18$; call this n

As there is one mole of barium chloride in the formula for the crystals the number of moles of water combined with this must be n/m (take this to the nearest whole number).

Exercise 9

Write down the M.M. of the following compounds using the table of relative atomic mass given at the back of the book and then work out *mentally* their percentage composition.

A (i) $CaCO_3$, (ii) $MgSO_4$, (iii) $KHCO_3$, (iv) HCHO.
B (i) $CaBr_2$, (ii) $NaHSO_4$, (iii) Na_3P, (iv) $(COOH)_2$.

Calculate the percentage composition of the following substances.

1 HNO_3.
2 $K_2SO_4 \cdot Al_2(SO_4)_3 \cdot 24H_2O$.
3 $NaNH_2$.
4 $MgCl_2 \cdot 6H_2O$.
5 $FeSO_4 \cdot 7H_2O$.
6 $KMnO_4$.
7 Mg_2SiO_4.
8 $Na_2S_4O_6$.

What mass of each element is present in the following?

9 7 g of sulphuric acid, H_2SO_4.
10 8 g of ethanoyl chloride, CH_3COCl.
11 20 g of sodium hydrogensulphite, $NaHSO_3$.
12 4 g of ammonium chloride, NH_4Cl.

13 10 g of ammonium hydrogensulphate, NH_4HSO_4.
14 12 g of ethanal (acetaldehyde), CH_3CHO.

Find the empirical formula for substances having the following percentage compositions.

15 P 20·19%, O 10·42%, C 69·39%.
16 Fe 72·42%, O 27·58%.
17 Ca 24·69%, H 1·23%, C 14·81%, O 59·27%.
18 Li 11·11%, S 50·80%, O 38·09%.
19 Fe 20·15%, S 11·51%, O 23·02%, H_2O 45·33%.

Find the molar formulae in the following cases.

20 A substance contains 39·6% C, 7·7% H and 52·7% O and has a vapour density of 91.

21 A substance containing water of crystallization has also 11·82% Mg and 34·98% Cl; its M.M. is 203.

22 A salt contains 45% of water of crystallization and the M.M. of the anhydrous salt is 111. How many moles of water does the salt contain?

23 A carbohydrate contains 42·11% C and 51·46% O and has a V.D. of 171.

24 A hydrocarbon of vapour density 33.5 contains 10.45% of hydrogen.

25 21 g of a compound A contain 8·4 g carbon, 1·4 g hydrogen and 11·2 g oxygen.

(i) What is the simplest formula A can have?
(ii) If its molecular mass is 180, what is the molecular formula of A? (C)

26 (a) a substance A has the following percentage composition by weight: carbon, 40·00%; hydrogen, 6·67% and oxygen, 53·33%. Calculate (i) its simplest formula, (ii) its molecular formula, given that its molecular mass is 60.

When A is dissolved in water, the solution formed is found to turn litmus paper red and to give carbon dioxide with sodium carbonate. What is the probable structure of the molecule?

(b) What is the maximum volume of carbon dioxide, measured at s.t.p., obtainable by treating 10·5 g of magnesium carbonate with dilute sulphuric acid

$$MgCO_3 + H_2SO_4 = MgSO_4 + H_2O + CO_2,$$

and what mass of hydrated magnesium sulphate,

$$MgSO_4 \cdot 7H_2O,$$

could be recovered from the solution by evaporation? Explain your working fully.

10
Equations and What They Tell Us

A chemical equation is more than a collection of formulae; when interpreted correctly it gives us a considerable amount of information about the reaction with which it is concerned. Let us consider the equation:

$$Zn + H_2SO_4 = ZnSO_4 + H_2$$

In a chemical reaction no matter is created and non is destroyed. Hence, the same number of the same atoms must exist after the change as before, but they are now combined in different ways. The left-hand shows the original arrangement and the right-hand side the new groupings. This equations tells us that:

(1) under some conditions not mentioned zinc acts on sulphuric acid to set the hydrogen free and to take its place;

(2) one mole of zinc acts on one mole of sulphuric acid (containing 2 moles of hydrogen atoms, 1 mole of sulphur and 4 moles of oxygen atoms) to form one mole of zinc sulphate (containing 1 mole of zinc, one mole of sulphur atoms and four moles of oxygen atoms) and one mole of hydrogen molecules containing 2 moles of hydrogen atoms;

(3) the valency of zinc is 2 (as one atom of it replaces 2 atoms of hydrogen) and therefore the valency of the sulphate group is also 2;

(4) if we know the relative atomic masses we can see that 65 units by masses of zinc react with 98 units of sulphuric acid to form 161 units of zinc sulphate and 2 units of hydrogen.

Information not given by the equation includes:
(a) whether it is necessary to apply heat or not;

(b) whether concentrated or dilute acid is necessary;
(c) whether the substances are solids, liquids, gases or in solution;
(d) whether heat is given out (an *exothermic* reaction) or taken in (an *endothermic* reaction) during the action;
(e) whether any solid is precipitated or not;
(f) whether the action is explosive or not;
(g) whether a catalyst is needed or not.

Some of these facts may be incorporated in the equation if desired. We can write:

$$Zn + dil. H_2SO_4 = ZnSO_4(aq) + H_2(g)$$

illustrating points (b) and (c). The (aq) stands for aqua (solution in water) and the (g) indicates a gas.

Some other equations in which additional information is given are:

1. $2HgO \xrightarrow{heat} 2Hg + O_2(g)$ or
   $2HgO \xrightarrow{\Delta} 2Hg + O_2(g)$ The sign Δ indicates that heat is required.
2. $NaCl + c.H_2SO_4 \xrightarrow{\Delta} NaHSO_4 + HCl(g)$
3. $AgNO_3(aq) + NaCl(aq) \rightarrow AgCl(s) + NaNO_3(aq)$

The (s) indicates that a precipitate is formed when two solutions or a gas and a solution are mixed.

4. $2SO_2 + O_2 \xrightarrow[Pt]{heat} 2SO_3$

Here a platinum catalyst is used and heat applied.

5. $3Fe(s) + 4H_2O(g) \xrightleftharpoons[\Delta]{\Delta} Fe_3O_4(s) + 4H_2(g)$
   $H_2CO_3(aq) \xrightleftharpoons[\Delta]{\Delta} H_2O(l) + CO_2(g)$

These are examples of reversible reactions. They can proceed in either the forward or the backward direction depending on the conditions and on which substances are present at the start. For instance, steam will react with heated iron to form tri-itron tetroxide (iron(II) diiron(III) oxide) (Fe_3O_4) and hydrogen, but hydrogen will react with heated iron oxide to form iron and water.

In many equations the *state* of the substance (solid, liquid, or gas) is added after each formula (s), (l) or (g): thus $H_2O(s)$ means ice, $H_2O(l)$ means water and $H_2O(g)$ is steam. The equation

$$CO_2(g) + 2KOH(aq) \rightarrow K_2CO_3(aq) + H_2O(l)$$

tells us that if gaseous carbon dioxide is passed into a solution of potassium hydroxide we get formed potassium carbonate also in solution in water. Where there is no doubt about the state of the substances they need not be mentioned, but notice the difference between HCl(g), hydrogen chloride and HCl(aq), hydrochloric acid.

Ionic Equations

In an earlier chapter you learned that most elements and groups form ions and join together forming electrovalent compounds. Most of the reactions which occur in solution are of this type and are termed *ionic reactions* and can be represented by *ionic equations*.

If a solution of sodium nitrate is mixed with one of potassium chloride we have present a mixture of sodium, potassium, nitrate and chloride ions, all remaining in solution. There is no visible reaction.

However, if solutions of silver nitrate and sodium chloride are mixed together a white precipitate of silver chloride is formed. The same result occurs with solutions of (a) silver nitrate and potassium chloride, (b) silver sulphate and sodium chloride and (c) silver sulphate and hydrochloric acid.

The common linkage between these pairs of substances is the presence of silver and chloride ions.

If two ions in a solution can join together to form an insoluble substance, then this reaction always happens and precipitation occurs (unless there is some inhibiting factor):

$$Ag^+(aq) + Cl^-(aq) \rightarrow AgCl(s)$$

This ionic equation represents what happens between *all* the pairs of substances named, the other two ions remaining in solution and taking no part in the reaction, e.g.:

$$Ag^+NO_3^- + Na^+Cl^- \rightarrow AgCl(s) + Na^+NO_3^-$$

Let us look at one or two other examples. Copper (II) hydroxide can be precipitated by adding aqueous sodium hydroxide to a solution of copper (II) chloride, nitrate or sulphate. Potassium hydroxide will do exactly the same thing. The following equations

$CuSO_4 + 2NaOH = Cu(OH)_2(s) + Na_2SO_4$
$CuCl_2 + 2NaOH = Cu(OH)_2(s) + 2NaCl$
$Cu(NO_3)_2 + 2KOH = Cu(OH)_2(s) + 2KNO_3$

can be replaced by an ionic equation:

$$Cu^{2+}(aq) + 2OH^-(aq) \rightarrow Cu(OH)_2(s)$$

If any soluble sulphate is added to a solution of barium chloride, nitrate or acetate a white precipitate of barium sulphate comes down. Some equations for this reaction are:

$BaCl_2 + H_2SO_4 = BaSO_4(s) + 2HCl$
$Ba(NO_3)_2 + Na_2SO_4 = BaSO_4(s) + 2NaNO_3$

The ionic equation would be:

$$Ba^{2+}(aq) + SO_4^{2-}(aq) \rightarrow BaSO_4(s)$$

How Much of X Makes How Much of Y?

A very common type of problem is to determine the theoretical masses of products formed as a result of a certain reaction. (Practical considerations and side-reactions would almost certainly reduce these in an actual preparation or industrial manufacture.) Conversely, it may be required to find out how much of a certain raw material will be needed to produce a definite amount of another substance.

These problems are best solved by a consideration of the chemical equations representing the reactions.

EXAMPLE (a). What mass of anhydrous zinc sulphate can be produced from 10 g of zinc?

In simple cases like this it is not necessary to put down the whole equation as it should be obvious that one mole of the sulphate will be formed from one mole of the metal.

i.e. $\quad\quad$ Zn $\rightarrow$ ZnSO$_4$
$\quad\quad\quad\quad$ 1 mole $\quad$ 1 mole

Now insert the necessary atomic and formula masses.

Zn: 65·5 g ZnSO$_4$: $\underbrace{65\cdot5 + 32 + (4 \times 16)}_{161\cdot5}$

65·5 g of zinc form 161·5 g of zinc sulphate

10 g of zinc form $\dfrac{161\cdot5}{65\cdot5} \times 10$

= $\underline{24\cdot7}$ g zinc sulphate

EXAMPLE (b). 1·28 g of copper(II) oxide are completely reduced by hydrogen to copper. What masses of copper and water are formed?

$$CuO + H_2 \rightarrow Cu + H_2O$$
$\quad$ 79·5 g $\quad\quad\quad$ 63·5 g $\quad$ 18 g

79·5 g of copper(II) oxide form 63·5 g of copper

∴ 1 g of copper(II) oxide forms $\dfrac{63\cdot5}{79\cdot5}$ g of copper

∴ 1·28 g of copper(II) oxide form $\dfrac{63\cdot5}{79\cdot5} \times 1\cdot28$ g

of copper = $\underline{1\cdot02}$ g

Similarly the mass of water formed will be

$\dfrac{63\cdot5}{79\cdot5} \times 18$ g = $\underline{0\cdot29}$ g

Exercise 10

Have tables of relative atomic masses and valencies in front of you and then try to answer the following problems mentally:

A. What mass of oxide can be formed by burning (i) 24 g Mg, (ii) 20 g Ca, (iii) 31 g P, (iv) 3 g H?
B. What mass of chloride can be formed from (i) 12 g Mg, (ii) 11·5 g Na, (iii) 3·9 g K, (iv) 108 g Ag?
C. What mass of sulphide can be formed from (i) 4 g H, (ii) 207 g Pb, (iii) 2 g Ca, (iv) 6·35 g Cu?

Using the methods illustrated in the examples try the following problems.

1 What mass of magnesium oxide can be obtained by burning 6 g of the metal in air.

2 2000 kg of titanium(IV) oxide are needed in a white paint. What mass of titanium would produce this amount?

3 How much zinc dissolved in hydrochloric acid will produce 0·15 g of hydrogen, and how much anhydrous zinc chloride will be produced?

4 What mass of sodium carbonate can be produced from 6·4 g of sodium hydroxide, and what mass of of carbon dioxide will be needed?

5 Some sodium hydroxide solution is neutralized by hydrochloric acid and 20 g of water are formed in addition to that already present. What masses of acid and alkali are used?

6 An impure specimen of sulphur weighing 12·5 g is burnt, and the mass of sulphur dioxide formed is 24 g. What is the percentage impurity in the sample?

7 Calculate the reacting mass of calcium hydroxide and ammonium chloride which will produce 25 g of ammonia. What mass of water is also produced?

8 3·60 g of copper(II) oxide are reduced by coal gas. How much copper and water are formed? If the water produced is converted to steam and passed over heated iron how much tri-iron tetroxide (Fe_3O_4) will be formed?

9 1·36 g of silver nitrate crystals were dissolved in water and the silver chloride precipitated with excess brine. If the precipitate had a mass of 1·06 g, what was the percentage of water of crystallization in the crystals?

10 3·70 g of magnesium carbonate are dissolved in dilute acid and the gas produced slowly passed through a tube containing carbon and heated to redness. What maximum mass of carbon monoxide can be obtained?

11 Write down as much information as you can get from the following equation:

$CaCl_2(aq) + 2AgNO_3(aq) \rightarrow Ca(NO_3)_2(aq) + 2AgCl(s)$

12 Write a balanced equation in each case to represent:
(a) the oxidation of sulphur dioxide in the Contact Process;
(b) the reduction of sulphur dioxide by hydrogen sulphide;
(c) the action of heat on potassium nitrate·
(d) the combustion of carbon monoxide.
(S)

13 How many moles of water of crystallization are present in one mole of crystalline sodium carbonate? In 100 kg of these crystals how many kg are contributed by the water of crystallization?

14 Calculate (a) the theoretical mass of chalk precipitated by the passage of carbon dioxide through a sample of lime water containing 0·37 g of calcium hydroxide, (b) the percentage gain in mass when magnesium is heated in oxygen. (S)

15 0·1 mole of calcium carbonate is heated strongly until it is completely decomposed: the product is then shaken with sufficient water to produce a clear solution and carbon dioxide is passed in until the precipitate which forms redissolves. This solution is then evaporated to dryness. By means of equations outline the course of the reactions and state how many moles of the final product are obtained. How many grams of solid will there be?

16 What mass of potassium chlorate(V) would have to be heated to yield 4 g of oxygen? What would be the mass of potassium chloride left? (V)

17 Complete the following equations and show the physical state of the products.

$Fe(s) + 2HCl(g) \xrightarrow{\Delta}$

$Cu(OH)_2(s) + 2HNO_3(aq) \longrightarrow$

$Fe_2(SO_4)_3(aq) + 6NaOH(aq) \longrightarrow$

$2NH_3(g) + 3CuO(s) \xrightarrow{\Delta}$

$Na_2CO_3(s) + 2HCl(aq) \longrightarrow$

$2Al(s) + 6HCl(aq) \longrightarrow$

$2NH_4Cl(s) + Ca(OH)_2(s) \xrightarrow{\Delta}$

$3Fe(s) + 4H_2O(g) \xrightarrow{\Delta}$

$PbO(s) + 2HNO_3(aq) \longrightarrow$

18 Calculate: (a) the amount of iron in 20 g of crystalline iron(II) sulphate; (b) the percentage of water of crystallization in the crystals; (c) the percentage loss of mass on heating the crystals to dull red heat, the final residue being iron(III) oxide. (L)

19 What is the theoretical mass of nitric acid obtainable from 0·6 mole of sodium nitrate, and how much sodium hydrogensulphate will be produced at the same time?

20 (i) 1 g of sodium is added in small pieces to cold water. The resulting solution is evaporated to dryness. Calculate the theoretical mass of the residue obtained.

(ii) A chloride of copper is found to contain 47·27 per cent of the metal. Find the simplest formula of the compound.

(iii) Calculate the volume of hydrogen at s.t.p. given off on dissolving 3·25 g of zinc in excess dilute sulphuric acid.

(iv) Calculate the percentage of water of crystallization in copper(II) sulphate, $CuSO^4.5H^2O$.
(1 dm³ of hydrogen has a mass of 0·09 g) (S)

21 Calculate:
(a) the mass of water used to react completely with 6·9 g of sodium.
(b) the percentage loss in mass on strongly heating chalk for some time. (S)

22 (a) Calculate the percentage composition of anhydrous sodium sulphate.
(b) A compound has the composition by mass: carbon 75 per cent; hydrogen 25 per cent. Find its formula. If all of the carbon present in the compound is converted on combustion into carbon dioxide, calculate the mass of carbon dioxide produced by burning 8 g of the compound. (S)

23 (a) Calculate the mass of ammonia gas which would be used in the formation of 1 kg of sulphate of ammonia.
(b) A hydrated salt on heating gently loses 36 per cent of its mass as steam. The anhydrous salt left is found to have the formula XSO_4, where X is a metal of atomic weight 64. Showing each step in your working, calculate the formula of the hydrated salt. (S)

24 (a) What mass of tri-iron tetroxide (Fe_3O_4) is produced by passing steam over 8·4 g of red hot iron filings until no iron remains?
(b) 10 g of a mixture of copper(II) oxide and copper is warmed with an excess of dilute sulphuric acid. When the reaction ceases, the hot mixture is filtered. The residue in the filter paper, after washing and drying, has a mass of 8·4 g. What is the mass of copper(II) sulphate crystals, $CuSO_4.5H_2O$, which could theoretically be obtained from the filtrate? (S)

11 Archimedes' Principle and Floating Bodies

It is an effect which most people have observed, that things seem to lose weight in water. You will have noted that in water, you can support easily, with one hand, a person who would otherwise be too heavy for you to lift. Indeed some things seem to lose all their weight in water, in which case we say that they float. You will be able to see from Fig. 11.1 that this is an effect of liquid pressure. It shows a block of solid material immersed in a liquid.

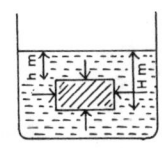

Fig. 11.1. Pressures on a solid immersed in a liquid

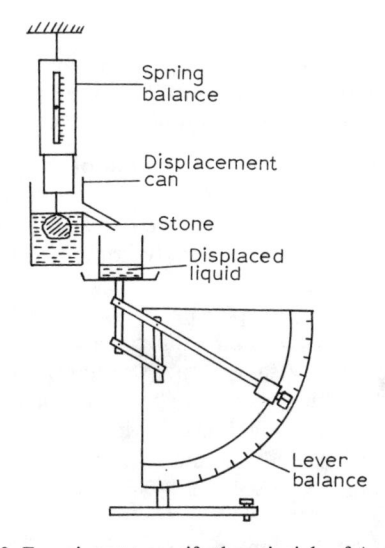

Fig. 11.2. Experiment to verify the principle of Archimedes

There is an upward pressure of $H \times d \times g$ N/m² on the bottom face and $h \times d \times g$ N/m² is the downward pressure on the top face. Since H is greater than h, the pressure on the bottom will be larger than that on the top. There will thus be a net upward force or upthrust on the body. Whether it floats or sinks depends on whether this upthrust is greater or less than its weight. We shall now try to find out by experiment what this upward force depends on.

Experiment 11.1. Weigh a stone by suspending it from a spring balance. Now fill a displacement can to the level of the spout with water and arrange it so that any displaced water may be collected in a beaker which rests on the pan of a lever balance (Fig. 11.2). Immerse the stone (still suspended from the spring balance) in the water contained in the displacement can. Note the new spring balance reading and the weight of the displaced water as indicated by the lever balance. Repeat the experiment with other convenient liquids in the displacement can, e.g. methylated spirit and other solids, e.g. a brass weight. Table 11.1 shows a typical set of results.

Table 11.1

	Water and stone	Meths and stone	Water and brass	Meths and brass
Weight of solid in air (N)	0·50	0·50	0·80	0·80
Weight of solid in liquid (N)	0·30	0·36	0·71	0·74
Apparent loss in weight (N)	0·20	0·14	0·09	0·06
Weight of liquid displaced (N)	0·20	0·14	0·09	0·09

With regard to the table, it should be pointed out that, although your spring and lever balances may be calibrated in grams or kilograms, they are in fact measuring forces in all cases and so the correct units are newtons.

The apparent loss in weight referred to in the table is the upthrust on the solid which we mentioned previously. The results show that the upthrust is equal to the weight of liquid displaced. This experiment illustrates the *Principle of Archimedes* which states:

When a body is wholly or partially immersed in a fluid the upthrust or apparent loss in weight is equal to the weight of fluid displaced.

You will see that this statement of the principle refers to a fluid. This implies that it works for a gas as well as for a liquid. We can show the existence of an upthrust on a body immersed in a gas.

Experiment 11.2. Suspend a hollow glass bulb (B) from one arm of a small balance which is mounted on a base plate attached to a vacuum pump. (Fig. 11.3). Adjust the position of the counter-mass (C) until a balance is achieved. Now cover the apparatus with a bell jar and switch the pump on. The balance beam will go down on the side of the bulb.

Before the air was removed from the bell jar both the bulb and the counter-mass experienced an upthrust or buoyancy effect. However Archimedes' principle tells us that since the bulb is bigger it will displace a larger weight of air and hence the upthrust on it will be greater.

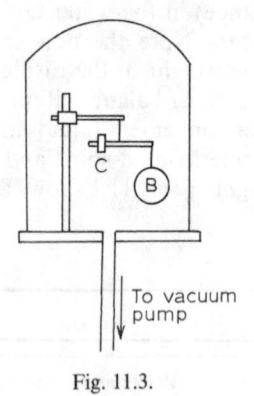

Fig. 11.3.

Floating

An object which is floating in a liquid (or in a gas, e.g. an airship or balloon) seems to have lost all its mass. The upthrust on it is equal to its mass so that it moves neither up nor down. Archimedes' principle tells us that upthrust is equal to the mass of fluid displaced. For a floating body this means that its mass is equal to the mass of fluid displaced. We say that 'a floating body displaces its own mass of the fluid in which it floats'. We may verify this application of Archimedes' principle by the following experiment.

Experiment 11.3. Take a wooden stick as shown in Fig. 11.4 (*a*). It should be of uniform cross section 1 cm², divided along its length into cm, and weighted at one end by a nail or screw. Float it vertically in various liquids contained in a tall vessel (gas jar or measuring cylinder).

In each case note the volume immersed. Finally dry it and find its mass.

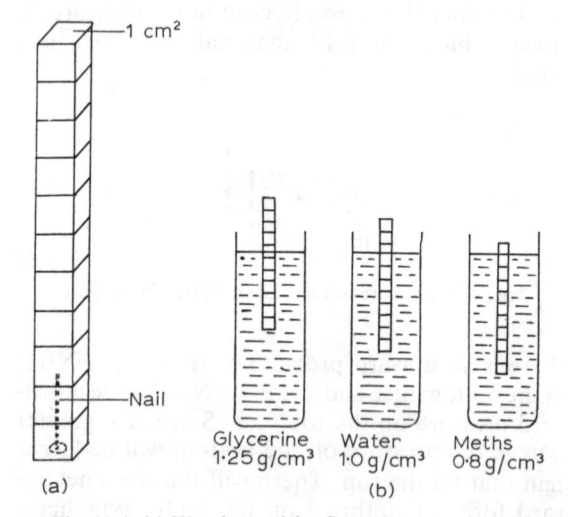

Fig. 11.4. (*a*) Wooden rod for floatation experiments
(*b*) Floatation experiments

The stick shown in Fig. 11.4(*b*) has a mass of 8 g and displaces:
(1) 6·4 cm³ of glycerine
   or since the density of glycerine is 1·25 g/cm³, 8 g of glycerine.
(2) 8 cm³ or 8 g of water.
(3) 10 cm³ of methylated spirits
   or, since the density of meths. is 0·8 g/cm³ 8 g of meths.

In each case the mass of liquid displaced is equal to the mass of the floating object.

You will see that the stick in the diagram has a volume of 11 cm³. Thus its mean density

is $\frac{8}{11}$ g/cm³ = 0·73 g/cm³

If we placed the stick in a liquid whose density

was 0.5 g/cm^3, even when it was totally immersed the mass of liquid displaced would only be 0.5×11

$$= 5.5 \text{ g}$$

Thus the upthrust would be less than the mass of the stick and it would sink. In fact a solid will float in a fluid only if its density is less than that of the fluid. In view of this, it may seem hard to account for a ship made of steel (density 7.7 g/cm^3) floating in water. However it is the *mean* density of the floating object we are really concerned with. The hull of a ship contains a great deal of air and its mean density will be much less than that of water. Of course if the hull were of solid steel it would sink.

You may have heard reference made to the 'displacement of a ship'. The *Queen Elizabeth*, for example, has a displacement of 84 000 t (1 t (tonne) = 1000 kg). This means that it displaces 84 000 t of water and hence from Archimedes' principle it weights this much. Although the weight of water displaced by a ship will always be the same, the volume displaced will depend on the density of the water. Thus a ship will sink a little deeper in fresh water than it does in salt water. We shall see later that temperature affects the density also, so the ship will sink deeper in warm water than in cold. Every ship must carry a set of lines marked on its hull indicating by the depth to which the ship sinks, the legal limit of mass of cargo which it may carry. These are known as Plimsoll lines (Fig. 11.5).

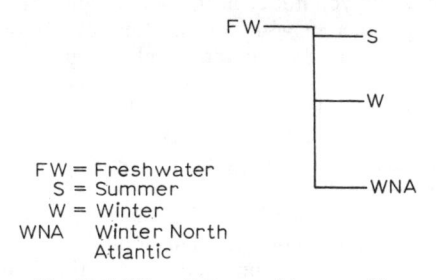

FW ─── S
     ─── W
     ─── WNA

FW = Freshwater
S = Summer
W = Winter
WNA = Winter North Atlantic

Fig. 11.5. Plimsoll line markings on ships

The Hydrometer

Experiment 11.3 may have suggested to you a way of measuring density easily. The stick used in that experiment could be marked off to read density by calibrating it in various liquids whose densities are known. The principal disadvantage of this simple device is that it is not sensitive enough. In the case illustrated in Fig. 11.4(*b*) a difference of 0.2 between the densities of water and meths. produces a difference in levels of 3 cm. In many practical applications it is necessary to find the densities of liquids to 3 significant figures. To do this, our stick would need to be very long and thin indeed. Fig. 11.6 shows how this is overcome in the design of a common hydrometer. It consists of a glass bulb containing lead shot to ensure that it floats in a vertical position, a larger bulb containing air and a stem of uniform, small cross section on which the scale of densities is marked. Hydrometers, in practice, usually cover a restricted range of densities. The narrow stem means that a small difference in densities produces a large difference in liquid levels when the hydrometer floats. The hollow bulb ensures that the hydrometer floats. Without it the hydrometer would need to be very long indeed if it was not to sink. The volume of the hollow bulb determines the range of the instrument. If it were to cover the range 0.8 to 1 g/cm^3 the bulb would be larger. You should note that the low density values are at the top of the scale and the scale is not evenly divided.

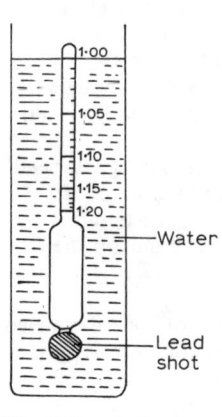

Fig. 11.6. A hydrometer

Balloons

In recent years much important research has been done in meteorology and in nuclear physics with the aid of high flying balloons. We showed in Experiment 11.2 the buoyancy effect of air on an air-filled glass bulb. If a container of thin rubber or sheet polythene is filled with a gas of very low density such as hydrogen then its mass may be less than the mass of the air it displaces. There will thus be a net upward force on it and it will rise. However, the air gets less dense at higher altitudes and the balloon will reach its maximum height when the mean density of the balloon envelope and the contained hydrogen is equal to the air density.

The balloons used by meterologists for weather research are usually of rubber and closed. They ascend with a load of scientific instruments on the end of a line. The balloons used for research in nuclear physics have been of sheet polythene and of the open necked type (Fig. 11.7). They carry large stacks of photographic plates to heights of 30 km or more. The plates record tracks of atomic particles in the cosmic radiation which falls on the earth's upper atmosphere from space. These balloons are not, of course, attached to a line and they have to be recovered, often at large distances from their launching point.

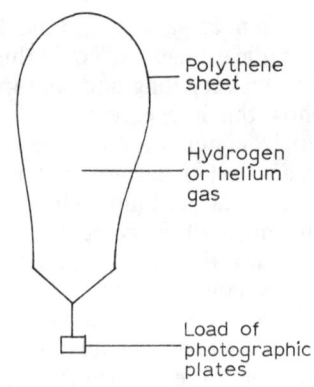

Fig. 11.7. 'Open' balloon used for cosmic ray research

Exercise 11

1 A rectangular block of metal measures 10 cm × 5 cm × 2 cm and has a mass of 500 g.
(a) What is its density?
(b) What is the greatest pressure it can exert when resting on a flat surface?
(c) What is the least pressure it can exert?
(d) If it was hung on a spring balance and then submerged in water, what reading would the balance show? (SR)

2 A solid of mass 0·8 kg, suspended by a string, is completely immersed in water. The tension in the string is 6 N. Calculate (a) the upthrust on the solid, (b) the volume of the solid, and (c) the density of the solid. Explain how you arrive at your answer in each case. (S)

3 State the principle of Archimedes and describe an experiment by which it may be verified.
Calculate the useful load which can be lifted by a helium balloon of volume 800 m³, if the envelope and carriage together weigh 100 kg. Assume that, under the existing conditions, the density of helium is 0·18 kg/m³ and that of air is 1·26 kg/m³. (J)

4 State Archimedes' principle as applied to a body which floats in a liquid. Draw a diagram of a commercial hydrometer. Point out the functions of its different parts and explain how Archimedes' principle is applied in its working.
An iceberg floats in sea water with $\frac{1}{11}$ of its volume showing. If the density of the sea water is 1·012 g/cm³, calculate the density of the ice. (L)

5 A certain common hydrometer has a mass of 25 g and the cross-sectional area of its stem is 1·0 cm². Calculate the distance between the 1·00 and 0·80 markings on the stem. (J)

6 State the principle of Archimedes and describe an experiment by which it may be verified.
A solid piece of glass of volume 23·8 cm³ is floated on mercury. Calculate the volume of glass submerged. Assume that the relative densities of glass and mercury are 2·5 and 13·6 respectively. (J)

7 State the principle of Archimedes and show how it leads to the principle of flotation. How would you test the latter experimentally?
A spring balance supporting a lump of material reads 300 g when the object is in air, 250 g when it is completely immersed in water and 260 g when completely immersed in paraffin. Calculate the relative densities of paraffin and the solid. (S)

8 The area of cross-section of a ship at the waterline is 2000 m². Find the change in the position of the waterline if the ship is loaded with 1000 t of cargo. Explain the steps in your working.
[Density of water = 1000 kg/m³
1 t = 1000 kg]

9 A block of brass is first weighed in air and then when it is completely immersed in water. What difference would you notice in the two weighings?
An experimenter with two identical blocks of brass places one block in water and it sinks. Then he attaches a big piece of cork to the other block and places it in the water; this block floats. Why does the first block sink while the second block floats?
What happens when a solid is placed in a liquid of the same density?
A metal cube of side 2 cm has a weight of 0·50 N in air. It is then completely immersed in a liquid of density 0·9 g/cm³. What is the volume and mass of the liquid displaced? Calculate the weight, in N, of the liquid displaced and deduce the apparent weight of the block. (C)

10 A piece of cork has a volume of 30 cm³ and a density of 0·25 g/cm³.
Calculate the mass of the cork.
Calculate the mass of water displaced when the cork is floating in water.
What is the mass of liquid displaced when the cork is floating in salt solution?
What is the volume of salt solution displaced when the cork is floating?
(The density of salt solution is 1·2 g/cm³. Assume that $g = 10$/s² or that 1 kg has a mass of 10 N.) (C)

12
Statics

In chapter 4 we talked about the sort of movement which is produced when forces act on bodies. It often happens, however, that a number of forces act on an object and yet produce no movement. We say that the body is 'in equilibrium' under the action of a number of forces. The study of this state of affairs is most important, for example, to the civil engineer. His job is to ensure that the bridges and buildings he erects will remain in equilibrium under the action of the various forces they may encounter. Some of the civil engineer's work may be experimental. For example, when a large bridge is being designed it is common practice to construct a model and study its behaviour in a wind tunnel in order to anticipate the behaviour of the completed bridge in high winds. On the other hand, much of the design work on a structure may be carried out by applying the simple physical principles which we shall discuss in this chapter.

Equilibrium under the Action of Two Forces

Figure 12.1(a) shows an object resting on a table, Fig. 12.1(b) an object suspended by a string. In both cases the body is in equilibrium under the action of two forces. In the first case there is the vertical downward attraction of the earth on the body which we call the weight of the body (W). In the absence of the table this force would cause the body to fall vertically with an acceleration of about 9·8 m/s². Since this does not happen there must be an upward force, equal to the weight. This we call the reaction of the table on the body (R).

In the case shown in Fig. 12.1(b), the force equal and opposite to the weight is called the tension in the string (T).

A body will be in equilibrium under the action of just two forces only if they are equal in size, opposite in direction and have the same line of action. The object shown in Fig. 12.2 for example will not be in equilibrium even though the forces F_1 and F_2 are equal in size and opposite in direction, since they do not have the same line of action. What kind of movement do you think the object will make in this case?

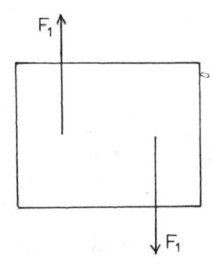

Fig. 12.2. Equal and opposite forces which do not produce equilibrium

We shall now consider cases of equilibrium under the action of more than two forces. There is really only one simple instrument which we can use to measure forces in our experiments and this is the spring balance. We discussed in Chapter 2, how we can use this instrument to measure weight and there is no reason why it should not be used to measure other forces. However, it should be noted that a spring balance is normally made to be used in a vertical position in which the weight of the spring itself and the hook attached to it will help to stretch it. Thus when used, say horizontally, it is likely to have a 'zero error'.

It should be noted also that, although the spring balances available to you may be calibrated in grams or kilograms, they really do measure forces and should be calibrated in newtons.

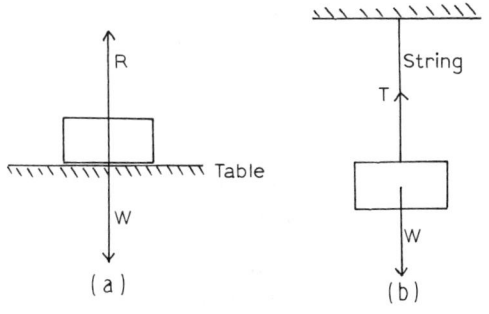

Fig. 12.1. Equilibrium

Parallel Forces

Experiment 12.1. Weigh a metre rule on a lever balance. Now suspend it (Fig. 12.3) from two suitable spring balances (S_1 and S_2). Hang 50, 100 and 200 g masses in turn by means of cotton or fine wire from various points on the rule noting each time the readings of S_1 and S_2. Tabulate your results as follows:

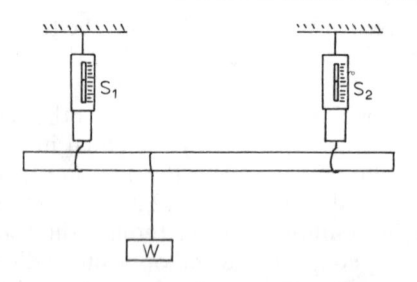

Fig. 12.3. Total upward force = total downward force

Weight suspended w (N)	Total downward force (N)	s_1 reading (N)	s_2 reading (N)	Total upward force (N)

The total downward force is the weight suspended + the weight of the rule. The total upward force is the sum of the tensions in the two spring balances (i.e. $S_1 + S_2$). Your table of results should show that in each case the total upward force equals the total downward force.

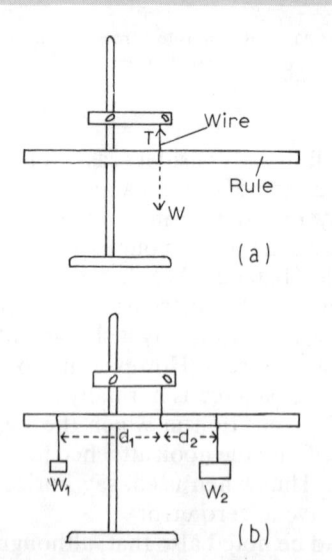

Fig. 12.4. (a) Rule in equilibrium under the action of two forces
(b) Clockwise and anticlockwise turning effects

Experiment 12.2. Suspend the metre rule by cotton or wire from a clamp, and adjust the point of suspension until the rule just balances in a horizontal position (Fig. 12.4(a)). This will probably happen with the ruler suspended at about its mid-point. We can now say that the ruler is in equilibrium under the action of two forces, the tension in the wire suspension and its own mass. These two forces must have the same line of action, hence we may regard the mass of the ruler as acting downwards through the point of suspension. Now hang 50 g (0·05 kg) and 100 g (0·1 kg) masses (W_1 and W_2 in Fig. 12.4(b)) from the two sides of the ruler. If we assume that g (the acceleration of gravity) is 10 m/s², then a 1 kg mass has a weight of 10 N. Our 0·05 and 0·1 kg masses will thus have masses of 0·5 N and 1 N respectively. Set the distance (d_1) of the 50 g mass from the point of suspension at 0·4 m and adjust the distance d_2 until the ruler again rests in a horizontal position. Repeat the experiment with d_1 0·3, 0·2 and 0·1 m, noting the value of d_2 for which balance is obtained each time. Repeat the experiments with $W_2 = 200$ g. Tabulate your results as follows:

W_1 N	d_1 m	$W_1 \times d_2$ Nm	W_2 N	d_2 m	$W_2 \times d_2$ Nm
0·5	0·4	0·2	1·0	0·2	0·2

You should see that the figures in columns 3 and 6 are approximately equal in each case. Now W_1 is pulling down on the left-hand side and tending to turn the ruler anticlockwise. W_2 is tending to turn the ruler clockwise. A condition of equilibrium will be reached when these two 'turning effects' are equal. The table of figures above shows what we mean by turning effect. $W_1 \times d_1$ is the turning effect or *moment* of the force W_1 about the point of suspension. The moment of a force is measured in newton metres (N m) if the force is in newtons and the distance in metres.

From the results of these two experiments we can give two general conclusions about a body in equilibrium under the action of a number of parallel forces.
(1) The total upward force equals the total downward force.
(2) The total clockwise moment about the point of suspension equals the total anticlockwise moment about the same point.

Moments and the Principle of Moments

The idea of the turning effect or moment of a force is a most important one and it needs to be given more attention. If a garage mechanic wishes to release a particularly rusty or tightly done up nut he will select a spanner with as long a handle as possible. He will probably say that it gives him more 'leverage'.

The long handle does not increase the force he is capable of exerting, but it does increase the turning effect or moment of the force about the axis (in this case the bolt). The moment here (Fig. 12.5) is $F \times D$, hence the larger D, the larger the moment.

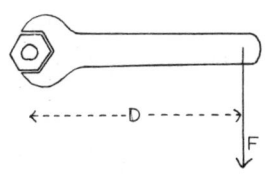

Fig. 12.5. The turning effect of the force is $F \times D$

The moment of a force about a point is defined as: force × the *perpendicular* distance from the point to the line of action of the force. The significance of the word 'perpendicular' in this definition may be seen by considering another simple example. Fig. 12.6(*a*), (*b*) and (*c*) represents three positions of a crank and pedal of a bicycle. If the cyclist exerts a force F on the pedal which is always vertically downwards, then in case (*a*) the turning effect of this force is zero. The line of action of the force passes through the axle, so the perpendicular distance (D) between them is zero. Thus $F \times D = 0$. In case (*b*) the moment of the force is $F \times d_1$ and in case (*c*) $F \times d_2$. Clearly a given force has a maximum moment when the force is applied at right angles to the crank as in case (*c*).

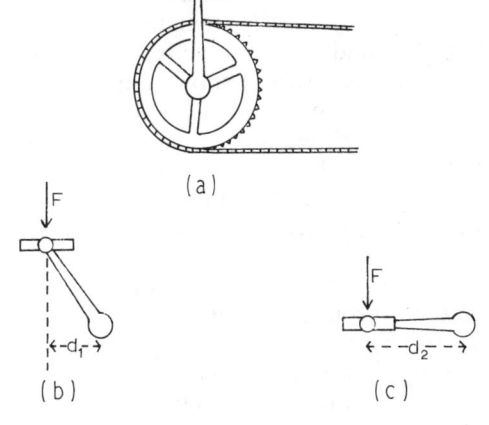

Fig. 12.6. (*a*) Force F has no turning effect
(*b*) Turning effect is $F \times d_1$
(*c*) Turning effect is $F \times d_2$

The second conclusion of experiments 12.1 and 12.2 may be generalized to include cases where the forces are not parallel and for moments taken about any point not just a point of suspension. The principle of moments states:

For a body in equilibrium under the action of a number of forces the total clockwise moment about any point is equal to the total anticlockwise moment about the same point.

Centre of Gravity

At the beginning of Experiment 12.2 we suspended a metre rule so that it was in equilibrium in a horizontal position. We then concluded that the weight of the ruler could be considered as acting vertically downwards through the point of suspension so that it could be just balanced by the tension in the wire suspension. We say that the point of suspension in this case is the *centre of gravity* of the ruler. Each atom of which the ruler is made up has its own weight represented by forces W_1, W_2, W_3, etc. in Fig. 12.7. To deal with each of these individually in solving a problem would be an impossibly complicated task. It is much more convenient to work with the weight of the whole ruler $W (= W_1 + W_2 + W_3 +$ etc.) which acts at a point C, the centre of gravity. Since the ruler has uniform cross section, the centre of gravity will be at the mid-point. If it happened, for example, to have a splinter of wood missing at one end, the centre of gravity would not be quite in the middle and would have to be found by balancing the rule.

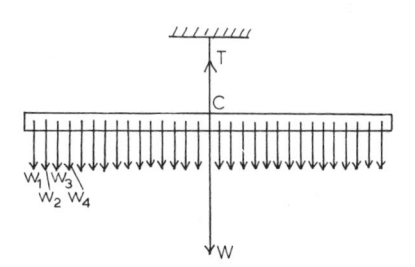

Fig. 12.7. The force W acting through C is equivalent to forces $W_1 W_2 W_3 W_4$ etc.

We have been regarding our metre rule as something having length but no other dimensions. How then can we find the centre of gravity of a body which has a substantial area? We could still try balancing it in a horizontal position when suspended by a string. There is an easier and more satisfactory method.

Experiment 12.3. To find the centre of gravity of a flat object (a lamina) of irregular shape.

A suitable object to illustrate this method is a piece of hardboard or thick cardboard with a number of small holes drilled through it. Attach a length of cotton with a metal weight (P) tied to the end, to a steel knitting needle. Pass the knitting needle through one of the holes in the lamina (A in Fig. 12.8(a)). The weight of the lamina (W) acts through G (the centre of gravity) whose position we are trying to locate. There will thus be a turning moment, $W \times d_1$, and the object will rotate until this moment is zero, i.e. until G is vertically below A (Fig. 12.8(b)). Mark the vertical through A on the lamina using the cotton and weight as a plumb line. We may now be sure that G is on the line we have marked. Place the needle through a hole at B (Fig. 12.8(c)). Once again the lamina will rotate until G is vertically below B. Rule the vertical through B on the hardboard. G is now at the intersection of the two ruled lines. In order to check on this, pivot the object at several other points and rule verticals through the point of suspension in each case. All the lines ruled should meet at a point. See if the lamina will balance on a pin point at the point you have found.

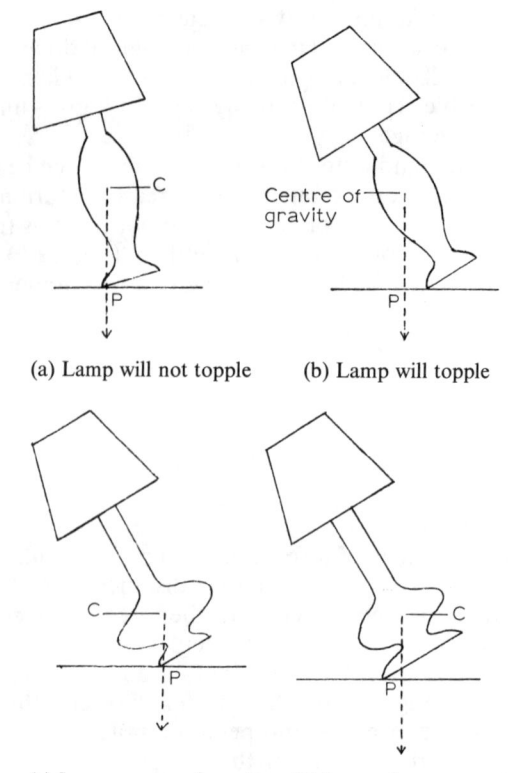

(a) Lamp will not topple (b) Lamp will topple

(c) Lower centre of gravity (d) Larger base

Fig. 12.9.

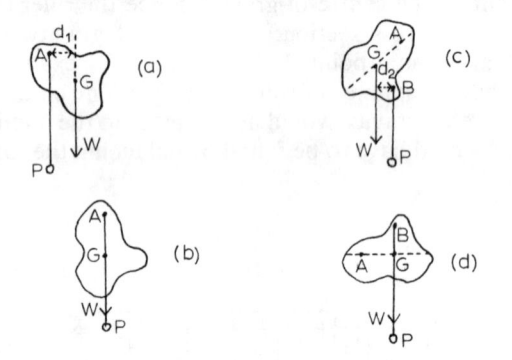

Fig. 12.8. Experiment to find the centre of gravity of an irregular lamina

Stability of Equilibrium

Designers of all sorts of things from buses and ships to teapots and table lamps must consider where the centre of gravity of the thing they are designing will be, since this will very much affect its stability. We say that an object is stable if, when it is displaced from its normal rest position, it returns to it instead of toppling over. We can see how the stability of an object may be increased by considering the designs of table lamp shown in Fig. 12.9.

Fig. 12.9(a) shows a lamp which has been displaced through a small angle from its normal rest position. If it were going to topple over, its axis would be the point P, where the edge of the base touches the ground. In this case the weight of the lamp w acts to the right of this point and the turning effect produced will restore the lamp to its normal rest position. However if the same lamp is tilted a little further (Fig. 12.9(b)) then the weight acts to the left of P and its turning effect will make the lamp topple. Fig. 12.9(c) shows a different lamp with a lower centre of gravity but the same base as the first. It is inclined at the same angle as (b) but since the centre of gravity is lower it will not fall over. Finally Fig. 12.9(d) shows a third design with a low centre of gravity like (c) but with a larger base than the other two. In this diagram it is clear that the lamp will have to be tipped a lot further before the vertical through the centre of gravity passes outside the base causing the lamp to topple over.

The lamps in Fig. 12.9 when resting normally on a horizontal surface would all be in a condition of *stable equilibrium*. This means that if slightly displaced from their rest position they would return to it. The cylinder shown in Fig. 12.10(a) is also in a condition of stable equilibrium. However when the plane on which it is resting is tipped as in Fig. 12.10(b) it reaches a point where it is on the verge of toppling. The vertical through the centre of gravity just passes through the edge of the base. Here, if the cylinder is tipped just a little further, its equilibrium will be disturbed. This is a condition of *unstable equilibrium*. Fig. 12.10(c) shows the same cylinder in a condition of *neutral equilibrium*. If it is rolled it will rest in any position with its curved surface in contact with the plane.

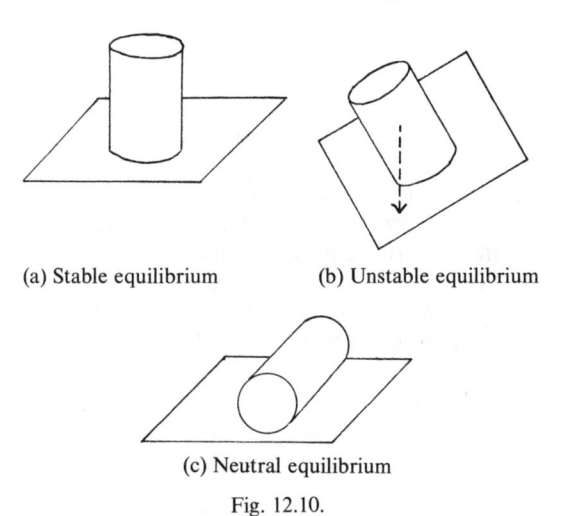

(a) Stable equilibrium   (b) Unstable equilibrium

(c) Neutral equilibrium

Fig. 12.10.

Application of the Principle of Moments

Before we leave these topics of moments and centres of gravity, we shall consider a simple problem based on them.

A uniform metre rule of mass 0·1 kg is supported by a vertical cord attached at the 35 cm mark. A 0·15 kg mass is suspended on the 5 cm mark on the rule. At what position on the rule must a mass of 0·06 kg be suspended so that the rule will come to rest in a horizontal position?   (*J*)

The arrangement is shown in Fig. 12.11. Let the 0·06 kg mass be suspended x m from the point of suspension of the rule.

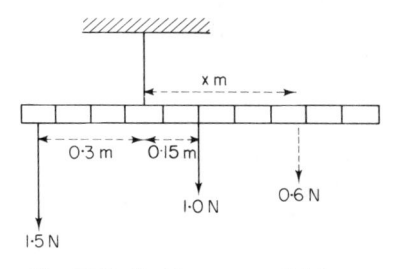

Fig. 12.11. Problem on parallel forces

If we consider moments about the point of suspension, the 0·15 kg is turning anticlockwise while the 0·06 kg and the weight of the rule itself (acting as the mid-point since the rule is uniform) are turning clockwise.

Assuming that the 0·1, 0·15 and 0·06 kg masses have weights of 1·0, 1·5 and 0·6 N respectively:

$1·5 \times 0·3 = 1·0 \times 0·15 + 0·6 \times x$

or $0·45 = 0·15 + 0·6x$

and $0·3 = 0·6x$

so $x = 0·5$ m

The 0·06 kg mass must therefore be placed 50 cm from the point of suspension to produce equilibrium, or on the 85 cm mark of the rule.

Non-parallel Forces

Most structures which engineers have to deal with involve forces which are not all parallel. Consequently the principle of moments alone is not sufficient for solving the problems which arise. We will consider now the fairly common case of a body in equilibrium under the action of three non-parallel forces.

Experiment 12.4 Fix two spring balances at points *P* and *Q* (Fig. 12.12) and join the hooks of the balances by a piece of string *RSU*. Attach another piece of string to it so that it can slide along and tie a mass *W* (say 0·5 kg) to this second string. Fix a drawing board with a piece of drawing paper behind the apparatus so that the positions of the strings may be marked on it. Set the suspended mass at *S* and mark the lines *RS*, *SU* and *SW*. Note the readings of the spring balances (T_1 and T_2). Remove the drawing paper. It will look like Fig. 12.13(a). Produce *AS* and draw a line *XY* parallel to *SC* to form a triangle *SXY*.

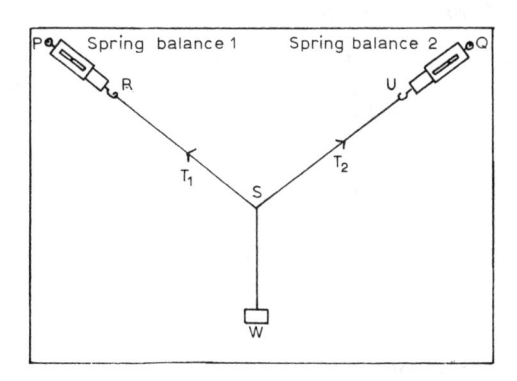

Fig. 12.12. Experiment to verify triangle of forces rule

(a)

(b)

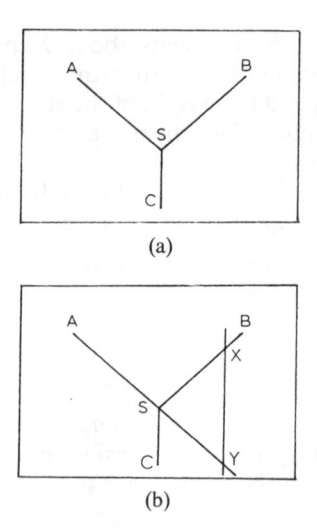

Fig. 12.13. Construction of triangle of forces

Now in the experimental arrangement the point S is in equilibrium under the action of three forces, the tension in RS (T_1 N), the tension in SU (T_2 N) and the tension in SW (Wg N). The sides of the triangle SXY represent these three forces in direction (SY represents T_1, SX represents T_9 and XY represents Wg). Measure the lengths of these three sides and write down their ratio in the form:

$$XY:SX:SY = 1:?:?$$

Write also the ratio of the three forces in the form:

$$Wg:T_9:T_1 = 1:?:?$$

Repeat the experiment with various suspended masses and with S in different positions. You should find always that the ratio of the lengths of the sides of the triangle obtained is the same as the ratio of the forces. A triangle obtained in this way is known as a triangle of forces.

The result of this experiment is summarized in the *Triangle of Forces rule* which states:

If a body is in equilibrium under the action of three forces, then those forces may be represented both *in size and direction* by the sides of a triangle taken in order.

The significance of the phrase 'taken in order' may be seen from Fig. 12.14. The three forces, represented in size and direction by the sides of the triangle ABC in Fig. 12.14(a), will produce equilibrium. However if, as in Fig. 12.14(b), we reverse one of the arrows, then all the forces are acting generally from left to right and will certainly not produce equilibrium.

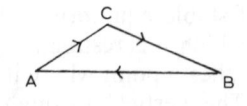

(a) Three forces in equilibrium

(b) Three forces not in equilibrium

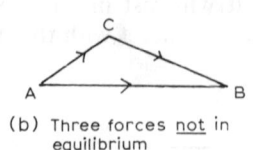

Fig. 12.14.

Application of the Triangle of Forces Rule

A simple problem will show how the rule we have demonstrated may be applied.

A load of 100 kg is suspended from a beam by two ropes, each of which makes an angle of 30° with the vertical. Find the tension in each rope. (C)

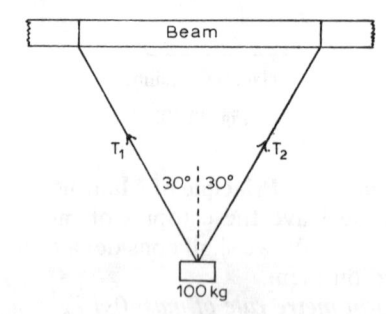

Fig. 12.15. Problem on the triangle of forces

The actual arrangement is as shown in Fig. 12.15. The weight of the 100 kg mass is 1000 N. We will represent it by a line 10 units long and in a vertical direction. Thus the scale of our diagram is, 1 unit of length = 100 N. The other forces T_1 and T_2 may be drawn in direction only, making 30° with the vertical (Fig. 12.16).

However our triangle of forces is complete and by measuring the lengths of the other two sides (both 5·7 units of length) we see that $T_1 = T_2 = 570$ N.

The Parallelogram of Forces

If we have two forces, P and Q, acting on a body as in Fig. 12.17(a) we know it cannot be in equilibrium. However there is a certain force, E, as in Fig. 12.17(b) which would reduce it to equilibrium.

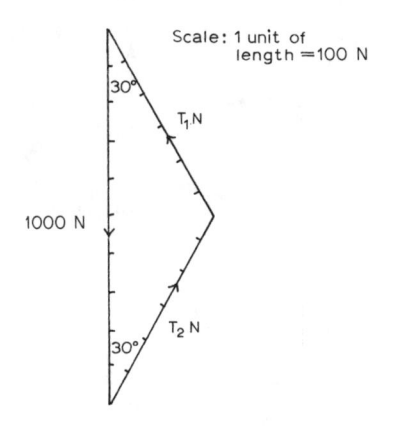

Fig. 12.16. $T_2 = T_2 = 570$ N

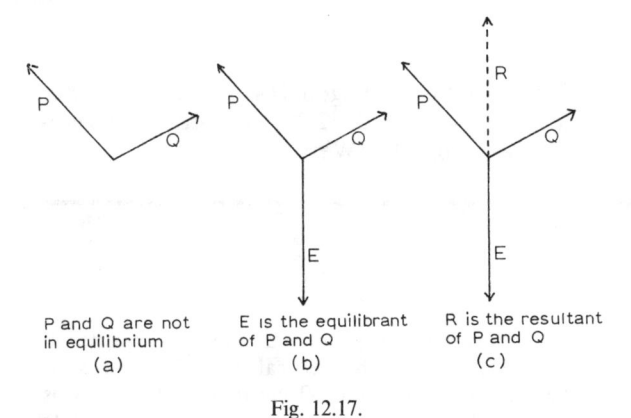

P and Q are not in equilibrium
(a)

E is the equilibrant of P and Q
(b)

R is the resultant of P and Q
(c)

Fig. 12.17.

E, which is known as the *equilibrant* of P and Q, may be found by means of a triangle of forces. E reduces to equilibrium the two forces P and Q. It would also reduce to equilibrium a single force R, equal in size and opposite in direction to E. It follows then that the single force R is producing the same effect as the two forces P and Q. R is known as the *resultant* of P and Q.

The resultant of two forces may be found by application of the *parallelogram of forces rule* which states:

If two forces (X and Y in Fig. 12.18) are represented in size and direction by the adjacent sides of a parallelogram, then their resultant (R) is represented in size and direction by the diagonal of the parallelogram passing through the point of intersection of the forces.

This is really only another way of stating the triangle rule, and it may be verified by an experiment similar to Experiment 12.4. In this case the resultant of the forces T_1 and T_2 (Fig. 12.12) will be a force W vertically upwards.

Resolving Forces

Just as we can find one force which has the same effect as any two (or more) forces, so we can split a force up into 'component' parts. This may be done by means of the parallelogram rule used in reverse. Fig. 12.19(a) shows an example of this. In driving home a nail a force (F) is applied at an angle to the nail. This force tends to drive the nail in and to bend it. Whether in fact it bends depends on how large the component X of the force is.

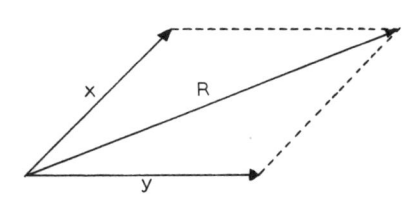

Fig. 12.18. A parallelogram of forces

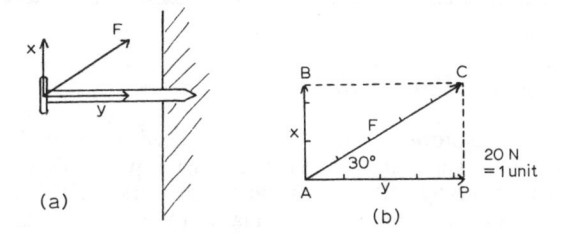

Fig. 12.19. Resolving a force into two components

Suppose F is 100 N and applied at 30° to the nail. We may find the components X and Y by construction. Let one unit of length in the diagram represent 20 N. AC is constructed 5 units long to represent F. AB and AD making 60° and 30° with F respectively are drawn and the rectangle is completed (Fig. 12.19(b)). AD and AB are measured and found to be 4·3 units and 2·5 units respectively. Thus $Y = 86$ N and $X = 50$ N.

In fact the scale drawing is not needed in cases like this where we are resolving a force into two components at right angles to each other since in this case:

$$\frac{X}{F} = \sin 30 \quad \text{or } X = F \sin 30$$

$$\frac{Y}{F} = \cos 30 \quad \text{or } Y = F \cos 30$$

It is very convenient to remember these results.

Vectors

Forces have both size and direction. They are known as *vector* quantities. Mass and volume have size but *no* direction. Such quantities are called *scalar* quantities. Speed is a scalar but velocity is a vector. The parallelogram rule which we have been using for adding and resolving forces may be used with all vectors. An important example of the application of the parallelogram rule is with velocities particularly as applied to aircraft. Instruments in an aircraft enable the pilot to measure his speed through the air, but this does not tell him his speed over the land as the air may well be moving quite quickly. The following problem shows how the parallelogram rule helps the pilot.

An aeroplane heading due north has an airspeed of 300 km/h. A wind is blowing at 60 km/h from the east. Find the velocity of the aircraft relative to the ground. (C)

The velocity over the ground is the resultant of the velocity through the air and the wind velocity.

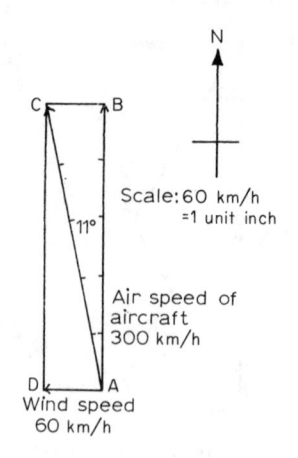

Fig. 12.20. Parallelogram of velocities

This resultant is the diagonal (*AC*) of our parallelogram of velocity (Fig. 12.20). It represents a velocity of 306 km/h 11° West of North.

Exercise 12

1 A uniform metre scale is balanced horizontally across a knife edge at the 20 cm mark with a 300 g mass hung by cotton from the 11 cm mark. Show in a diagram the three forces acting on the scale and calculate its mass. (J)

2 Describe an experiment to find the position of the centre of gravity of a thin, flat, irregularly-shaped piece of tin plate. (J)

3 What conditions must be fulfilled so that three forces, in the same plane and acting on the same body, may be in equilibrium (*a*) if the forces are parallel, (*b*) if the forces are not parallel?

A 10 kg mass hangs by a string (assumed weightless) from a hook in the ceiling. The string is maintained in a position 40° to the vertical by a force acting from the mass in a direction at right angles to the string. Find EITHER graphically OR by calculation the value of this force. (J)

4 A uniform metre rule of mass 175 g is supported in a horizontal position by vertical weightless cords attached at the 10 cm and 80 cm marks. Each cord is suspended from the hook of a spring balance. Calculate the load registered by each spring balance. (J)

5 (*a*) Distinguish between *stable* and *unstable* equilibrium.

Show how the stability of a body is affected by the position of the centre of gravity and by the size of the base.

(*b*) What is meant by a *vector quantity*? Give TWO examples.

A porter is pulling a trolley by exerting a force of 50 N which is inclined at 30° to the horizontal. What is the equivalent of this force in a horizontal direction? (C)

13
Sulphur, the Yellow Solid, and the Sulphides

A number of minerals containing sulphur as one of the combining elements are found in the earth's crust, e.g. lead(II) sulphide or galena, PbS, iron pyrites, FeS_2 and calcium sulphate or gypsum, $CaSO_4.2H_2O$. The only country where sulphur occurs to a large extent as the element is the U.S.A. which produces over 90% of the world's total supply and twenty-five times as much as the next producer, Sicily. However, much sulphur is now obtained from areas where crude oil is found and refined and France now supplies much of Britain's requirements.

Extraction

The sulphur is present in limestone rocks about 150 m below ground and practical difficulties make ordinary mining methods unsuitable. However, a very simple and economical method of obtaining the sulphur was devised by Herman Frasch in 1904. A hole is bored down to the sulphur-bearing rocks and is lined with steel. A Frasch pump (Fig. 13.1) is lowered through this. The pump consists of three concentric tubes, the widest being only about six inches diameter. Water heated under pressure to about 180°C is blown down the outermost tube to liquefy the sulphur near the bottom of the pump. This liquid sulphur enters the pump and is forced up the middle tube as a mixture of sulphur, air and water by compressed air blown down the centre tube. The mixture is kept warm as it flows away into vats where it solidifies.

Sulphur finds widespread use in the manufacture of sulphuric acid, perhaps the most important of all heavy industrial chemicals. Much is also used in making fireworks, matches, gunpowder, certain ointments and medicines (sulphonamides, penicillin, etc.) and in the vulcanizing of rubber, a process which removes the stickiness and renders the rubber more pliable.

Forms of Sulphur

There are a number of elements which can exist in different physical forms, each of which, however, has exactly the same chemical properties as the others. Such forms may take up different crystalline shapes or may be amorphous. These several forms are called *polymorphs* or *allotropes* and quite a number have been described for sulphur. The most important are rhombic and monoclinic (also called prismatic) and an amorphous variety produced during certain chemical reactions. Fig. 13.2 shows the usual appearance of the two crystalline forms.

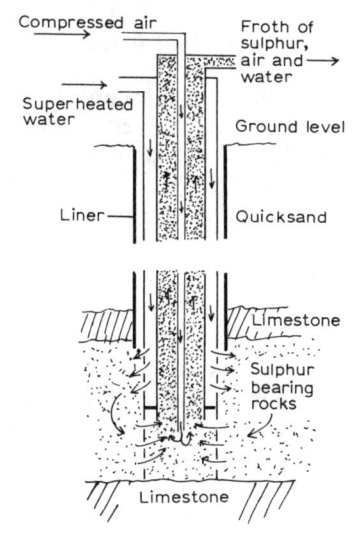

Fig. 13.1. Frasch pump

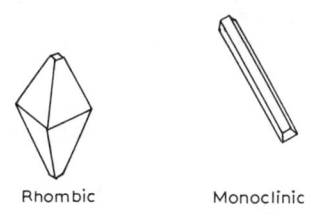

Fig. 13.2. Sulphur crystals

Experiment 13.1. To make various allotropes of Sulphur. (See also p. 317)
(i) Add powdered roll sulphur to about 20 cm³ of carbon disulphide in a test-tube and shake to dissolve as much as possible. Pour through a filter

paper into a dish and cover with a papier mâché mat or another filter so that the solvent evaporates slowly. (*Keep all flames away during this experiment as carbon disulphide is very inflammable.*) When the liquid has evaporated you should find fairly large octahedral crystals of rhombic sulphur.

(ii) Fit up a flask and condenser as in Fig. 13.3 and in the flask place some powdered roll sulphur and just cover it with methylbenzene (toluene). Boil until all the sulphur goes into solution. The upright (reflux) condenser prevents loss of the volatile liquid. Have ready a hot boiling tube standing in a beaker or tin and firmly packed round with cotton-wool. Pour most of the sulphur solution into this tube and leave to cool. Long needle-like crystals of monoclinic sulphur will then be seen in both the tube and the flask.

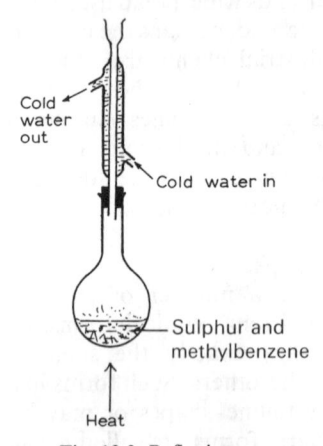

Fig. 13.3. Refluxing

If these are examined under a microscope from time to time (about 20 × magnification) they will be seen to lose their smooth outline as they slowly change into small rhombic crystals.

(iii) Similar crystals can be formed if molten sulphur is allowed to cool. Melt some sulphur in a test-tube and when it becomes liquid pour it into a wet filter paper in a funnel. When the sulphur has *nearly* set remove the paper from the funnel and carefully unfold it. Examine the needle-shaped crystals formed (Fig. 13.4).

(iv) Make a fairly strong solution of sodium thiosulphate(VI) (hypo) and add dilute hydrochloric acid to it. A yellowish-white suspension and amorphous sulphur will be formed.

$$Na_2S_2O_3(aq) + 2HCl(aq) \rightarrow$$
$$2NaCl(aq) + H_2O(l) + SO_2(g) + S(s)$$

(v) If a thin stream of liquid sulphur nearly at its boiling point is poured into cold water it forms plastic sulphur, an unstable rubbery form which soon hardens and changes to the rhombic variety.

Flowers (i.e. 'flour') of sulphur is a mixture of amorphous and rhombic sulphur, only the latter part being soluble in carbon disulphide: roll sulphur is a mass of very small rhombic crystals. Rhombic sulphur melts at 113°C but if kept at a temperature just above 96°C it gradually changes to the monoclinic form; the latter melts at 119°C and below 96°C changes to the rhombic form: this temperature of 96°C is called the transition temperature.

When sulphur is heated in a test-tube interesting changes occur due to alterations in the molecular structure, i.e. in the way in which the atoms making up the sulphur molecule are joined together. The solid melts at 113°C and forms a mobile amber liquid; it soon darkens and becomes so viscous that it will not pour out of the tube, but at a higher temperature it again becomes mobile and almost black; finally it boils at 444°C.

Sulphur burns with a blue flame (lilac colour in oxygen) and the sulphur dioxide formed can be recognised by its acrid smell and by its ability to turn a piece of filter paper dipped in a solution of potassium chromate(VI) from yellow to green.

$$S + O_2 \rightarrow SO_2$$

All forms of sulphur can be shown to be identical chemically by burning a fixed mass of each in oxygen. In each case the same mass (or volume under identical conditions) of sulphur dioxide and nothing else is formed.

Structure

At room temperature the sulphur molecule consists of 'puckered' rings of eight atoms. When it is heated and melts some of these rings break apart to form long chains of eight atoms. At this stage the liquid is runny, but as it is heated further, more rings break, the chains get tangled together and the liquid is then viscous. When the sulphur boils, the vapour consists of simpler molecules containing only two atoms.

Fig. 13.4. Monoclinic crystals formed from molten sulphur

Sulphides

Sulphur will join when many other elements, both metals and non-metals, when heated with them.

> *Experiment 13.2. Synthesis of Metallic Sulphides.*
>
> Mix intimately together (but do not grind) some powdered zinc and sulphur and place a layer about 2 cm deep in the bottom of a test-tube. Clamp this almost vertically and put a large sheet of asbestos underneath. Place a lighted bunsen below the tube and do not stand too close. In a few minutes there will be a flash of light and clouds of white zinc sulphide appear. The tube gets extremely hot. Alternatively the mixture may be placed on a folded strip of asbestos paper and heated or a cone of it may be piled on to an asbestos slab and lit with a bunsen flame.
>
> $$Zn + S \rightarrow ZnS$$
>
> If iron is used instead of zinc it should be in the form of a very fine filings ground with the sulphur in the correct proportions (7 g Fe : 4 g S). The tube should be heated strongly and a glow spreads throughout the mixture as the iron(II) sulphide forms. *Do not use powdered magnesium in this experiment in a test-tube.* It is dangerously explosive, especially when ground with the sulphur.

It should be apparent that the behaviour of these metals is in accordance with their position in the electrochemical series. It might therefore be expected that hydrogen would show little affinity for sulphur and this is so. (Copper and mercury, however, join up quite readily with sulphur.) The two elements can be made to combine together in small amounts if the gas is passed through boiling sulphur in the apparatus shown in Fig. 13.5. The lead (II) ethanoate (acetate) solution placed *after* the reaction tube turns a brownish-black although neither hydrogen nor sulphur has any affect on this substance. A new compound, hydrogen sulphide, has been formed.

$$H_2 + S \rightarrow H_2S$$

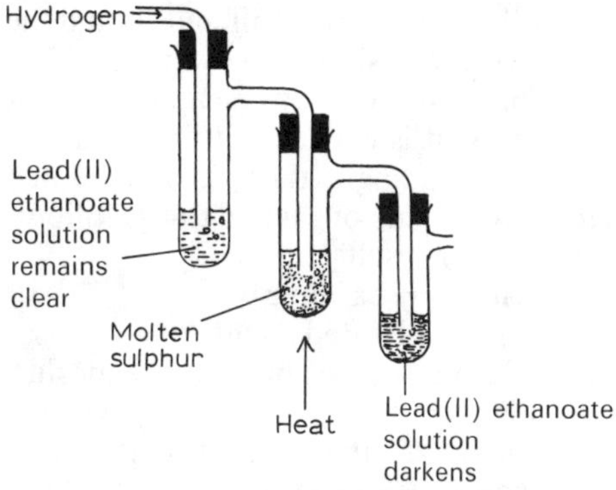

Fig. 13.5. Synthesis of hydrogen sulphide

14
The Oxides and Acids of Sulphur

Sulphur Dioxide

We have already seen that sulphur burns with a blue flame to form the gas sulphur dioxide. Although this method gives a pure gas and is used in industry, laboratory methods are concerned with its production from sulphur-containing compouds. You know from earlier work that sulphur dioxide dissolves in water to form a weak acid called sulphurous acid, H_2SO_3. This possesses the normal properties of an acid and so forms salts called sulphites and hydrogensulphites, (or bisulphites), which in some respects are similar to carbonates and hydrogencarbonates, the salts of carbonic acid, H_2CO_3. These latter are easily decomposed by dilute acids to yield carbon dioxide.

$Na_2CO_3(g) + 2HCl(aq) \rightarrow 2NaCl(aq) + H_2O(l) + CO_2(g)$

$NaHCO_3(s) + HCl(aq) \rightarrow NaCl(aq) + H_2O(l) + CO_2(g)$

Sulphites and hydrogensulphites are similarly decomposed to give sulphur dioxide.

$Na_2SO_3(s) + 2HCl(aq) \rightarrow 2NaCl(aq) + H_2O(l) + SO_2(g)$

$NaHSO_3(s) + HCl(aq) \rightarrow NaCl(aq) + H_2O(l) + SO_2(g)$

Experiment 14.1. To investigate the gas obtained from Sodium Hydrogensulphite.

Set up the apparatus of Fig. 14.1 and add the acid slowly. A gas is given off, but should the action slow down the flask may be warmed. The trap is used to prevent excess gas escaping into the air. If the experiment is carried out in a fume cupboard with a good draught this will not be necessary. Collect several jars of gas and test as follows. Make a note of your results.
  (i) Sniff *very* cautiously.
  (ii) Insert a lighted splint.
  (iii) Hold a strip of burning magnesium in the jar.
  (iv) Invert a jar of gas in a trough of water containing blue litmus.
  (v) Add some dilute sulphuric acid to a solution of potassium manganate(VII) and pour it into the jar.
  (vi) Acidify a solution of potassium chromate(VI) or dichromate(VI) in the same way and add this to a jar of gas.
  (vii) Warm a little purple-brown lead(IV) oxide in a deflagrating spoon and place it in a jar.
  (viii) Pass the gas into bromine water.

Having made your observations see how closely they agree with what follows here.

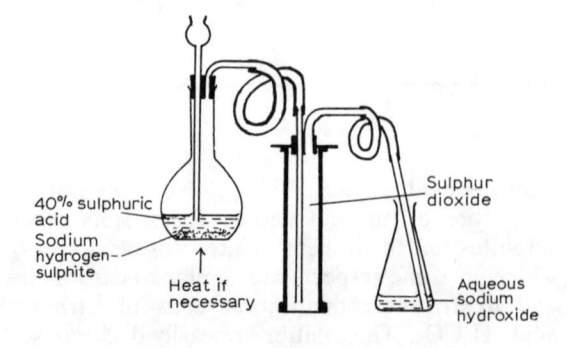

Fig. 14.1. Preparation of sulphur dioxide

(i) The gas smells like that formed when sulphur burns. It is sulphur dioxide.
(ii) The splint goes out as sulphur dioxide does not readily support combustion; the gas does not burn.
(iii) The magnesium continues to burn and a yellow deposit is formed (look up the action of burning magnesium on carbon dioxide). The sulphur dioxide is reduced to sulphur by the magnesium, the gas acting as an oxidizing agent.

$$\overset{\text{oxidation}}{2Mg + SO_2} \rightarrow 2MgO + S \text{ (reduction)}$$

(iv) The water rises in the jar showing that the gas is fairly soluble and forms an acid solution.

$$H_2O + SO_2 \rightarrow H_2SO_3$$

(v) The solution is decolorized.
(vi) The solution turns green.
(vii) The lead(IV) oxide glows and changes to white lead(II) sulphate—at least on the surface.

$$PbO_2 + SO_2 \rightarrow PbSO_4$$

(viii) The bromine water is decolorized as it is reduced to hydrobromic acid, the sulphur dioxide in solution forming sulphuric acid. (See p. 73 for the method of testing for this.)

$$SO_2 + H_2O \equiv H_2SO_3$$
$$Br_2 + H_2O \equiv 2HBr + O$$
$$H_2SO_3 + O \equiv H_2SO_4$$

Add:

$$SO_2 + 2H_2O + Br_2 \rightarrow 2HBr + H_2SO_4$$

A common alternative laboratory method of preparing sulphur dioxide makes use of the reaction between metals such as zinc, iron and copper and hot concentrated sulphuric acid. The action with copper is represented by

$$Cu + 2H_2SO_4 \rightarrow CuSO_4 + 2H_2O + SO_2$$

but it is really more complex than this. The blue colour of the copper(II) sulphate cannot be seen unless the final solution is very much diluted. Some black copper(I) sulphide is also formed.

Sulphur dioxide can easily be liquefied by applying pressure to the gas and it is stored in this form in glass siphons. When the pressure is released the liquid vaporizes and the gas is given off. Sulphur dioxide is over twice as dense as air. It can be dried if necessary after preparation by passing it through silica gel or concentrated sulphuric acid.

Uses of Sulphur Dioxide

In the presence of water sulphur dioxide behaves as a bleaching agent. It is milder in its action than is chlorine and is used for bleaching straw (for hats), newspaper, wool (for flannel trousers) and sponges. The dyes present are reduced to colourless compounds, but the yellow colour may be restored on prolonged exposures to the air. Substances that have been bleached by chlorine are often soaked in vats containing a solution of sulphur dioxide (known as an 'antichlor') as this combines with any excess chlorine

$$H_2SO_3 + H_2O + Cl_2 \rightarrow H_2SO_4 + 2HCl$$

and the two dilute acids formed can easily be washed away. Fruit, fruit juice and fruit pulp are preserved during transport and storage by the presence of sulphur dioxide which prevents fermentation occurring.

If a suspension of calcium hydroxide is saturated with sulphur dioxide calcium hydrogensulphite is formed.

$$Ca(OH)_2 + H_2O + SO_2 \rightarrow Ca(HSO_3)_2$$

If wood chips are strongly heated with this solution the wood is broken down into fibres and bleached; it is then wood pulp and is used in the manufacture of paper. The main use of sulphur dioxide, however, is in the production of sulphuric acid.

Sulphuric Acid

It has been said that a country's industrial importance can be measured by the amount of sulphuric acid it uses, as this acid plays a vital part in the manufacture of such substances as fertilisers (ammonium sulphate and 'superphosphate'), titanium(IV) oxide (white paint), rayon, soap, detergents and dyestuffs, in the refining of petroleum and it is the electrolyte in all the batteries so necessary for our cars. A considerable amount is also used in the cleaning or pickling of steel and tinplate.

There are two methods by which the acid is manufactured, (i) the Contact Process and (ii) the Lead Chamber Process. The former method predominates and can quite easily be simulated in a laboratory experiment.

Experiment 14.2. The Action of Sulphur Dioxide and Oxygen in the presence of a Catalyst.

If sulphur dioxide and oxygen are mixed together at room temperature—or even heated—they do not combine, only a mixture is formed. But if a catalyst such as platinum is present they will join together at about 450°C to form sulphur (VI) oxide (sulphur trioxide).

$$2SO_2 + O_2 \xrightarrow{Pt} 2SO_3$$

This is a white crystalline solid at low temperatures, but can only be collected in a flask surrounded by ice.

The apparatus for carrying out this experiment is shown in Fig. 14.2. Oxygen is obtained from a

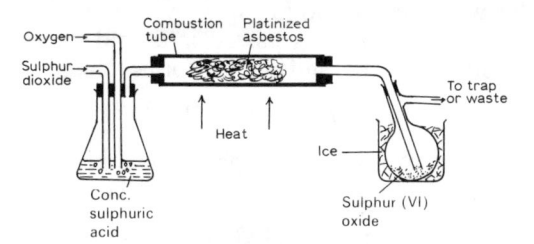

Fig. 14.2. Catalytic oxidation of sulphur dioxide

cylinder (or by heating oxygen mixture) and sulphur dioxide from a siphon (or as previously described). The gases are passed through a drying bottle containing concentrated sulphuric acid and the rate of delivery of the gases arranged so that there is two or three times as much oxygen as sulphur dioxide. The mixed dry gases then pass through a combustion tube containing a platinum catalyst. This is made by soaking some asbestos or mineral wool in platinum chloride solution and heating it strongly. Platinum is left dispersed on the wool and the substance is then known as platinized asbestos.

$$PtCl_4 \rightarrow Pt + 2Cl_2$$

The tube and catalyst must be dried thoroughly by heating before the experiment. The tube is now heated gently and white fumes should soon be apparent and pass into the flask (previously dried). When a small amount of sulphur(VI) oxide has been formed stop the experiment and uncork the flask. Addition of a drop of water should cause a hissing sound as the oxide dissolves to form sulphuric acid.

$$SO_3 + H_2O \rightarrow H_2SO_4$$

An acidic oxide which dissolves in water to form an acid is called an *acidic anhydride*.

The industrial 'Contact Process' follows the same lines. Sulphur dioxide is made either by burning sulphur (from the U.S.A.) or by roasting (i) anhydrite, $CaSO_4$, (ii) 'spent' oxide from the coal gas works (having absorbed hydrogen sulphide) or (iii) iron pyrites.

$$4FeS_2(s) + 11O_2(g) \rightarrow 2Fe_2O_3(s) + 8SO_2(g)$$

The gas has to be purified, washed and dried if made by any process except the burning of sulphur or the catalyst may be spoilt. It is then passed along with excess air through a steel cylinder carrying the catalyst on perforated shelves (vanadium pentoxide is used as it is cheaper than platinum). Heat is given out in the reaction and is sufficient to maintain the temperature at the required value. The sulphur(VI) oxide made is not dissolved in water as this would cause too much spray and acid would be lost. Instead it is passed into cold 98% acid already prepared and with it forms 'oleum'.

$$H_2SO_4 + SO_3 \rightarrow H_2S_2O_7$$

This is converted into sulphuric acid of the right concentration by the addition of water.

$$H_2S_2O_7 + H_2O \rightarrow 2H_2SO_4$$

Properties of the Acid

Dilute sulphuric acid is very similar in its properties to dilute nitric or hydrochloric acids, but when we consider the three concentrated acids we find their properties are very different.

(A) Dilute Acid

In making a dilute solution from concentrated sulphuric acid always be very careful to do so by adding the acid very cautiously to the water and *never the other way*. Great heat is given out in this mixing and were the water added to the acid and the initial amount of water small it would possibly be turned immediately to steam, and this expansion to about 1700 times the original volume might blow the acid about. If the acid is poured into water with stirring the water will absorb this heat safely.

Experiment 14.3. To examine the properties of dilute Sulphuric Acid.

By carrying out the following tests it can be shown that dilute sulphuric acid has the properties usually associated with an acid.

(a) Dip a piece of blue litmus into the acid; the paper turns red. If Universal indicator is used this also turns red showing the acid to be quite a strong one.

(b) Add a small piece of magnesium ribbon to the acid and test any gas issuing from the test-tube with a lighted splint. It should burn with a slight pop as it is hydrogen.

$$Mg(s) + H_2SO_4(aq) \rightarrow MgSO_4(aq) + H_2(g)$$

(c) Add some sodium carbonate, notice the effervescence and test the gas given off with a drop of lime water held on the end of a glass rod. It is carbon dioxide.

$$Na_2CO_3(s) + H_2SO_4(aq) \rightarrow Na_2SO_4(aq) + H_2O(l) + CO_2(g)$$

(d) Warm a little copper(II) oxide in a test tube with some of the acid and a blue solution of copper (II) sulphate will be formed. This is one of the methods of producing a salt using an acid and a base.

$$CuO + H_2SO_4 \rightarrow CuSO_4 + H_2O$$

The formula for sulphuric acid, H_2SO_4, shows that there are two atoms of hydrogen, each of which can be replaced by an atom of a metal with a similar electronic structure, i.e. with one electron only in the outermost shell.

Such an acid is said to be dibasic and can form two different series of salts, e.g. $NaHSO_4$ and Na_2SO_4, sodium hydrogensulphate and sodium sulphate.

Experiment 14.4. To make two different Salts from Sulphuric Acid.

(i) Put 20 cm³ of dilute sodium hydroxide in a dish with a few drops of phenolphthalein and run in dilute sulphuric acid from a burette until the indicator becomes colourless, stirring all the time, Make a note of the volume of acid used, discard the solution just made and repeat without using any indicator. Use exactly the same volumes as before. Place the dish on a water bath and evaporate until the solution left is ready to crystallize. This can be determined by dipping a glass rod in it and blowing on the rod to cool it. If crystals form cover the dish and put aside. The salt formed will be sodium sulphate, a normal salt.

$$2NaOH + H_2SO_4 \rightarrow Na_2SO_4 + H_2O$$

(ii) In order to make the hydrogensulphate 10 cm³ only of sodium hydroxide solution are taken and sulphuric acid added equal in volume to that used in (i). Evaporate and crystallize as before to get sodium hydrogensulphate (bisulphate).

$$NaOH + H_2SO_4 \rightarrow NaHSO_4 + H_2O$$

Examine the crystals under a low-power microscope or hand lens and notice any differences.

Hydrogensulphates containing as they do an excess of acid, are not neutral to litmus and are known as 'acid' or hydrogen salts, the sulphates being known as normal salts. It should be clear from the above method of making the salts that they can be converted into each other by methods represented by the following equations.

$$Na_2SO_4 + H_2SO_4 \rightarrow 2NaHSO_4$$
$$NaHSO_4 + NaOH \rightarrow Na_2SO_4 + H_2O$$

A solution of sodium hydrogensulphate will effervesce with carbonates and dissolve magnesium, but all 'acid' salts do not behave in this way. Sodium hydrogencarbonate is actually alkaline to litmus as the acid from which it is derived, carbonic acid, is so much weaker than the alkali, sodium hydroxide. It is still, however, an 'acid' salt. The term 'acid' used here refers to the hydrogen remaining, not to the reaction of the salt towards an indicator.

Metals higher than hydrogen in the electrochemical series liberate this element from dilute acids. Sodium and calcium are too active for this to be really safe, but the action of dilute sulphuric acid on zinc will be recalled as the usual method of preparing hydrogen. Let us try the experiment with iron and make iron(II) sulphate.

Experiment 14.5. To prepare Crystals of Iron(II) Sulphate, $FeSO_4.7H_2O$

Take some dilute sulphuric acid and add powdered iron to it with stirring until hydrogen ceases to be evolved.

$$Fe(s) + H_2SO_4(aq) \rightarrow FeSO_4(aq) + H_2(g)$$

Some iron should be left undissolved. Filter the solution into a dish and add two or three more drops of dilute sulphuric acid (this helps to prevent iron(III) sulphate being formed by oxidation). Now evaporate to crystallization point as previously explained and put aside to cool. Pale green crystals should be formed.

One other acidic property of the dilute acid is its ability to liberate carbon dioxide from carbonates.

(B) *Concentrated acid*

The concentrated acid has properties which may be classified as: (i) dehydrating, (ii) oxidizing. These will now be considered.

(i) Dehydrating properties

Most crystals contain water of crystallization and this attachment of molecules of water to a molecule of an anhydrous substance may cause a rearrangement of its structure with the 'production of colour. Many coloured crystals on gentle heating lose this water of crystallization and become white. Concentrated sulphuric acid which has a strong affinity for water dehydrates crystals in a similar way.

Experiment 14.6. The action of concentrated Sulphuric Acid on Copper(II) Sulphate crystals

Take a large crystal of copper(II) sulphate (blue vitriol) and place it in a test-tube. Pour in sufficient concentrated sulphuric acid (with care!) to cover the bottom half only of the crystal. After a short time this part will become white on the surface. The blue colour can be restored by placing the crystal in water. This action can be represented by the reversible equation:

$$CuSO_4 \cdot 5H_2O \rightleftharpoons CuSO_4 + 5H_2O$$

copper(II) sulphate crystals anhydrous copper(II) sulphate

There are many organic compounds which do not actually contain water of crystallization, but in which the hydrogen and oxygen atoms are present in the same ratio as in water. Concentrated sulphuric acid attacks some of these to remove the hydrogen and oxygen as water leaving behind carbon; the substances are said to be charred.

Experiment 14.7. The action of concentrated Sulphuric Acid on Sugar and Paper.

(a) There are numerous sugars represented by such formulae as $C_6H_{12}O_6$ (glucose) or $C_{12}H_{22}O_{11}$ (cane sugar). Place a little cane sugar in a test-tube and cover it with the concentrated acid. Warm; you will notice the sugar gradually becomes brown and then black. Do not heat too strongly or acrid fumes will be given off. A similar result is obtained using starch.

(b) Dissolve a little sugar in water in a test-tube and carefully add some concentrated acid. The blackening now takes place quite rapidly, helped by the heat generated as the acid mixes with the water. The substance left is sugar charcoal.

$$C_{12}H_{22}O_{11} \xrightarrow{c.H_2SO_4} 12C + 11H_2O$$

(c) Dip a glass rod in the concentrated acid and mark a piece of paper with it; this soon chars, especially if warmed. Repeat with dilute acid and allow to dry naturally; the mark should now be invisible. Warm the paper and as the water from the dilute acid evaporates a black mark will appear. (This method can be used for 'secret' writing.) The concentrated acid has a very similar effect on flesh as both this and paper contain cellulose: so be very careful how you handle it.

There are some other organic compounds where the hydrogen and oxygen atoms are not in the proportions of two to one and yet, when acted on by concentrated sulphuric acid, these elements are removed to form water.

*Experiment 14·8. The Dehydration of Methanoic (Formic) and Ethanedioic (Oxalic) Acids.

(a) Add a little concentrated sulphuric acid to some methanoic acid in a test tube. A gas is evolved which can be made to burn with a blue flame; it is carbon monoxide† and forms carbon dioxide on burning.

$$HCOOH \xrightarrow{c.H_2SO_4} H_2O + CO$$

(b) Repeat using a little solid ethanedioic acid instead of methanoic and warm. A mixture of carbon monoxide and carbon dioxide is given off.

$$(COOH)_2 \xrightarrow{c.H_2SO_4} H_2O + CO + CO_2$$

The two gases can be separated by passing them through sodium hydroxide solution which absorbs the dioxide.

Notice that the formulae of the acids are not written in the same way as the mineral acids you have met. In the latter, hydrogen comes first in the formula; in organic acids it is the hydrogen at the end of the formula which is replaceable by metals. Methanoic acid can be obtained from ants (when they sting they inject it into their victim) and ethanedioic acid, a poison, is present to some extent in both rhubarb and in spinach.

(ii) Oxidizing reactions

Experiment 14.9. The Action of Heat on concentrated Sulphuric Acid.
Take a small hard-glass or silica test-tube and in it place concentrated sulphuric acid to a depth of about 2·5 cm. Now add small pieces of mineral wool until all the liquid is soaked up. Small pieces of porous pot are then put in to fill about 5–7 cm of the tube.
Arrange the apparatus as in Fig. 14.3. Heat the porous pot strongly about 5 cm from the closed end and occasionally move the flame for a moment to the mineral or asbestos wool. A gas comes off which on testing can be shown to be oxygen. The

†Carbon monoxide is *extremely* poisonous and no attempt should be made to smell it.

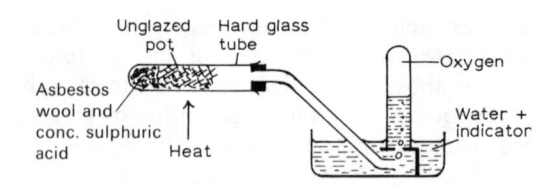

Fig. 14.3. Action of heat on sulphuric acid.

water in the small trough becomes acidic due to dissolved sulphur dioxide.

$$2H_2SO_4 \rightleftharpoons 2H_2O + 2SO_2 + O_2$$
reduction

In many of its reactions concentrated sulphuric acid, when heated with metals or non-metals, behaves in the above manner. The oxygen set free brings about oxidation.

If it is heated with a piece of charcoal the latter is oxidized to carbon dioxide; both this and the sulphur dioxide can be identified using the apparatus of Fig. 14.4. The potassium chromate(VI) solution becomes green and the lime water milky.

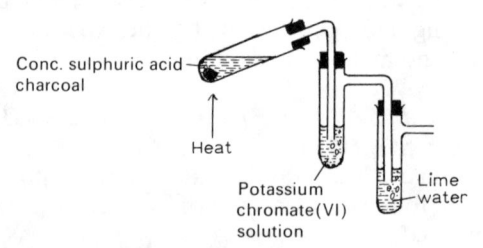

Fig. 14.4. Reduction of sulphuric acid with charcoal

$$2H_2SO_4 \equiv 2H_2O + 2SO_2 + \cancel{O}_2$$
$$C + \cancel{O}_2 \equiv CO_2$$
$$\overline{C + 2H_2SO_4 \rightarrow CO_2 \uparrow + 2H_2O + 2SO_2}$$

Some other non-metals (e.g. sulphur) behave similarly, but with metals the action usually goes farther and a salt (a sulphate) is formed. The reaction with copper can be represented by the equations:

$$2H_2SO_4 \equiv 2H_2O + 2SO_2 + \cancel{O}_2$$
$$2Cu + \cancel{O}_2 \equiv 2\cancel{CuO}$$
$$\underline{2\cancel{CuO} + 2H_2SO_4 \equiv 2CuSO_4 + 2H_2O}$$
$$Cu + 2H_2SO_4 \rightarrow CuSO_4 + SO_2 + 2H_2O$$
(adding and dividing by two)

As mentioned earlier in the chapter this method is frequently used in the laboratory for preparing sulphur dioxide. With zinc the reaction would be:

$$Zn(s) + 2H_2SO_4(aq) \rightarrow ZnSO_4(aq) + SO_2(g) + 2H_2O(l)$$

Experiment 14.10. The Action of concentrated Sulphuric Acid on Copper.

Place a few pieces of copper in a test-tube and just cover with concentrated sulphuric acid. Notice that there is no action at all. Cold concentrated sulphuric acid has very little action on metals other than those near the top of the electrochemical series (and then only if a little water is present). Fit the tube with a cork and a delivery tube dipping into some potassium chromate(VI) solution, acidified with a little dilute acid. Heat the copper and concentrated acid; a gas is evolved which turns the solution green. Take the delivery tube out of the solution, allow the test-tube to cool down. (If it is not done this way water will be sucked back on to the hot acid with disastrous results.) Pour the black contents into a large beaker of water when the blue colour due to copper(II) sulphate should be noticeable.

Although a metal sulphate is formed in this and similar reactions, this method is NOT used for making sulphates.

Identification of a Sulphate

All common sulphates other than those of lead, barium and calcium are soluble in water. If a solution containing a sulphate is added to one containing barium ions, i.e. a soluble barium salt such as the chloride, a white precipitate of barium sulphate is formed.

$$BaCl_2(aq) + Na_2SO_4(aq) \rightarrow BaSO_4(s) + 2NaCl(aq)$$

or

$$Ba^{2+}(aq) + SO_4^{2-}(aq) \rightarrow BaSO_4(s)$$

A similar white precipitate is obtained with a soluble sulphite or carbonate, but barium sulphite and barium carbonate are both soluble in dilute hydrochloric acid, whereas the sulphate is insoluble. In short:

addition of barium chloride solution to a solution of a sulphate acidified with dilute hydrochloric acid will produce a white precipitate of barium sulphate.

Exercise 13–14

1 Give an account of the Frasch process for the extraction of sulphur.

Describe how sulphuric acid is manufactured on a commercial scale by the contact process. (Technical details of the plant are not required.)

Name *two* gases for which concentrated sulphuric acid is a suitable drying agent, and *two* for which it is unsuitable. (*OC*)

2 Describe in outline the contact process for the manufacture of sulphur trioxide.

What is the action of concentrated sulphuric acid upon (*a*) sugar, (*b*) sulphur? What is the action of dilute sulphuric acid on (*c*) zinc, (*d*) sodium carbonate? (*L*)

3 How is sulphur found in nature? Give *three* different sources and indicate briefly in each case how the sulphur is obtained reasonably pure or is otherwise made use of.

Under what conditions is sulphur converted into pure sulphuric acid in industry?

How, and under what conditions, does sulphuric acid react with (*a*) hydrogen sulphide, (*b*) sodium hydroxide? (*O*)

4 (*a*) Write equations for the main reactions that occur, and state the conditions under which they take place, when sulphuric acid is manufactured from sulphur.

(*b*) Describe and explain *one* reaction in which concentrated sulphuric acid acts as an oxidising agent, and *one* in which it acts as a dehydrating agent.

(*c*) What reactions take place between (i) dilute sulphuric acid and excess sodium carbonate solution, (ii) sodium hydroxide solution and excess dilute sulphuric acid? (*C*)

5 Mention *four* uses of sulphur. What chemical changes occur when:

(*a*) Sulphur is heated in the air.

(*b*) A mixture of iron filings and sulphur is heated in a test tube.

(*c*) Potassium nitrate is heated.

(*d*) Sulphur is added to molten potassium nitrate?

6 (*a*) Name the compound formed when sulphur burns in oxygen.

(*b*) Give the equation which shows the reaction which occurs.

(*c*) How much of the compound would be made if 5 kg of sulphur were burnt in oxygen?

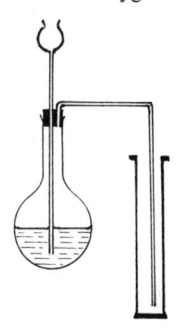

(d) The drawing shows the apparatus which could be used to prepare the same compound in a different way. Name the reagents in the flask.

(e) Why must the end of the funnel be kept under the liquid?

(f) Sulphur is the raw material from which sulphuric acid is made. Give *brief* notes explaining the process. (SR)

7 (a) Give *two* reactions in which sulphur dioxide is a product. State the conditions in each case.

(b) Draw a clear labelled diagram to show how you would prepare and collect a few jars of sulphur dioxide.

(c) How could unwanted sulphur dioxide be absorbed, and what reaction would occur in the process?

(d) Give *two* chemical tests by which you could distinguish a jar of sulphur dioxide from one of hydrogen chloride. (C)

8 Draw labelled diagrams of apparatus you would use to prepare and collect (a) sulphur dioxide, (b) hydrogen chloride.

Describe (i) *two* tests (other than the use of indicators) by which you could show solutions of these gases to be strongly acid; (ii) *two* tests, other than smell, by which you could distinguish between them. (O)

9 Describe, with the aid of a diagram, how you would obtain several gas jars of sulphur dioxide, starting with concentrated sulphuric acid.

Give a brief account of the physical properties of sulphur dioxide.

What is seen to happen, and why, when separate gas jars of sulphur dioxide are shaken with (a) litmus solution, (b) acidified potassium dichromate(VI) solution, (c) chlorine water, afterwards adding a little barium chloride solution? (S)

10 Given ordinary rhombic (octahedral) sulphur, describe how you would prepare a sample of monoclinic (prismatic) sulphur.

Draw the apparatus you would use, and explain how you would show that equal weights of these two allotropes produce equal weights of oxide. (L)

11 Name a process by which sulphuric acid is made.

Mention three uses of the acid which cause it to be so important.

Describe briefly two simple experiments you would do to show that sulphuric acid is in fact an acid. Say whether you are using the concentrated or the dilute acid in the experiments you describe, and if the experiments result in the formation of any gas, give a simple test by which you could identify the gas.

What is meant by saying that sulphuric acid is a dehydrating agent?

Draw a clearly labelled diagram to show that sugar is dehydrated by sulphuric acid.

Why is it dangerous to pour water into concentrated sulphuric acid? (EM)

12 Describe (a) how you would prepare a sample of monoclinic sulphur, starting from rhombic sulphur; (b) what is observed when sulphur is heated from its melting-point to its boiling-point.

Under what conditions does sulphuric acid react with (i) copper, (ii) sodium nitrate, (iii) sodium chloride? Write an equation for each reaction. (OC)

13 Describe and explain the action of concentrated sulphuric acid on (i) a strong solution of cane sugar, (ii) sodium chloride.

Describe carefully how you would dilute some concentrated sulphuric acid, and how you would proceed to use it in preparing some dry crystals of sodium sulphate.

What mass of sulphur is theoretically needed to produce 1000 kg of pure sulphuric acid?

14 State what is observed and what happens when dilute sulphuric acid is added to (a) copper(II) oxide, (b) sodium hydrogen carbonate, (c) lead(II) nitrate solution? Write the equations representing the reactions. (J)

15 Each of the gases chlorine and sulphur dioxide is a bleaching agent.

(a) Explain what happens to each of them when used in this way.

(b) Describe what you would do to make them react with each other, and explain the reaction which would then take place.

(c) Draw a labelled diagram showing how you would prepare and collect ONE of them. (C)

16 What do you understand by the term *allotropy*? Given rhombic sulphur, describe how you would prepare samples of (a) monoclinic sulphur, (b) plastic sulphur.

State briefly how you could obtain sulphur dioxide, staring from sulphur. What is the vapour density of sulphur dioxide?

Describe and explain ONE chemical test for sulphur dioxide.

17 Sulphur(VI) oxide is made *in industry* by the catalytic oxidation of sulphur dioxide under certain conditions.

(a) (i) What is a catalyst?
  (ii) Name the catalyst used in this reaction.

(b) State *two* conditions, other than the use of a catalyst, necessary to obtain a reasonable yield of sulphur trioxide.

(c) From which natural sources are (i) the sulphur dioxide, (ii) the oxygen obtained?

(d) Give the equation for the reaction, showing the physical states for the reactants and product.

(e) The reaction between the reactants and product is in equilibrium and the equilibrium position lies towards the product. What would be the effect upon the equilibrium position if
  (i) the sulphur(VI) oxide were removed as fast as it is formed,
  (ii) more oxygen were added to the equilibrium mixture?

(f) How is the sulphur(VI) oxide formed converted into sulphuric acid?

(g) What would be observed, and which new substance would be formed, when concentrated sulphuric acid is added to
  (i) copper(II) sulphate crystals,
  (ii) sodium chloride,
  (iii) sugar (L)

18 (a) You are provided with powdered sulphur and xylene. How would you prepare (i) monoclinic and (ii) rhombic sulphur using these substances? Suggest any safety precaution you would take.
(b) What do you observe when powdered sulphur is gently heated to boiling in a test tube?
(c) When hot molten sulphur is poured into cold water, the physical properties of the sulphur change. What are the physical properties of this rapidly cooled sulphur?
(d) Draw a neat labelled diagram for the preparation and collection of sulphur dioxide in the laboratory.
(e) Write a balanced equation for the reaction of sulphur dioxide with sodium hydroxide?
(f) Write a balanced equation for the reaction of sulphur dioxide with oxygen.
(g) What is the industrial importance of this reaction? (C)

15
Machines, Work, Energy and Power

When a wheel on a motor car has to be changed, it is necessary to lift the car so that the wheel in question is clear of the ground. This involves exerting a force of several kilonewtons and is well beyond the capabilities of most people. However with the aid of a simple machine, namely a 'screw jack', it is possible to do the job quite easily by exerting a fairly small force. Other examples of machines like this may come to your mind. You may have seen, for example, a motor mechanic lifting the engine from a car by a chain passing round a system of pulleys. He can do this almost effortlessly with one hand although the engine being lifted is very heavy. You may have seen a lock keeper opening a massive pair of lock gates without too much effort with the aid of a system of gear wheels.

We shall consider a number of machines like these later in this chapter. Before we do so we must consider some of the principles involved and define some new terms which will enable us to talk about them more easily.

Work

In the three examples quoted above something is being moved by a force. In the first two a load is being raised and in the third the lock gates are being opened or closed. In all three cases we say that work is being done. The amount of work is defined as *the force × the distance moved in the direction of the force*. To take a simple example, if a 10 kg weight is to be lifted vertically without the aid of a machine an upward force of 100 N must be exerted. If the weight is lifted through 2 m (Fig. 15.1), then the work done is: 100×2 newton-metres = 200 newton-metres.

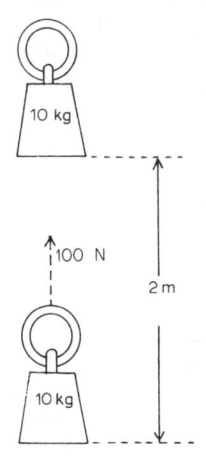

Fig. 15.1. $20g$ joules of work done

A newton metre is known as a 'joule' (abbreviation J).

The significance of the phrase 'in the direction of the force' in the definition of work done may be seen by considering the case of a vehicle free wheeling down a hill (Fig. 15.2). It is moving under its own weight (W), which acts vertically downwards. If it moves a distance s down the slope, we may *not* say that the work done is $W \times s$. The distance moved in the direction of the force is h, the vertical height through which the vehicle drops and the work done is $W \times h$.

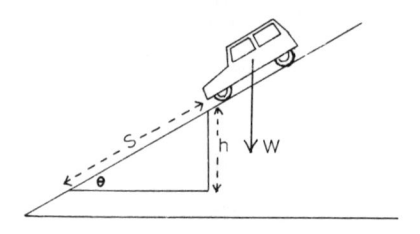

Fig. 15.2. Work done is $W \times h$

Energy

Something (or someone) capable of doing work is said to possess energy. The 10 kg mass referred to in the last section might be raised by a human being or an animal, or it might be raised by a rope attached to an electric motor or to a steam engine. The energy or ability to do work possessed by the person or animal is obtained from chemical changes undergone by the food consumed. Work is done by the electric motor as it consumes electrical energy and by the steam engine as it takes heat energy from the steam. We have mentioned here three forms of energy, chemical energy, electrical energy and heat energy. We shall, in later chapters, be discussing these and other forms (light or radiation, sound and nuclear energy) all of which are capable, either directly or indirectly, of doing work. Until nuclear energy was utilised all the energy available to man originated as radiation from the sun. The following table shows the principal sources.

Energy source	Origin
Food	Solar radiation needed to produce photosynthesis in vegetation.
Coal	Decayed vegetation.
Oil	Decayed animal matter.
Gas	From coal or oil.
Water power (hydro-electric)	Heat from the sun evaporates sea water which condenses and falls as rain on high ground. Rivers so formed drive turbines.
Wind power	Convection currents in the atmosphere are caused by heat from the sun.

Conservation of Energy

When we use an electric motor in a crane to lift a load, a certain amount of electrical energy is used up. Is this energy completely lost? If the load which has been raised is released, it will fall back to the ground. Its mass will cause it to move and so work is done. Thus the load in its elevated position possesses energy due to that position. This is known as potential energy. So some, at least, of the electrical energy consumed reappears as potential energy in the load. Some of it will reappear as heat in the electric motor, pulleys and other moving parts of the crane. In fact all of the electrical energy used up will be converted into other forms of energy.

This example is an illustration of a most important general law, *the Law of Conservation of Energy*. This states: *Energy cannot be created or destroyed but only converted from one form to another.*

Many of the devices we shall discuss later in the book are designed just to convert one form of energy into another. Energy, like work, is measured in joules. The efficiency of any device for converting energy from form A to form B is defined as:

$$\frac{\text{Energy of form } B \text{ obtained}}{\text{Energy of form } A \text{ supplied}} \times 100\%$$

We shall see in Chapter 19 that if form B happens to be heat, then the efficiency may be 100%. In other words an electric heater may convert all the electrical energy supplied into heat. If form B happens to be anything other than heat, then the efficiency will be less than 100%. For example with an electric filament lamp, for every 100 J of electrical energy supplied perhaps only 5 J will become light. The rest becomes heat and is thus 'wasted', We should say that the efficiency of the lamp is 5%.

Potential and Kinetic Energy

We have mentioned that if a mass of M kg is lifted to a vertical height h m above the ground (Fig. 15.3) it is given potential energy. It is able to

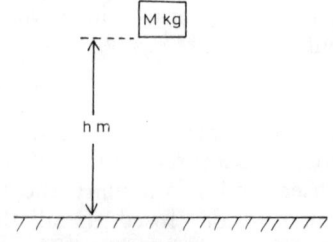

Fig. 15.3. Potential energy possessed by mass M kg is Mgh joules

do a certain amount of work (namely to fall back to the gound). You may be able to think of cases where this potential energy due to height is put to use. A counter-mass is used for example with most lifts. It will do much of the work when the lift is going up (Fig. 15.4). Slowly falling masses attached to chains are used to drive some types of clock.

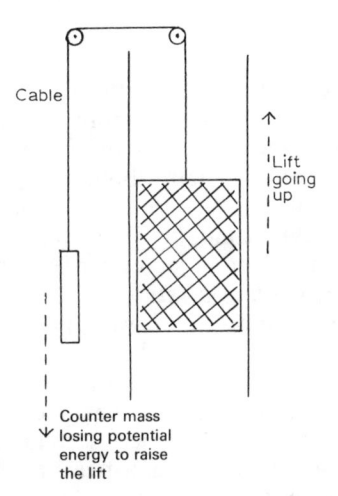

Fig. 15.4. A lift

The force on the mass, M kg, in Fig. 15.3 is Mg N. Thus the work which it may do or the energy which it stores is Mgh J.

If the mass is connected as the lift counter-mass in Fig. 15.4, most of its potential energy will be transferred to the lift when it falls. If, on the other hand, it falls quite freely its potential energy apparently disappears. It will however acquire considerable speed. This leads to the conclusion that a moving mass must have some energy by virtue of its motion. This idea is supported by the fact that when a moving vehicle is brought to rest by application of the brakes a considerable amount of heat is produced at the brakes. The energy due to motion (*the kinetic energy*) is here converted to heat energy. The freely falling mass also will produce heat (and a little sound energy) when it is brought to rest on impact with the ground.

The amount of kinetic energy possessed by a mass, m kg, moving with a speed v m/s will be the amount of work which has been done in accelerating the mass from rest to this speed. Supposing that a uniform acceleration of a m/s^2 is brought about by a constant force, f N, acting over a distance, s m.

$$\text{The work done} = f \times s \text{ J}$$
$$\text{Now } v^2 = 0^2 + 2as \quad (1)$$

$$\text{and } f = ma \quad (2)$$
$$\text{From (1) } a = \frac{v^2}{2s}$$

and substituting in (2)

$$f = \frac{mv^2}{2s}$$

$$\text{Thus work done} = \frac{mv^2}{2s} \times s \text{ J}$$
$$= \frac{mv^2}{2} \text{ J}$$

Hence the kinetic energy possessed by a mass m kg having a speed v m/s is $\frac{1}{2}mv^2$ J.

In the case of the mass M in Fig. 15.5, if it falls freely through the height h m, its acceleration will be g m/s^2. Its velocity just as it hits the ground will then be given by:

$$v^2 = 2gh. \quad (v = \sqrt{2gh})$$

Fig. 15.5. Energy changes as a body falls from rest

and the kinetic energy on impact

$$\tfrac{1}{2}Mv^2 = \tfrac{1}{2}M \times 2gh$$
$$= Mgh$$

Thus the mass in position 1 has potential energy Mgh J and no kinetic energy. Just before impact in position 3 it has lost all the potential energy but has kinetic energy Mgh J. Thus the total energy remains unchanged. At any intermediate position 2, the falling mass has both kinetic and potential energy, the total of which is Mgh J. In practice, some work will be done by the falling mass in overcoming air resistance, with the result that heat will be produced during the fall and the final kinetic energy will be a little less than the original potential energy.

Potential energy does not arise only from height. A compressed or stretched spring stores energy as does a piece of stretched elastic. If the latter is used in a catapult, the stored potential energy is rapidly changed into kinetic energy when the elastic is released.

The following example will show how some of these ideas may be applied.

A metal ball of mass 4 kg is released at the top of a building and takes 2 s to reach the ground. (a) What is the height of the building? (b) Calculate the energy of the ball just before it hits the ground. (c) What would be the potential energy after falling for 1 s? (d) What would be the kinetic energy at that time? (L)

(a) In this case:
$$s = \tfrac{1}{2}gt^2$$
$$= \tfrac{1}{2} \times 9\cdot 8 \times 4$$
$$= 19\cdot 6 \text{ m}$$

Height of building is 19·6 m.

(b) Velocity before impact is given by:
$$v = 0 + gt$$
$$= 9\cdot 8 \times 2$$
$$= 19\cdot 6 \text{ m/s}$$

thus K.E. just before impact
$$= \tfrac{1}{2}mv^2$$
$$= \tfrac{1}{2} \times 4 \times 19\cdot 6^2$$
$$= 768 \text{ J}$$

P.E. just before impact is zero hence 768 J is the total energy.

(c) After 1 s the height fallen through
$$s = \tfrac{1}{2}gt^2$$
$$= \tfrac{1}{2} \times 9\cdot 8 \times 1$$
$$= 4\cdot 9 \text{ m}$$

thus height above the ground
$$= 19\cdot 6 - 4\cdot 9$$
$$= 14\cdot 7 \text{ m}.$$

So P.E. after 1 s is:
$$mgh \text{ J}$$
$$= 4 \times 9\cdot 8 \times 14\cdot 7$$
$$= 576 \text{ J}$$

and (d) Velocity after 1 s is given by:
$$v = gt$$
$$= 9\cdot 8 \times 1$$
$$= 9\cdot 8 \text{ m/s}$$

and the K.E. at this instant is
$$\tfrac{1}{2}mv^2 = \tfrac{1}{2} \times 4 \times 9\cdot 8^2$$
$$= 192 \text{ J}$$

Power

Just one more important term needs to be defined before we consider in more detail the type of machine we mentioned at the beginning of this chapter. When we talk about the *power* of an engine, we mean *the rate at which it does work*. Power is thus the amount of work done per second. It will be measured in joules per second. A device which does work at a rate of *1 joule per second* is said to have a power of *1 watt*. You are probably familiar with the use of this unit in defining the power of electrical appliances. A 100 W light bulb, for example, is one which consumes 100 J of electrical energy per second. The unit may equally well be used as a measure of the power of any device which changes one form of energy into another.

You can easily obtain a value for the power at which you are capable of working.

Experiment 15.1. Time yourself running up a flight of stairs. Measure the vertical height of the stairs. You must know also your mass. The following typical set of values will show you how to calculate your power.

$$\text{Time} = 15 \text{ s}$$
$$\text{Height} = 7 \text{ m}$$
$$\text{Mass} = 80 \text{ kg}$$
$$\text{Work done} = \text{Force} \times \text{distance}$$
$$= (80 \times 10) \text{ N} \times 7 \text{ m}$$
$$= 5600 \text{ J}$$
$$\text{power} = \text{rate of doing work}$$
$$= \frac{5600}{15} \text{ J/s}$$
$$= 373 \text{ } W$$

Machines

(*a*) *Pulleys*

On almost any building site you will see a single pully being used to pull buckets and other fairly light loads to the top of the building as shown in Fig. 15.6. In a case like this the effort E, which the workman must apply in order to lift the load L, has to do two things. It has to work against the weight

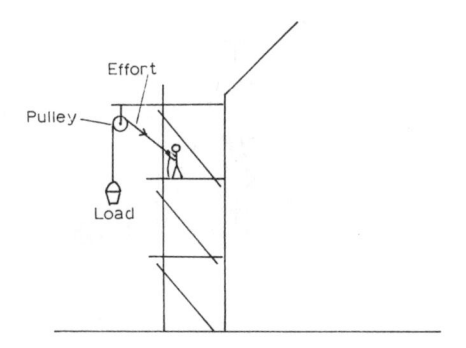

Fig. 15.6. Use of a single pulley

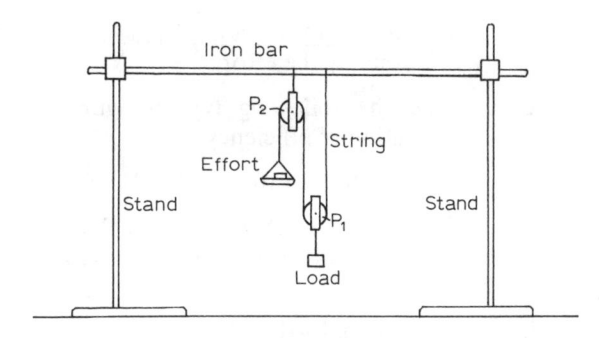

Fig. 15.7. A two pulley system

of L; it also has to work against the friction at the axle of the pulley. This work done in overcoming friction is, of course, work wasted, and the effort E is greater than would be necessary to raise the load directly. Why then use the pulley? Of course it is a convenient arrangement, but there is another good reason for doing it like this. With the pulley the workman has to exert a downward force and he can use his own weight to help him in this.

For heavier loads more complicated pulley systems must be used. An experiment will help you to understand the principles involved in these.

Experiment 15.2. Set up a two pulley system as shown in Fig. 15.7. A string is fixed to the cross bar. It passes round the two pulleys and a scale pan is attached to the other end. The effort to raise the load will be applied by placing weights in this scale pan. The total effort applied will, of course, be the weight of the scale pan itself together with masses placed in it.

Make the load L, say, 200 g. Add weights to the scale pan until, when it is given a slight downward push, it continues to move down at a more or less steady speed. Note the values of load and efforts. Repeat the experiment with various larger values of load and tabulate your results as follows.

Load (N)	Effort (N) (including weight of scale pan)	Load/Effort

Now measure the distance through which the load rises when the scale pan goes down a certain distance, say 20 cm. This is something you could find out just by thinking about the arrangement, without doing an experiment. You will see that the effort moves twice as far as the load.

You now know two important facts about the machine.

You know:

$$\frac{\text{the distance moved by the effort}}{\text{the corresponding distance moved by the load}}$$

This ratio, which in this case is 2, is known as the *velocity ratio* of the machine.

You will probably have found that the values in the third column of the table $\left(\frac{\text{load}}{\text{effort}}\right)$ are approximately constant. This ratio is known as the *mechanical advantage* of the system. The actual value of the mechanical advantage will depend on the type of pulleys you are using, but it may be about 1·3. It will certainly be less than 2.

This experiment shows that with a machine consisting of two pulleys it is possible to exert a force greater than the effort which is applied. We shall see that with more pulleys it is possible to exert a force many times greater than the applied effort. Fig. 15.8. shows an application of a two pulley system which makes it easier to pull in the sail of a sailing boat.

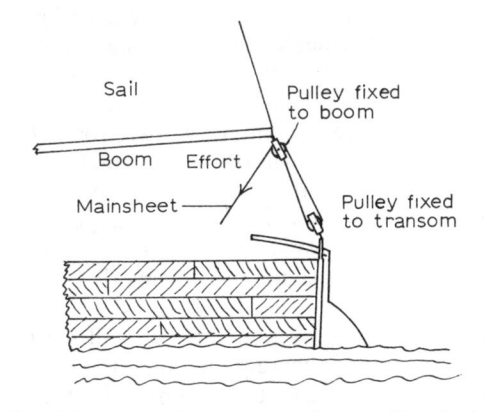

Fig. 15.8. A two pulley system used on a sailing dinghy

The results of your experiment will enable you to calculate also the efficiency of the machine. This is defined as:

$$\frac{\text{Work done on the load}}{\text{Work done by the effort}} \times 100\%$$

Let us take the following typical figures to show the calculation of efficiency.

Load = 2 N   Effort = 1·3 N

If the effort (scale pan) goes down 2 m we know that the load rises through 1 m. (The velocity ratio is 2.)

Thus work done on load = 2 × 1 J
and work done by effort = 1·3 × 2 J

$$\text{Hence the efficiency} = \frac{2 \times 1}{1 \cdot 3 \times 2} \times 100\%$$
$$= 77\%$$

Now $\frac{2}{1 \cdot 3}$ is the mechanical advantage of the system and $\frac{1}{2}$ is $\frac{1}{\text{velocity ratio}}$. Hence efficiency may be calculated from $\frac{\text{M.A.}}{\text{V.R.}}$

This result means that 23% of the work done by the effort is wasted. As we pointed out for a single pulley some of this work will be done against friction at the axles of the pulleys. There is an additional waste in this case. Besides raising the load we are raising the lower pulley (P_1 in Fig. 15.7). In fact the results of Experiment 15.2 will only yield reasonably constant values for mechanical advantage and efficiency if the weight of this pulley is small by comparison with the mass of the load.

If we were able to make a pulley system of this kind in which there was no friction and the pulleys had no mass, then the machine would be 100% efficient and the mechanical advantage would have the same value (2) as the velocity ratio. In general for all machines as the friction is reduced the value of the mechanical advantage will approach that of the velocity ratio without ever reaching it. In any case, if one requires to lift a heavy load (i.e. one wants a large mechanical advantage) then the machine must have a large velocity ratio. For ordinary pulley systems this means that one must have more pulleys. Fig. 15.9 shows a six pulley system. If you think about it you will see that if six metres of rope were pulled in at the effort, the load would rise one metre. The machine thus has a velocity ratio of 6 and the mechanical advantage may be about 5. A pulley system of this type is referred to as a 'block and tackle'. When such a system is used in practice it will not look quite like the arrangement in Fig. 15.9. The three pulleys in

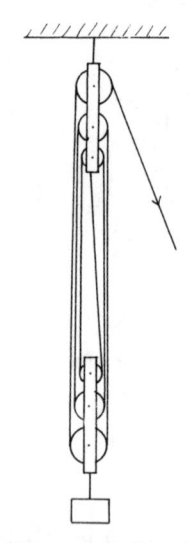

Fig. 15.9. A six pulley system

each block will all be the same size and mounted side by side as in Fig. 15.10. The pulleys in Fig. 15.9 are arranged simply to show more clearly how the apparatus is set up.

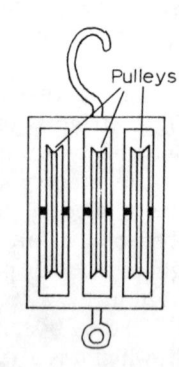

Fig. 15.10. A pulley block

Fig. 15.11 shows a pulley system with a still larger velocity ratio. It is known as a Weston differential pulley. In the top block there are two pulleys which are fixed together. They therefore rotate together whereas those in Fig. 15.9 rotate independently. The lower block to which the load is attached contains just one pulley. A long continuous loop of chain passes around the pulleys as shown in Fig. 15.11. The load is raised by applying an effort E to the free loop of chain. We can see how such a system can have such a large velocity ratio as follows:

Suppose the larger of the two top pulleys has a diameter D and the smaller, d. If chain is pulled in at E such that these two pulleys make just one revolution, then chain of length πD is taken in on

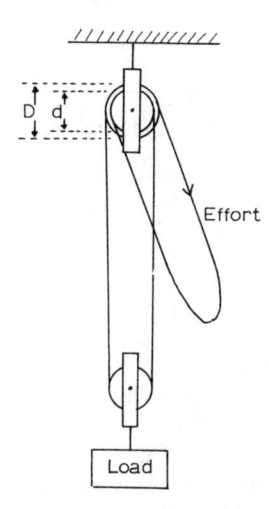

Fig. 15.11. A Weston differential pulley

chain. It is prevented from falling by the frictional forces.

(b) The wheel and axle

It is not an easy matter to apply a large turning moment directly to an axle (e.g. the shaft connecting to the steering mechanism of a motor vehicle (Fig. 15.12)). In order to do this the simplest method is to attach a large diameter wheel to the axle and apply a force at the rim of the wheel. This will give a much larger turning effect than if the same force were applied directly to the shaft.

the left hand side of the loop supporting L and a length πd is paid out on the right hand side of this loop. Thus the loop is altogether shortened by an amount $\pi D - \pi d$. The distance moved by the load is then:

$$\frac{\pi(D-d)}{2}$$

At the same time the effort moves a distance

$$= \pi D$$

Hence the velocity ratio =

$$\frac{\text{Distance moved by effort}}{\text{Corresponding distance moved by load}}$$

$$= \frac{\pi D}{\frac{\pi(D-d)}{2}}$$

$$= \frac{2D}{(D-d)}$$

In order to make a large velocity ratio one can make D very large or $(D-d)$ very small. The second alternative is easier to achieve. In other words one makes d very nearly equal to D. If the two top pulleys are equal in diameter the load will not move however fast the operator pulls in chain at E. If the difference is very small, L will move very slowly. A large velocity ratio gives a large mechanical advantage, so this system may be used for very heavy loads. You will find one in most engineering workshops suspended from a heavy girder. The efficiency of this machine is not high since friction is large. This is to some extent an advantage since it means that a heavy load may remain hanging even if the operator lets go of the

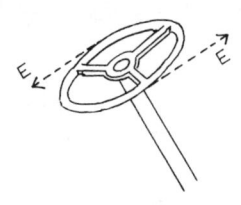

Fig. 15.12. The steering column of a car. A 'wheel and axle'

Fig. 15.13 shows a simple application of this 'wheel and axle' machine. This device which used to be used for raising water from wells is known as a windlass. The velocity ratio here is easily calculated. The effort applied to the handle moves a distance $2\pi R$ for one complete revolution. (R is the length of the handle.) At the same time a length of rope $2\pi r$ is taken in on the 'axle' (r is the radius of the axle) and the load moves through this distance.

Thus velocity ratio =

$$\frac{\text{Distance moved by effort}}{\text{Corresponding distance moved by load}}$$

$$= \frac{2\pi R}{2\pi r}$$

$$= \frac{R}{r}$$

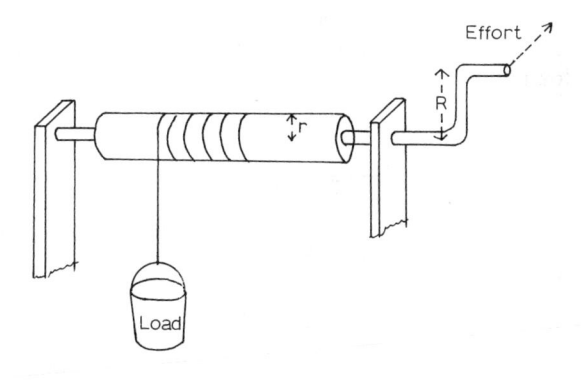

Fig. 15.13. A windlass

(c) The screw

We mentioned at the beginning of this chapter the need to lift part of a motor vehicle clear of the ground in order, for example, to change a wheel. This is usually done by means of a 'jack'. This may be a type of hydraulic press (see page 40), but it is more often based on a screw. A typical screw jack is shown in Fig. 15.14. Such a machine will have a very high velocity ratio, but in general a rather low efficiency. We can best see how the high velocity ratio comes about by giving some typical values to the quantities involved for our screw jack. The effort will be applied at the end of a handle, say 20 cm long (see Fig. 15.14). If

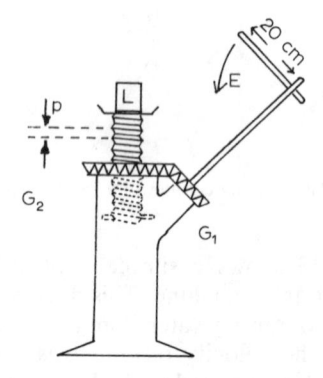

Fig. 15.14. A screw jack

this makes one rotation, then the effort moves a distance $2\pi \times 20$ cm. The bevel gear wheel G_1 also makes one rotation, and causes the other gear wheel G_2 to turn also. If G_2 has 4 times as many teeth as G_1 then one rotation of G_1 produces a quarter revolution of G_2. This in turn produces a quarter revolution of the screw. Now one revolution of a screw makes it move forward or back a distance p (Fig. 15.15) which is known as the pitch of the screw. Let us suppose that the pitch of the screw in our screw jack is 5 mm. It follows then that one revolution of the handle, which turns the screw through a quarter revolution, raises (or lowers) the load by $\frac{1}{4} \times \frac{1}{2}$ cm.

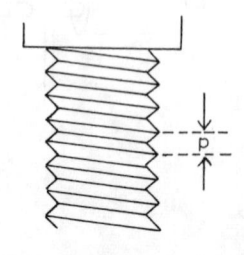

Fig. 15.15. p is the pitch of the screw

Thus:

$$\text{Velocity ratio} = \frac{\text{Distance moved by effort}}{\text{Distance moved by load}}$$

$$= \frac{2\pi \times 20}{\frac{1}{4} \times \frac{1}{2}}$$

$$= 40 \times 8\pi$$

The value of the velocity ratio then may be over 1000. Even if the machine is only 10% efficient, the mechanical advantage will be about 100.

There are many other machines which employ screw threads to give a large velocity ratio. Even a simple wood screw may be regarded as such a machine. The value of the velocity ratio here will depend not only on the pitch of the thread but also on the diameter of the handle of the screwdriver.

(d) The inclined plane

When heavy loads have to be man-handled upwards, say on to the back of a lorry, the simplest way of doing the job is with the aid of a ramp or inclined plane, (Fig. 15.16). The work output of such a 'machine' is:

$$\text{Load} \times \text{vertical height}$$
$$= L \times h$$

Fig. 15.16. An inclined plane

The work input is:
  Effort $\times$ distance along the inclined plane
$$= E \times s$$

The mechanical advantage $= \dfrac{L}{E}$

and velocity ratio $= \dfrac{s}{h}$

$$= \frac{1}{\sin \theta}$$

where θ is the inclination of the plane.

It follows then that the velocity ratio is larger the smaller the inclination of the plane.

As in all previous cases some of the work put in will be 'wasted' in overcoming friction, and the longer the plane the more will be the wasted work.

However a long gentle slope will also require a smaller effort for a given load.

(e) Levers

In the previous chapter we mentioned the use by engineers of long handled spanners to loosen particularly stubborn nuts. This is an example of the simplest and probably most widely used of all machines which we call the lever. Fig. 15.17 shows a crowbar being used to lift a paving stone. This is another example of a lever in use. The crowbar consists of a rigid iron bar. The effort is applied at one end, a distance d from the pivot or *fulcrum*. The load is at the other end, a much smaller distance from the fulcrum. The effort will have to move quite a long way to produce a small movement of the load. Hence there is a large velocity ratio and consequently a large mechanical advantage.

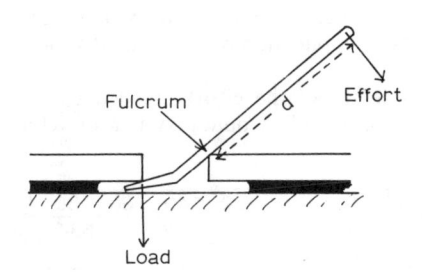

Fig. 15.17. A crowbar being used to lift a paving stone

A crowbar used in this way is typical of a class of levers in which the fulcrum comes between the load and the effort. A pair of pliers (Fig. 15.18) is another example of this class and you can, no doubt, think of many others. A crowbar may also be used as in Fig. 15.19. In this case the fulcrum is the ground instead of a neighbouring stone. Hence the load and effort are on the same side of the fulcrum. Another example of this second class of lever is the wheelbarrow shown in Fig. 15.20. Here also a velocity ratio and mechanical advantage greater than 1 is obtained.

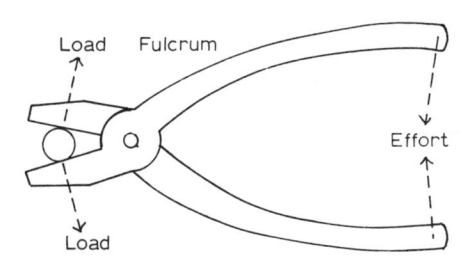

Fig. 15.18. Pliers

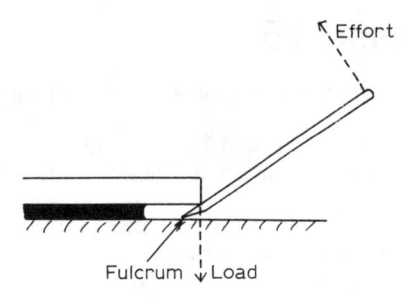

Fig. 15.19. An alternative way of using a crow bar

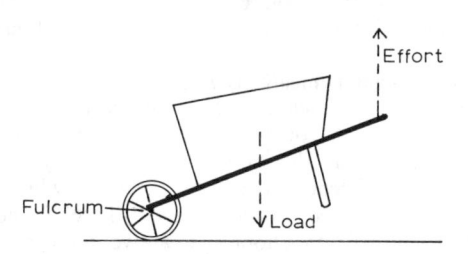

Fig. 15.20. A wheelbarrow

Fig. 15.21 shows a type of crane in which the load is raised by a rigid arm which may be regarded as a lever. Here load and effort are on the same side of the fulcrum but since E is closer to the fulcrum than L, E must be greater than L to lift it. Thus the mechanical advantage and velocity ratio of this third class of lever will be less than 1. The movement of the human forearm by the biceps muscle (Fig. 15.22) is a second example of this type of lever.

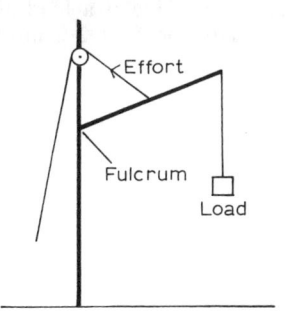

Fig. 15.21. A jib crane

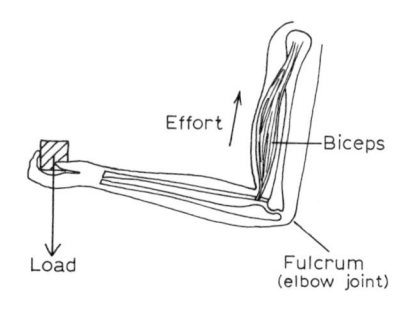

Fig. 15.22. The human arm

Exercise 15

1 A bicycle has a mass of 30 kg. How much work is done:
   (i) in lifting it 2 m above the ground?
   (ii) in pushing it up a 4 m long slope with a force of 150 N.
   (iii) in overcoming friction if, after pushing it up the slope, it is 2 m above the ground. *(SR)*

2 The diameters of a wheel and axle are 20 cm and 5 cm respectively. If an effort of 200 N is needed to raise a mass of 60 kg, calculate (i) the mechanical advantage, (ii) the velocity ratio, and (iii) the efficiency of the machine. *(S)*

3 State what is meant by *mechanical advantage* and *efficiency* of a machine. Describe an experimental method by which you would determine the values of these quantities in the case of a given pulley system.
Find the efficiency of a 5 kw escalator which can carry ten persons, of average mass 70 kg each, slowly from ground level to a floor 10 m above in one minute. *(O)*

4 Explain the terms *force, work, energy*.
A motor scooter with its rider together weighing 200 kg is driven up a hill whose vertical height is 40 m in 4 minutes; calculate (*a*) the work done against gravity, (*b*) the power developed in working against gravity, (*c*) the power of the scooter, assuming its efficiency to be 20%. What energy changes are involved from the time the motor scooter begins to move until it reaches the top of the hill? *(O)*

5 Define the terms *work, energy*, and *power*.
Draw a diagram of a pulley system having a velocity ratio of 4. Find its efficiency if a load of 90 N can just be raised by a force of 25 N applied to the machine.
At what power is a machine working if a load of 90 N is being raised at a steady speed of 2·2 m/s? *(O)*

6 Draw a labelled diagram of the machine known as a wheel and axle. Include in your diagram suitable dimensions to give the machine a velocity ratio of 5. An effort of 10 N turns this machine for 3 minutes and is sufficient to raise the load a distance of 20 m. Neglecting all frictional forces, find the power applied. *(J)*

7 Define *energy*. What is meant by the conservation of energy?
A 20 kg mass is raised to a height of 10 m above the ground. What quantity and type of energy has been given to it? It is then allowed to fall. What changes of energy have occurred when it has come to rest on the ground? *(L)*

8 State two types of energy possessed by a satellite orbiting above the Earth's surface with constant speed and briefly explain why the satellite has these types of energy. *(L)*

9 Make a diagram of a pulley system with a velocity ratio (i.e. displacement ratio) of 5, indicating clearly the points of application of the load (*W*) and the effort (*E*).
The table shows the efforts needed slowly to raise various masses with a pulley system of velocity ratio 5.

Effort in N (*E*)	120	140	160	180	195	210
Mass lifted in kg	30	35·3	43	53·3	65	80
Load in N (*W*)						

Work out the load for each case and plot a graph of *W* against *E*. From the graph estimate the minimum effort needed to raise a mass of 47·5 kg. Calculate the efficiency of the machine when this load is being raised. Why is it less than 100%? *(C)*

16 Expansion of Solids and Liquids. Temperature

Experiment 16.1. Set up the apparatus shown in Fig. 16.1. The roller may be a piece of wooden dowel or a length of glass tube plugged at the end with cork. A straw is pinned with a drawing pin into the end of the roller. Use a Bunsen burner to heat the middle of the retort stand. You will observe that the straw, which is acting as a pointer, moves to the right. This indicates that the iron bar of the retort stand has got longer. Allow it to cool. The pointer will return to its original position indicating that the bar has contracted.

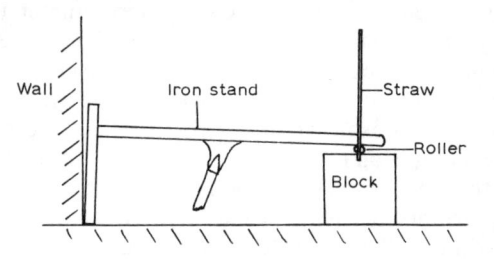

Fig. 16.1. Expansion of a metal rod

Experiment 16.2. Fill a flask right to the top with water and push in a rubber bung fitted with a glass tube. Some of the water will rise up the glass tube (Fig. 16.2). Mount a piece of paper on the tube and make a pencil mark against the water level. Now heat the apparatus and watch the water level carefully. You may see it fall a little at first but then it will rise steadily. Again allow it to cool and you will see the level return to its original position.

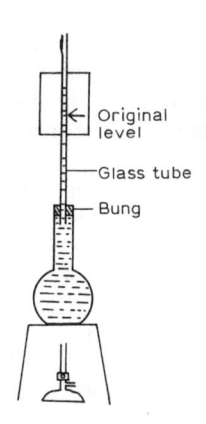

Fig. 16.2. Expansion of a liquid

These two experiments illustrate the very important general principle that solids and liquids expand when heated and contract when cooled. Of course we have not proved that all solids and liquids always do this. In fact we shall mention later one important instance in which things behave rather differently. Two other points arise from these experiments. In the case of the solid the amount of expansion is very small and special arrangements are made in order to detect it. Water expands much more and the increase in volume is clearly visible. Also, in the case of the solid we show an increase in length while with a liquid an increase in volume is shown. When we come to try to measure expansion, we shall find that it is convenient to deal with 'linear' expansion of solids and 'volume' or 'cubical' expansion of liquids.

The initial drop in water level in Experiment 16.2 may puzzle you. It is important to remember here that the glass vessel expands as well as the liquid contained in it. Thus the total amount of expansion indicated by the change in water level in the tube is less than actually occurs. And since heat reaches the glass before it reaches the water there will be first a drop in level. In fact if glass expanded more than water this drop in level would continue. But this is not the case.

Temperature. We shall study expansion in more detail later on. Now we must consider one of the most important applications of expansion, namely its use in measurement of temperature. We can tell whether one thing is hotter than another by touching them in succession. Just how hot things are we cannot tell without the aid of an instrument which we call a thermometer. Thermometers use properties of materials which change as they get hotter. We might try measuring temperature by measuring the length of the iron rod or the level of the water in experiments 1 and 2. Neither of these is very suitable. The change in length of the iron is too small. The reason for the unsuitability of water in a thermometer will appear later. The most common type of thermometer uses the expansion of mercury in a glass vessel. Other thermometers use the change with temperature of the volume or pressure of a gas and the change in the electrical resistance of a coil of platinum wire.

The mercury-in-glass thermometer consists of a narrow bore glass tube with a bulb containing mercury at one end. The other end is sealed after the air over the mercury has been pumped out. Before this or any other thermometer may be used for measuring temperature it must be calibrated. First the fixed points must be marked. The bulb is placed in pure melting ice as shown in Fig. 16.3 and the mercury level in the stem of the thermometer marked.

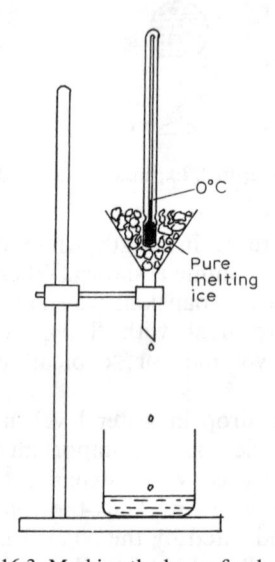

Fig. 16.3. Marking the lower fixed point

The other fixed point is the temperature of pure boiling water at normal atmospheric pressure (76 cm of mercury). We shall see later that impurities and variations in atmospheric pressure alter the boiling point. The upper fixed point may be marked on the thermometer by placing it in the vessel shown in Fig. 16.4 which is known as a hypsometer. It will be seen that the bulb of the thermometer is placed in the steam, the temperature of which is not affected by impurities which may be present in the water. A mercury manometer attached to the vessel measures the difference (if any) between the pressures inside and outside. If atmospheric pressure is known the steam pressure may be calculated. A small correction must be applied if this is not 76 cm of mercury.

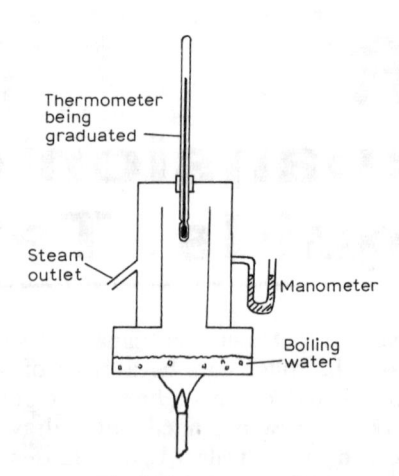

Fig. 16.4. A hypsometer

The Celsius Scale of Temperature. The temperature scale most widely used for everyday practical purposes is the Celsius scale. You may sometimes hear degrees Celsius referred to as degrees Centigrade (°C). On a celsius thermometer the lower fixed point, the temperature of pure melting ice, is labelled 0°C. The upper fixed point, the temperature of steam over boiling water at normal atmospheric pressure, is called 100°C. The calibration of a mercury-in-glass thermometer on the Celsius scale may be completed by dividing the distance between the marks on the stem corresponding to 0 and 100°C into 100 equal divisions.

Thermometric Liquids. Mercury thermometers are very widely used for ordinary scientific work. They do have certain limitations however. Mercury freezes at about −40°C which makes it unsuitable for use in the very coldest climates. Also it is rather expensive. Alchohol is commonly used in inexpensive thermometers for everyday use in measuring atmospheric temperature. It has a freezing point of −115°C. The boiling point of alcohol is 78°C which makes it unsuitable for most scientific work. Mercury boils at 357°C.

Special Thermometers

(a) Clinical

The normal human body temperature is 36·9°C. Accurate measurement of body temperature is very valuable to a doctor diagnosing and treating illness. The clinical thermometer used for this purpose is shown in Fig. 16.5. The essential features are its limited range which leads to high sensitivity and the constriction. The liquid in the bulb of the thermometer can expand past the constriction but the thread of liquid in the stem

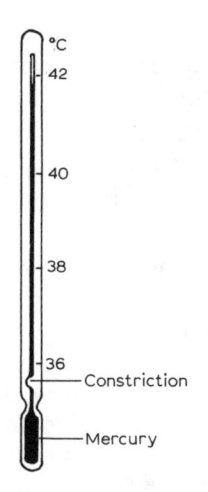

Fig. 16.5. A clinical thermometer

will not pass it when it cools and a break in the thread will occur. The thermometer is thus a maximum reading type. When a reading has been taken with a clinical thermometer, the thermometer is shaken to get the thread of mercury in the stem back into the bulb.

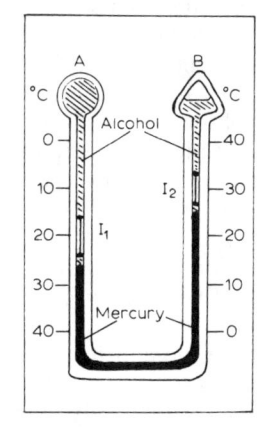

(b) Maximum and Minimum Thermometers

In many cases it is convenient to know the maximum and minimum air temperatures over a certain period of time. Workers in the building industry, for example, require to know the lowest overnight temperature, as recently laid concrete might break up if the temperature fell below freezing. Meteorologists keep statistical records of maximum and minimum temperatures of each 24-hour period. These values may be obtained from a thermometer of the type shown in Fig. 16.6.

At the beginning of the period over which maximum and minimum temperatures are being taken the two steel indexes I_1 and I_2 are moved

Fig. 16.6. Maximum and minimum thermometer

with the aid of a magnet so that they rest on the mercury thread. If the temperature rises, the alcohol in bulb A expands and the mercury thread moves down on the left hand side and up on the right. Index I_1 is left in position while I_2 will be, by the end of the period, left at the highest level in the right hand limb. I_1 will mark the lowest temperature during the period. The lower end of the index in each case indicates the reading. It should be noted that since the mercury is only a fine thread in the tube its expansion is negligible by comparison with the quite large volume of alcohol in the bulb.

Other Applications of Expansion

(a) Bimetal strips

Strips of two different metals, for example, iron and brass, pressed or riveted together make what is known as a bimetal strip (Fig. 16.7(*a*)). Since, for a certain rise in temperature, the brass will expand more than an equal length of iron, a bimetal strip will bend when heated as in Fig. 16.7(*b*). This effect is used in a number of devices including thermostats, fire alarms, time delay switches in electronic circuits and flashing direc-

Fig. 16.7. A bimetal strip

tion indicators in motor vehicles. A bimetal thermostat is illustrated in Fig. 16.8. The heating element might, for example, be an immersion heater in a tank of water. When a certain water temperature has been reached the bimetal strip

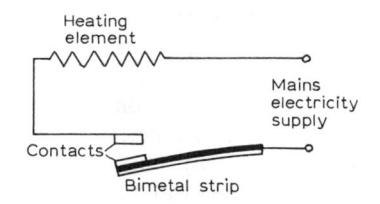

Fig. 16.8. A bimetal thermostat

bends sufficiently to separate a pair of contacts switching off the current in the heating element. A drop in temperature results in the bimetal strip straightening, completing the current at the contacts and so switching on the current again. In this way a constant temperature is maintained.

(b) Riveting

Fig. 16.9 shows the process of riveting two metal plates, very widely used particularly in shipbuilding. The rivet is first heated to red heat and then put through holes in the two plates to be fixed together (Fig. 16.9(a)). The end of the rivet is hammered flat (Fig. 16.9(b)). While this is being done the plates are as far as possible in contact and the diagrams exaggerate the gap between them. However when the rivet cools it contracts and pulls the plates hard together (Fig. 16.9(c)). A row of such rivets along an overlap between two metal plates produces a strong watertight join.

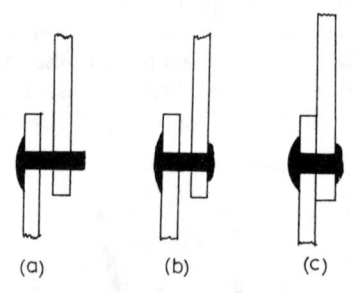

Fig. 16.9. Riveting. (a) Rivet hot (b) End hammered over (c) Rivet cold

Effects of Expansion

The change in the size of an object with variations in temperature is an undesirable effect which has to be allowed for in almost all engineering projects. These effects are most obvious either when large structures like bridges or railway tracks are being built or when large temperature differences occur, for example, in the large turbogenerators used in power stations.

In building the recently completed Forth road bridge in Scotland allowance was made for a total expansion of the road deck of about 2 m. Four deck expansion joints of the type shown in Fig. 16.10 were incorporated. The road at these joints is on rollers and on expansion some of the surface moves under a fixed portion.

Expansion and contractions arising from changes of atmospheric temperature also must be allowed for when railway lines are laid. In this case, the lengths of line are laid with a small gap between them so that expansion may take place

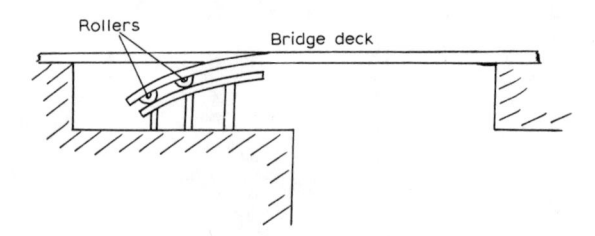

Fig. 16.10. Bridge expansion joints

without causing them to buckle. The lines are joined together by means of flat metal plates, known as fish plates, which have slotted bolt holes so that expansion may take place freely (Fig. 16.11).

The technique has recently been developed of laying lines in 400 m lengths without expansion joints. Here the lines have to take the strains set up by temperature rises without buckling. This track has the advantage that it gives a much smoother ride.

A temperature change of several hundred degrees Celsius may occur when a power station turbine generator set is brought into use (see Chapter 46). Thus, although its overall length does not compare with the other two examples quoted, a substantial increase in length will occur. To allow for this, one end will be rigidly fixed while the other is once again secured through slotted bolt holes.

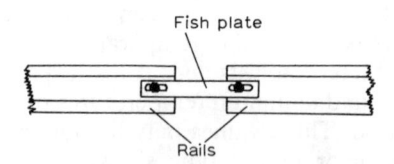

Fig. 16.11. Railway line expansion joints

The expansion of metals also causes problems to the makers of clocks and watches. You may recall that in Chapter 3 we discussed how the length of a pendulum determines its time of swing. In fact if a pendulum gets longer, its time of swing gets longer also. It follows then that a rise of temperature causing a pendulum to expand may in turn cause the clock to go slow. Several elaborate methods have been adopted to try to eliminate errors of this type. Perhaps the simplest is to choose a metal for a pendulum which expands very litte. A metal with this property is known as invar.

In a watch, the time control is provided not by a pendulum, but by a balance wheel (Fig. 16.12). Temperature changes are likely to lead to a watch going fast or slow also. This happens not only

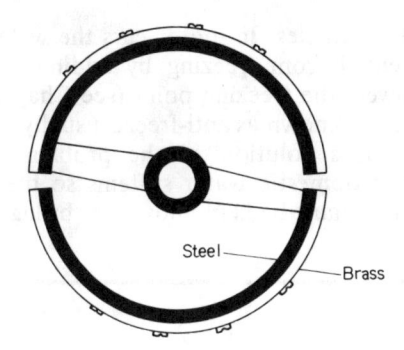

Fig. 16.12. Balance wheel

because of the expansion or contraction of the balance wheel, but also on account of changes which occur in the elasticity of the spring controlling the oscillations. Both of these sources of error are compensated for by making balance wheels of bimetal strip. Thus when the temperature rises the rim of the wheel tends to bend inwards and keep the average diameter approximately constant.

Density Changes and Convection

Since the volumes of solids and liquids generally increase when they are heated without changes in mass, there must be changes in density.

$$\text{Density} = \frac{\text{Mass}}{\text{Volume}}$$

Thus when volume increases, density decreases. In solids these density changes are not very large but in liquids (and gases) they have very important consequences. Suppose a saucepan of a liquid is being heated on a hotplate. The liquid at the bottom closest to the hotplate will be heated first, its density will become lower than that of the rest and it will 'float' to the surface. Colder, denser liquid from above will sink to the bottom where it in turn will be heated. In this way a circulation may be set up as shown in Fig. 16.13. This process, known as convection, will be discussed in more detail in Chapter 18. However one consequence of the effect is that if the liquid in a vessel is not all at the same temperature, then the warmest will be at the surface and the coldest at the bottom.

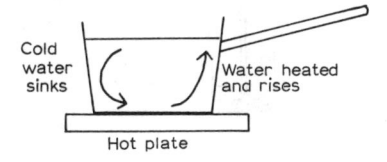

Fig. 16.13. Convection in water

Unusual Behaviour of Water

There is one important exception to the general principles stated in the section above. If, in the case of water, the coldest is always at the bottom, then one would expect freezing to start there. However it is common experience that a layer of ice forms first on the surface of water. Clearly, at temperatures close to 0°C water behaves differently from most liquids. In fact, as water is cooled from room temperature it contracts, as expected, until 4°C is reached. After this it starts to expand as the temperature is reduced (Fig. 16.14) and as freezing takes place, a substantial increase in volume occurs. If the ice produced is further cooled below 0°C it contracts normally.

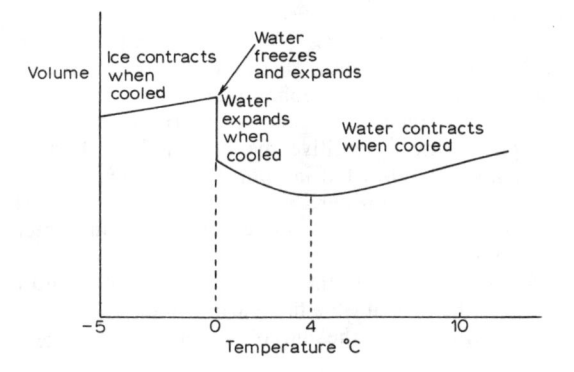

Fig. 16.14. Volume changes as water is cooled

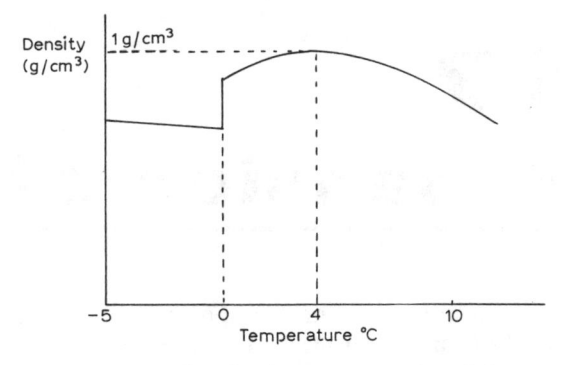

Fig. 16.15. Density changes as water is cooled

Fig. 16.15 shows the variation in the density of water over the same temperature range.

There are several important consequences of this unusual behaviour of water. If it behaved like other liquids, when air temperatures fall in winter to 0°C or below, water on the surface of lakes, rivers and oceans would be cooled, sink to the bottom and be replaced by warmer water from below. In this way all the water would quite quickly be cooled to freezing point. In fact the coldest water remains on the surface, and unless a body of water freezes right through there will always be some at the bottom at 4°C. In this way fish can survive even in water with a thick layer of ice above them.

The sudden increase in volume as water freezes is the cause of bursts in water systems in buildings and motor vehicles. In motor cars the water may be prevented from freezing by adding a liquid which lowers the freezing point (see Chapter 21). Such liquids, known as anti-freeze, usually contain glycerol. This solution to the problem is not possible in domestic water systems so the water temperature must be kept above 0°C by lagging or heating.

Exercise 16

1 Describe experiments, ONE in each case, by which you would test the accuracy of a mercury-in-glass thermometer (a) the lower fixed point, (b) the upper fixed point.

State THREE advantages of mercury compared with alcohol when used in a thermometer and state ONE instance in which an alcohol thermometer must be used in preference to a mercury thermometer.

Explain why a sensitive mercury-in-glass thermometer first records a fall in temperature when its bulb is placed in boiling water. (J)

2 (a) Why does a milk bottle crack if boiling water is put into it?

(b) Describe with the aid of a diagram how EITHER a thermostat OR a fire alarm works.

3 Describe, with the aid of a diagram, how you would determine experimentally the coefficient of linear expansion of brass.

State TWO cases in which the expansion of solids is a disadvantage and indicate in each case how the disadvantage is overcome. (L)

4 A block of ice at $-10°C$ is heated until the temperature reaches $+10°C$. Draw a sketch graph showing how the volume changes over this range of temperature. Mark on the graph the temperature at which the *density* is a maximum.

After a severe frost the following observations were made: (a) a water-pipe had burst; (b) a neighbouring water-pipe which was lagged with felt had not burst; (c) the surface of a pond was covered with ice but the water at the bottom was above freezing-point. Give a clear explanation of each observation. (C)

17
Expansion of Gases

Experiment 17.1. Set up the apparatus shown in Fig. 17.1 and warm the flask very gently with a Bunsen burner. You will observe bubbles coming to the surface of the water in the beaker from the end of the tube. The air contained in the flask is expanding. After heating for some while, allow the flask to cool. Water will rise in the tube and eventually partially fill the flask. The air left at the end of the heating process has contracted again.

This experiment has shown a gas behaving in the same sort of way, when heated, as we observed

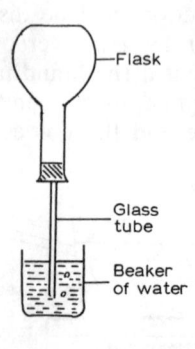

Fig. 17.1. Expansion of air

with solids and liquids in the last chapter. There is one important difference. The forces involved in the linear expansion and contraction of a solid are very large indeed and in the case of liquid it is almost impossible to prevent it expanding when heated. We say that liquids are almost incompressible. With a gas, on the other hand, it is possible to store it in a metal cylinder which keeps its volume almost constant in spite of temperature changes. A rise in temperature of the gas contained in a cylinder, since it cannot increase the volume, will increase the pressure instead.

We believe that a gas consists of a large number of particles (molecules) flying about in rapid, random motion. The collisions which these molecules have with the walls of the containing vessel are responsible for the pressure of the gas. When the temperature of the gas rises the molecules move faster. Clearly, for a given container, the faster the molecules are moving the greater will be the gas pressure.

In many practical problems it is important to be able to calculate how the volume and pressure of a gas will change as the temperature changes. In order to deal with a problem like this in which there are three variables (Pressure, Volume and Temperature), one must, at first, keep one of them constant, and investigate the variation of the other two. We will first investigate how the volume of a fixed mass of gas varies with temperature when the pressure remains constant.

Effect of Temperature on the Volume of a Gas

Experiment 17.2. The apparatus shown in Fig. 17.2 consists of a length of capillary tube, sealed at one end and mounted on a length of cm scale. A small mass of air is trapped in the tube by the thread of concentrated sulphuric acid, which also ensures that the air is free of water vapour. Clamp this apparatus in a vertical position with the open end of the tube uppermost and with the trapped air entirely below the level of water contained in a beaker. Note the length of the column of trapped air and the reading on a mercury thermometer whose bulb is in the water. Now heat the water bath and note the length of air column at various temperatures up to 100 °C.

It will be seen that the pressure on the gas is constant throughout this experiment. It is atmospheric pressure plus the small pressure due to the sulphuric acid thread. The cross section of the tube is assumed to be constant for the purposes of the experiment. Thus the volume of trapped air is

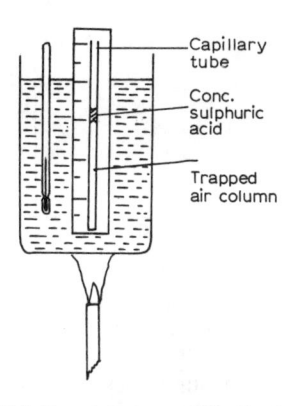

Fig. 17.2. Experiment to verify Charles Law

proportional to the length of the air column and when we wish to plot our results to show how volume varies with temperature, we may justifiably plot length against temperature.

Fig. 17.3 shows a typical graph for an experiment of this sort. The points lie on a straight line, but since it does not pass through the origin, we are *not* justified in saying that the volume is proportional to the temperature in °C. If we did say this it would imply that at 0 °C the volume of the gas is zero. This would be a most surprising statement,

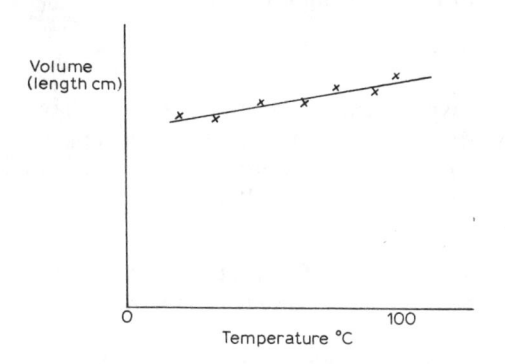

Fig. 17.3. Volume–temperature graph for air (constant pressure)

particularly since 0 °C is a quite arbitrary temperature, that of pure melting ice, which we have chosen to make our zero on a particular scale. How then can we summarize our results? If we produce back our straight line graph we can find a temperature at which the volume would apparently become zero (Fig. 17.4). You should find that this temperature is around −270 °C if you produce back the graph obtained from your experiment. Suppose we devise a new scale of temperatures on which this is our zero. Now, our straight line graph will pass through the origin of axes and we may say that the volume of the gas is proportional to the tempera-

91

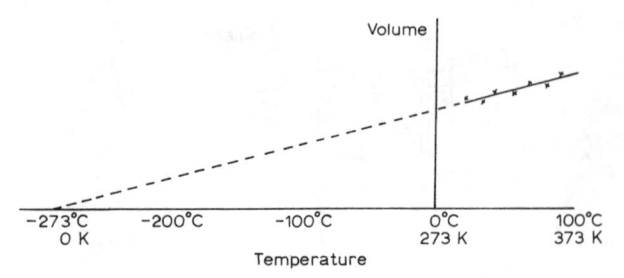

Fig. 17.4. Volume–temperature graph to show absolute zero of temperature

ture measured on our new scale. The zero on this scale is, accurately, a little lower than −273 °C and is known as the absolute zero of temperature. The scale is known as the 'Kelvin' scale. (Lord Kelvin proposed the system in 1848.) Temperatures on this scale are measured in kelvins (abbreviation K). In this book the symbol T will be reserved for temperatures on the Kelvin scale while t will be used temperatures on the Celsius scale. Our experiment shows then for the gas used (air), over the range of temperatures considered (room temperature to 100 °C) the volume of the gas is proportional to the temperature in kelvins.

$$V \alpha T$$
or $V = $ a constant $\times T$
or $\dfrac{V}{T} = $ a constant

The experiment illustrates a general law, followed closely by all gases under appropriate conditions of temperature and pressure, which is known as *Charles' Law*. It states:

For a fixed mass of gas, at constant pressure, the volume is proportional to the temperature in kelvins.

We have defined the zero on our Kelvin scale of temperature, but in order that the size of the degree shall be known we must have at least one other fixed point. In fact the size of the degree on the Kelvin scale is the same as that on the Celsius scale so that as −273 °C ≡ 0K then 0 °C ≡ 273K, 100 °C ≡ 373K and t °C ≡ $(273 + t)$K.

We can now see how Charles' Law may be used to deal with problems involving the expansion of gases at constant pressure. Fig. 17.5 shows a cylinder in which some gas is maintained at a constant pressure P by a piston. The gas occupies 1 dm³ at 17 °C. If the gas is heated it will expand and push the piston back. Suppose it is heated to 162 °C. What will be its volume then? Charles' Law may be written:

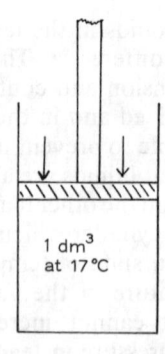

Fig. 17.5. Gas in a cylinder at constant pressure

$$\dfrac{V}{T} = \text{a constant}$$

or $\dfrac{V_1}{T_1} = \dfrac{V_2}{T_2}$ (Pressure constant)

if V_1 and V_2 are the volumes of our fixed mass of gas at temperatures of T_1 and T_2 respectively. Thus in this case:

$$\dfrac{1}{(273 + 17)} = \dfrac{V_2}{(273 + 162)}$$
$$\dfrac{1}{290} = \dfrac{V_2}{435}$$
$$V_2 = \dfrac{435}{290}$$
$$= 1 \cdot 5 \text{ dm}^3$$

One other point needs to be emphasized about the graph in Fig. 17.4. It is not suggested that by cooling a gas sufficiently it is possible to make it occupy no volume at all. In fact all gases become first liquids and then solids before a temperature of −273 °C is reached. We have merely produced back a graph of volume against temperature at constant pressure. Since the graph at these very low temperatures does not represent physical reality, you may feel that our absolute zero is just something we have invented to simplify the mathematics of problems dealing with the expansion of gases at higher temperatures. However, research over the last 40 or 50 years has shown that the absolute zero of temperature is very much a matter of physical reality. Scientists have been able to produce temperatures down to within a small fraction of a degree of absolute zero. But the closer they approach to it, the more difficult the next step becomes. The difficulties in producing and measuring the very low temperatures and the extraordinary behaviour of some substances near absolute zero have given rise to a large amount of research in the fields of Low Temperature Physics and Chemistry. This, until now, has been mostly pure research. That is, it has been carried out only to find out more about nature. However it seems likely that important physical applications will arise in the future.

Effect of Pressure on the Volume of a Gas

We have mentioned previously that exerting a pressure on a solid or liquid will have only a very small effect on the total volume (although it may result in a change of shape). With a gas this is not so. If a large enough pressure is applied a gas may be compressed into a very small fraction of its original volume. However, when a gas is compressed, in addition to a decrease in the volume of the gas, there is usually an increase in temperature. You will have noticed this probably when pumping up a bicycle tyre. If you pump vigorously enough, the end of the pump may become quite hot. If we wish to carry out an experiment to see how the volume of a gas varies with pressure alone, we must ensure that the temperature remains constant.

Fig. 17.6 shows an apparatus we may use to investigate this relationship. Once again we are using a column of air trapped in a capillary tube, this time by a thread of mercury. Again we will assume the tube to have uniform cross section, so we may measure the length of the air column instead of its volume. The pressure on the trapped air we shall measure by means of the pressure gauge and this pressure may be increased by pumping air by means of a bicycle pump into the hollow vessel V.

Pressure gauge reading (kN/m^2)	Length of air column (cm)	$\dfrac{1}{\text{length}}$ (cm^{-1})
0	29.4	0.034
50	24.2	0.041
100	18.1	0.055
150	14.5	0.069
200	12.1	0.083
250	10.4	0.096

Note that in the above table $1 kN/m^2$ is 1000 newtons per square metre.

The figures in the table indicate the results of an experiment of this kind. It is clear that when the pressure increases the volume decreases so we have not got a case of simple proportionality. In fact if you plot a graph of pressure gauge reading against length you will obtain a curve which you would find difficulty in interpreting precisely. But try plotting pressure gauge reading against $\dfrac{1}{\text{length}}$. Fig. 17.7 shows the sort of graph you should obtain in this case. It is now a straight graph but it does not pass through the origin. This leads again to the question 'What does the zero reading on the pressure gauge indicate?'. It means, of course, that the pressure inside the vessel V (and so that on our trapped air) is the same as atmospheric pressure. Now atmospheric pressure is generally about $100 \, kN/m^2$ so our zero on the pressure axis should in fact show about $100 \, kN/m^2$ and the true zero is at about -100 on the axis in Fig. 17.7. If we make this our origin of axes our straight line graph passes through it. We may thus conclude from our experiment that the pressure is proportional to $\dfrac{1}{\text{length of the column}}$

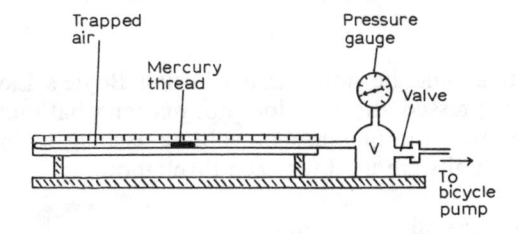

Fig. 17.6. Experiment to verify Boyles Law

Experiment 17.3. Note the length of the air column with the pressure gauge reading zero (does this mean that the pressure of the trapped air is zero?). Pump some air into the apparatus and note the new length and pressure gauge reading. Obtain a series of pairs of values over the whole range of the gauge and tabulate your results as below:

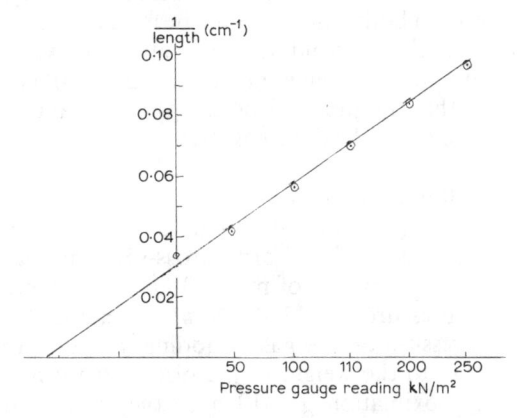

Fig. 17.7. Results of experiment to verify Boyle's Law

93

(and hence proportional to $\frac{1}{\text{Volume of gas}}$).

$$P \propto \frac{1}{V}$$

We say pressure is *inversely* proportional to volume. When we say $x \propto y$, we mean that if x is doubled, y is doubled, etc. If we say x is *inversely* proportional to y ($x \propto \frac{1}{y}$) we mean that if x is doubled y is halved, etc.

This experiment illustrates *Boyle's Law* which states:

For a fixed mass of gas at constant temperature, the pressure is inversely proportional to the volume.

The constant temperature condition will have been fulfilled in this experiment, since the heat produced by compressing the trapped air in the tube will have been quickly lost to the surroundings, and the temperature will have stayed that of the atmosphere.

We can rewrite Boyle's Law:

$$P = K\frac{1}{V}$$

or $PV = K$ (K a constant)

or $P_1 V_1 = P_2 V_2$

where V_1 and V_2 are the volumes of a certain mass of gas at pressures P_1 and P_2 respectively, the temperature being constant.

Pressure on the Gas and Pressure of the Gas

In our experiment to verify Charles' Law we pointed out that the pressure *on* the gas was constant (atmospheric + pressure due to the thread of sulphuric acid). In the Boyle's Law experiment the pressure gauge measured the pressure in V which was the pressure on the air column. In both cases it should be noted that, if the gas is neither expanding nor contracting while a measurement is being made, we are justified in saying that the pressure on the gas is equal to the pressure exerted by the gas itself.

Use of Boyle's Law

Fig. 17.8(*a*) shows 1000 cm³ of gas contained in a cylinder of uniform cross-sectional area 100 cm² by a piston of mass 10 kg. If the atmospheric pressure A is 100 kN/m² we can find the total pressure on the gas by adding to A the pressure due to the weight of the piston. If we make the approximation $g = 10$ m/s² the mass of the piston is 100 N and the pressure it causes is

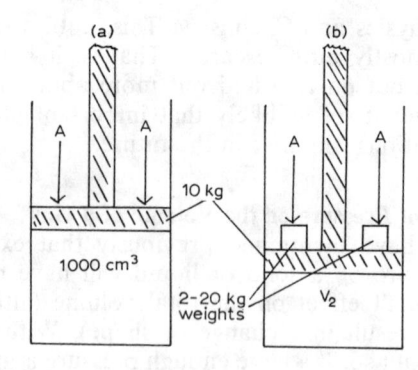

Fig. 17.8. Gas at constant temperature. Pressure increased

$\frac{100}{0\cdot 01} = 10\,000$ N/m² (0·01 m² is the area of the piston). So the total pressure on the gas is 110 kN/m².

Now we wish to find the new volume (V_2) occupied by the gas when a 40 kg mass is placed on the piston (Fig. 17.8(*b*)) assuming there is no change in temperature.

The mass added increases the pressure by $\frac{400}{0\cdot 01} = 40\,000$ N/m² Hence the new total pressure on the gas is 150 kN/m².

From Boyle's law

$$P_1 V_1 = P_2 V_2$$
$$110 \times 0\cdot 001 = 150 \times V_2$$

(0·001 m³ is the initial volume of the trapped gas).

Hence the new volume V_2

$$= \frac{110 \times 0\cdot 001}{150}$$
$$= 0\cdot 00073 \text{ m}^3$$
$$\text{or } 730 \text{ cm}^3$$

It should be noted that in using Boyle's Law as expressed above it does not matter what units are used for pressure or volume as long as the same units are used for each throughout.

The General Gas Equation

The two gas laws we have dealt with so far refer to the special cases of constant pressure and constant temperature. In general if the pressure on a gas is changed there will be changes to both its volume and its temperature. In order to deal with this kind of problem we need to combine the two laws.

Suppose we have a fixed mass of gas at pressure P_1, temperature T_1K and volume V_1 which is going to be compressed so that its volume changes to V_2, its temperature rises to T_2K, and its pressure increases to P_2. This change may take the place in two stages (Fig. 17.9). First it may be heated at constant pressure to the final temperature T_2. It will expand to volume V^1 and the pressure stays at P_1. For this Charles' Law applies.

$$\frac{V_1}{T_1} = \frac{V^1}{T_2}$$

$$\text{or } V^1 = \frac{T_2 V_1}{T_1} \quad (1)$$

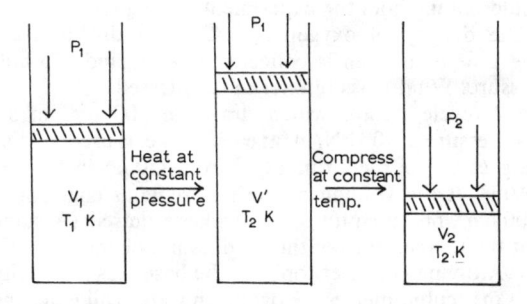

Fig. 17.9. The general gas equation: $\frac{P_1 V_1}{T_1} = \frac{P_2 V_2}{T_2}$

Now it may be compressed at constant temperature so that the pressure and volume go to their final values of $P_2 V_2$ while the temperature stays at T_2. Applying Boyle's Law to this:

$$P_1 V^1 = P_2 V_2$$

or, substituting for V^1 from equation (1)

$$\frac{P_1 T_2 V_1}{T_1} = P_2 V_2$$

$$\text{or } \frac{P_1 V_1}{T_1} = \frac{P_2 V_2}{T_2}$$

This is known as the general gas equation. It may be used, for example, if the volume of a fixed mass of gas is known under certain conditions of temperature and pressure to find its volume under any other specified conditions. In particular, if the mass of some gas is required, it may be found from its volume and density. The density is normally quoted in books of tables at s.t.p. (standard temperature and pressure) which means 0 °C and 76 cm of mercury pressure. In order to use this value of density the volume of gas must be found at 0 °C and 76 cm of mercury pressure. We can best illustrate this by means of an example: The density of air at s.t.p. is 1·3 kg/m³. What is the mass of air in a room of volume 40 cubic metres if air temperature is 20 °C and atmospheric pressure 78 cm of mercury?

Initially
$\begin{cases} \text{Volume of air} = 40 \text{ m}^3 \\ \text{Temp.} = 293 \text{ K} \\ \text{Pressure} = 78 \text{ cm of mercury} \end{cases}$

Finally
$\begin{cases} \text{Volume} = V \text{ m}^3 \\ \text{Temp.} = 273 \text{ K } (0\,°\text{C}) \\ \text{Pressure} = 76 \text{ cm of mercury} \end{cases}$

$$\frac{P_1 V_1}{T_1} = \frac{P_2 V_2}{T_2}$$

$$\frac{78 \times 40}{293} = \frac{76 V}{273}$$

$$\therefore V = \frac{78 \times 40 \times 273}{76 \times 293}$$

$$= 38 \cdot 2 \text{ m}^3 \text{ at s.t.p.}$$

$\therefore$ Mass of air = Volume (at s.t.p.) × density (at s.t.p.)
$= 38 \cdot 2 \times 1 \cdot 3$ kg
$= 50 \cdot 1$ kg

The preparation of gases in chemistry often involves the use of heat or heat may be produced during the reaction so the conditions under which they are collected will vary considerably. The temperature is unlikely ever to be 0 °C and the pressure may not be 760 mm. In order to compare the volumes collected at different times or under differing conditions each must be corrected to some standard set of conditions, viz. 0 °C and 760 mm. The example above shows how this is done.

Most apparatus now used in chemistry for measuring volumes of gases and liquids is marked or graduated in millilitres (ml) or in cubic centimetres (cm³). As 1 litre = 1000 cm³, 1 ml = 1 cm³, so the two measurements are the same. You should learn to use both as, in problems, volumes may be expressed in either way. The term ml is not used, however, for volumes of solids. One litre is usually expressed as one dm³.

Exercise 17

1 Draw a fully labelled diagram of the apparatus you would use to verify Boyle's Law. State clearly the measurements you would take, and describe how you would use these to verify the Law. A vertical cylinder, whose area of cross section is 5 cm², is fitted with a piston of negligible mass, and encloses 1 dm³ of gas at an atmospheric pressure of 100 kN/m². If a mass of 1 kg is now placed on the piston, calculate the new volume of the gas. Assume the temperature remains constant. *(S)*

2 A flask of capacity 1 dm³ is slowly exhausted of air by means of a pump. The cylinder of the pump has an internal area of 5 cm² and the length of the stroke is 20 cm. If the initial air pressure in the flask is 77 cm of mercury what will be the pressure after (*a*) one stroke, (*b*) two strokes? *(J)*

3 One dm³ of air at a pressure of one atmosphere is compressed to a quarter of its original volume. Calculate the additional pressure required, assuming the temperature remains the same.

Name the Law used in this calculation. *(S)*

4 Describe experiments, ONE in each case, to show that gases and liquids expand on heating.

A balloon, which contains 10 dm³ of gas 20°C and 76 cm of mercury pressure, rises until it reaches a height where the pressure of gas is 60 cm of mercury and the temperature is 5°C. What volume does the gas now occupy?

Why are balloons which are used for high altitude observations not fully inflated before being released? *(C)*

5 Describe an experiment you could perform to show that the volume of a mass of dry air at constant pressure varies with the temperature. Explain how the air is kept dry and how the pressure is kept constant. Draw a sketch graph to show the type of results you would obtain from the experiment.

The density of oxygen is 1·43 g per dm³ at s.t.p. One dm³ of oxygen is collected at 25°C and 765 mm pressure. What mass of oxygen is collected? *(C)*

6 A cycle pump, which draws in 100 cm³ of air at a pressure of 100 kN/m² at each stroke, is used to raise the pressure of the air in a tyre from 150 kN/m² to 250 kN/m². If the volume of the tyre remains constant at 1000 cm³, find the number of strokes required. (Assume that the temperature of the air does not change.) *(C)*

7 A diving chamber, open at the base, has a capacity of 9 m³ (cubic metres). How much water will enter the chamber when it is lowered to a depth of 2·6 metres? (Assume that the pressure of the atmosphere is equal to the pressure exerted by a column of water 10·4 m long.) *(C)*

18
How Heat gets from One Place to Another

One of the most important engineering problems, cropping up day after day in all sorts of industries, is how to get heat from one place to another as quickly as possible and with a minimum of waste. Possibly the biggest of these heat transfer problems arises in the designing of nuclear power stations where heat has to be transferred from the fuel (usually rods of uranium) to steam which drives the turbines and generators (see Chapter 44). We must try to understand the processes by which heat travels before we can deal with complicated problems of this sort.

Convection

Experiment 18.1. Drop a few crystals of potassium manganate(VII) very carefully down the side of a beaker containing water. Now heat the bottom of the beaker gently beneath the crystals. You will observe that the mauve colour produced by the potassium manganate(VII) moves in the direction indicated by the arrow in Fig. 18.1. It is a movement of the water itself that you observe. The crystals simply colour the water close to them and make its movements visible.

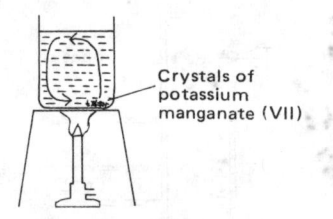

Fig. 18.1. Experiment to show convection in water

We have already suggested an explanation for this effect in the section dealing with the expansion of liquids in Chapter 16. The water above the bunsen flame is heated and expands. Thus its density gets less and it 'floats' to the surface. Its place is taken by colder water from the surface, which in turn is heated and returns to the surface. In due course all the water in the beaker is heated. In this experiment heat travels through a liquid by the actual movement of water molecules.

This process of heat transfer is know as *convection* and it will be clear that it might take place equally well in a gas where the atoms or molecules are even more free to move about. On the other hand, in a solid the atoms have only a very limited amount of movement and heat transfer by convection is virtually impossible. In liquids and gases, convection is generally the principal process by which heat is transferred. We can show this in the following experiment.

Experiment 18.2. The apparatus shown in Fig. 18.2 consists of a loop of glass tube with an opening at the top through which it may be filled with water. Drop a few crystals of potassium manganate(VII) through this opening into the water and heat the tube as in Fig. 18.2(a) under the crystals. You will observe the coloration to spread across the top limb of the tube and the water in this limb may ultimately boil, but no great amount of colour will spread to the rest of the apparatus and the water in it will remain quite cold. Now empty and clean the apparatus, refill with fresh water and add some potassium manganate(VII) crystals. Heat this time as in Fig. 18.2(b). You will see a circulation of water set up as shown by the arrows in the diagram and all the water in the apparatus is quickly coloured by the crystals. In due course all the water is made warm.

In the first part of this experiment convection cannot take place, since water which has been heated will not travel downwards through colder water. As the water in the bottom of the apparatus remains cool it is clear that no other method of heat transfer is very effective in water.

Convection in Gases

Experiment 18.3. Fig. 18.3 shows a glass fronted wooden box fitted with two glass chimneys. Place a lighted candle under one of these chimneys and hold a bundle of smouldering corrugated cardboard over the other. The smouldering cardboard produces a lot of smoke which normally rises. However when you bring it near to the top of the chimney you will see much of the smoke drawn down it into the box. In due course smoke will start to emerge from the other chimney. In this experiment the smoke serves the same purpose as the potassium manganate(VII) in the previous ones. It makes movements of air visible. The air close to the candle flame is heated, its density gets less and so it rises. Colder air comes in through the other chimney to take its place, and in this way a steady convection current is set up.

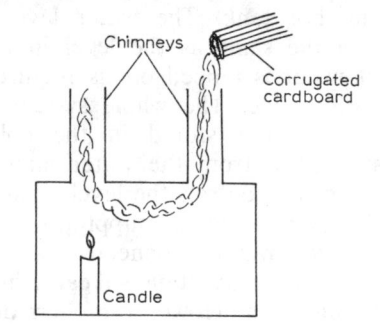

Fig. 18.3. Experiment to show convection in air

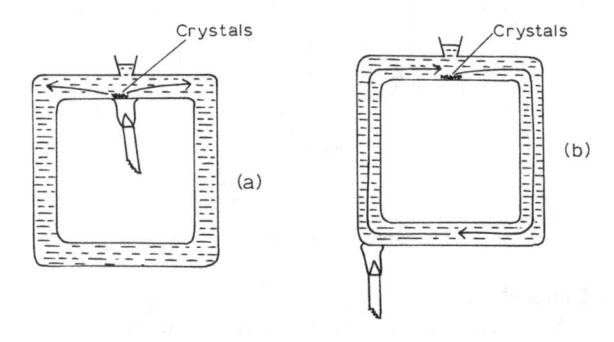

Fig. 18.2. Experiment to show convection in water

Applications of Convection

1. Domestic Heating

With an ordinary open coal fire almost all the hot air produced by the burning coal rises up the chimney with the gases produced by combustion. Thus most of the heat getting into the room in this case does so by heat transfer processes other than convection. Other solid fuel burners which produce convection currents going into the room and are therefore much more efficient are widely used today. Fig. 18.4 shows one such arrangement. Of the cold air which is drawn in at the bottom of the fire, some goes through the hot coal of the fire and up the chimney but the rest goes around the back and sides of the metal firebox in which the coal is contained. There it is heated and comes out through louvres at the front. (For the chemical reactions occurring see Chapter 26.) Gas burning convector heaters operate on a similar principle.

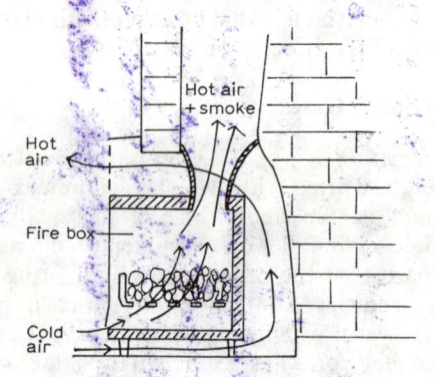

Fig. 18.4. A solid fuel convector fire

2. Hot water systems and central heating

Fig. 18.5 shows a typical domestic hot water system based on a solid fuel boiler. Water in the boiler is heated and rises to the hot tank while cold water from the bottom of the hot tank falls to the boiler to take its place and be heated in turn. The hottest water in the system is in the expansion pipe which comes from the top of the hot tank. The water level in this will remain the same as the level in the cold tank. Hot water is tapped off as required from the expansion pipe. The whole system is kept filled by the water stored in the cold tank which is supplied from the water mains via a float valve which controls the level. A system of this kind may be used to supply hot water to radiators which will be connected as shown in Fig. 18.5. Natural convection will cause hot water to flow through them. However, modern domestic central heating systems are usually based on narrow bore copper pipes and the rate of flow of water by natural convection is not fast enough. It is therefore common to connect a pump in the water circuit.

3. Motor car engine cooling systems

A great deal of heat is produced in the engine of a motor car due to the 'burning' of the petrol fuel. In order that the engine shall not get too hot a circulation of water through it is arranged. Fig. 18.6 shows a greatly simplified diagram of such a system. Water, heated in the engine block, passes out through a hose to the top of the radiator. As it falls through the radiator it is cooled by cold air coming in through the radiator grille (the flow of air is assisted by a fan driven by the engine). The cold water enters the bottom of the engine block to begin its cycle again. In modern cars designers wish to have lower bonnets to improve the streamlining of the body. This makes a natural convection circulation of the type described very difficult to achieve, hence it is usual to fit water circulation pumps.

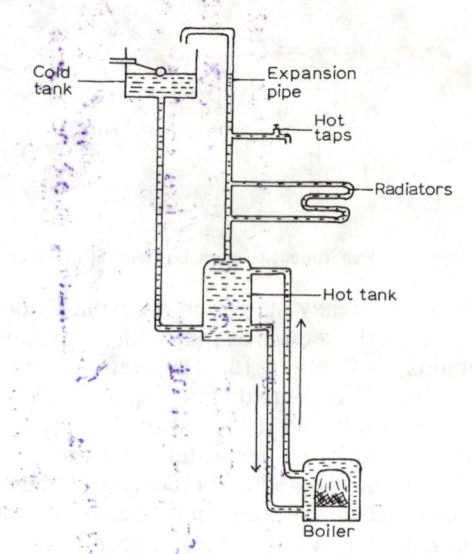

Fig. 18.5. Domestic hot water system

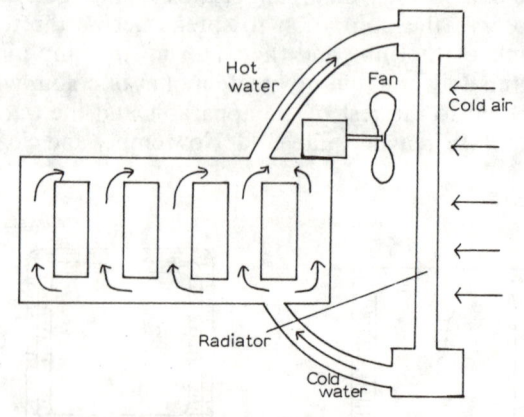

Fig. 18.6. Motor car engine cooling system

Conduction

We have pointed out that, since the atoms in a solid material have only limited movements available to them, convection in solids is almost impossible. On the other hand, we are all aware that heat travels very well through metals. The handle of a poker which has been left in the fire soon becomes very hot. The process by which this kind of heat transfer takes place is known as conduction. It involves heat being passed from one atom to the next in the material. Experiment 18.2 showed that only by convection does heat travel well in liquids. In other words liquids are bad conductors of heat. Gases are even worse conductors. This is readily understood if you consider that in gases the atoms are quite large distances apart and it is therefore only with the utmost difficulty that heat may be transferred from one atom to a neighbouring one. The same is true to a lesser extent of liquids. Among solids, plastics, wood and glass are very poor conductors by comparison with metals.

Bad Conductors and their Uses

The technique of surrounding something with a bad conductor in order to prevent heat getting into or out of it is sometimes known as lagging. It has very wide industrial and domestic application. Cold pipes are lagged to prevent them freezing in winter, hot pipes are lagged to reduce heat losses. The walls of refrigerators and ovens are packed with heat insulation to stop heat getting in and out. We wear thick woollen clothing to keep ourselves warm in winter. In all these cases porous materials which contain a lot of air are very suitable since air is a bad conductor of heat. Typical materials of this sort are cotton wool, glass fibre, expanded polystyrene, knitted woollen materials and cork. Fig. 18.7 shows some of the steps taken in modern house building to keep a house warm in winter. A layer of granulated cork or glass fibre is laid between the joists in the loft, the brickwork consists of a double wall with an air cavity and two panes of glass are used in each window with a small gap between them.

Good Conductors and their Uses

Experiment 18.4. Place an iron or copper gauze on a tripod stand with a Bunsen burner beneath it. Turn on the gas and light it well above the gauze (Fig. 18.8(*a*)). You will find that the gas will burn for some while above the gauze without igniting below it. Now turn off the gas, give the gauze ample time to cool off and try lighting the gas below it. Again you will find that it will burn for some time without the flame appearing above the gauze (Fig. 18.8(*b*)).

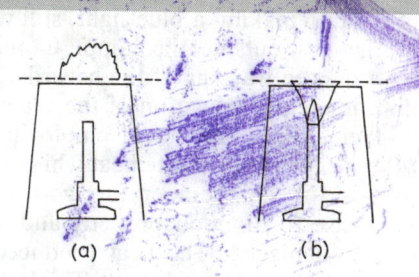

Fig. 18.8 Conduction in metals

At first one is inclined to think that this experiment shows that the gauze is a bad conductor of heat since it takes a long while for the other side to become hot enough to ignite the gas. However a little thought will show that the opposite is true. If it were a bad conductor, the part of the gauze in contact with the flame would quickly become very hot while the rest of it would remain quite cool. In fact, since it is a good conductor the heat is dispersed throughout the whole gauze and thence to the surrounding air. Consequently it takes a considerable while for any part of it to become hot enough to ignite the gas.

This ability of metals to disperse heat has a number of practical applications. Sir Humphrey Davy (1778–1829) was presented with the problem of designing a miner's oil burning lamp that would prevent the explosions of inflammable gas which caused great loss of life in those days. Fig. 18.9 shows his solution to the problem. The flame could not be completely enclosed, since air had to get in, and the gaseous combustion products out, in order that the oil should burn. Davy decided that if he covered the lamp with two or three tall caps of wire gauze, the heat produced by the combustion of the oil would be dispersed without the temperature of the outside of the lamp or the escaping

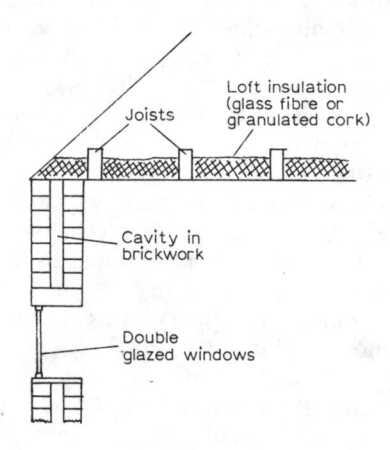

Fig. 18.7. Thermal insulation of a house

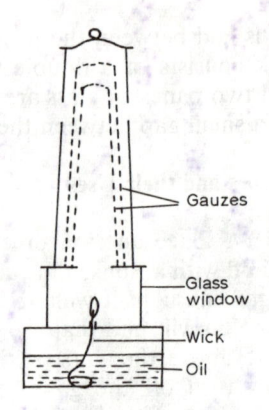

Fig. 18.9 The 'Davy' safety lamp

gases getting high enough to set off an explosion. It was found that the explosive gases could burn inside the lamp making a blue light, still without igniting the gas outside the lamp. In this way miners were given warning of the hazard.

A more modern application of the same effect occurs in electronic equipment employing transistors. These devices generate heat while they are operating, but if their temperature gets too high they will cease to function properly and may be permanently damaged. The heat produced must therefore be dispersed in some way. This is done by mounting them in good thermal contact with a sheet of metal which causes the heat to be dispersed and lost to the surroundings. Also when soldering a transistor into a circuit it is essential that the soldering iron does not make it too hot. To prevent this the wire from the transistor is gripped with a pair of pliers as in Fig. 18.10 and the heat is thus conducted away. In both these cases we are using a piece of metal as a 'heat sink'.

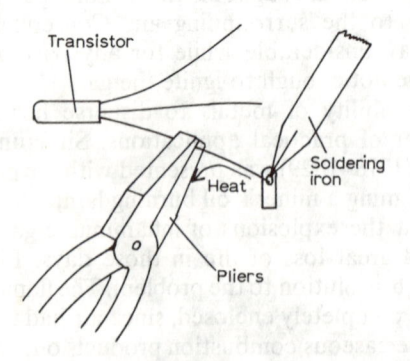

Fig. 18.10. A heat 'sink'

Radiation

Both of the heat transfer processes we have discussed so far require a 'medium'. In the case of convection it will be a liquid or gas, for conduction, a solid. However, the vast quantities of heat received on earth from the sun travel through a long distance in a vacuum. A third type of heat transfer must be involved here which we call 'radiation'. As is mentioned in Chapter 32, besides visible light the sun (and other light sources) emit so called 'invisible radiation' which we call ultra violet and infra red. When any of this radiation is absorbed in matter heat is produced. Most of the heat radiated by the sun and other hot bodies is in the infra red and it is this we usually mean when we refer to radiant heat.

In order to investigate the properties of radiant heat we need a source and a sensitive detecting device. A convenient source of radiant heat is a small electric heating element of the type shown in Fig. 18.11. A most sensitive detecting device is a thermopile. The structure of this instrument is rather beyond our scope here. However, the radiant heat energy falling on the cone shaped collector is converted to a small electric current which is measured by a sensitive galvanometer connected to the thermopile.

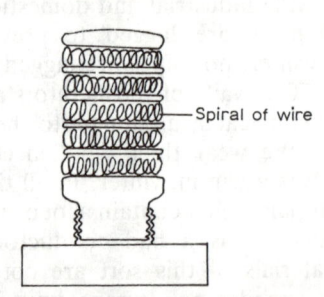

Fig. 18.11. Radiant heat source

Experiment 18.5. Place the heating element at such a distance in front of the thermopile as to cause a suitable deflection of the galvanometer. Now place a piece of wood between source and detector (Fig. 18.12). You will observe that all the radiation is blocked. Try a plate of glass. The deflection of the galvanometer will indicate that a little of the radiation gets through but most of it is stopped. So glass, which is almost completely transparent to visible light, is transparent only to a very limited extent to infra red. On the other hand air is quite transparent to both.

Now arrange the apparatus as in Fig. 18.13 (which shows a plan view). Place a piece of wood between heat source and thermopile to prevent radiation directly between them. By adjusting the position of the piece of polished metal, you will find it possible to get a substantial deflection of the

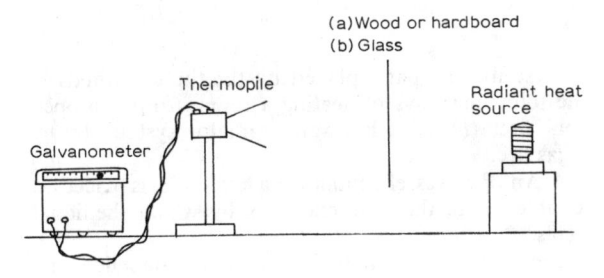

Fig. 18.12. Absorbtion of radiant heat

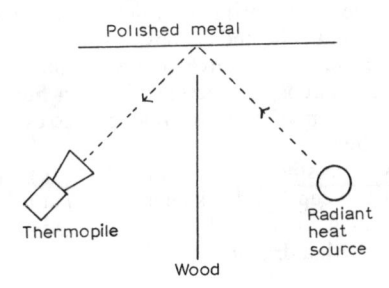

Fig. 18.13. Reflection of radiant heat

galvanometer. Thus radiant heat is reflected in the same way as light (see Chapter 29). If a rough black surface replaces the polished metal, the galvanometer deflection is greatly reduced.

Radiating Surfaces

The amount of radiation given off by a hot body depends on its temperature and also on the nature of its surface. An apparatus for showing the effect of the surface was invented by a Scottish scientist, Leslie (1777–1832). It consists of a hollow metal cube with the four vertical faces treated differently. One is highly polished, one painted matt black and the other two in different colours.

Experiment 18.6. Fill a Leslie cube with hot water and place it with the polished surface facing a thermopile at a measured distance from it (Fig. 18.14). Note the galvanometer deflection. Turn the cube so that each face in turn faces the thermopile and is the same distance from it. Note in each case the deflection. Your results should show that the matt black surface is the most effective radiating surface while the polished one is the least effective. It should be emphasised that, since the cube is of metal (a good conductor), all the surfaces should be at the same temperature.

To summarize the results:

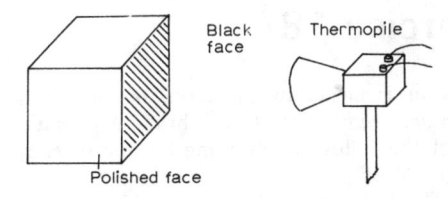

Fig. 18.14. Radiation from black and polished surfaces

Black surfaces are bad reflectors and so are good absorbers of radiant heat. They are also good emitters of radiant heat.

Polished surfaces are good reflectors and so are bad absorbers of radiant heat. They are also bad emitters.

The Vacuum Flask

A device which embodies many of the conclusions reached in this chapter is the vacuum flask (Fig. 18.15). We are familiar with this vessel for keeping tea or coffee hot. It was invented by Sir James Dewar (1842–1923) for keeping liquid gases cold. It consists of a double-walled glass vessel with a vacuum space between the walls. No conduction or convection can take place between the walls. Radiation is kept to a minimum by silvering the inside of the outer wall and the outside of the inner wall. Convection in the air over the liquid is reduced by a cork in the neck of the flask.

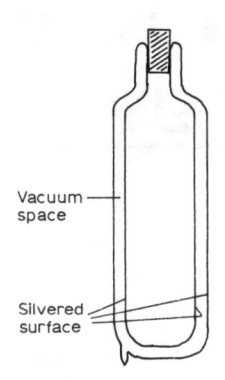

Fig. 18.15. A vacuum flask

Exercise 18

1 What is meant by the terms *thermal conduction, convection, radiation*? Give brief explanations of each of the following, pointing out the physical principles involved:

(*a*) The hot-water cylinder in the hot-water system in a house is often covered with glass wool contained in a white plastic case.

(*b*) The wall above a hot-water radiator is much dirtier than the wall below it.

(*c*) The radiant type of electric fire has a brightly polished metal sheet behind the heating coil.

(*d*) The freezing compartment inside a refrigerator is placed at the top. (*C*)

2 Explain with the aid of a diagram, how convection currents may be used to cool a motor car engine. (*L*)

3 Describe simple experiments, ONE in each case, to show the transference of heat by (i) conduction, (ii) convection, (iii) radiation.

Describe the parts played by the three methods in the following ways of heating a room: (*a*) by an open coal fire, (*b*) by a hot-water radiator system, (*c*) by a gas fire. (*C*)

4 An open vessel containing a hot liquid, is placed on a table. State the different ways by which the liquid cools.

Describe the vacuum flask (a clearly-labelled diagram will be accepted), and explain why a liquid placed in such a flask remains hot for long periods. (*L*)

5 Describe an experiment which shows that a gas expands when it is heated at constant pressure.

It is often said that convection is a result of thermal expansion. What is convection? Explain how it takes place in a room of a house that possesses a central heating system.

How is it possible for a vacuum flask (*a*) on certain occasions to keep hot drinks hot, yet (*b*) at other times to keep chilled food cold? Give a detailed answer and illustrate it with a drawing. (*C*)

19
Heat and Mechanical Energy

In Chapter 15 it was suggested that heat is one form of energy and that heat is usually produced when other forms of energy (for example kinetic or electrical energy) disappear. This idea may be familiar to you, for you may have noticed how a piece of metal becomes hot when it is drilled or filed. In this case the mechanical energy applied to the tool is being converted into heat energy. You will also be familiar with electric heaters whose function is to change electrical energy into heat.

The idea of heat as a form of energy has not always been widely accepted. Some scientists in the 18th century thought heat was a fluid which they called caloric. This fluid, they believed, could not be created or destroyed but flowed into things when they became hot and out when they cooled.

An American called Benjamin Thompson living in the early 19th century disputed this idea.

He was involved at one time in drilling cannon barrels and he succeeded in boiling a kettle with heat produced in the drilling process. Thompson concluded that as long as he went on drilling he could go on producing heat. On the caloric theory this implied that either the cannon barrel or the drill had an inexhaustible supply of caloric. He rejected this possibility and suggested instead that the motion of the drill (what we should call its kinetic energy) was being changed into some sort of motion within the metal itself. He was suggesting that, in modern terms, heat is the kinetic energy of the atoms or molecules of a material.

This idea was taken up by an English scientist, James Prescott Joule (1818–1889), who devoted a large part of his life to answering these questions:

Is it possible to convert some mechanical

energy *entirely* into heat energy? If it is, will a unit of mechanical energy always produce the same quantity of heat? Joule and a number of other nineteenth century scientists devised experiments in which mechanical work was done in many different ways. In each experiment they tried to measure the mechanical energy supplied and the heat energy produced.

In many of Joule's experiments the heat produced was supplied to water and the amount of heat energy produced was calculated in terms of the rise in temperature of a certain mass of water.

Most of these experiments are not easily carried out and in order to obtain consistent results many corrections need to be applied. However as a result of them, by the middle of the last century the answers to Joule's questions had emerged quite clearly.

It was possible to convert mechanical energy entirely into heat, and when this conversion was carried out a unit of mechanical energy always produced the same amount of heat.

We shall not describe one of Joule's experiments. Instead you could carry out a simple rough experiment which shows the basic principles of all such experiments.

Experiment 19.1. Note the temperature of a quantity of lead shot and place it in a long cardboard tube which is sealed at one end with a bung (Fig. 19.1). Place over the other end a tight fitting cup of cardboard or plastic. Hold the tube in a vertical position and invert it fifty times so that with each inversion the lead shot falls the length of the tube. Now take the lead shot out of the tube in the cup and measure its temperature again. It should show a small increase. Finally measure the length of the tube.

The following are the results of an experiment of this kind:

Initial temperature of lead shot = 17 °C
Final temperature of lead shot = 19·7 °C
Number of inversions = 50
Length of tube = 1 m

When the tube is inverted, the lead shot is raised to the top of the tube where it possesses Mgh J of potential energy (see Chapter 15), where M kg is its mass and h m the length of the tube. As it falls this potential energy is converted into kinetic energy and when it hits the bottom of the tube the kinetic energy is converted into heat. Now with 50 inversions and a tube 1 m long:
Total potential energy lost

$$= 50 \times 9·8 \times 1 \times M \text{ J}$$
$$= 490 \, M \text{ J}.$$

Thus:

490 M J heat M kg of lead by 2·7°C

and

490 J heat 1 kg of lead by 2·7°C

also

$\frac{490}{2·7}$ J (181) heat 1 kg of lead by 1°C

The results of our experiment together with the conclusions arrived at by Joule mentioned earlier in this chapter lead us to believe that if we supply 1 kg of lead with 181 J of energy of any kind which is completely converted into heat, then the temperature of the lead will rise by 1 °C.

It should be remembered that the experiment we have been discussing is a very rough one and more accurate experiments give for lead a value of 116 J.

As we pointed out earlier, in most of the experiments of Joule, energy was supplied to water, and the general conclusion reached was that to heat 1 kg of water through 1 °C we must supply 4200 J of energy.

This quantity, the amount of energy needed to raise the temperature of 1 kg of a substance by 1 °C is known as the specific heat capacity of the substance. This definition and the fact that the specific heat capacity of water is about 4200 J, will be the starting point for Chapter 20 which deals with measuring quantities of heat energy.

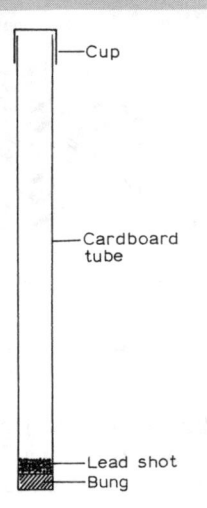

Fig. 19.1. Simple experiment to investigate the conversion of mechanical to heat energy

Conversion of Heat into Mechanical Energy

We have indicated that mechanical energy may be converted entirely into heat, but the reverse process is possibly of more practical importance. Heat is the form of energy most readily available to us. Nature provides large quantities of fuels; coal, oil and natural gas which when burnt provide heat. Devices whereby this heat can be made to do mechanical work have been the concern of engineers for hundreds of years. Although the efficiency of these machines is steadily improving it is now generally believed impossible to convert *all* the heat produced by burning fuel into mechanical energy.

Steam Engines

Fig. 19.2 shows what is probably the earliest form of heat engine. Its design is attributed to Hero of Alexandria and its moving part is a metal sphere with two outlet pipes as shown in the diagram The sphere is mounted on a horizontal axle which also supplies it with steam generated in the boiler. The steam emerges from the outlet pipes under some pressure and the reaction of the sphere causes it to spin. This type of engine is highly inefficient and of no practical importance.

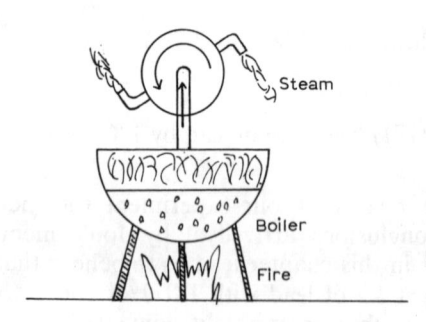

Fig. 19.2. Hero's steam engine

The basic features of a steam engine which has been of very great practical importance are shown in Fig. 19.3. Steam enters through opening 1 in Fig. 19.3(a) and its pressure pushes the piston to the right. When the piston is nearly at the right hand end of the cylinder the slide valve moves so that steam is admitted through opening 2 (Fig. 19.3(b)). At the same time opening 1 is connected to an exhaust pipe. Thus the piston moves towards the left again and the 'spent' steam to the left of the piston is pushed out. The backwards and forwards movement of the piston is converted to a rotation by means of a crankshaft to which the slide valve is also attached so that the two are correctly synchronised.

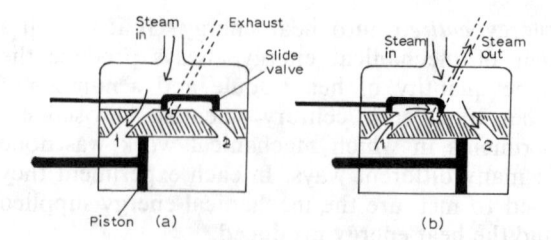

Fig. 19.3. A steam engine

For many applications the steam engine just described has been replaced by the steam turbine which was developed during the last quarter of the last century. The turbine may be regarded as a sort of windmill. High pressure steam is directed against a large number of vanes arranged around a wheel, so causing the wheel to rotate. In a practical steam turbine of the kind used, for example, in power stations to drive the generators, there are several wheels of rotating vanes with sets of fixed vanes so as to produced maximum turning effect. Fig. 19.4 shows a greatly simplified diagram of this sort of engine.

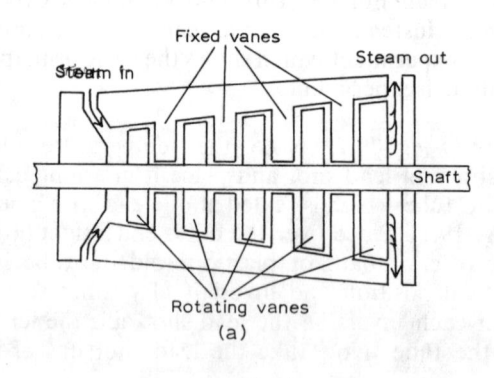

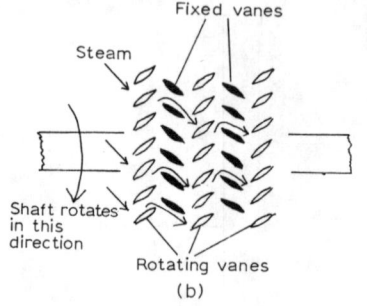

Fig. 19.4. (a) A steam turbine
(b) Flow of steam through turbine vanes

An engine thought to be of great importance for the future is the gas turbine. In this, instead of steam, the high pressure products of the combustion of an air-liquid fuel mixture are directed against the turbine blades. This system is used in turbo-prop aircraft engines in which a propeller is attached to the turbine shaft.

We shall describe briefly just one more type of heat engine. This is the four stroke petrol engine which is used in most light motor vehicles. In its modern form this is a rather complicated system and we shall only be able to give the barest outline here. Fig. 19.5(a) shows the piston going down and a mixture of air and petrol vapour being drawn into the cylinder through valve 1 which is open. In Fig. 19.5(b) the piston is moving up again with both valves closed and this air-gas mixture is being compressed. Just before the piston reaches the top of the cylinder a large electrical potential difference is applied to the spark plug and a spark crosses the gap. The compressed air-gas mixture is ignited and the expanding gases push the piston down again (Fig. 19.5(c)). The fourth and last stroke of the cycle (Fig. 19.5(d) consists of the gas left in the cylinder being pushed by the rising piston through the open valve V_2. The up and down motion of the piston is converted to a rotation by linking it by the connecting rod to a crankshaft. You will notice that only one stroke in the four described is a 'power' stroke. Thus motor car engines usually have four or six cylinders which fire in sequence so that a steadier supply of power is provided. Each of the valves opens only once every other complete rotation of the crankshaft. The valves are then operated by a camshaft geared to the crankshaft so that it makes half as many revolutions.

In all the engines we have described, hot exhaust gases are taken out. Thus much of the heat energy put in is wasted. In fact even in the most efficient of power stations only about 30–40% of the heat energy put in is obtained as electrical energy. So although we can convert all forms of energy entirely into heat, the reverse process can never be 100% efficient.

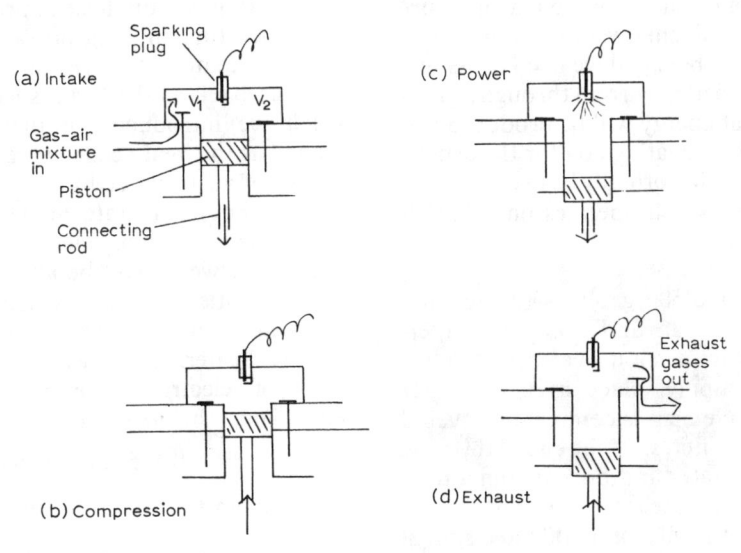

Fig. 19.5. 4-stroke petrol engine

Exercise 19

1 Show that heat is a form of energy (i) by describing an example where heat is produced as a result of work done, (ii) by describing an example where work is done because heat has been supplied. (C)

2 Describe TWO simple experiments you could perform to show that heat is a form of energy.

Describe, with the help of a diagram, the action of an internal combustion engine. Explain how the energy of the fuel is converted into mechanical work. (C)

3 Calculate the rise in temperature resulting from water falling down a waterfall 100 m high. (g = 9·8 m/s² specific heat capacity of water = 4200 J/kg°C.)

20 Measuring Quantities of Heat Energy

In the last chapter we described how the experiments of Joule established that the specific heat capacity of water (the amount of energy needed to raise the temperature of 1 kg of water by 1°C) is 4200 J. In this chapter we shall be concerned with measuring the specific heat capacity of other substances and with measuring quantities of heat energy in general.

We shall find that the most convenient way of supplying heat energy to a liquid is by electrical means. We shall not discuss how we may measure quantities of electrical energy until later in this book, but for the time being all we need to assume is that if we pass a constant current through a given coil of wire then heat energy will be produced in it at a constant rate. Our heating coil for the experiments in this chapter will normally be in the form of an immersion heater which operates on a 12 volt supply of electricity.

Experiment 20.1. Place 500 g of cold water in a light aluminium saucepan and note its temperature. Connect the immersion heater to the low voltage electricity supply, place it in the water so that the heating element is completely covered and switch on for 5 minutes. At the end of this time switch off, stir the water and note the final temperature of the water. Repeat the experiment with 1 kg of water and with 500 g of methylated spirits.

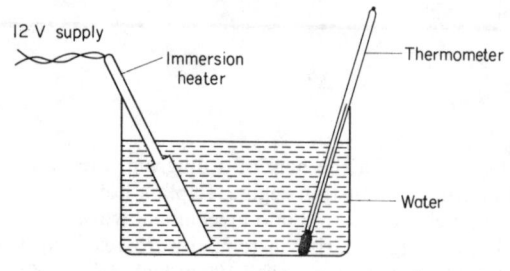

Fig. 20.1. Specific heat capacity of water

A typical set of results might be as follows:

0.5 kg of water, rise in temperature = 7.0°C

1.0 kg of water, rise in temperature = 3.5°C

0.5 kg of meths, rise in temperature = 12.0°C

Remember that we must assume that in each of these experiments we were supplying the same amount of electrical energy and we must therefore be producing the same amount of heat energy. If we consider the first two of these results we see that if we double the mass, we halve the temperature rise. The first and third of the results indicate that it requires more heat to produce a certain rise in temperature in 1 kg of water than to produce the same temperature rise in 1 kg of meths. This leads us back to the term 'specific heat capacity' which we introduced in the last chapter. It is defined as the heat needed to raise the temperature of 1 kg of a substance by 1°C.

If we knew the rate at which the immersion heater was consuming electrical energy in Experiment 20.1, we should be able to find the specific heat capacities of water and meths from the results. Supposing, for example, that we are told that it is a 50 W heater. This means that it is converting 50 J of electrical energy every second into heat. Thus the heat supplied in 5 minutes is:

$$50 \times 5 \times 60 = 15\,000 \text{ J}$$

Hence from the result of the second experiment:

1 kg of water is heated through 3.5°C by 15 000 J

and

1 kg of water is heated through 1°C

by $\frac{15\,000}{3.5}$ J

= 4300 J

This by definition is the specific heat capacity of water. In the last chapter we indicated that the value is, in fact, closer to 4200 J/kg°C. From the result of the third experiment we may say:

0.5 kg of meths is heated through 12.0°C by 15 000 J

and

$$1 \text{ kg of meths is heated through } 12 \cdot 0\,°C \text{ by} \\ 15\,000 \times 2 \text{ J}$$

and

1 kg of meths is heated through 1 °C

$$\text{by } \frac{15\,000 \times 2}{12} \text{ J}$$

$$= 2500 \text{ J}$$

Thus the specific heat capacity of meths is about 2500 J/kg °C.

We shall see later in this book how it is possible to measure the electrical energy supplied to an immersion heater. For the time being it is sufficient to connect the heater in circuit with a joulemeter during the experiment. This is an instrument very like the normal domestic electricity meter which you have at home to record for the electricity company the energy you consume. The joulemeter will register the number of joules supplied to the heater and, in this way, Experiment 20.1 becomes an experiment to measure the specific heat capacity of a liquid.

Specific Heat Capacity of Metals

Experiment 20.2. A very similar experiment to 20.1 may be used to determine the specific heat capacity of a metal, only in this case it is convenient to use a specially prepared metal specimen for the purpose. It will be a cylinder, drilled with two holes, one to take the immersion heater and the other the bulb of a thermometer. (Fig. 20.2).

It is quite important that both heater and thermometer make a good 'thermal contact' with the block so they should be a good fit in their holes. A little oil in the thermometer hole will help in this respect also. The calculation is made much simpler if the metal sample is of a mass 1 kg. Note the initial temperature of the block. Now connect the heater to a joulemeter and the 12 V electricity supply, and switch on for 5 minutes. Record the highest temperature indicated by the thermometer and the electrical energy supplied as indicated by the joulemeter.

The following is a typical set of results for such an experiment:

Mass of aluminium block	1 kg
Initial temperature	20 °C
Final temperature	34 °C
Energy supplied	15 000 J

Thus:

1 kg of aluminium is heated through

$$14\,°C \text{ by } 15\,000 \text{ J}$$

and

1 kg of aluminium is heated through

$$1\,°C \text{ by } \frac{15\,000}{14} \text{ J}$$

$$= 1100 \text{ J}$$

Thus, from our experiment, the specific heat capacity of aluminium is 1100 J/kg °C.

You will note that this value is only about one quarter of that for water. It is a fact that liquids generally have much higher specific heat capacities than solids. This has many important practical consequences. For example, in the field of climate, the fact that the sea heats up and cools down much more slowly than the land, causes places near the sea to have much more moderate climates than places far inland.

You will realise that the experiments we have described in this chapter are not very accurate ones. You may, in fact, be able to think of ways to improve their accuracy. One suggestion would be to wrap the metal block in a cloth jacket. This would help to prevent heat produced by the immersion heater from escaping to the surrounding air.

The Heat produced by Fuels and Foods

Many experiments of this kind are carried out in industrial laboratories today to investigate the heat produced by burning fuels. For example, most coal burning power stations will employ a quality control scientist, part of whose job it is to find the calorific value of the fuel supplied. He will carefully select a sample from each consignment of

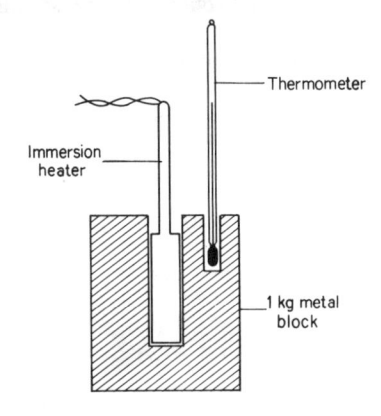

Fig. 20.2. Specific heat capacity of a metal

coal, grind it to a powder and burn it in a 'bomb calorimeter' (Fig. 20.3). The fuel is ignited by an electric heating coil in a pressure vessel which contains also oxygen gas. The heat produced causes a rise in temperature of the calorimeter which contains water. If the mass of the water and of the calorimeter and fittings are known, the amount of heat produced may be calculated.

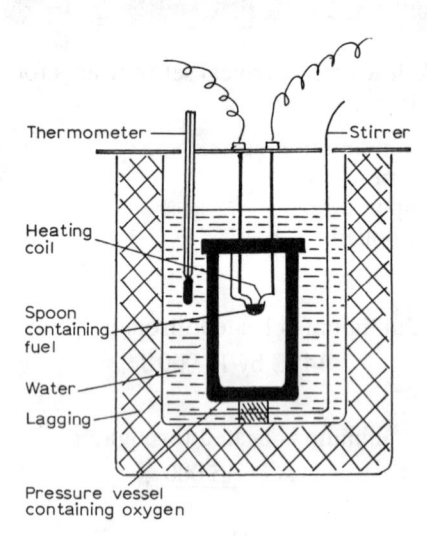

Similar techniques are also used to find the calorific values of gas and other fuels. The gas supply company, for example, is responsible for ensuring that the gas supplied to consumers keeps up to a certain specified calorific value. This may be about 20 MJ/m^3. 1 MJ is one million joules. (See also Chapter 26.) If you look at a gas bill you will probably find that the charge has been calculated in terms of the amount of heat energy consumed. However the gas meter measures the volume of gas used. The calorific value of the fuel must therefore be known to calculate the cost.

The food eaten by humans and animals may be regarded to some extent as fuel and the chemical changes involved in digestion result in the production of heat. The calorific value of foods then is of interest, particularly to dieticians. These are people whose job it is to advise doctors and their patients on matters of diet. Although scientists generally have agreed to measure heat energy in joules, dieticians tend to still measure the calorific value of foods in 'calories' or more usually 'kilocalories'. A kilocalorie is the heat needed to raise the temperature of 1 kg of water by 1 °C. We have already seen that this is equal to 4200 J.

Exercise 20

1 Define specific heat capacity.

An electric heater is rated at 12 V, 24 W. It is fitted into a metal block having a mass of 1 kg and a specific heat capacity of $4 \cdot 0 \times 10^2$ J/kg °C.

Calculate
 (i) the number of joules produced by the heater in one second,
 (ii) the number of joules produced by the heater in five minutes,
 (iii) the greatest possible temperature rise in five minutes. (C)

2 Define 'specific heat capacity'.

A small immersion heater is placed in 510 g of water and raises its temperature 20 °C in 5 min. The same heater is then placed in 400 g of turpentine and produces a rise of 34 °C in 3 min. Find the specific heat capacity of turpentine. (L)

3 A gas water heater raises the temperature of 100 kg of water from 10 to 60 °C in 1 hour. The cost of gas is 1 p for 10 MJ, and 75% of the gas used actually heats the water. Find the cost of using the heater for one hour. (The specific heat capacity of water is 4200 J/kg°C.)

4 Describe how you would find the specific heat capacity of a metal.

It was found that during an experiment that when 1·5 g of coal was burnt the heat produced raised the temperature of 500 g of water from 15 °C to 35 °C. Find the calorific value of the coal.

21
Change of State and Latent Heat

We pointed out in the last chapter that supplying heat to boiling water produces no temperature change. Instead a 'change of state' (from liquid to vapour) is caused. The heat which must be supplied to a boiling liquid to change it into a vapour is known as *latent heat*. When the reverse occurs and steam condenses to water, this latent heat is given out. We can show that latent heat is also involved in a change from solid to liquid and the reverse.

Experiment 21.1. Place some naphthalene in a test-tube mounted in a clamp (Fig. 21.1). Heat it very gently until the naphthalene melts and then put in a thermometer. Allow the apparatus to cool and record the temperature every minute until a few minutes after the naphthalene has become completely solid again. Plot a graph of temperature against time.

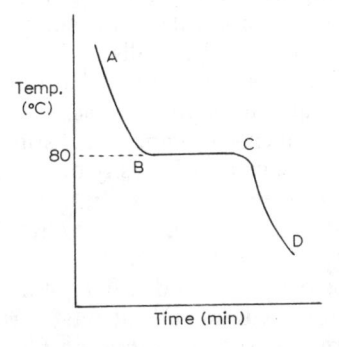

Fig. 21.2. Cooling curve for naphthalene

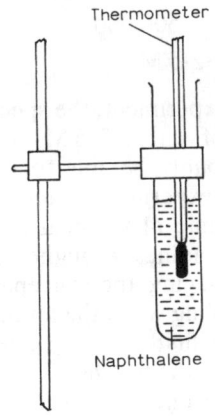

Fig. 21.1. Experiment to plot a cooling curve for naphthalene

Fig. 21.2 shows the sort of graph you will probably obtain. The portion AB of the graph shows the temperature of the liquid naphthalene falling as it loses heat to the surrounding air. Then at B the graph levels out and the temperature stops falling. Since the naphthalene is still well above room temperature we must not suppose that it has ceased to lose heat. Between B and C then it is losing latent heat as it solidifies. At C when all the naphthalene is solid again, loss of heat causes a fall in temperature once more.

This experiment, besides showing the existence of latent heat, provides a simple but reliable method of finding the *freezing or melting point* of a substance, that is, the temperature at which the change of state between liquid and solid occurs. If the experiment were carried out with a substance like paraffin wax which gradually softens before becoming liquid, no flat portion like BC on the graph would be obtained.

The reason why heat must be supplied in order to bring about melting or vaporization may be appreciated by considering the nature of solids, liquids and gases. The atoms in a solid are closely packed together, as in Fig. 21.3(a). Although

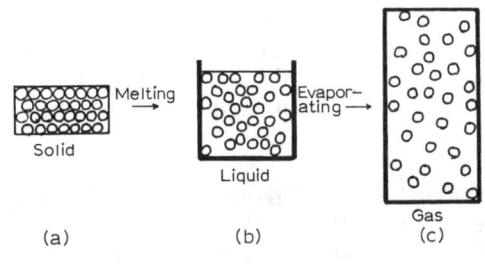

Fig. 21.3.

they are vibrating they do not move any substantial distance and they are held together by the forces between them (possibly electrostatic forces, see Chapter 47). In a liquid (Fig. 21.3(b)) the atoms or molecules are relatively free to move about, bounded for the most part only by the liquid surface. Thus to change solid to liquid, work has to be done to overcome the forces of attraction between atoms in the solid state. This work is done by supplying heat energy. In a gas (Fig. 21.3(c)) the atoms or molecules are completely free, bounded only by the walls of the containing vessel. Hence more heat energy has to be supplied to change liquid to vapour or gas.

An experiment designed to stimulate this change of state can fairly easily be set up.

Experiment 21.2. To show a Connection between Energy supplied and the Movement of Particles.
Connect a large cylindrical tap-funnel via the stem to a foot-bellows and put ants' eggs, pith balls or rice krispies into the funnel to fill about one-third of it. The particles are stationary, representing the molecules in a solid. Now open the tap and get a pupil to start pumping the bellows. The particles will start moving around just as the molecules of a solid do when heat is applied and liquefaction takes place; the odd one or two may even leave the funnel. Increase the rate at which air is pumped in until all the particles are in rapid motion as in a gas and many will escape into the air. The pupil working the bellows will be very conscious of the energy he has put into the experiment and the result of this input of energy will be obvious to the class.

Latent Heat of Vaporization and its Measurement

As always, we would like to be able to measure the quantities of heat involved in these changes of state. Before we can do this we must give a more precise definition of latent heat than we have so far. In particular we must specify the mass of substance we are melting or vaporizing, and we must exclude heat which causes changes of temperature. Thus *Specific Latent Heat of Vaporization* is defined as the amount of heat needed to change 1 kg of liquid at the boiling point into vapour without change of temperature. It will be measured in kJ/kg. We shall now describe an experiment by which it is possible to find the value of this quantity for water.

Experiment 21.3. Determination of the Specific Latent Heat of Vaporization of Water.
Place about 50 g of water in a light aluminium saucepan and heat with an immersion heater of known power until boiling is reached. If you use the standard 12 V 50 W heater, this will take a long time so it is better to use a mains (240 V) heater of about 500 W power. When boiling is reached, switch off and remove the heater. Now quickly weigh the saucepan and its contents on a lever balance, replace and switch on the immersion heater and boil for a set period of time, say 5 minutes. Finally switch off and remove the heater and reweigh the saucepan.

A typical set of results is shown below:

Power of heater = 500 W
Time of boiling = 300 s
Initial mass of saucepan
 + water = 530 g
Final mass of saucepan
 + water = 470 g

Thus, during our experiment, 500×300 J of energy are supplied to the boiling water and cause the evaporation of 60 g of it. Now we have defined specific latent heat of vaporization as the heat energy needed to vaporize 1 kg of liquid at the boiling point. In this case:

60 g are vaporized by 500×300 J and 1 g is vaporized by $\dfrac{500 \times 300}{60}$ J and 1 kg is vaporized by

$$\dfrac{500 \times 300}{60} \times 1000 \text{ J}$$

$$= 2\,500\,000 \text{ J}$$

$$= 2 \cdot 5 \text{ MJ}$$

Thus from our experiment, the specific latent heat of vaporization of water is $2 \cdot 5$ MJ/kg.

You will probably be able to see some of the reasons why this experiment is not an accurate one. For example, some of the heat produced will be lost to the surroundings throughout. This could be minimised by lagging the saucepan. Also, there will be some cooling when the immersion heater is removed for the first weighing and so when it is switched on again the first heat provided will have to bring the water once again to boiling point. This explains the suggestion that the weighing should be carried out quickly.

The generally accepted value for the latent heat of vaporization of water at 100 °C is $2 \cdot 27$ MJ/kg. We shall see later that evaporation and boiling may take place at temperatures other than 100 °C and the latent heat will vary with the temperature.

The value for latent heat of vaporization of water which these experiments give has at least one important practical consequence. If we consider 1 kg of steam at 100 °C first condensing and then cooling to normal room temperature (say 20 °C), 2·2 MJ approximately are given out during condensation and only 0·34 MJ in the subsequent cooling. This shows how much more effective steam is for heating things than hot water. Steam heating is used in industry. You may have come across it elsewhere. It is used, for example, in coffee shops for heating milk quickly. We can also see how much more serious are scalds produced by steam than those produced by boiling water.

Evaporation and Boiling

If you leave a little water in a shallow dish for a day or two you will find that some or all of the water disappears. We should say that the water had evaporated or changed into a vapour. So this change of state does not occur only when the water is boiling or at 100 °C. In fact it can occur at any temperature. At any temperature some of the water molecules will be travelling sufficiently fast to burst through the surface of the liquid and become molecules of water vapour. Evaporation then is something which takes place at the surface of a liquid only and it occurs at any temperature. Boiling, on the other hand, consists of a change from liquid to vapour taking place through the body of a liquid and it occurs, provided the atmospheric pressure is normal, at a particular temperature.

Cooling by Evaporation

Although evaporation may take place 'spontaneously', without an external heat source, the latent heat must still be supplied to bring about the change of state. Thus when evaporation does take place, the liquid remaining and its surroundings are cooled. Cooling by evaporation is a most important effect which can be demonstrated by the following simple experiments.

Experiment 21.4. Soak a piece of blotting paper in water and wrap it around the bulb of a thermometer. Wave the thermometer gently in the air. You may observe a drop in reading of several degrees. To be certain that this is really a case of cooling by evaporation of water from the blotting paper and not caused by the water being below room temperature, it may be necessary to leave the water standing in the room for some hours before use.

Experiment 21.5. Place a little ether in a metal vessel and stand it on a drop of water on the bench (Fig. 21.4). Cause a draught of air to blow across the ether surface. Quite quickly you will observe that the water under the vessel has frozen and it is stuck to the bench. Ether, which is in any case volatile, has been made to evaporate rapidly by the draught over its surface. It has extracted latent heat from its surroundings including the water which was cooled to freezing point.

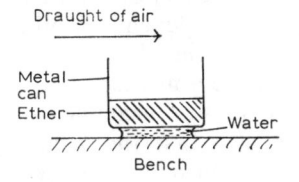

Fig. 21.4. Cooling by evaporation

Cooling by evaporation is the basis of almost all systems of refrigeration. Fig. 21.5 shows a typical small domestic refrigerator. A gas which is easily liquefied, (usually called 'freon', but ammonia has been used) is in the sealed cooling system of the refrigerator. The gas (or liquid) circulates in the direction indicated by the arrows on the diagram. Cooling is caused in the freezer compartment by the evaporation of liquid freon which must extract its latent heat of vaporization from the contents of this compartment. The vapour thus produced passes on to a motor driven pump which compresses it into the condenser section. Here the vapour is liquefied. Some heat is generated in this process, but since this part of the circulation is outside the box of the refrigerator and fitted with cooling fins, the heat is dissipated to the atmosphere. The liquid produced in the condenser passes

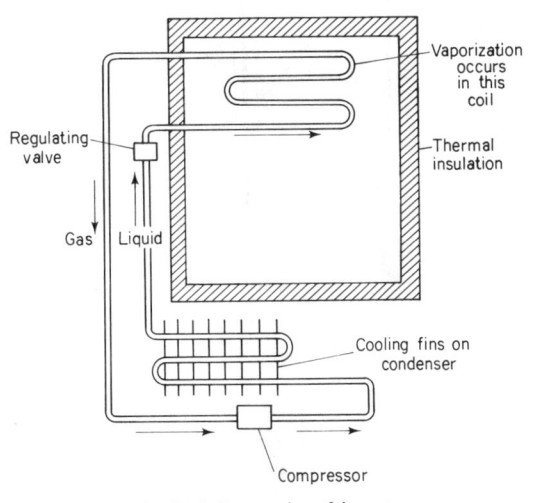

Fig. 21.5. Domestic refrigerator

111

on to the evaporation coil again via a regulating valve. This makes it possible for a high pressure to be maintained in the condenser while there is a much lower pressure in the evaporation coil.

Water Vapour in the Atmosphere

The fact that evaporation is taking place all the time from the surfaces or rivers, lakes and oceans means that there is always a certain amount of water vapour in the atmosphere. The precise quantity of vapour will depend on other atmospheric conditions. However, for many purposes it is important to have a measure of the 'humidity' of the atmosphere. For example, some industrial processes require that the humidity should be controlled. The most common used instrument for measuring humidity is the wet and dry bulb hygrometer (Fig. 21.6). It consists of two ordinary thermometers mounted side by side. One of the thermometers has a piece of muslin wrapped around the bulb and dipping into a small reservoir of water. Water will be continually evaporating from this 'wet bulb' and producing cooling of it. The rate of evaporation will depend on how much water vapour there is already in the atmosphere and also on the air temperature. Thus a large difference between the temperatures indicated by the two thermometers shows that evaporation is taking place quickly and the air is fairly dry. A small difference tells us that the air is already rather damp. No difference at all between the two readings indicates that no evaporation is taking place as the air is already saturated with water vapour. A wet and dry bulb hygrometer is normally supplied with a set of tables, from which, knowing the dry bulb reading and the difference in readings, one can read off the 'relative humidity'. This is a percentage figure, 100% relative humidity meaning that the atmosphere is saturated. This will happen, of course, when condensation is taking place on walls, windows and any other slightly cool surfaces.

Atmospheric Pressure and Boiling Point

Experiment 21.6. Boil some water in a stout, round bottomed flask. When it is boiling steadily, remove the heat and when the boiling has subsided cork down the flask. Now hold the flask under a running cold water tap (Fig. 21.7). You will see that the water starts to boil again quite vigorously and continues to do so.

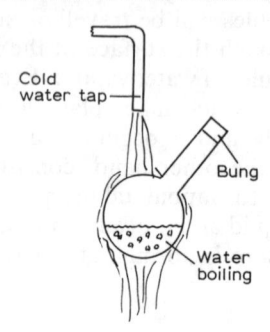

Fig. 21.7. Boiling under reduced pressure

When the water in the flask first boiled some of the air in it was driven out by the steam formed. The flask was then corked, so that the expelled air could not return, and cooled. Some of the steam in the flask condensed and left a partial vacuum inside. Under this reduced pressure the boiling point of water is considerably less than 100°C so the water started to boil again.

The boiling points of all liquids depend on the pressure over the surface. At normal atmospheric pressure water boils at 100°C.

At half this pressure (about 38 cm of mercury) it boils at 82°C. This variation of boiling point with pressure has many applications. For example, mercury at normal atmospheric pressure boils at 357°C and this would normally be the upper limit for any mercury thermometer. However by having a gas at a pressure well above 76 cm of mercury over the mercury in a thermometer it is possible to extend the range upwards. If the pressure is 8 atmospheres (8 × 76 cm of Hg) the boiling point

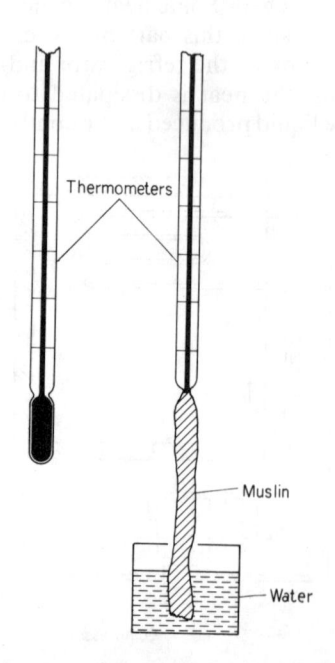

Fig.21.6. Wet and dry bulb hygrometer

is raised to 500 °C. You may be familiar with the use of pressure cookers. These are large heavy saucepans fitted with a tight fitting lid, so that when the contents start to vaporize, an increased pressure is built up inside. Thus the boiling point of water is raised. In this way food may be cooked at a higher temperature and so more quickly than usual.

Specific Latent Heat of Fusion and its Measurement

Specific latent heat of fusion is defined as the amount of heat needed to change one kilogramme of solid at the melting point into liquid without a change in temperature. The following experiment is one to measure this quantity for ice changing to water.

Experiment 21.7. Set up the apparatus shown in Fig. 21.8. It consists of a funnel containing ice in small pieces packed around an immersion heater. Leave the apparatus for a measured period of time, say 5 minutes, and collect the water that comes from the melting ice. Note the volume (or the mass) of this water. Now switch on the immersion heater and collect the water emerging in a further 5 minutes.

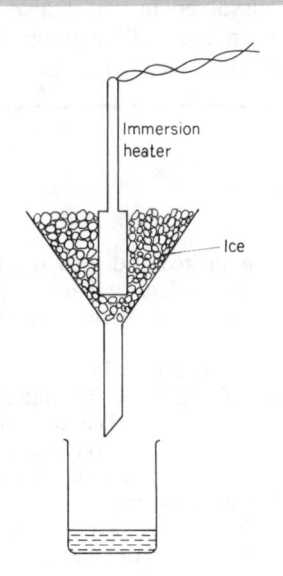

Fig. 21.8. Specific latent heat of fusion experiment

The following might be the results of such an experiment:

Power of immersion heater = 50 W
Water collected in first 5 minutes = 12 g
Water collected in second 5 minutes = 62 g

Now the ice melting in the first 5 minutes gets its latent heat of fusion simply from the surroundings. During the second 5 minutes heat is supplied by the surroundings and by the immersion heater. We may thus say that the heat generated in the heater melts $62 - 12 = 50$ g of ice. The 50 W heater produces 50 J of heat energy each second or 50×300 J in 5 minutes.
Thus:

50 g of ice are melted by 15 000 J

and

1 g of ice is melted by $\dfrac{15\,000}{50}$ J

and

1 kg of ice is melted by $\dfrac{15\,000 \times 1000}{50}$

$= 300$ kJ

Thus the specific latent heat of fusion of ice is 300 kJ/kg, from the results of this experiment. The generally accepted value is 336 kJ/kg.

The Effect of Pressure on Melting Points

We have discussed how the boiling point of a liquid varies with the pressure over its surface. A similar, but much smaller variation in melting point is demonstrated in the following experiment.

Experiment 21.8. Mount a large slab of ice as in Fig. 21.9 and place over the top of it a thin copper wire with weights attached to each end. Over a period of a few hours you will observe that the copper wire passes right through the slab but leaves it as one block. The ice re-freezes after the wire has passed.

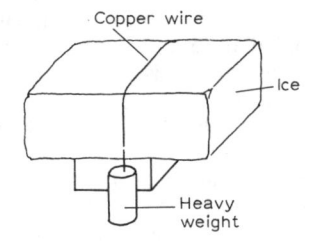

Fig. 21.9. The regelation experiment

We must call upon several of the conclusions we have reached in this and previous chapters to explain this rather surprising effect which is called *regelation*. First, in view of the large downward force of the weights and the small area under the wire over which this force is acting, we may say that there is a large pressure on the ice under the wire. This pressure *lowers* the melting point of the ice in

113

this region. The depression of the freezing point will probably be less than a centigrade degree, but if the ice is at 0°C and the freezing point is −1°C, then the ice will melt. Thus the wire is allowed to move a small way into the ice. The ice immediately below the wire will continue to melt, but as it does so it must extract its latent heat of fusion from its surroundings. It will extract heat through the wire from the water above it (Fig. 21.10) causing this water to freeze again. It is important then that the wire is a good conductor of heat. If you tried the experiment with say a nylon thread, you would probably not succeed in leaving the block in one piece.

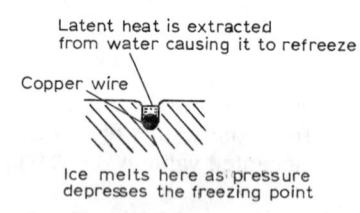

Fig. 21.10. Explanation of the regulation experiment

This depression of the freezing point and consequent melting of ice which is under a high pressure has an application in ice skating. The whole of the skater's weight is acting on the ice through the knife edge blades. The ice beneath these blades is therefore under a high pressure. This causes some melting as in the regelation experiment and a film of water is formed between the ice and the blades. The water acts as a lubricant and enables the skater to move smoothly over the ice.

A high pressure does not lower the freezing point of all substances. Ice contracts when it melts, consequently an increase of pressure may be regarded as 'helping' the melting process so a depression of the freezing point does occur. For those substances which expand when melting occurs, the pressure is opposing melting and so the freezing point is made higher.

Effect of impurities on Freezing and Boiling points

During a hard winter hundreds of tonnes of common salt are spread on British roads after snowfalls. Sodium chloride and many other salts have the effect of lowering the freezing point of water, and so cause ice and snow to melt. An ice-salt mixture will produce temperatures well below 0°C since as the ice melts it absorbs its latent heat of fusion from its surroundings so producing cooling. The ice and salt constitute what is called a freezing mixture.

Impurities dissolved in water also have a substantial effect on the boiling point. A saturated brine, for example, boils at 125°C.

Exercise 21

1 When 500 g of water, originally at 34°C are heated, boiling-point is reached in 4 min. but the kettle is left for a further 1 min. before being switched off. It is then found that 15 g of water have been lost as steam. What value does this indicate for the specific latent heat of steam?
[You may neglect heat losses to the kettle and the surroundings.] *(O)*

2 What is (i) evaporation, (ii) boiling?
Describe an experiment to show that the temperature at which water boils is lowered when the pressure on the surface of the water is lessened. *(L)*

3 State what is meant by the *melting-point* of a solid.
Describe how you would find the melting-point of naphthalene by plotting a cooling curve. Draw the shape of the curve you would expect to obtain and state what each part of the curve represents.
What is the effect of an increase of pressure on (a) the melting-point of ice, (b) the boiling-point of water? Describe briefly a practical application of boiling water at an increased pressure. *(C)*

4 Ice at 0°C is melted and then warmed at 10°C. Describe the changes in density that take place.
Define the terms *specific latent heat of fusion of ice* and *joule*.
A calorimeter, fitted with a 75 W heating coil and a thermometer, all of negligible thermal capacity, contains exactly 1 kg of water. The calorimeter and its contents are cooled to 0°C, surrounded by a felt jacket and 0.20 kg of ice at 0°C is added. The heater is then switched on. After gently stirring with the thermometer for 14 minutes, it is found that the last particles of the ice have just melted and the thermometer still reads 0°C. Calculate the specific latent heat of fusion of ice from this experiment in kJ/kg.
Why was a felt jacket employed?
The answer obtained is a little lower than it should be. Suggest what modification in procedure might have ensured a more accurate result.
What temperature rise will occur if the heating coil is switched on for a further period of 7 minutes?
[Specific heat capacity of water = 4·200 kJ/kg K.] *(L)*

5 A calorimeter contains 200 g of a liquid of specific heat 3360 J/kg deg. C and is at a temperature of 20 °C when 30 g of ice at 0 °C are added to it. Find the lowest temperature reached, neglecting any heat losses by the calorimeter and by the surroundings.

6 An iron ball of mass 100 g is placed in a furnace and allowed to remain there until its temperature is the same as that of the furnace. It is then removed and quickly placed in a dry hole in a large block of ice at 0°C. The mass of ice melted is 120 g. What is the temperature of the furnace? Assume that the specific heat capacity of iron is 462 J/kg °C and the specific latent heat of fusion of ice is 336 kJ/kg.

22
The Air We Breathe

Air, the commonest substance known to man, was considered by the ancient philosopher to be one of the four fundamental elements, the others being earth, fire and water. As time progressed and man's store of knowledge increased, he became more and more interested in the exact nature of the air and in those reactions for which air seemed essential e.g. burning.

During the seventeenth and eighteenth centuries the idea developed that all substances contained something in common which they gave up *to* the air on burning, but it was proved by Lavoisier that, in fact, they took something *from* the air in the process, viz. oxygen. It had been suggested in 1684 by Mayow that air contained at least two gases, but it was left to Lavoisier to show that one of these two was essential for combustion and that it occupied about one-fifth of the atmosphere. He suggested the name, *oxygen*, for it. By the end of the eighteenth century quite a number of different gases had been prepared and were considered to be various kinds of air, e.g. inflammable air (hydrogen), active air (oxygen), inactive air (nitrogen) and fixed air (carbon dioxide).

The greater part of the air is nitrogen, but in 1785 Henry Cavendish, a son of the then Duke of Devonshire, carried out several experiments on air in which he removed the oxygen, nitrogen and carbon dioxide; always some part of the air remained. His notebook recording this was put aside for more than a century and only in 1894 was the presence of other gases definitely recorded. These gases, argon, neon, krypton and xenon are called the rare gases because they occur in such small amounts in the air. The volume of xenon in one cubic kilometre of air is only 6 m^3. Oxygen makes up about 20% of the air by volume and is, of course, essential to life on earth. You will have learnt already that carbon dioxide is continually being produced by the burning of carbon-containing fuels in fires, in heating equipment, in internal combustion engines and in jet engines. Every time we breathe out we add to the amount of carbon dioxide in the air. Most of this gas is taken in by growing plants or is absorbed in the sea, but there is always a certain amount present in the atmosphere. This volume, about 0·03–0·04%, is small, but plays a significant part in our lives (see p. 151). The constituent of the air which varies most is the water vapour content. On damp or muggy days that is fairly high; on dry days it is low. It varies from day to day and from place to place. In Britain it averages about 1%, but in hot desert areas it may be only one-tenth of this, while in humid districts (such as equatorial rain forests) it may be as much as 6–7%. Because of this variation, figures relating to the composition of air are usually given referring to dry air; a typical analysis would be:

	% by volume
nitrogen	78
oxygen	21
rare gases	<1
carbon dioxide	0·03

Let us look at some of the facts that support the idea that air is a mixture of gases and not a compound of some of them.

(i) The gases in the air can easily be separated and if the different gases comprising it are mixed together in the proportions shown in the table, then a substance of exactly the same characteristics as air is formed.

(ii) The properties of a compound are usually very different from those of the elements from which it is made, but air behaves as though it is a mixture of oxygen and nitrogen, e.g. oxygen will relight a *glowing* splint and nitrogen will extinguish a *burning* one, whereas air will do neither; it will allow a *burning* splint to go on burning.

(iii) If air is shaken with water more oxygen dissolves than does nitrogen; if this air is now boiled out of the water it is found to contain about 35% of oxygen.

(iv) All compounds can be given a formula, but none can be found to fit in with the proportions of the gases in the air, especially as the actual composition is not constant but varies slightly according to physical conditions.

(v) If a formula N_4O is adopted to fit in with the four volumes of nitrogen to one of oxygen present such a compound would have a molecular mass of $(4 \times 14) + 16 = 72$. The relative vapour density of a gas is half its molecular mass, in this case 36. However, the density of a *mixture* of four volumes of nitrogen with one of oxygen would be $72/5 = 14 \cdot 4$, the actual density of air.

Water vapour. If air is drawn through a U-tube containing a deliquescent or hygroscopic substance such as anhydrous (fused) calcium chloride or silica gel the water vapour is removed. It can be obtained separately by distilling the solid which has absorbed it.

Carbon dioxide. Carbon dioxide is removed when air is passed through a sodium of potassium hydroxide in a Drechsel or Woulfe bottle.

$$2KOH + CO_2 \rightarrow K_2CO_3 + H_2O$$

It can be re-obtained from the carbonate formed by the addition of a dilute acid.

$$K_2CO_3 + H_2SO_4 \rightarrow K_2SO_4 + CO_2 + H_2O$$

Oxygen. Air is passed through aqueous potassium hydroxide (why?) and then over heated copper (see Fig. 22.1) which changes to copper(II) oxide.

$$2Cu + (4N_2 + O_2) \rightarrow 2CuO + 4N_2$$

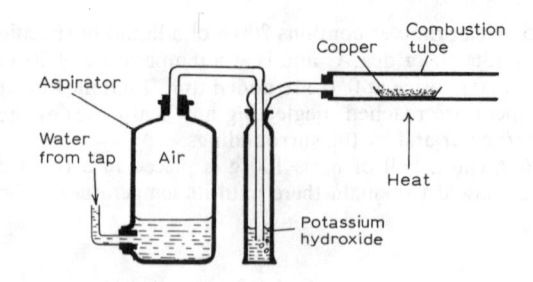

Fig. 22. 1. Removal of Oxygen and Carbon Dioxide from the air

The copper(II) oxide is then warmed in a stream of hydrogen or coal gas when it is reduced to copper and water is formed

$$CuO + H_2 \rightarrow Cu + H_2O$$

The water can then be electrolyzed and the oxygen collected.

Nitrogen. It is not possible to remove the nitrogen and leave the other gases, but it is easy to remove the others and have a sample of nitrogen left. The nitrogen not used in the experiment above (when air is passed over heated copper) can be collected in gas jars over water. It will, of course, contain the rare gases as well. There are other methods available. Phosphorus burned in air will leave nitrogen, but a convenient method by which the volume of oxygen in the air can be estimated at the same time is as follows.

Experiment 22.1. To Determine the Approximate Percentage by Volume of Oxygen in Air.

(a) Take a corked graduated tube (a eudiometer) and put in it a little pyrogallol dissolved in water. Add a pellet of sodium hydroxide, cork the tube again and invert several times. This solution absorbs oxygen and carbon dioxide, turning brown in the process. Note the volume. Leave for a few moments and then remove the cork with the tube held inverted under water in a tall jar. Adjust the tube so that the water is at the some level inside it as in the jar. The water will rise about a fifth of the way up the tube. Nitrogen will be left.

(b) A modification of the apparatus shown in Fig. 22.1 is shown in Fig. 22.2 using gas syringes. Place some copper metal (preferably 'pencil lead' type) in a hard glass or silica tube, c, filling the space at each end with a loose-fitting glass rod. Have 100 cm³ of air in (a) or (b) and connect the syringes to the tube. Heat the oxide and pass the air over it several times slowly, rotating the pistons a little as you do this. Continue until there is no

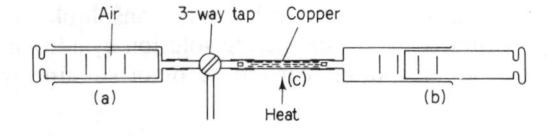

Fig. 22.2.

further contraction in the volume. Allow the apparatus to cool down, pass all the remaining gases into one syringe and read off the final volume. This should be about 79–80 cm³.

(c) Other methods which are reasonably satisfactory include
  (i) allowing yellow phosphorus to smoulder in a long tube of air,
  (ii) sprinkling iron filings in a long damp tube.

In both cases the tube is stood in a jar of water and the volume of air noted at the beginning and after some days. As mentioned already, all volumes must be measured after raising the tube so that the water levels inside and out are the same. Why do you think this is necessary?

Industrial Methods for Obtaining Oxygen and Nitrogen

Laboratory methods for making oxygen and nitrogen are not suitable on the industrial scale where the cost of the raw material is of vital importance. There is an abundant supply of both these gases in the air which is available free.

Although the apparatus used is complicated the principle of the method can easily be understood. Air is cooled down to about 200°C below zero, at which temperature both oxygen and nitrogen have become liquefied. The temperature is then allowed gradually to rise and at −196°C the liquid nitrogen begins to boil and comes off as a gas and can be collected. The temperature continues to rise and the liquid oxygen left begins to vaporize at −183°C and is collected as the gas. The rare gases present in the atmosphere can also be obtained if required by taking similar advantage of their different boiling points.

Oxygen and its Properties

The preparation of oxygen in the laboratory should already be familiar to you and will only be mentioned briefly here. It is usually obtained by heating 'oxygen mixture' which contains white potassium chlorate(V) and black manganese(IV) oxide, (manganese dioxide), the latter forming only about a fifth of the total bulk. The oxygen comes from the potassium chlorate(V).

$$2KClO_3(s) \rightarrow 2KCl(s) + 3O_2(g)$$

and the manganese(IV) oxide acts as a catalyst. Remember the definition of a catalyst.

A catalyst is a substance which alters the rate of a chemical reaction and itself remains unchanged chemically at the end.

If our final mixture of potassium chloride and manganese(IV) oxide is shaken with water and filtered the manganese(IV) oxide will be left on the filter paper.

Other methods of preparing oxygen (in a purer state) include heating potassium nitrate or potassium manganate(VII) (only a low temperature is needed for the latter) or allowing hydrogen peroxide to react with manganese(IV) oxide or potassium manganate(VII). The manganese(IV) oxide again acts as a catalyst, all the oxygen coming from the hydrogen peroxide.

$$2H_2O_2(aq) \rightarrow 2H_2O(l) + O_2(g)$$

The apparatus shown in Fig. 22.3 is suitable for the reaction between the peroxide and the manganate(VII). The filter flask contains about 30 cm³ of 20-volume hydrogen peroxide and the solution in the dropping funnel is made by pouring 20 cm³ of concentrated sulphuric acid into twice its volume of water and dissolving 3 g of potassium manganate(VII) in it when cooled. About one dm³ of gas can be obtained using these amounts.

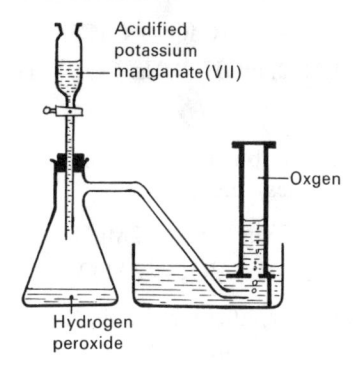

Fig. 22.3. Preparation of Oxygen

Oxygen is a vigorous supporter of combustion and allows many metals and nonmetals to burn in it forming oxides,* e.g.

magnesium $2Mg + O_2 \rightarrow 2MgO$
phosphorus $4P + 5O_2 \rightarrow 2P_2O_5$

*Details can be found in an elementary textbook such as *Basic Chemistry* by the author (with C. F. Dingle and R. A. Southcott).

The oxides formed from metals are termed 'basic oxides' and those from non-metals are 'acidic oxides'. Many compounds also burn in oxygen, particularly liquids or gases which are used as fuels, e.g. petrol, alcohol, butane (calor gas), turpentine and fats such as paraffin wax or lard. In all these cases the only products are carbon dioxide and water as can be shown using the apparatus set up as in Fig. 26.8.

Classification of Oxides

There are several different kinds of oxides, each with some distinctive properties by which it can be recognised. Some of these classes with examples are listed below.

Acidic Oxides

Acidic oxides are obtained when a nonmetal is burnt in air or in oxygen and most of them, but not all, dissolve in water to form acids.

$P_2O_5 + H_2O \rightarrow 2HPO_3$
(meta)phosphoric acid

$CO_2 + H_2O \rightarrow H_2CO_3$ carbonic acid

Some of these oxides, such as silicon dioxide (silica), SiO_2, are insoluble in water and the corresponding acid has to be obtained by a different method. These oxides all form salts by combining with bases or alkalis (soluble bases). Carbon dioxide joins up with calcium oxide or sodium hydroxide to form carbonates.

$CO_2 + CaO \rightarrow CaCO_3$
$CO_2 + 2NaOH \rightarrow Na_2CO_3 + H_2O$

Basic Oxide

Metals which burn in air form basic oxides (also called bases), e.g.

$2Mg + O_2 \rightarrow 2MgO$
$4Al + 3O_2 \rightarrow 2Al_2O_3$

Mostly, however, these oxides are produced by heating metal carbonates, nitrates or hydroxides strongly. Make a note of the following typical reactions.

$2Cu(NO_3)_2(s) \rightarrow 2CuO(s) + 4NO_2(g) + O_2(g)$
$ZnCO_3(s) \rightarrow ZnO(s) + CO_2(g)$
$Pb(OH)_2(s) \rightarrow PbO(s) + H_2O(g)$

The nitrates are made by dissolving the metal in dilute nitric acid and evaporating carefully to dryness or to crystallization point over a water bath; further heating causes decomposition. The carbonates and hydroxides, being insoluble, are precipitated from the nitrate solution by adding a solution of sodium carbonate or of sodium hydroxide.

$Zn(NO_3)_2(aq) + Na_2CO_3(aq) \rightarrow ZnCO_3(s) + 2Na_2CO_3(aq)$
or $Zn^{2+}(aq) + CO_3^{2-}(aq) \rightarrow ZnCO_3(s)$
white

$Cu(NO_3)_2(aq) + 2NaOH(aq) \rightarrow Cu(OH)_2(s) + 2NaNO_3(aq)$
or $Cu^{2+}(aq) + 2OH^-(aq) \rightarrow Cu(OH)_2(s)$
pale blue

Note carefully that the oxides of sodium and potassium can NOT be produced by heating the nitrates, carbonates or hydroxides.

The few basic oxides which are soluble in water (sodium, potassium, lithium and barium oxides with calcium and magnesium oxides showing a slight solubility) form alkaline solutions of the corresponding hydroxides—these also are called bases.

$Na_2O + H_2O \rightarrow 2NaOH$
$BaO + H_2O \rightarrow Ba(OH)_2$

A basic oxide is best defined as one which reacts with an acid to form a salt and water only.

$CuO + H_2SO_4 \rightarrow CuSO_4 + H_2O$
$MgO + 2HCl \rightarrow MgCl_2 + H_2O$

Amphoteric Oxides

There are some oxides which show the properties of both acidic and basic oxides and these are called amphoteric ones (Gk. = both kinds). The most familiar example is zinc oxide which is soluble both in acid and in alkali.

With hydrochloric acid it forms zinc chloride:

$ZnO + 2HCl \rightarrow ZnCl_2 + H_2O$

and with excess sodium hydroxide solution it dissolves to form sodium zincate, Na_2ZnO_2

$ZnO + 2NaOH(aq) \rightarrow Na_2ZnO_2(aq) + H_2O$

(notice that the formula is made up of those of the two oxides, Na_2O and ZnO.)

If aqueous sodium hydroxide is added to a solution of zinc sulphate or chloride a precipitate of zinc hydroxide is first produced:

$ZnSO_4(aq) + 2NaOH(aq) \rightarrow Zn(OH)_2(s) + Na_2SO_4(aq)$
$Zn^{2+}(aq) + 2OH^-(aq) \rightarrow Zn(OH)_2(s)$

and this then dissolves in excess alkali to form a solution of sodium zincate.

$$Zn(OH)_2 + 2NaOH \rightarrow Na_2ZnO_2 + 2H_2O$$

Oxidation and Reduction

The terms 'oxidation' and 'reduction' are used quite frequently in connection with seemingly different reactions, so it is necessary to understand clearly exactly what they mean.

Simple changes in which oxygen is added to an element by combustion are obviously ones involving oxidation.

$$2Mg + O_2 \xrightarrow{\Delta} 2MgO$$

as is:

$$C + 2H_2SO_4(conc.) \xrightarrow{\Delta} CO_2 + 2SO_2 + 2H_2O$$

Reactions in which oxygen is removed from compounds are example of reduction, e.g.

$$2KClO_3 \xrightarrow{\Delta} 2KCl + 3O_2$$
$$CuO + H_2 \xrightarrow{\Delta} Cu + H_2O$$
$$PbO + C \xrightarrow{\Delta} Pb + CO$$

Oxygen and concentrated sulphuric acid (hot) are oxidizing agents; hydrogen and carbon are reducing agents. In some ways hydrogen can be looked upon as the chemical opposite of oxygen and so the removal of hydrogen from a compound is also a case of oxidation. Ammonia is oxidized to nitrogen when it is passed over heated copper oxide.

$$2NH_3 + 3CuO \rightarrow 3Cu + 3H_2O + N_2$$
(reduction / oxidation)

You should notice from this equation that oxidation and reduction go on together; the substance that brings about the oxidation is itself reduced. Conversely, the reducing agent in a reaction becomes oxidized, e.g. in the production of iron in the blast furnace the ore, iron(III) oxide, is reduced by carbon monoxide which is itself oxidized to carbon dioxide.

$$Fe_2O_3 + 3CO \rightarrow 2Fe + 3CO_2$$
(reduction / oxidation)

such actions are called REDOX reactions (REDUCTION—OXIDATION).

The change of iron(II) oxide → iron(III) oxide:

$$FeO \rightarrow Fe_2O_3$$

is oxidation. In the former there is one atom of oxygen to one of iron, but in the latter the proportion is $3:2$ or $1\frac{1}{2}:1$. This increase in the proportion of the non-metallic part of a compound is typical of oxidation and occurs also in such changes as:

$$FeS \rightarrow FeS_2 \quad (1:1 \rightarrow 2:1)$$
$$FeCl_2 \rightarrow FeCl_3 \quad (2:1 \rightarrow 3:1)$$

and these also are regarded as examples of oxidation. The last change is brought about by chlorine:

$$2FeCl_2 + Cl_2 \rightarrow 2FeCl_3$$

and thus chlorine is an oxidizing agent.

Similarly, a decrease in the proportion of the non-metallic part is reduction: e.g.

$$CuO \rightarrow Cu$$

Here the proportion of oxygen to copper changes from $1:1$ to $0:1$, hence reduction. The change from phosphorus pentachloride to the trichloride:

$$PCl_5 \rightarrow PCl_3$$

is also reduction although there is no 'metallic' part in these compounds. The change, however, is similar to that in:

$$FeCl_3 \rightarrow FeCl_2$$

In the preparation of chlorine by the action of manganese(IV) oxide on hydrochloric acid:

$$MnO_2 + 4HCl \rightarrow MnCl_2 + 2H_2O + Cl_2$$

the acid loses hydrogen to form chlorine and so is oxidized. The manganese(IV) oxide is reduced.

In the above cases the valency of the phosphorus changes from 5 to 3, that of the iron from 3 to 2 and that of the manganese from 4 to 2, a reduction in each case.

Notice that the loss of water is neither oxidation or reduction.

Definition. Oxidation is a change in which the proportion of the non-metallic part of a compound is increased and reduction is a change in which this proportion is decreased.

In modern theory oxidation and reduction are connected with the change in the number of electrons attached to a particular atom. An atom of iron has 26 planetary electrons, but on joining with chlorine to form iron(II) chloride this number is reduced to 24 and in iron(III) chloride it is only 23. This *loss of electrons* is typical of oxidation. Conversely, sulphur has 16 electrons attached to each atom, but in the compound sodium sulphide it has received 2 more from the two sodium atoms and now has 18. A *gain of electrons* represents reduction.

Remember this in the following way:
Loss of **E**lectrons is **O**xidation. (**LEO,** the lion)
Gain of **E**lectrons is **R**eduction. (**GER,** the lion's roar—just try shouting it!)

An oxidizing agent or oxidant can be defined as a substance which will remove electrons from another, whereas a reducing agent or reductant will supply these electrons. In the reaction

$$2FeCl_2 + Cl_2 \rightarrow 2FeCl_3$$

each chlorine atom removes one electron from each iron(II) ion which is thus oxidized to an iron(III) ion and itself is reduced to a chloride ion

$$Fe^{2+} - e^- \rightarrow Fe^{3+}$$
$$\text{and} \quad Cl + e^- \rightarrow Cl^-$$

A metal going into solution in water or in an acid (a chemical change in each case) loses electrons to become a positive ion, e.g.

$$Na - e^- \rightarrow Na^+ \text{ and } Mg - 2e^- \rightarrow Mg^{2+}$$

and so the metal is oxidized. You will meet many more examples.

Typical oxidizing agents are oxygen, chlorine, ozone, hydrogen peroxide, potassium manganate (VII), manganese(IV) oxide, most metal dioxides and nitrates. Typical reducing agents are hydrogen, carbon, hydrogen sulphide, carbon monoxide, sulphur dioxide, ammonia and most metal powders.

Most oxidizing agents will:
(i) liberate iodine from a solution of potassium iodide containing a little dilute sulphuric acid. If a piece of starch paper (or a little starch itself) is added this will go blue. Test papers containing both starch and potassium iodide are available for testing for oxidizing agents. Note that ordinary molecular oxygen will not give this result.
(ii) liberate chlorine from concentrated hydrochloric acid. Heating is sometimes necessary and the chlorine is recognised by its ability to bleach moist litmus paper.

Most reducing agents (but not molecular hydrogen) will remove the colour from potassium manganate(VII) solution acidified with dilute sulphuric acid or will turn acidified potassium chromate(VI) solution green.

Experiment 22.2. The Action of Reducing Agents.
Take three test tubes containing acidified potassium manganate(VII) solution and treat as follows, noticing any changes.

(a) Pass in sulphur dioxide. (If this gas is not available it can be produced in the test tubes by adding a little sodium hydrogensulphite.)
(b) Pass in hydrogen made by the action of zinc on dilute sulphuric acid in another tube or flask.
(c) Add a piece of granulated zinc.

You should find that the colour is removed by the sulphur dioxide and the zinc but not by the hydrogen. This is rather curious, but we can learn something from it. In case (c) the zinc acts on the acid present to produce hydrogen on the spot or 'in situ'. This hydrogen at the moment of its production is 'nascent', i.e. in the form of single atoms which may either join together to form molecules or act instead on the manganate(VII). The latter action requires the less energy and so takes place and the colour is removed. In (b) the atoms have already used up much of their energy in joining together to form molecules and can no longer decolorize the solution. Most elemental gases in the nascent state are much more active than in the form of molecules, e.g. nascent oxygen will bleach many coloured materials (this is produced when chlorine reacts with water and is the reason why moist chlorine, but not the dry gas, is a good bleaching agent.

The process of reduction is used in industry in the extraction of many metals from their ores, e.g. iron and zinc are obtained in this way.

$$Fe_2O_3(s) + 3CO(g) \rightarrow 2Fe(s) + 3CO_2(g)$$
$$ZnO(s) + C(s) \rightarrow Zn(s) + CO(g)$$

Many aluminium articles are given a coating of the oxide ('anodizing') to protect them from corrosion. This layer may be coloured to give an attractive appearance. Most paints have oil or water as a base mixed with colouring matter. The oil is usually linseed oil which is oxidized in the air to a resin which causes the paint to dry or harden. Sugars that are eaten or formed in the body are oxidized there to carbon dioxide and water:

$$C_6H_{12}O_6 + 6O_2 \rightarrow 6CO_2 + 6H_2O + heat$$

and the heat given out in the reaction maintains our bodies at a fairly uniform temperature in both summer and winter (see also p. 151).

Nitrogen

It is not always convenient to prepare nitrogen from the air by the method used in Experiment 22.1. Like oxygen, it is best obtained by heating certain chemicals; in this case, ammonium nitrite,

NH_4NO_2. This, however, is an unstable substance and is not usually available in the laboratory. Instead, a mixture of sodium nitrite and ammonium chloride, which behaves in the same way, is used.

Experiment 22.3. The preparation of chemical nitrogen.

Place a mixture of about 8 g of sodium nitrite and 6 g of ammonium chloride in a 250 cm³ round-bottom flask and cover with water. Arrange to collect the gas over water. Warm, and when you think all the air has been displaced collect several jars of the gas. Carry out the following tests:
 (a) Shake with water containing neutral litmus, methyl orange or Universal indicator.
 (b) Apply a lighted splint—what happens?
 (c) Add a little lime water and shake—is there any change?

You will find that nitrogen is a most uninteresting gas—all its properties seem to be negative ones. It has no effect on indicators, extinguishes flames (as does carbon dioxide) and does not turn lime water milky (here it differs from carbon dioxide). However, negative tests such as these are important, they enable us to distinguish nitrogen from other common gases. Consider the properties listed here.
 (i) No colour (not chlorine or nitrogen dioxide).
 (ii) No action on litmus (not ammonia, hydrogen chloride, sulphur dioxide or carbon dioxide).
 (iii) Does not burn (not hydrogen or carbon monoxide).
 (iv) Does not support combustion (not oxygen or dinitrogen oxide).
 (v) No effect on lime water (not carbon dioxide).
 (vi) No smell (not ammonia, sulphur dioxide or hydrogen chloride).
 (vii) No brown colour on mixing with air (not nitrogen monoxide).

A gas which has the negative properties listed above must be nitrogen.

However, nitrogen is not quite devoid of activity as can be shown by plunging a piece of burning magnesium into the gas where it continues to burn. Although most of the product formed here is magnesium oxide (from the air present) there is also some magnesium nitride. The following experiment illustrates this in a more satisfactory way.

Experiment 22.4. The Action of Nitrogen on heated Magnesium.

Half fill the boiling tube (Fig. 22.4) with a mixture

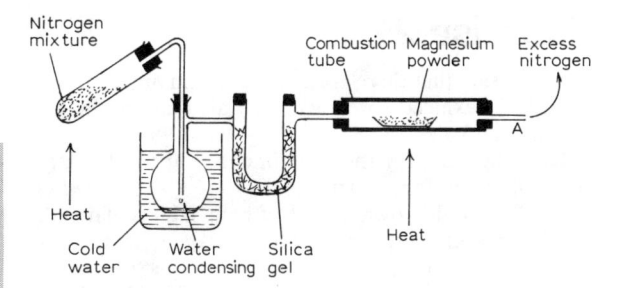

Fig. 22.4. Formation of Magnesium Nitride

of ammonium chloride and sodium nitrite and heat gently. A liquid condenses in the flask which can be shown to be water (test with anhydrous copper(II) sulphate). The reaction which takes place on heating 'nitrogen mixture' is shown by:

$NH_4Cl(s) + NaNO_2(s) \rightarrow$
$\qquad NaCl(s) + N_2(g) + 2H_2O(l)$

The gas is then dried from any water still carried along and passed over magnesium powder in a silica or hard glass tube. When the gas issuing at A puts out a lighted splint (what does this tell us?) the magnesium should be strongly heated. It should burn or smoulder and after cooling (with the dry gas still passing through) the surface of the powder will be yellow-green magnesium nitride.

This experiment shows that at a high temperature nitrogen will support the combustion of magnesium.

If a *little* of the green powder is dropped into about 50 cm³ of warm water containing a drop of Nessler's reagent (see p. 129) a yellow-brown colour should develop. This is a very sensitive test for the gas ammonia which is produced by the action of water on magnesium nitride.

$Mg_3N_2(s) + 6H_2O(l) \rightarrow 3Mg(OH)_2(aq) + 2NH_3(g)$

Uses of Nitrogen

A supply of nitrogen is essential for animal and vegetable life and this subject will be discussed in detail in the section dealing with nitric acid. Most of the nitrogen obtained from the air is used in the manufacture of ammonia, an intermediate step in the production of both ammonium sulphate, a very good fertilizer, and of nitric acid, which finds considerable use in many industries.

Exercise 22

1 It is said that there is less oxygen and more carbon dioxide in exhaled (breathed out) air than in inhaled (breathed in) air.

Describe carefully the experiments you would carry out to find out if this is true. (MR)

2 (a) A well-known brand of bath-salts is claimed, by the manufacturer, to 'radiate oxygen' when it is placed in hot water. How would you test this statement?

(b) Some clothing is left to air near a good fire. After a short time, the clothing catches fire and burns. Explain briefly what has happened. (WR)

3 What are (a) acidic oxides, (b) basic oxides? Give one example of each to illustrate your answer.

How would you prepare specimens of two oxides from each of the following compounds: (i) lead(II) nitrate, (ii) calcium carbonate?

How would you show that the solid oxide from (ii) was free from the original substance? (L)

4 In each of the reactions represented below, give the name of the reducing agent:
(a) $SO_2 + Cl_2 + 2H_2O = H_2SO_4 + 2HCl$
(b) $SO_2 + 2H_2S = 2H_2O + 3S$
(c) $CuO + H_2 = Cu + H_2O$
(d) $Pb + KNO_3 = PbO + KNO_2$ (S)

5 Describe an experiment to show that air contains about 20 per cent. oxygen by volume.

Name (a) TWO other constituents of the atmosphere, (b) TWO substances usually present in the air over an industrial town but not in the open country.

Give TWO pieces of evidence indicating that air is a mixture. (J)

6 Describe experiments (a) to obtain the air that is dissolved in water, (b) to show what fraction of this air is oxygen. Comment on the result of experiment (b).

What mass of anhydrous copper(II) sulphate crystals would you expect to produce by heating 100 g of ($CUSO_4.5H_2O$)?

7 Give the names of TWO substances which yield oxygen as the only gaseous product when heated, and write equations for their decomposition. Describe with the aid of a sketch how you would determine experimentally, for any ONE of these substances, the volume of oxygen liberated (expressed in cm^3 at s.t.p.) when one gram of it is heated.

How would you find the approximate percentage by volume of oxygen in air? (OC)

8 Describe an experiment to show that air contains approximately 20 per cent. of oxygen by volume.

Give TWO reasons for believing that air is a mixture.

Describe and explain what happens when (a) crystals of washing soda, (b) sticks of sodium hydroxide (caustic soda) are exposed to air. (J)

9 (a) (i) Give the names and formulae of ONE acidic and ONE basic oxide.

(ii) What are the characteristic reactions of each of these classes of oxide?

For each oxide you have named give ONE reaction that illustrates your answer.

(b) Name and give the formula of ONE oxide that behaves in some ways as a basic oxide and in some ways as an acidic oxide, and give TWO reactions that illustrate this behaviour.

(c) Give the name and formula of ONE oxide that readily acts as an oxidizing agent, and of ONE that acts as a reducing agent. (C)

10 State TWO experimental facts that support the view that air is a mixture and not a compound.

Two identical graduated bell-jars are set up so that 5 dm^3 of air are trapped, at atmospheric pressure, in each above dilute solutions of sodium hydroxide in glass troughs. In one bell-jar a candle is allowed to burn, in the other white (yellow) phosphorus is left to smoulder. The candle gradually goes out after a few minutes. When this bell-jar has cooled once again to room temperature, more dilute sodium hydroxide solution is poured into the trough so that the levels inside and outside are the same. It is found that the volume of gas left inside the bell-jar has been reduced to 4.50 dm^3. Some time later the phosphorus ceases to smoulder and the volume of gas left after using the same procedure as with the first bell-jar is found to be 4.0 dm^3. Little or no carbon dioxide or carbon monoxide can be detected in either bell-jar. What is the composition of the air left in the first bell-jar? Why was sodium hydroxide solution used in the trough, instead of water?

Samples of each of the gases remaining in the bell-jars were passed over separate pieces of copper foil, heated in glass tubes. State the changes, if any, that you would expect to observe in the appearance of the foil. Give reasons for your answer. (C)

11 Give reasons for the view that air is mainly a mixture of oxygen and nitrogen and not a gaseous compound of formula N_4O.

Describe how you would find the exact proportion of oxygen in a sample of air if you were given a narrow glass tube, sealed at one end and graduated in cm^3, there being a full range of laboratory chemicals and common laboratory equipment available.

Describe how oxygen and nitrogen are obtained on the large scale for commercial use. (C)

12 Hydrogen, obtained by reacting a solid X with a liquid Y, is passed through some fused calcium chloride and then over some copper(II) oxide (Z) being heated in a horizontal combustion tube. At the far end of the tube is some anhydrous copper(II) sulphate and excess hydrogen passes out from the tube through a jet B where it can be burnt.

(a) (i) What chemicals (X and Y) would you use to generate hydrogen?

(ii) Write a balanced equation for the reaction.

(b) What is the colour of the solid Z (i) at the beginning of the experiment and (ii) at the end of the experiment?

(c) What happens to the anhydrous copper sulphate during the experiment? What does this show?

(d) What is the function of the fused calcium chloride?

(e) Why is it necessary to pass hydrogen through the apparatus for a while before igniting the gas at B?
(f) What is the name given to the reaction between copper(II) oxide and hydrogen?
(g) How is hydrogen produced on an industrial scale? Only the essential details should be given.
(h) Carbon and hydrogen form many different compounds. What is the general name given to such compounds?
(i) One such compound is propane. What is the formula of propane?
(j) Write an equation for the reaction between propane and oxygen.
(k) Name one use of propane. (C)

Other questions or problems concerning nitrogen and carbon dioxide will be found in Exercises 23, 24 and 26.

23
Some Simple Chemical Laws

In 1774, Lavoisier—as a result of much careful measuring and weighing—put forward what is now known as the Law of Conservation of Mass. This states:

In any chemical reaction matter is neither created nor destroyed.

This has been verified by use of the most accurate and refined balances available; within the limits of the chemical balance you have in your laboratory you also may show it to be true.

Experiment 23.1. To illustrate the Law of Conservation of Mass.
Place some dilute sulphuric acid in a clean conical flask and suspend in the flask a small tube containing barium chloride solution (Fig. 23.1). Weigh the whole unit as accurately as possible. Tip the flask so that the two liquids mix together. A white precipitate of barium sulphate is formed by double decomposition.

$$BaCl_2(aq) + H_2SO_4(aq) \rightarrow BaSO_4(s) + 2HCl(aq)$$
$$Ba^{2+}(aq) + SO_4^{2-}(aq) \rightarrow BaSO_4(s)$$

On reweighing the unit no change in mass will be found. A similar result can be obtained using solutions of silver nitrate and hydrochloric acid or lead nitrate and potassium iodide.

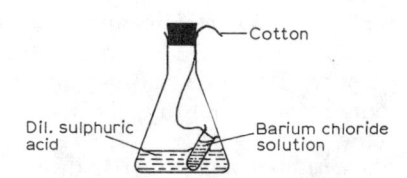

Fig. 23.1. Illustrating the Law of Conservation of Mass

When a substance burns it usually gets smaller and appears to lose mass, but this is because some of the products of combustion are lost; e.g. a candle forms carbon dioxide and steam, both of which escape into the air. We know that, in fact, elements such as magnesium, carbon, phosphorus, etc., gain in mass on burning because they *take in* oxygen from the air. If a piece of phosphorus is burned in a sealed flask the reaction:

$$4P + 5O_2 \rightarrow 2P_2O_5$$

takes place and the phosphorus gains in mass, but *the total mass remains the same*, showing that the air must have lost in mass by the same amount.

In radioactive changes, which involve the *nucleus* of the atom, this law does not hold good in the form in which it is stated. Some of the mass is converted into energy which is released, usually as heat. The enormous amounts of heat produced by the fission of heavy elements such as uranium into smaller ones (as in the atom bomb) or by the fusion of light elements (hydrogen) into heavier ones (as in the hydrogen bomb) are due to this conversion of a very small amount of matter into an enormous quantity of energy. The control of such changes is making nuclear or atomic energy available to the human race.

Law of Constant Composition

If a specimen of *pure* chalk from anywhere in the world is analyzed it will be found to contain 40% of calcium, 12% of carbon and 48% of oxygen. Similarly, every sample of pure water contains 11·2% of hydrogen and 88·8% of oxygen by mass. These facts are expressed in the Law of Constant Composition (also called the Law of Constant—or Definite—Proportions) which states:

All pure samples of the same compound, no matter how made, always contain the same elements joined together in the same proportions by mass.

This law is one that can easily be illustrated by a simple experiment in which one particular compound is made by several different methods.

Experiment 23.2. To analyze various samples of Copper(II) Oxide and to Deduce its Formula.

Black copper(II) oxide can be obtained from other copper compounds by the methods described below.

(i) Take a sample of copper(II) nitrate and heat it strongly in an evaporating basin in a fume cupboard until it is all black and then weigh it after it has cooled. Heat it again, cool and reweigh. This should be done until the mass no longer shows any alteration; the residue will now be pure copper(II) oxide.

(ii) Repeat the above experiment using copper(II) carbonate instead of the nitrate, again heating to constant mass. Once more copper(II) oxide is left.

(iii) Add some sodium hydroxide solution to one of copper(II) sulphate in a beaker and a light blue precipitate of copper(II) hydroxide is formed.

Filter and wash the precipitate on the paper. Scrape some of it into an evaporating dish and heat strongly to constant mass as above:

You now have three different samples of pure copper(II) oxide which have to be reduced to the metal. Weigh three porcelain or fireclay boats and add to each about 1 g of one of the oxide samples. Weigh again to determine the exact mass of each sample. Let these masses be a, b and c g respectively. Place the boats in a long combustion tube connected to a source of hydrogen (or coal gas). After the gas has passed for a short time to drive out all the air, light a Bunsen under the boats and heat at a moderate temperature (air-hole half-open and flame about 3 cm high) for about twenty minutes (Fig. 23.2). Allow the tube to cool with the hydrogen still passing and again weigh the separate boats. As was done in the preparation of the oxide samples, further heating, cooling and re-weighing must be carried out until the mass is constant.

Suppose the masses of copper formed are respectively x, y and z g. Work out the percentage of copper in each sample of copper(II) oxide. You should find that the values of:

$$\frac{100x}{a}, \frac{100y}{b} \text{ and } \frac{100z}{c}$$

are the same or very nearly so (allowing for experimental error).

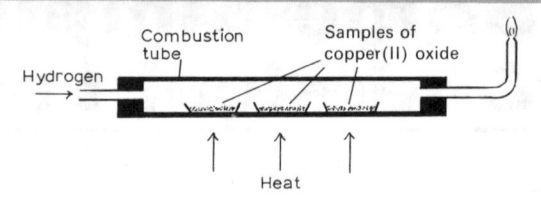

Fig. 23.2. Reduction of copper oxide

An average value of all the percentages obtained by the class should be worked out and used in deducing the formula. Let us suppose this is 80% of copper and 20% of oxygen.

Then, in 100 g of the oxide there will be $\frac{80}{63 \cdot 5}$ moles of copper and $\frac{20}{16}$ moles of oxygen atoms. Both these expressions equal 1·25 and this means that there are the same number of atoms of copper as of oxygen in the oxide and this fits in with the formula CuO.

The equations for the various reactions involved are:

(i) $2Cu(NO_3)_2(s) \xrightarrow{\Delta} 2CuO(s) + 4NO_2(g) + O_2(g)$

(ii) $CuCO_3(s) \xrightarrow{\Delta} CuO(s) + CO_2(g)$

(iii) $CuSO_4(aq) + 2NaOH(aq) \rightarrow Cu(OH)_2(s) + Na_2SO_4(aq)$
and $Cu(OH)_2(g) \xrightarrow{\Delta} CuO(g) + H_2O(g)$

Finally, $CuO(s) + H_2(g) \xrightarrow{\Delta} Cu(s) + H_2O(g)$

Let us now look at one or two examples of calculations based on this law.

EXAMPLE (a). 8·8 g of a metal M join with 2·2 g of oxygen in one experiment, while in another the reduction of 3·5 g of the oxide leaves 2·8 g of M. Show that these figures are in accordance with the Law of Constant Composition.

(i) 11 g of oxide are formed from 8·8 g of M:
the oxide contains $\dfrac{8\cdot 8 \times 100}{11} = 80\%$ M

(ii) 3·4 g of oxide leave 2·8 g of M:
the oxide contains $\dfrac{2\cdot 8 \times 100}{3\cdot 5} = 80\%$ M

EXAMPLE (b). *Two specimens of a mineral ore are analyzed: 5·76 g of the first contain 3·84 g of the metal, 0·48 g of sulphur and 1·44 g of oxygen. 9·24 g of a second sample give 0·79 g of sulphur, 3·16 g of oxygen and 5·29 g of metal. Are the two samples identical?*

Let us find out the masses of the various elements which would be contained in 9·24 g of the *first* sample.

$$\text{mass of metal} = \dfrac{9\cdot 24 \times 3\cdot 84}{5\cdot 76} = 6\cdot 16 \text{ g}$$

$$\text{mass of sulphur} = \dfrac{9\cdot 24 \times 0\cdot 48}{5\cdot 76} = 0\cdot 77 \text{ g}$$

$$\text{mass of oxygen} = \dfrac{9\cdot 24 \times 1\cdot 44}{5\cdot 76} = 2\cdot 31 \text{ g}$$

As these are not the same as the masses found in the *second* sample we must conclude that the two samples are not the same compound.

Law of Multiple Proportions

Under differing conditions some elements will combine with varying amounts of another element, e.g. carbon monoxide CO and the dioxide CO_2; (b) iron(II) chloride $FeCl_2$ and iron(III) chloride $FeCl_3$; (c) copper(I) oxide Cu_2O and copper(II) oxide CuO; (d) dinitrogen oxide (nitrous oxide) N_2O, nitrogen monoxide (nitric oxide) NO and nitrogen dioxide NO_2.

In (a) 12 g of carbon combine with 16 or 32 g of oxygen.

In (b) 56 g of iron combine with 71 or 106·5 g of chlorine.

In (c) 16 g of oxygen combines with 128 or 64 g copper.

In (d) 14 g of nitrogen combine with 8, 16 or 32 g of oxygen.

You should be able to see that the figures representing the different masses bear a simple relationship to one another; 1:2; 2:3; 2:1; 1:2:4.

The law of chemical combination concerned with these ratios is called the Law of Multiple Proportions and states:

If two elements combine together to form more than one compound, then the different masses of the one element combining with a fixed mass of the other bear a simple relationship to each other.

(It should be noted that this law does not hold in the case of alloys or of organic carbon compounds. Thousands of such compounds, mostly also containing hydrogen, are known; formulae such as $C_{70}H_{142}$ and $C_{80}H_{162}$ are encountered in which no simple relationship exists between the masses of hydrogen united with a fixed mass of carbon in the different molecules.)

EXAMPLE (c). *Carbon monoxide contains 57·14% of oxygen, whereas the reduction of 5 g of the dioxide leaves 1·363 g of carbon. Do these figures illustrate the Law of Multiple Proportions?*

In the monoxide 42·86 g of carbon combine with 57·14 g of oxygen.

In the dioxide 1·363 g of carbon combine with 3·637 g of oxygen, i.e. 42·86 g of carbon join with 114·3 g of oxygen.

The values 57·14 g and 114·3 g being in the ratio of 1:2, illustrate the law.

EXAMPLE (d). *Compounds of nitrogen and oxygen contain the following percentages of nitrogen: 63·66, 46·67, 38·85, 30·44, 25·93, 22·58. Show that these figures conform to the Law of Multiple Proportions.*

%Nitrogen:
　63·66　46·67　36·85　30·44　25·93　22·58
% Oxygen:
　36·34　53·33　63·15　69·56　74·07　77·42

If the mass of oxygen combined with, say, 63·66 parts of nitrogen is calculated in each compound we get these figures:

Oxygen:

36·34　71·70　109·05　145·40　181·90　218·25

These amounts are in the simple ratio of:

$$1:2:3:4:5:6$$

As with the Law of Constant Composition, it is quite easy to illustrate this other law in the laboratory. One method would be to take samples of two or three different oxides of a metal (e.g. lead) and reduce them in hydrogen as in Experiment 23.2. Another method is described here.

Experiment 23.3. To illustrate the Law of Multiple Proportions.

Take two pieces of clean copper sheet, one weighing about 0·5–0·7 g and the other about

1–1·4 g. Cut the smaller into strips (for easier solution) after weighing it and place in a test-tube. Cover with concentrated nitric acid in a fume cupboard (to form copper(II) nitrate). The metal soon dissolves with evolution of brown fumes (nitrogen dioxide).

$$Cu + 4HNO_3 \rightarrow \underset{\text{green}}{Cu(NO_3)_2} + \underset{\text{brown}}{2NO_2(g)} + 2H_2O$$

This must now be converted into the oxide by heating strongly. Loss of liquid by spitting can be prevented by adding a few pieces of porous pot to it and by placing a loose plug of glass wool in the mouth of the tube. When *all* the green-brown colour anywhere in the tube has been replaced by a black one

$$2Cu(NO_3)_2(aq) \rightarrow 2CuO(s) + 4NO_2(g) + O_2(g)$$

add sufficient concentrated hydrochloric acid to convert all the oxide into copper(II) chloride on gentle warming.

$$CuO(s) + 2HCl(aq) \rightarrow CuCl_2(aq) + H_2O(l)$$

The solution will be a greenish or brownish yellow, depending on its concentration.

Now add the larger piece of (weighed) copper—in one piece—to the solution and continue heating for above five minutes to convert the copper(II) chloride to copper(I) chloride. This reaction is complete when a drop of the solution poured into a beaker of water produces a milkiness (copper(I) chloride is insoluble in water).

$$Cu + CuCl_2 \rightarrow 2CuCl$$

When this happens, pour all the solution into water and wash the copper remaining in the tube with (a) water and then (b) a little methylated spirits. Dry between filter papers and hold for *one moment* about twelve inches above a very small non-luminous bunsen flame or put under an infra-red lamp. Now reweigh.

If the masses obtained are:

Small piece of copper = x g
Large piece of copper = y g
Copper remaining = z g

Then the mass of copper in the first chloride is x gm. Mass of copper in the second chloride is $(x + y - z)$g

The ratio $\dfrac{x + y - z}{x}$ should be found. What value would you expect to find for it?

Exercise 23

Without putting any figures down on paper say if the following figures support the law of constant composition:

(a) One piece of marble contains 40% Ca; 3 g of another sample are found to contain 1·2 g of the metal.

(b) Two boys analyze the same oxide. One takes 4 g of it and finds it loses 1·6 g as oxygen on heating. The other boy reports that after heating 6 g of his sample 0·36 g remain.

(c) 33·3% of a certain mineral sulphide is sulphur; 6·5 g of a sample from another source yield 2·5 g of sulphur.

1. Iron forms an oxide containing 77·8% of iron, and another oxide containing 70·0% of iron. Show that these two oxides conform to the Law of Multiple Proportions.
(OC)

2 Black copper(II) oxide contains 79·87% copper, and 2·062 g of the red copper(I) oxide contain 0·2308 g of oxygen. Do these compounds illustrate the Law of Multiple Proportions?

3 Copper forms two chlorides, containing respectively 35·8 and 52·7% of chlorine. Show that these figures conform to the Law of Multiple Proportions.
(OC)

4 When 9·54 g of black copper(II) oxide was heated in a current of hydrogen, 2·16 g of water was formed and 7·62 g of copper was left. Calculate the proportion of oxygen to hydrogen, by mass, in water. (WR)

5 Copper(II) oxide is formed in the following three ways; show that all the figures bear out the Law of Definite Proportions:

(i) 0·2133 g of the metal dissolved in nitric acid and heated gives 0·2667 g of the oxide.

(ii) 0·9143 g of the metal is converted via the hydroxide to 1·143 g of the oxide.

(iii) 1·631 g of the oxide are left after heating 2·528 g of the carbonate which contains 51·62% of copper.

6 (a) Calculate the formula of the oxide of lead which contains 7·4% of oxygen.

(b) Indicate very briefly how you would obtain metallic lead from its oxide. (S)

7 An element X forms two oxides containing respectively 40% and 50% of oxygen. Do these proportions agree with the Law of Multiple Proportions? If the formula of the first is XO what is that of the second?

8 A nitride is a compound containing only two elements, one of which is nitrogen. A metal nitride X was prepared in the laboratory and analyzed. As a result of two experiments it was found that

(a) 5 g of X contained 3·6 g of the metal
(b) 5·4 g of the metal combined with 2·1 g of nitrogen.
See if these figures support the idea of constancy of composition and if they do, try to deduce the formula of X given that the metal has an A.M. of 24 and nitrogen of 14.

9 State the laws of (a) definite proportions and (b) multiple proportions.
Choosing copper(II) oxide as an example, describe an experiment by which you would illustrate the former law.

10 B is a compound containing only the elements X and Y. 4 atoms of X weigh as much as 3 atoms of Y. In an experiment to synthesize B it was found that the mass of Y combining with X was exactly twice that of X. Deduce a suitable formula for B.

11 State (a) the Law of Conservation of Mass (indestructibility of matter), (b) the Law of Definite Proportions.
Describe an experiment you could do in the laboratory to illustrate the first of these laws.
Assuming that coal contains only the elements carbon, hydrogen, oxygen and sulphur, explain what takes place when a piece of coal is burned, applying the principle of conservation of mass in giving your explanation. (*J*)

12 Sketch and label the apparatus you would use to prepare several jars of dry chlorine, without using electrolysis. Name the starting materials, and write the equation for the reaction involved.
Write equations for any THREE reactions involving chlorine as one of the reactants, and state the conditions under which these reactions take place.
A metal forms two chlorides, containing respectively 55·9% and 65·6% by mass of chlorine. Show that these chlorides conform to the Law of Multiple Proportions.
(*OC*)

24
Ammonia and the Oxides of Nitrogen

Although nitrogen is not a very active element it is present in all living matter, being a constituent of all proteins. The number of nitrogen compounds used in industry is very considerable; many of them are organic substances, but two inorganic ones play a large part in the manufacture of essential commodities. These are ammonia and nitric acid.

Ammonia NH_3

Ammonia, NH_3, is a very pungent gas which you may already have met. It was first prepared by Joseph Priestley in 1774 and collected over mercury, although its solution in water had been previously known. It is always obtained in the laboratory from one of its compounds; these are called ammonium salts and the one generally used is ammonium chloride (sal ammoniac), NH_4Cl.

Experiment 24.1. Preparation of Ammonia
Make a mixture of calcium hydroxide and ammonium chloride in the proportions of approximately 2:1 and sniff it cautiously. What do you notice? Hold a piece of red litmus over it; is there any change? What does this suggest?

Now place the mixture in a boiling tube and set up the apparatus shown in Fig. 24.1. The tube should slope slightly downwards so that water

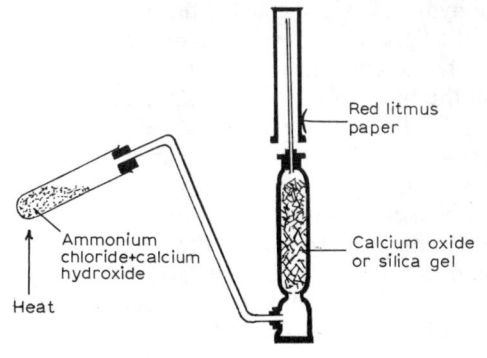

Fig. 24.1. Preparation of Ammonia.

produced in the reaction can easily be driven out. Heat on the mixture produces ammonia and the water given off is absorbed in a quicklime or silica gel tower. The gas is collected by downward displacement of air, the jars being judged full when a piece of moist red litmus on the outside of the jar turns blue. (Which two properties of the gas can be deduced from this method of collection?)

$$2NH_4Cl(s) + Ca(OH)_2(s) \rightarrow 2NH_3(g) + CaCl_2(s) + 2H_2O(l)$$

The normal drying agents, calcium chloride and concentrated sulphuric acid, cannot be used with ammonia as the gas reacts with them. It has no action on quicklime which remains dry even after being slaked by the water—or on silica gel.

$$CaO + H_2O \rightarrow Ca(OH)_2$$

(See Chapter 50 for a small-scale method of preparing ammonia, suitable for a class experiment.)

Properties of Ammonia

This gas has a characteristic pungent smell which makes it easy to recognise—however, it should only be sniffed with caution as heavy concentrations can be dangerous. It is less dense than air and extremely soluble in water. You may have seen the fountain experiment performed with another soluble gas, hydrogen chloride, but it works even better with ammonia, one volume of water dissolving about 1200 volumes of the gas.

*Experiment 24.2. The Fountain Experiment.

Take a fairly large flask, 500–1000 cm³ and fit it with a bung carrying two tubes, A and B. A is drawn to a point and reaches well into the bulb of the flask, the lower end being fitted with a pinch-clip. The shorter tube B, carries a small teat (Fig. 24.2). Prepare ammonia as in the previous experiment (or by heating a concentrated solution of ammonium hydroxide) and pass it through the tube A, into the flask for some time in order to displace all the air, the cork fitting loosely in the neck. Next push the bung in tightly and after passing the gas for a further minute or so close the clip on A. Squeeze the teat so that a drop of water enters the flask and swirl it around. Invert the flask in a large beaker of water containing some colourless phenolphthalein and open the clip on A. Water should rise slowly up the tube and then spray over like a fountain to fill most of the flask. Notice the change in colour of the indicator; what does this suggest?

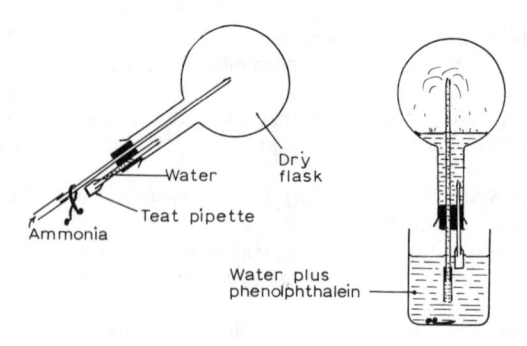

Fig. 24.2. The Fountain experiment

When common salt dissolves in water the change is a physical one, no new substance being formed. When ammonia dissolves in water both physical and chemical changes occur. Most of the gas just forms a solution, but some combines with the water to form ammonium hydroxide, NH_4OH. That this chemical change does take place is shown by the change in colour of the indicator. Ammonium hydroxide is alkaline (like sodium hydroxide).

$$NH_3 + H_2O \rightarrow NH_4OH$$

If ammonia is passed through a delivery tube directly into water there is every likelihood that the water may be sucked back into the tube in which the gas is being prepared. A funnel should be fitted to the end of the delivery tube with the rim just below the surface of the water in a beaker. If the water rises up the funnel the level of it in the beaker drops and soon the funnel is out of the liquid in the vessel. This causes any water sucked up to fall back into the beaker.

If a light is applied to a jet of ammonia the gas attempts to burn, but once the light is removed the reaction stops. However, if pure oxygen is used (Fig. 24.3), the gas burns with a yellow-brown flame to form nitrogen and steam. (This is *not* a suitable method for preparing nitrogen from ammonia.)

$$\overset{\text{oxidation}}{4N\overset{\longrightarrow}{H_3} + 3\underset{\longleftarrow}{O_2} \rightarrow 2\overset{\longrightarrow}{N_2} + 6H_2O}$$
$$\text{reduction}$$

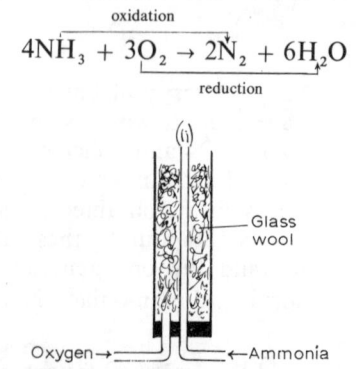

Fig. 24.3. Oxidation of Ammonia

The glass-wool serves to spread the oxygen around the jet of ammonia.

This is an obvious example of oxidation of the ammonia (by loss of hydrogen) which itself reduces the oxygen to water (gain of hydrogen).

The reducing action of the gas can best be shown by passing it over heated copper(II) oxide, using the apparatus of Fig. 24.4. The oxide is reduced to copper, steam is formed and condenses to water in the cooled tube (how recognised?) and nitrogen is collected over water.

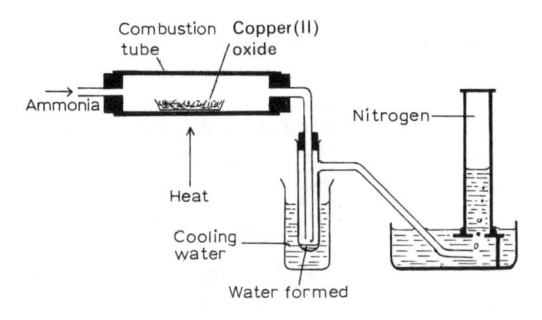

$$\underset{\text{oxidation}}{2NH_3} + \underset{\text{reduction}}{3CuO} \rightarrow N_2\uparrow + 3Cu + 3H_2O$$

Fig. 24.4. Reduction of Copper Oxide by Ammonia

Ammonia, unless completely dried, is an alkaline gas and reacts with acids to form salts. It is a basic anhydride.

$$NH_3 + HCl \rightarrow NH_4Cl$$
$$2NH_3 + H_2SO_4 \rightarrow (NH_4)_2SO_4$$

White fumes are noticeable in these reactions, particularly when the ammonia comes into contact with hydrogen chloride or its solution.

Experiment 24.3. To identify Ammonia Gas. (see p. 319).

(a) Take two rods dipped respectively into bottles of concentrated hydrochloric acid and ammonium hydroxide and hold them near each other, but not touching. Dense white fumes of ammonium chloride immediately form. If a drop of the concentrated acid is held on a glass rod at the mouth of a tube in which an ammonium salt is being heated with sodium hydroxide these same fumes appear.

(b) Collect jars of ammonia and hydrogen chloride and place in contact. Notice the fumes. Alternatively put a little of the acid into a gas jar and a little of the ammonium hydroxide into another. Swirl around and then pour out the liquids.

Now place the jars with their mouths in contact. Again the same reaction is observed.

$$NH_3 + HCl \rightarrow NH_4Cl$$

(c) Put a little concentrated ammonia solution in a gas jar and swirl it round. Have a filter paper with a drop of copper sulphate solution in the centre (this will be almost colourless). Put this paper on the jar in place of the cover plate. Very soon the spot will become deep blue in colour.

When chlorine is passed into ammonia solution (or chlorine water is added) white fumes are formed and bubbles of some gas may be seen rising in the liquid. The gas is nitrogen, produced by the oxidation of the ammonia.

$$\underset{\text{reduction}}{2NH_3} + \underset{\text{oxidation}}{3Cl_2} \equiv N_2 + 6HCl$$

However, the acid produced immediately combines with excess ammonia to form ammonium chloride, hence the white fumes.

$$6NH_3 + 6HCl \equiv 6NH_4Cl$$

The final reaction is therefore:

$$8NH_3 + 3Cl_2 \rightarrow N_2 + 6NH_4Cl$$

Experiment 24.4. To identify Ammonia in Solution.
Place about 10 cm³ of concentrated ammonium hydroxide solution in a graduated cylinder (A) and dilute with water to 100 cm³. Transfer 10 cm³ of this dilute solution to another cylinder (B) and again make up to 100 cm³ with water. Repeat once again in a third cylinder (C). To each of the three cylinders add three drops of Nessler's reagent (a mercury compound). A brown colour will develop in cylinder A and a pale yellow one in cylinder C. This is a very sensitive test and can be used to detect as little as one part of ammonia in about one million parts of water.

The Industrial Preparation of Ammonia

As ammonia is composed only of nitrogen and hydrogen it might be expected that it would be comparatively easy to make it from its elements. This is done industrially (the Haber process), but it is not suitable for the laboratory preparation because of the necessary conditions.

Hydrogen, prepared mainly from petroleum or natural gas, is mixed with nitrogen from the air in the proportions of 3:1 by volume and passed at a pressure of 350 atmospheres over a catalyst of

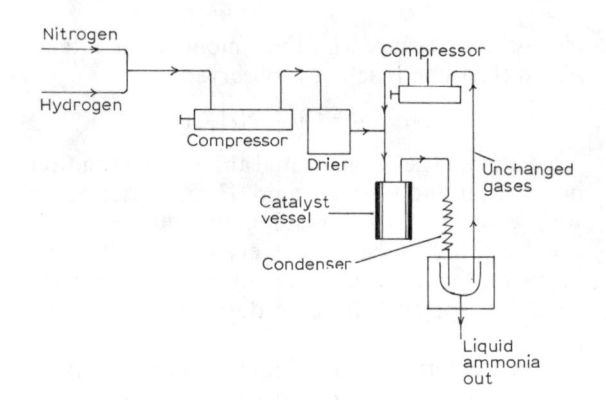

Fig. 24.5. Flow diagram for Haber Process

iron at a temperature of 350°–400°C. About 10–15% of the gases combine to form ammonia.

$$N_2 + 3H_2 \rightarrow 2NH_3$$

This is liquefied and removed. Fresh gases are added to the unchanged nitrogen and hydrogen and the whole lot is recirculated. This is a continuous process, ammonia being drawn off all the time.

Uses of Ammonia

The main use for most of the ammonia produced in the world today is the preparation of fertilizers. The direct application of ammonia for this purpose is now growing rapidly, but in the main, ammonium compounds such as the sulphate are employed. Next in importance comes its conversion to nitric acid and explosives while other uses include the manufacture of synthetic fibres, plastics and glues. It is also used in large refrigerating plants and for softening water.

Ammonium Compounds

We have already spoken of ammonium chloride, ammonium sulphate, etc., and now let us have a look at them again and compare their formulae with those of the corresponding sodium compounds and of the acids from which they are derived.

Ammonium chloride:
NH_4Cl NaCl   HCl
Ammonium nitrate:
NH_4NO_3 $NaNO_3$ HNO_3
Ammonium sulphate:
$(NH_4)_2SO_4$ Na_2SO_4 H_2SO_4
Ammonium carbonate:
$(NH_4)_2CO_3$ Na_2CO_3 H_2CO_3
Ammonium hydroxide:
NH_4OH NaOH

You will see the group NH_4 persists as such throughout all the compounds and takes the place of one atom of hydrogen in the acid (as does one atom of sodium). You are accustomed to such acid radical groups as nitrate, sulphate and carbonate; the ammonium group is a similar but basic radical (the only one you will meet) and its salts form ions similar to those of other salts. The ammonium ion carries a unit *positive* charge and is written NH_4^+.

Ammonium Hydroxide NH_4OH

This compound is only known in solution, any attempt to drive off the water by heating decomposes the solution.

$$NH_4OH \rightarrow NH_3 + H_2O$$

This suggests a suitable method for obtaining the gas when required. It is not, however, a method of *preparing* the gas.

If ammonia is passed into water the density of the solution gradually *decreases* until the water is saturated and it is then 0·88; i.e. one dm^{3*} of the solution has a mass of 880 g. This is usually called '880 ammonia'.

Although the compound, ammonium hydroxide is present, most of the liquid can be regarded simply as a solution of ammonia in water.

Ammonium hydroxide is a caustic alkali like sodium hydroxide, but it is not so strong. It has similar properties and forms salts with acids.

$$NH_4OH + HCl \rightarrow NH_4Cl + H_2O$$
$$NH_4OH + HNO_3 \rightarrow NH_4NO_3 + H_2O$$

It also precipitates insoluble metal hydroxides from solutions of their salts.

$$FeCl_3(aq) + 3NH_4OH(aq) \rightarrow Fe(OH)_3(s) + 3NH_4Cl(aq)$$
or $Fe^{3+}(aq) + 3OH^-(aq) \rightarrow Fe(OH)_3(s)$
red-brown
$$CuSO_4(aq) + 2NH_4OH(aq) \rightarrow Cu(OH)_2(s) + (NH_4)_2SO_4(aq)$$
or $Cu^{2+}(aq) + 2OH^-(aq) \rightarrow Cu(OH)_2(s)$

This pale-blue copper(II) hydroxide precipitate dissolves in excess ammonium hydroxide to give a deep-blue solution and this colour change can be used as a test for ammonia in solution (or as a test for copper(II) ions). Other hydroxides precipitated are those of lead, zinc and aluminium.

'Household Ammonia' is a solution of ammonia and it is used for cleaning brass and copper articles, for softening hard water and it helps to prevent

*One litre = 1000 cm^3 = 1 dm^3 (one cubic decimetre). The term dm^3 is used in this book in calculations.

woollen articles matting together when they are washed. Water containing calcium hydrogencarbonate is hard owing to the presence of calcium ions, but the addition of ammonia solution removes these ions by precipitating insoluble calcium carbonate.

$$2NH_3(aq) + Ca(HCO_3)_2(aq) \rightarrow CaCO_3(s) + (NH_4)_2CO_3(aq)$$

Ammonium Chloride NH_4Cl

The chloride, 'sal ammoniac', is the commonest of the ammonium salts and is found naturally along with the sulphate in volcanic districts. It exhibits the property of sublimation; i.e. when heated it changes directly to a vapour and is re-formed as a solid on cooling. In this process the ammonium chloride actually decomposes into the gases ammonia and hydrogen chloride.

$$NH_4Cl(s) \underset{cool}{\overset{heat}{\rightleftharpoons}} NH_3(g) + HCl(g)$$

This is an example of *thermal dissociation* which may be defined as a process in which a compound is decomposed by heat into simpler substances which recombine on cooling. Notice that this type of action differs from simple decomposition as represented by heat on potassium chlorate(V).

$$2KClO_3(s) \rightarrow 2KCl(s) + 3O_2(g)$$

in that it is always reversible. It can easily be illustrated using a test tube as shown in Fig. 24.6.

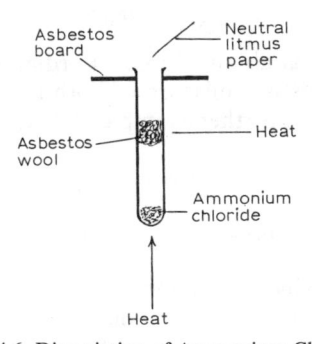

Fig. 24.6. Dissociation of Ammonium Chloride

Experiment 24.5. The Thermal Dissociation of Ammonium Chloride

A 15 cm test-tube containing some slightly moist ammonium chloride has a plug of asbestos wool about two inches above the solid and this is heated strongly from the side. The chloride is then heated from below and a strip of damp neutral litmus held at the mouth of the tube. This first turns blue and later becomes red.

The molecules of ammonia are smaller than those of hydrogen chloride and move about more rapidly. When the sal ammoniac is split up the ammonia molecules are able to pass through the asbestos wool plug the more easily and reach the litmus first, turning it blue. These are followed by a mixture of both ammonia and hydrogen chloride molecules and their recombination to form ammonium chloride is shown by white fumes appearing. At the end the final hydrogen chloride molecules turn the litmus red.

Ammonium chloride is constituent of Leclanché cells and is used as a flux in soldering.

Ammonium Nitrate NH_4NO_3

This salt is a very powerful fertilizer, containing a high proportion of nitrogen, and is also used in the preparation of certain explosives such as ammonal (in which it is mixed with aluminium powder). When used as a fertilizer its explosive nature is nullified by mixing it with chalk. It is made by passing ammonia gas into dilute nitric acid.

$$NH_3 + HNO_3 \rightarrow NH_4NO_3$$

It dissolves in water endothermically, i.e. heat is absorbed and the temperature falls very considerably, so the salt solution can be used as a 'freezing mixture'. The action of heat on the nitrate is referred to on p. 138.

Ammonium Sulphate $(NH_4)_2SO_4$

This compound is of very considerable value as a fertilizer; it is one of the cheapest and best and is the most widely used synthetic nitrogenous fertilizer in the world. It is made either by the reaction of ammonia with sulphuric acid or by passing the gas with carbon dioxide over the mineral anhydrite ($CaSO_4$) in the presence of hot water.

$$2NH_3 + H_2O + CO_2 \rightarrow (NH_4)_2CO_3$$
$$(NH_4)_2CO_3 + CaSO_4 \rightarrow (NH_4)_2SO_4 + CaCO_3$$

The ammonium sulphate is crystallized and used as such, or mixed with phosphates and potassium salts to make complete fertilizers.

Some is also obtained using the ammoniacal liquor from gas works.

The Oxides of Nitrogen

Nitrogen does not combine very readily with oxygen although a spark passed through a mixture of the gases will produce some nitrogen oxide (this happens during thunderstorms). However there are quite a number of oxides which can be formed.

Nitrogen Dioxide NO$_2$ (see also p. 317):

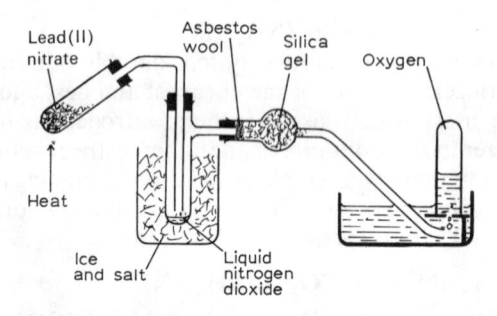

Experiment 24.6. The Preparation of Nitrogen Dioxide.

(a) Nitrogen dioxide is readily prepared by heating lead(II) nitrate. This is chosen as it does not contain any water of crystallization which would affect the result.

The solid is heated in a *dry* test-tube connected to a *dry* receiver surrounded by a freezing mixture of ice and salt. From the side-arm of this receiver a delivery tube passes via a silica gel tube to a collecting jar (a boiling tube is suitable) standing in a trough of water. The drying-tube serves to keep water vapour out of the receiver (Fig. 24.7).

Gentle heat causes the white nitrate crystals to melt to an orange liquid (which later solidifies). They decompose noisily (this is called decrepitation) and evolve a brown gas which liquefies in the freezer to a yellow-brown liquid. A colourless gas can be collected in jars over the water and shown to be oxygen (how?).

$$2Pb(NO_3)_2(s) \rightarrow 2PbO(s) + 4NO_2(g) + O_2(g)$$

(Note carefully this method of separating two gases which liquefy at two widely different temperatures.) The tube containing the liquefied nitrogen dioxide should be corked and the side-arm closed with a piece of rubber tubing and a clip.

Fig. 24.7. Preparation of Nitrogen Dioxide.

(b) Set up a flask with a thistle funnel and with a delivery tube dipping into a gas jar almost closed by a piece of card (as in preparing carbon dioxide). Put some copper turnings in the flask and place the apparatus in a *fume cupboard* with a good draught. Add concentrated nitric acid and notice the red-brown fumes immediately evolved. Collect and cover several jars of gas, warming the flask if necessary.

$$Cu(s) + 4HNO_3(aq) \rightarrow Cu(NO_3)_2(aq) + 2NO_2(g) + 2H_2O(l)$$

This experiment can be carried out on the demonstration bench if a Frankland cover is used on the collecting jars as was done in the preparation of chlorine (see p. 23).

Properties

Nitrogen dioxide has a pungent smell and is very harmful if inhaled. It dissolves in water to give a solution containing both nitric and nitrous acids.

$$2NO_2 + H_2O \rightarrow HNO_3 + HNO_2$$

Consequently if the gas is passed into sodium hydroxide solution it forms a mixture of sodium nitrate and sodium nitrite.

$$2NO_2 + 2NaOH \rightarrow NaNO_3 + NaNO_2 + H_2O$$

For this reason it is classified as a *mixed acid anhydride*. The gas does not burn and its ability to allow substances to burn in it will be discussed later. It is an oxidizing agent and will convert carbon monoxide to the dioxide. In this case the nitrogen dioxide becomes nitrogen oxide.

$$CO + NO_2 \rightarrow CO_2 + NO$$

The liquid collected in the receiving tube is much paler than the gas and, in fact, in it the molecules of the gas are joined in pairs (associated) to give the liquid a formula N_2O_4. It is named dinitrogen tetroxide. If the tube is warmed the liquid starts to vaporize and becomes darker in colour until the change represented by:

$$N_2O_4 \rightarrow 2NO_2$$

is complete at about 150°C. At ordinary temperatures the gas is a mixture of both kinds of molecules. This is a further example of thermal dissociation.

$$N_2O_4 \underset{\text{cool below 22°}}{\overset{\text{heat above 22°}}{\rightleftharpoons}} 2NO_2$$

yellow   dark brown

Nitrogen Oxide (Nitric Oxide) NO

We have just seen that concentrated nitric acid is changed by copper to the brown gas, nitrogen dioxide. The equation can be written in two parts:

$$\overset{\text{oxidation}}{Cu + 2HNO_3} \equiv CuO + 2NO_2 + H_2O$$

$$CuO + 2HNO_3 \equiv Cu(NO_3)_2 + H_2O$$

Adding:

$$Cu + 4HNO_3 \rightarrow Cu(NO_3)_2 + 2NO_2 + 2H_2O$$

the partial equation (a) shows that the copper is oxidized and therefore the acid must be reduced. Other reduction products can be obtained by using less concentrated acid as the following experiment shows.

Experiment 24.7. Preparation of Nitrogen Oxide
Fit a flat-bottom flask with a thistle funnel and a delivery tube leading to a trough of water. In the flask place some copper turnings covered by water and add concentrated nitric acid through the funnel. Effervescence occurs and brown nitrogen dioxide fills the flask. This dissolves in the water and is replaced by a colourless gas which can be collected in jars over the water. The solution in the flask becomes green as copper(II) nitrate is again formed. The gas formed is nitrogen oxide (also called nitric oxide).

Again reduction of the acid has taken place, but the equation for this reaction is rather complicated: it is $3Cu + 8HNO_3 \rightarrow 3Cu(NO_3)_2 + 2NO + 4H_2O$

Properties
Nitrogen oxide (nitric oxide) is a colourless gas, almost insoluble in water, and with a density almost the same as that of air. It is neutral to litmus, does not burn and only with difficulty will anything burn in it (see p. 134). In contact with air or oxygen it *immediately* forms brown fumes of nitrogen dioxide (this explains the colour noticed in the preparation flask).

$$2NO + O_2 \rightarrow 2NO_2$$
colourless   brown

The reaction serves to identify the gas and, conversely, could also be used as a test for oxygen. Because of this property the smell of nitrogen oxide is unknown.

One special property of the gas is that when it is passed into a solution of iron(II) sulphate it is absorbed with the formation of a blackish-brown compoung, nitroso iron(II) sulphate, $FeSO_4.NO$. As this compound readily splits up again on heating it can be used to obtain pure nitrogen oxide from a mixture of the gas with any other insoluble ones. If the mixture is passed through the iron(II) sulphate only nitrogen oxide is removed and it can be regained on heating.

If a few drops of carbon disulphide are added to a jar of nitrogen oxide and a light applied (care!) a brilliant lilac flame passes down the jar.

Dinitrogen Oxide (Nitrous Oxide)N_2O
Both nitrogen oxide and dinitrogen oxide can be prepared from nitric acid by making the acid more dilute and using zinc as the reacting metal. The gas obtained in this way, however, is usually mixed with other oxides of nitrogen and so a different method of preparation is employed. Dinitrogen oxide was discovered by Joseph Priestley in 1772 and first prepared in 1799 by Sir Humphrey Davy. Its more common name is nitrous oxide.

Experiment 24.8. Action of Heat on Ammonium Nitrate.
Take a little ammonium nitrate and heat it in a test-tube. As the amount gets smaller, small explosions occur with the production of light and the emission of 'smoke' rings.

If the temperature is kept below 260°C the reaction is:

$$NH_4NO_3(s) \rightarrow N_2O(g) + 2H_2O(g) \text{ (steam)}$$

but at higher temperatures and in close confinement a different change occurs.

$$2NH_4NO_3(s) \rightarrow 2N_2(g) + 4H_2O(g) + O_2(g)$$

All the solid is converted into gases (some nitrogen dioxide is also formed) and the expansion is very great. One gram of ammonium nitrate under such circumstances produces over four dm^3 of gas. You can imagine the effect if the solid were heated in a flask with a very small exit; the reaction is explosive! It is, however, quite safe when performed in an open test-tube. For preparing dinitrogen oxide it is better to use a mixture of sodium nitrate and ammonium chloride (compare the preparation of nitrogen) or ammonium nitrate mixed with sand.

Experiment 24.9. Preparation of Dinitrogen Oxide (Nitrous Oxide).
Set up the apparatus of Fig. 24.8. Heat the mixture gently and take care that the amount of solid does not become very small. Water is formed as one product and condenses in the receiver (how can it be identified?) and a colourless gas with a sweet smell collects in the gas jar.

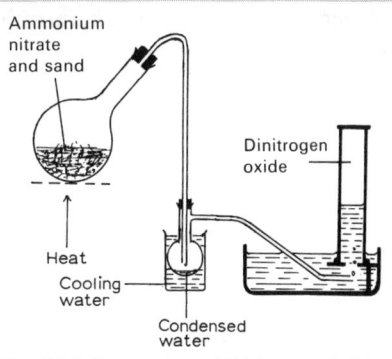

Fig. 24.8 Preparation of Dinitrogen Oxide

$$NaNO_3(s) + NH_4Cl(s) \rightarrow NaCl(s) + N_2O(g) + 2H_2O(g)$$

Properties and Uses

Dinitrogen oxide is used as an anaesthetic in dentistry and in minor operations. If breathed alone, the patient, on regaining conciousness, is likely to have fits of hysterical laughter (hence its name, 'laughing gas'). This after-effect is prevented by using a mixture of the gas with oxygen with a trace of carbon dioxide.

Dinitrogen oxide is more easily decomposed into its elements than either of the two other oxides and so it will support the combustion of feebly burning objects. It will even relight a glowing splint, being the only common colourless gas other than oxygen to do this. It can be distinguished from oxygen in the following way.

Experiment 24.10. Dinitrogen Oxide and Oxygen distinguished.

Place a jar of water on a shelf in a trough and displace half the water by nitrogen monoxide. Now pass in some dinitrogen oxide. The volume of gas increases, but no other change is noticeable. Repeat using oxygen instead of the dinitrogen oxide. Brown fumes of nitrogen dioxide form and dissolve in the water which then rises in the jar.

General properties of the Oxides of Nitrogen

Nitrogen oxide is the least soluble and nitrogen dioxide the most, the latter giving an acidic solution. Dinitrogen oxide is sparingly soluble in water and gives a neutral solution. Nitrogen oxide is the lightest oxide, being 15 times as heavy as the same volume of hydrogen and the dioxide is the densest (23); dinitrogen oxide has the same density as carbon dioxide (22).

None of the gases will burn, but all will support the combustion of such substances as are hot enough to decompose them and liberate oxygen. Dinitrogen oxide does this the most and nitrogen oxide the least readily. The decomposition:

$$2N_2O \rightarrow 2N_2 + O_2$$

occurs at 600° and heat is given out in the process, energy being lost by the gas. This helps the combustion of other substances and even a glowing splint will relight. With nitrogen dioxide the change:

$$2NO_2 \rightarrow 2NO + O_2$$

is complete at 620° and so combustion should readily occur. But heat is *taken in* during the decomposition (i.e. energy is absorbed) and so burning becomes more difficult unless the substance is already quite hot. A warm freshly-made sample of the dioxide will relight a glowing splint, but if the gas has stood for some time this may not happen. Nitrogen oxide only decomposes at a really high temperature and so substances must be very hot before they will burn in the gas.

Magnesium and phosphorus burn in the gases; feebly burning sulphur continues to do so only in dinitrogen oxide. If it is strongly alight it will go on burning in the dioxide but not in nitrogen oxide. The equations for the reactions with phosphorus are:

$$2P + 5N_2O \rightarrow P_2O_5 + 5N_2$$
$$4P + 10NO \rightarrow 2P_2O_5 + 5N_2$$
$$8P + 10NO_2 \rightarrow 4P_2O_5 + 5N_2$$

It should be obvious from these equations that all the gases are oxidizing agents. They are all readily reduced to nitrogen if passed over heated copper.

$$N_2O + Cu \rightarrow N_2 + CuO$$
$$2NO + 2Cu \rightarrow N_2 + 2CuO$$
$$2NO_2 + 4Cu \rightarrow N_2 + 4CuO$$

This is the best method of preparing nitrogen from one of the oxides.

Questions and exercises on this chapter will be found after chapter 25.

25
Nitric Acid

Nitric acid is one of the three common mineral acids and is used in large quantities in many industries. It was formerly known as *aqua fortis* because of its property of dissolving all common metals (but it has no action on gold or platinum).

Industrial Preparation

Nitric acid is made industrially by oxidizing ammonia in the presence of a catalyst. This gas, mixed with air, is passed through a convertor containing a platinum/rhodium gauze at about 550°C. The gaseous mixture forms nitrogen monoxide.

$$4NH_3 + 5O_2 \rightarrow 4NO + 6H_2O \text{ (plus heat)}$$

The monoxide with more air or oxygen (to form nitrogen dioxide) is passed up a tower fitted with baffles. Water or dilute nitric acid (already prepared) flows down from the top and absorbs the gases to form more concentrated acid; it is then run off.

$$4NO + 3O_2 + 2H_2O \rightarrow 4HNO_3$$

Any excess nitrogen dioxide formed is absorbed in aqueous sodium hydroxide to produce sodium nitrite and nitrate. Fig. 25.1 shows in outline part of the necessary plant. There may be several towers, the gases from the first passing to the second and so on.

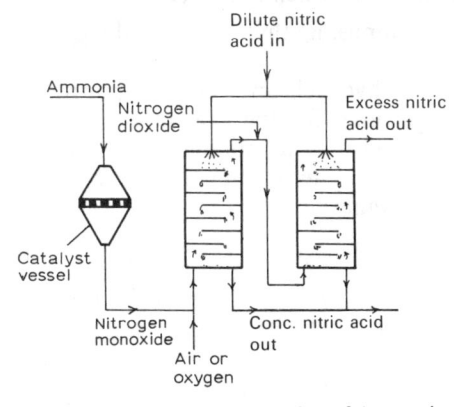

Fig. 25.1. Flow diagram for conversion of Ammonia to Nitric Acid

Experiment 25.1. Catalytic Oxidation of Ammonia.
The reactions which occur in the above industrial process can be illustrated by a laboratory experiment.

The apparatus depicted in Fig. 25.2 is set up and air (from an aspirator) or oxygen carries along with it ammonia from the concentrated ammonia solution. This mixture passes over a platinized asbestos catalyst heated by a low flame and forms nitrogen dioxide as brown fumes in the receiver. Any unchanged ammonia or moisture is removed by the calcium chloride tube. The nitrogen dioxide may be passed into a small amount of water and the solution tested with litmus to show that it is acidic. The formation of nitric acid may be verified by adding a little copper when brown fumes should again be obtained. Thus, nitric acid has been formed.

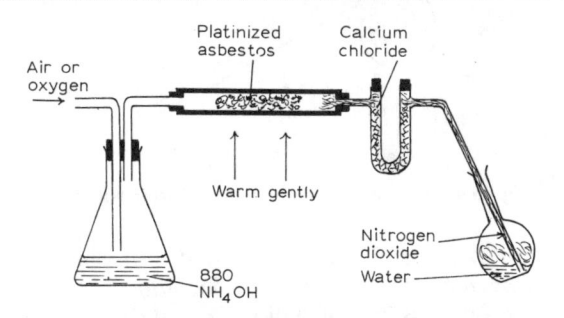

Fig. 25.2. Catalytic oxidation of Ammonia

Formerly much nitric acid was made from potassium or sodium nitrate, but this method is now almost obsolete; in fact, nitric acid is today used for making these nitrates. It was in 1648 that Glauber showed that nitric acid could be obtained from potassium nitrate (nitre) and sulphuric acid and it was much used by jewellers, gilders and brass finishers because of its action on copper and its use in the separation of silver and gold.

Experiment 25.2. Laboratory preparation of Nitric Acid.
The usual laboratory method of making nitric acid follows the obsolete industrial one just

mentioned. The apparatus used must consist of glass as nitric acid attacks rubber or cork.* The small-scale apparatus used in sixth-form work for the preparation of organic compounds is very suitable and should be fitted up as shown in Fig. 25.3. If this is not available a glass retort (with glass stopper) may be used with the neck right inside a flask being cooled by running water (Fig. 25.4). The nitric acid given off in the reaction:

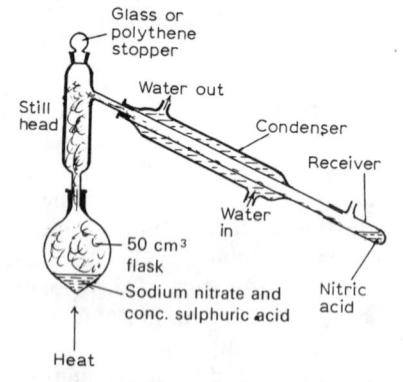

$$NaNO_3 + H_2SO_4 \rightarrow NaHSO_4 + HNO_3$$

Fig. 25.3. Small scale preparation of Nitric Acid

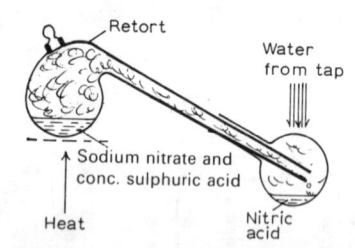

Fig. 25.4. Conventional method for preparing Nitric Acid

vaporizes, condenses again and runs into the receiver. Some of the vapour decomposes to form the brown gas, nitrogen dioxide, and this dissolves in the acid to give it a yellow colour which can be removed by blowing air from a hand-bellows through the liquid. At the temperature of the experiment in the laboratory sodium hydrogensulphate (not the sulphate) is obtained. You may recall that this was the by-product also in the preparation of hydrogen chloride.

Dilute the acid obtained with about eight times its volume of water and then carry out these tests on separate portions.
(a) Notice its action on (i) litmus paper, (ii) Universal indicator.

(b) Add a little (iii) sodium carbonate or marble, (iv) magnesium. Try to write equations if you can. What are the brown fumes formed?
(c) Dilute some of the solution further to about four times its volume so making it very dilute and again add magnesium. Are brown fumes still given off? If so, repeat with even more dilute acid until no coloured gas is seen although effervescence is still apparent. Hold your thumb over the tube for a while and then bring a light near the mouth of the tube. What happens? Can you identify the gas?

Properties

Nitric acid has a lower boiling point than sulphuric and so can be displaced from its salts by the latter. The acid is very corrosive and destroys organic matter (such as paper and flesh) turning it yellow. It should be handled carefully. A mixture of one part by volume of concentrated nitric acid with three parts of concentrated hydrochloric acid is 'aqua regia' and will dissolve gold.

(i) *Acidic properties*

Nitric acid is a mineral acid and it will

(a) neutralize bases to form salts:

$$CuO + 2HNO_3 \rightarrow Cu(NO_3)_2 + H_2O$$
$$NaOH + HNO_3 \rightarrow NaNO_3 + H_2O$$

(b) attack carbonates:

$$CaCO_3(s) + 2HNO_3(aq) \rightarrow Ca(NO_3)_2(aq) + CO_2(g) + H_2O(l)$$
$$Na_2CO_3(s) + 2HNO_3(aq) \rightarrow 2NaNO_3(aq) + CO_2(g) + H_2O(l)$$

The methods of (a) and (b) are used for preparing nitrates, all of which are soluble in water.

(c) turn litmus, methyl orange and Universal indicator red.

Unlike hydrochloric and sulphuric acids it *cannot* be used for the preparation of hydrogen. This gas can only be obtained from nitric acid by using a very dilute solution and a very active metal such as magnesium.

$$Mg(s) + 2HNO_3(aq) \rightarrow Mg(NO_3)_2(aq) + H_2(g)$$

The action of other metals is discussed later in this chapter.

(ii) *Oxidizing properties*

The laboratory preparation of nitric acid showed that the acid is decomposed by heat, one of the

---

*A polythene stopper may be used in place of a glass one, or a cork if the bottom half is covered with aluminium foil (or with a milk bottle top).

products being the brown gas nitrogen dioxide. This action can be better demonstrated using the apparatus of Fig. 25.5.

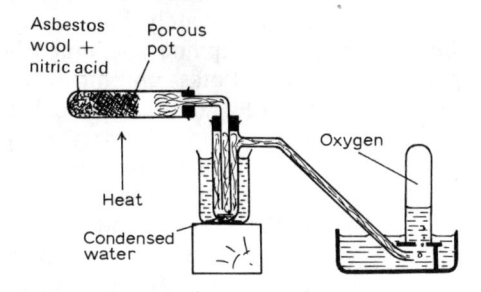

Fig. 25.5. Thermal decomposition of Nitric Acid

Experiment 25.3. The Thermal Decomposition of Nitric Acid.

Some concentrated nitric acid is poured into a hard test-tube and a few wisps of asbestos wool added until all the liquid is absorbed. On top of the wool are placed some pieces of broken porous pot or pumice and the tube arranged as shown.

Heat is first applied to the porous pot and when this is red-hot the Bunsen is shifted a *little* towards the closed end. The acid vaporizes and the vapour is 'cracked' by the heated pot into smaller molecules of water (steam), nitrogen dioxide (seen as red-brown fumes which dissolve in the water) and oxygen (collected over water).

$$4HNO_3 \rightarrow 2H_2O + 4NO_2 + O_2$$

Because of the relative ease with which concentrated nitric acid supplies oxygen it behaves as a very strong oxidizing agent. The equations for these reactions can be worked out using the above basic reaction (simplified):

$$2HNO_3 \equiv H_2O + 2NO_2 + O$$

and combining it with others. Let us look at some examples.

(a) Sulphur

If some powdered sulphur is put into concentrated nitric acid (plus a little bromine to act as a catalyst) and the tube stood in hot water the sulphur is converted into sulphur(VI) oxide and then into sulphuric acid.

Oxidation: $\quad S + 3O \equiv SO_3$

Solution: $\quad SO_3 + H_2O \equiv H_2SO_4$

3 × basic equation:
$$6HNO_3 \equiv 3H_2O + 6NO_2 + 3O$$

Final equation:
$$S + 6HNO_3 \rightarrow H_2SO_4 + 6NO_2 + 2H_2O$$

The solution should be diluted, filtered from any unchanged sulphur and then added to some barium chloride solution. If a sulphate has been formed what would you expect to see?

(b) Carbon

A reaction similar to the above occurs, but carbonic acid is not obtained as this is unstable. A small piece of charcoal should be heated with the nitric acid.

$$C + 2O \equiv CO_2$$
$$4HNO_3 \equiv 2H_2O + 4NO_2 + 2O$$
$$\overline{C + 4HNO_3 \rightarrow 2H_2O + 4NO_2 + CO_2}$$

This same reaction can be brought about by taking a bundle of wooden splints and fastening them together. The splints are lit, the flame blown out and the still glowing mass dipped into fuming nitric acid on a watch glass. Brown fumes are seen and the splints relight in the oxygen liberated.

(c) Phosphorus

$$2P + 5O \equiv P_2O_5$$
$$P_2O_5 + 3H_2O \equiv 2H_3PO_4$$
$$10HNO_3 \equiv 5H_2O + 10NO_2 + 5O$$

Add and divide by two:
$$P + 5HNO_3 \rightarrow H_3PO_4 + 10NO_2 + 5H_2O$$

If yellow phosphorus is used the action may be quite violent; it is safer to use the red variety and warm the mixture.

(d) Copper

The reaction of copper with concentrated nitric acid has already been considered. In the form of partial chemical equations the reactions can be set out:

$$Cu + O \equiv CuO$$
$$CuO + 2HNO_3 \equiv Cu(NO_3)_2 + H_2O$$
$$2HNO_3 \equiv H_2O + 2NO_2 + O$$
$$\overline{Cu + 4HNO_3 \rightarrow Cu(NO_3)_2 + 2H_2O + 2NO_2}$$

The copper here is oxidized to copper(II) nitrate; i.e. copper atoms become copper(II) ions. So the change:

$$Cu \rightarrow Cu^{2+}$$

is one of oxidation. Metal atoms on becoming ions lose electrons; this Loss of Electrons is Oxidation.*

With different concentrations of acid different products can be obtained. Concentrated acid gives nitrogen dioxide and moderately dilute acid nitrogen monoxide. Even the dilute acid is a good oxidizing agent.

(e) Potassium iodide

A little nitric acid added to a solution of potassium iodide oxidizes it to iodine (this is a common test for an oxidizing agent).

$$2HNO_3 \equiv H_2O + 2NO_2 + \emptyset$$
$$2KI + \emptyset \equiv K_2O + I_2$$
$$K_2O + 2HNO_3 \equiv 2KNO_3 + H_2O$$
$$\overline{2KI + 4HNO_3 \rightarrow 2KNO_3 + 2NO_2 + 2H_2O + I_2}$$

(f) Organic Matter

If concentrated nitric acid is dropped on a piece of rubber tubing there is much effervescence and nitrogen dioxide is evolved.

Uses

Many explosives such as nitroglycerine and trinitrotoluene (TNT) are produced by reactions between nitric acid and organic substances. Nitrates such as sodium nitrate and ammonium nitrate are good fertilizers. The latter is usually mixed with chalk (calcium carbonate) to form 'nitro-chalk' before use. Straw or cotton (cellulose) and the acid form nitrocellulose and solutions of this with pigments are used as cellulose paints; nitrophosphate fertilizers are made from phosphate rock and nitric acid.

Nitrates

The salts of nitric acid are called nitrates. Their preparation has been indicated on p. 118.

When nitrates are heated they all decompose: according to the manner in which this happens they can be allotted to one of four classes.

(i) Ammonium nitrate: this behaves differently from other nitrates and, as you have already seen, dinitrogen oxide is liberated on heating.

$$NH_4NO_3(s) \rightarrow N_2O(g) + 2H_2O(l)$$

(ii) Alkali nitrates, i.e. sodium and potassium nitrates: these form nitrites and evolve oxygen.

---

*Some metals such as iron and aluminium are only attacked by concentrated nitric acid for a short time; they then become 'passive' and no further action takes place.

$$2NaNO_3(s) \rightarrow 2NaNO_2(s) + O_2(g)$$

If a sample of one of these nitrates is melted and a little charcoal or sulphur added to the molten mass these elements immediately burn away to form the carbonate or sulphide and much heat and light are evolved. Potassium nitrate is a constituent of gunpowder (with charcoal and sulphur) and it provides the necessary oxygen for combustion. The volume of gas formed (mainly nitrogen and carbon dioxide) is immensely larger than the volume of the powder and results in enormous expansion—hence the propulsive force of gunpowder. (Sodium nitrate cannot replace the potassium salt as it is deliquescent.)

(iii) 'Heavy' metal nitrates, i.e. the nitrates of most other metals such as calcium, magnesium, zinc, lead and copper: these form the oxide and nitrogen dioxide along with oxygen.

$$2Mg(NO_3)_2(s) \rightarrow 2MgO(s) + 4NO_2(g) + O_2(g)$$
$$2Cu(NO_3)_2(s) \rightarrow 2CuO(s) + 4NO_2(g) + O_2(g)$$

(iv) The nitrates of mercury and silver behave like other 'heavy' metal nitrates and form the oxide, but further heating decomposes this to give the metal.

$$4AgNO_3 \equiv 2Ag_2O + 4NO_2 + O_2$$
$$2Ag_2O \equiv 4Ag + O_2$$

Simplifying: $2AgNO_3 \rightarrow 2Ag + 2NO_2\uparrow + O_2\uparrow$

Notice that the metals of group (ii) are those at the top of the electrochemical series and are strongly electro-positive and active. The metals in group (iv) are at the bottom of the table, are feebly electro-positive and not very reactive. Those in group (iii) occupy intermediate positions in the series.

All nitrates behave in the same way towards certain reactions and these can be employed as tests for the nitrate (NO_3^-) group.

(a) A solution of a nitrate is mixed with one of iron(II) sulphate and placed in a test-tube held at an angle (Fig. 25.6) while concentrated sulphuric acid is slowly and carefully poured down

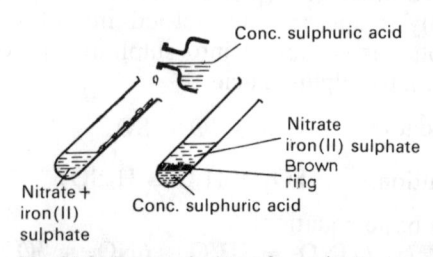

Fig. 25.6. 'Brown ring' test for a nitrate

the side of the tube so that it sinks to the bottom without mixing. Where the two liquids meet a brown ring appears. This test is known as the 'brown ring test'.

The explanation is fairly simple; the sulphuric acid and the nitrate form nitric acid and this is reduced by some of the iron (II) sulphate to nitrogen oxide. The oxide then reacts with more of the iron (II) salt to form nitroso iron (II) sulphate ($FeSO_4.NO$) as a brown ring. If the mixture is shaken the brown colour disperses with evolution of heat.

(b) When warmed with concentrated sulphuric acid and a little copper brown fumes of nitrogen dioxide are given off. There are two familiar reactions here:

a) the formation of nitric acid from the nitrate and the sulphuric acid.

b) the action of nitric acid on copper.

The Nitrogen Cycle in Nature

Unless certain elements are present in the soil much of its fertility is lost. These are mainly potassium, phosphorus and nitrogen with traces of such others as sulphur, magnesium, calcium, iron, etc. In this section we are concerned only with nitrogen.

When an element occurs in its natural elemental state we describe it as 'free'; if it is combined with other elements in a compound we say it is 'fixed'. Most plants and animals cannot take in 'free' nitrogen and make use of it, but have to obtain this element from the combined nitrogen in minerals present in the soil (either nitrates or ammonium salts). The plants extract this nitrogen, grow and then either die or are consumed by animals or humans. In the latter case the nitrogen is passed on as a constituent of the proteins in the body. When eliminated from the animal or when the plants die the nitrogen is returned to the soil as manure or compost, but in human settlements most of it passes away as sewage and is lost (some 'mud' from sewage works is used on allotments to restore the nitrogen). Unless nitrogen can be supplied to the soil, modern intensive cropping methods result in decreased amounts of crops.

Some plants such as peas, beans, clover, vetch and mosses are able to utilise 'free' nitrogen by means of bacteria on their roots which convert it into nitrates. Some nitrogen monoxide is produced during thunderstorms and this eventually reaches the ground as nitric acid and acts on minerals present to form nitrates. The decay of plants and the waste products of animals return proteins to the ground where they are acted upon by soil bacteria to give ammonia and ammonium compounds and by nitrifying bacteria to give nitrites and nitrates. By the removal of crops (often eaten far from where they are grown) or the washing away of nitrates the amount of fixed nitrogen becomes very low unless artificial means of replacing it are taken. Nitrogenous fertilizers such as ammonium sulphate or nitrate and sodium nitrate are added to the soil. Although some of these are obtained from other sources, the majority of such fertilizers depend on the fixation of atmospheric nitrogen as ammonia and on its subsequent conversion into nitric acid.

This supply and removal of nitrogen is called the 'Nitrogen Cycle' and is shown in a simplified form in Fig. 25.7.

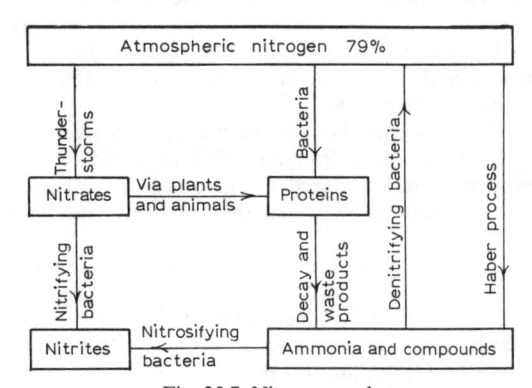

Fig. 25.7. Nitrogen cycle

Exercise 24–25

1 Give details (no diagrams) of the preparation of nitrogen (i) from the air, (ii) from a nitrogen compound. In what way would the samples possibly differ?

2 Draw a clearly labelled diagram to show how you would prepare ammonia in the laboratory. You must not start from ammonia solution. You must show how the ammonia is collected.

Write the word equation for the preparation, and construct underneath it the balanced chemical equation.

Describe what you would see if some ammonium chloride were heated in a dry test tube. Give a simple explanation of what is happening. (*EM*)

3 Give the chemical name of a compound containing nitrogen in each case which:

(*a*) is used in Leclanché cells;

(*b*) imparts an alkaline reaction to water;

(c) is a coloured gas, and heavier than air;
(d) is a salt used in testing for chlorides. (S)

4 Describe in outline the manufacture of ammonia, indicating how the starting materials are obtained. (Technical details are not required.)

Give a brief account of ONE experiment which shows that ammonia is very soluble in water. What ions are present in an aqueous solution of this gas?

Mention ONE use for each of the following salts: ammonium chloride; ammonium sulphate; ammonium nitrate. (OC)

5 (a) How would you *collect* a sample of ammonia gas in the laboratory?

(b) How and why does this method differ from the method of collection of (i) oxygen, (ii) dry carbon dioxide?

(c) How is ammonia manufactured on a large scale? (OC)

6 Describe the laboratory preparation of dry ammonia.

State THREE physical properties of ammonia.

Briefly describe the reaction between ammonia and (i) hydrogen chloride, (ii) heated copper(II) oxide.

Describe and explain what happens when ammonia solution is added to iron (III) chloride solution. (S)

7 (a) Give the reaction by which you would prepare nitric acid in the laboratory, say what conditions are necessary and draw a diagram of the apparatus required.

(b) How does dilute nitric acid react with: (i) copper (II) oxide, (ii) sodium carbonate?

(c) How does concentrated nitric acid react with copper? (C)

8 Draw labelled diagrams showing how you would prepare and collect, in the laboratory, (a) ammonia, (b) nitrogen monoxide.

Indicate briefly how ammonia is prepared on a large scale and converted into nitric acid. No sketch of technical plant is required, but give a careful account of the conditions under which the reactions take place. (O)

9 Draw a labelled diagram of the apparatus used in the laboratory preparation of nitric acid from sodium nitrate. Give the equation for the reaction.

Describe the action of concentrated nitric acid on (1) copper, (2) sulphur, (3) skin. (L)

10 Given lead(II) oxide (PbO) and dilute nitric acid, describe with essential experimental detail how you would prepare a dry crystalline specimen of lead nitrate. Describe the action of heat on this substance.

Give the names and formulae of TWO nitrogenous fertilizers, and explain briefly why the production of such fertilizers is so important in modern life. (OC)

11 (a) Two substances often used in gardening are lime and ammonium sulphate. Why is each used?

(b) (i) How would you prepare some ammonium sulphate crystals from ammonium hydroxide solution? (Colour due to indicator is not important in this case.)
(ii) What measurements, if any, would you make so that someone with no knowledge of chemistry could make the same crystals from your instructions?

(c) A bag of garden lime has been partly used over a number of years. The analysis on the bag reads:
Calcium oxide (CaO) 20%
Calcium hydroxide ($Ca(OH)_2$) 80%
yet on heating a sample strongly it gives off a gas which is found to turn lime-water milky. A second sample treated with dilute hydrochloric acid effervesces and this gas also turns lime-water milky. What is the name of the gas and how would you account for its presence in the two samples? (NR)

12 Describe a laboratory method of preparing and collecting jars of nitrogen monoxide.

What happens when a jar of this gas is opened to the air?

Give the names and formulae of TWO other common oxides of nitrogen. Write the equations representing the usual laboratory methods of preparing them. (J)

13 Describe how you would prepare a specimen of nitric acid, starting from potassium nitrate as the source of nitrogen.

Explain as fully as you can why the large-scale production of nitric acid is so important.

Given a sample of lead(II) nitrate, say briefly how you would prepare nitrogen dioxide. (OC)

14 (a) Describe how you would prepare and collect a few gas-jars of dry ammonia.

(b) What reactions occur between dry ammonia and (i) copper(II) oxide, (ii) hydrogen chloride? In EACH case describe any visible changes, name the products and state the conditions.

(c) Write equations for the action of heat on (i) copper(II) nitrate, (ii) potassium nitrate. (C)

15 Describe a laboratory method for the preparation and collection of dinitrogen oxide.

Give TWO tests by means of which the gas can be distinguished from oxygen.

Name and write the formulae of TWO other oxides of nitrogen.

What volume of nitrogen would be obtained by passing 150 cm³ of dinitrogen oxide over heated copper, all volumes to be measured at the same temperature and pressure? (J)

16 Explain how you could produce a fairly pure sample of atmospheric nitrogen in the laboratory.

Why is it difficult to produce pure nitrogen from the air?

Why do gardeners find it necessary to add ammonium sulphate to the soil at regular intervals?

What is the name of the nitrogen compounds which occur in the tissues of animals and plants?

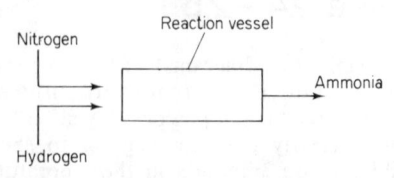

17 The diagram above is a simple flow diagram showing the essentials of the industrial production of ammonia.

(a) What are the conditions used for the industrial production of ammonia?
(b) Write a balanced equation for the reaction.
(c) What are the conditions necessary for the production of nitric acid from ammonia? Write the equations of the reactions for each stage of this production.
(d) What chemicals would you use to prepare nitric acid in the laboratory? Write a balanced equation for the reaction.
(e) Give an example of an oxidation reaction of nitric acid.
(f) Write a balanced equation for the action of heat on sodium nitrate.
(g) Give two uses each of ammonia and nitric acid.

(C)

26
Carbon, Carbon Monoxide and Fuels. Sources of Energy

You should be familiar with carbon dioxide, the gas produced by burning carbon, but may know very little about the element itself. Carbon, in combination with other elements, is present in every living thing and in everything that once had life and now forms part of the rocks and earth around us. Although it cannot be regarded as a very active element, it does have the unique property of linking its atoms together to form chains of almost any length and of twining these into branched and ring structures to form the 'skeletons' of an almost limitless number of substances which we call organic compounds. There are more of these compounds of carbon than there are compounds of all other elements put together.

Carbon occurs naturally in two very distinct forms, diamond and graphite, both of which may be classified as *mineral carbon*, although it is possible that they had a vegetable origin in the remote past.

Mineral Carbon

Diamonds are mined, chiefly in South Africa, and are so rare that the method used to obtain them is long and tedious. The soft rock (blue ground) in which they occur is exposed to the weather to cause it to crumble. It is then washed by a stream of water over a bed of grease, to which the diamonds (and other stones) adhere and they can afterwards be picked off. A good yield is one gram of diamond from every ten megagram of earth!* Diamonds are sometimes found as octahedral crystals (eight faces, see Fig. 26.1) and they all have this crystalline structure; they are not all colourless, many being black and opaque.

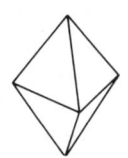

Fig. 26.1

Graphite occurs naturally in Siberia and Sri Lanka and was once mined in Cumberland under the name of plumbago. It is occasionally found as large hexagonal crystalline plates, dark grey in colour with an almost metallic lustre and greasy to the touch (Fig. 26.2). It is also manufactured in large quantities in the electric furnace from coke.

An element which exists in more than one physical form is said to show *allotropy* (polymorphism), or to be *allotropic* (polymorphic); graphite and diamond are both *allotropes* (polymorphs) of carbon. The physical properties of these are very different, but their chemical properties are identical. This is always so with allotropes. Few elements have such varied uses as carbon, and

*A megagram = 1 tonne.

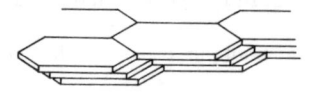

Fig. 26.2

these uses depend on the physical properties of the variety concerned. Figure 26.3 shows the arrangement of the carbon atoms in diamond, each atom being at the centre of a tetrahedron made up of four other carbon atoms. The interlocking structure is hard and rigid and renders diamond the hardest natural substance known; it will scratch all others.

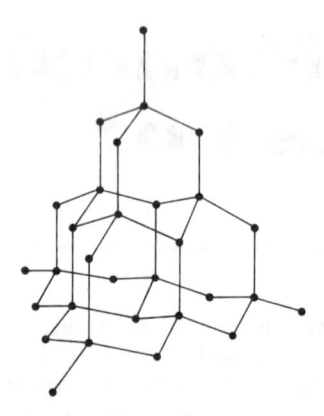

Fig. 26.3

Transparent diamonds are used as gems, but less valuable specimens, often discoloured, are used for cutting glass, for drilling hard materials such as rocks, and for drawing wire. For the latter purpose, a hole of the required diameter is made through a diamond by repeated tapping with a smaller diamond: the wire, of slightly greater diameter, has one end thinned and is then drawn through the hole so that the surplus metal is scraped off and a wire of uniform diameter is obtained. An improved method developed in London pushes the wire through the hole by an extremely high hydrostatic (i.e. fluid) pressure: wires of virtually unlimited length can be extruded without breaking—as may occur when they are pulled.

The flat crystals of graphite readily slide over each other (try pushing the top book of a pile of about six) giving it excellent 'dry' lubricating properties. It is also used as a 'wet' lubricant suspended in oil ('oildag') or in water ('aquadag'). Its crystal structure is shown in Fig. 26.4 and you will readily notice how different it is from that of diamond. The bonds between the carbon atoms in different

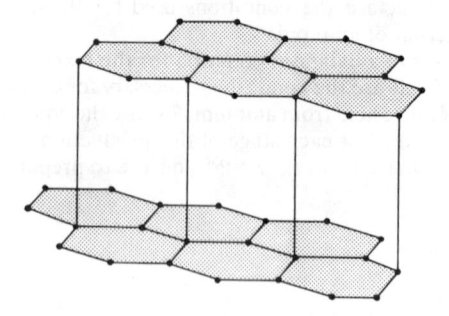

Fig. 26.4. Arrangement of carbon atoms in graphite.

layers are much weaker than those joining the atoms in any one horizontal layer and are easily 'broken'; hence the sliding of the crystals. Graphite is a good conductor of electricity along each layer, but a bad one in a direction perpendicular to this. Mixed with varied proportions of clay, pencil leads of different grades of hardness are obtained. It is also used as 'black-lead' for polishing ironwork, leather, etc. This name and the old name, 'plumbago', indicate a resemblance to lead—which also marks paper—but the combustion of graphite in oxygen (it does not burn in air), whereby carbon dioxide is the sole product, proves that no such relationship exists.

Diamond and graphite can be shown to be the same element by heating a weighed amount of each strongly in separate combustion tubes. A stream of oxygen is passed through each tube and the issuing gases are bubbled through a weighed quantity of potassium hydroxide solution. In each case the carbon burns to form carbon dioxide and this is absorbed by the aqueous potassium hydroxide with an increase in mass. A final weighing of the solid left in each tube will indicate how much carbon has been used and it will be found that 1 g of either form produces exactly the same mass of carbon dioxide, 3·667 g.

Vegetable Carbon

One form of carbon—charcoal—has been used in previous experiments. There are many kinds of charcoal which can be made by heating wood or other carbon-containing substances in the absence of air, or—at any rate—in the absence of sufficient air to burn it completely.

Experiment 26.1. The Preparation and Combustion of Charcoal.

(i) Place a few chips of wood in a crucible, fill up with clean sand and cover with a lid. Arrange it in a pipe-clay triangle on a tripod, and heat fairly strongly for about ten minutes. Note whether any fumes escape, and whether the underside of the crucible lid becomes blackened. Allow to cool, then tip out the contents into a basin and collect the charcoal. This may be placed in a deflagrating spoon, heated strongly and plunged into a jar of oxygen; it will burn away, leaving a little white ash (wood-ash) in the spoon.

If the ash is stirred into a little water containing a few drops of colourless phenolphthalein, the indicator turns pink showing that an alkali has been formed; the ash contains potassium carbonate, K_2CO_3. This and other minerals were taken up by the roots of the tree (from which the wood came) in the soil-water.

(ii) Place a few wooden splints in a narrow test-tube fitted with a short delivery tube and heat strongly. The wood does not burn, but a gas issues from the tube and may be lit, burning with a luminous flame. The charcoal left retains the shape and appearance, except for the colour, of the wood.

The old method (still used in some parts of the world) of making charcoal is to light a slow wood fire and cover it with turves to ensure slow combustion. Some of the wood burns and heats the rest; the latter is converted into charcoal. This method is extremely wasteful.

The modern method is to place the wood in iron retorts, heated by fuel gas. By this means, besides charcoal (left in the retorts), a watery liquid (containing valuable chemicals), wood tar (used for timber preservation) and an inflammable gas, available for helping to heat the wood, are produced. All these products are lost when wood is burnt.

Lamp black is a kind of soot made by the incomplete combustion of a mineral oil such as tung oil in the presence of sacking which catches the smoke. It has a very fine texture and, suspended in suitable liquids, is used for printer's ink, Indian ink and black paint. It gives colour and toughness to rubber (as in motor tyres) and is an ingredient used in the manufacture of gramophone discs.

Soot, itself, varies in composition according to the fuel which is burnt to produce it. The black deposit in a chimney, under which a coal fire has been burning, contains compounds of nitrogen and, possibly, of sulphur in addition to carbon.

Gas carbon, a hard, brittle substance, is left on the walls of the retorts when coal is heated to make coal-gas by a method similar to that just described for wood. A similar substance forms gradually on the piston crowns and cylinder heads of internal-combustion engines, due to the partial combustion of lubricating oil; this necessitates occasional 'decarbonizing'. It is the hardest and most compact form of carbon next to diamond. It is also a fair conductor of electricity, so that it is used for electrodes in dry and other batteries, and for the 'carbons' of arc lights.

Left in the retorts at gas works where coal is still heated is *coke*, the familiar smokeless fuel which is also much used in industry in the manufacture of iron, phosphorus and other substances.

If bones, blood and other animal refuse are heated in the absence of air, *bone black* (or animal charcoal) is left. This obviously contains bone-ash, which consists mainly of calcium phosphate(V) $Ca_3(PO_4)_2$, and in fact only a small part of animal charcoal is actually carbon.

The forms of carbon mentioned above may appear to be amorphous, but modern physical methods of analysis (by means of X-rays) have shown that they are composed of very tiny crystals of graphite (micro-crystalline), often interspersed with impurities. Some of the properties of the two allotropes of carbon and of charcoal are summarized in the table on p. 144.

Charcoal

The properties and uses of charcoal are so varied and interesting that we are going to study this 'variety' of carbon in more detail.

Charcoal is used as a fuel by blacksmiths and workers with wrought iron, etc., and as a reducing agent. Certain forms are very porous, capable of exposing enormous surfaces to gases and solutes, some of which are adsorbed* in large amounts. For this reason they are used in gas-masks, in the escape pipes of sewers and in charcoal biscuits for the relief of indigestion or flatulence. Charcoal is also used by artists for drawing purposes.

Experiment 26.2. The porous Nature of Charcoal.
Take a piece of compressed block charcoal and show that it floats on water. Now heat it very strongly for about ten minutes and place it under the water in a beaker. It should now sink (if not, reheat for a further five minutes) as the air originally 'trapped' in the 'pores' of the charcoal has been driven out by the heat.

Animal Charcoal is also used as a powerful adsorbent, as the following experiment will illustrate:

*Absorption and adsorption are not quite the same; blotting paper absorbs water which is taken up and penetrates right through to the inside of the paper: in adsorption, on the other hand, the substance taken up forms a very thin layer, possibly only one molecule thick, on the *surface* of the adsorbent.

Table 26.1

	DIAMOND	GRAPHITE	CHARCOAL
Crystalline form	Octahedral	Hexagonal plates	Apparently amorphous (micro-crystalline)
Colour	Often colourless	Dark-grey	Black
Optical properties	Often transparent. High refractive index	Opaque Lustrous	Opaque Dull
Hardness	Hardest natural substance known	Soft, slippery	Brittle. Hardness varies
Approximate density	3·5	2·5	1·5
Heated strongly in oxygen	All burn (if the temperature is high enough) to form carbon dioxide $C + O_2 \rightarrow CO_2$		

*Experiment 26.3. *Animal Charcoal used as an Adsorbent*

(a) *This experiment should be carried out by the teacher.* Place a few drops of bromine (CARE) in a gas jar by means of a teat-pipette; cover with a lid and shake. Brown fumes soon fill the jar. Drop in some activated charcoal (previously heated) and again shake; the fumes disappear.

(b) Place about 30 cm³ of a weak solution of methyl violet dye in a 100 cm³ stoppered cylinder, add about 5 cm³ in bulk of crushed activated charcoal. Shake well and filter. Wash the charcoal on the filter until the filtrate is colourless. Now add a few cm³ of alcohol to the charcoal; a violet filtrate is obtained.

(c) Boil some ordinary ink with about a quarter of its bulk of coarse animal charcoal for ten minutes. Filter and notice that the filtrate is colourless.

Draw your conclusions from these experiments.

Activated charcoal is made from animal charcoal by a process which involves heating it in a limited amount of air or in steam. This kind of Charcoal was sometimes used at coal gas works for removing the benzene from the coalgas.

Chemical Properties of Carbon

So far in this chapter we have confined ourselves almost exclusively to the *physical* properties of carbon. Carbon is not very active chemically; the exception, you will remember, is its power to undergo combustion in oxygen if heated strongly:

$$C + O_2 \xrightarrow{\Delta} CO_2$$

This property renders the element a powerful reducing agent at high temperatures. (You have met the term 'reduction' before in connexion with hydrogen which also had the power to remove oxygen from many substances containing it in combination.) The chief oxidizing agents in this process are the metallic oxides, of which many ores consist.

Experiment 26.4. To reduce Copper(II) Oxide with Charcoal.

Make an ignition tube out of a piece of ordinary glass tubing about four inches long. Two-thirds fill the bulb with an intimate mixture of powdered charcoal and excess copper(II) oxide. Heat, gently at first, then strongly, and test for the evolution of carbon dioxide by the method you are already familiar with. Remove some gas by means of the teat-pipette and bubble through lime-water. After heating for several minutes, allow to cool; then empty the contents into a test-tube containing a little water. Shake the tube, allow to stand for a few seconds, then pour off the water. Add more water and repeat the process. What does the water carry with it? What is the residue in the tube? If you are still uncertain (but the colour should be a guide) add a few drops of concentrated nitric acid to the residue. If any brown fumes (nitrogen dioxide) are seen, then copper has been formed. Copper(II) oxide does not cause these fumes to be produced.

Test for Carbon. The charcoal (carbon) used in this experiment may be replaced by almost any compound of this element and similar results obtained. This provides an important test for the element carbon in a compound; the latter, if a solid or liquid, is mixed with the copper(II) oxide and heated. If it is a gas it is passed over the heated copper(II) oxide. If carbon dioxide is detected by lime-water, carbon is present. You should repeat the experiment using sugar instead of charcoal.

Competition for Oxygen

In any oxide, there must be an attraction between the oxygen and the other element or the substance would readily decompose. The strength of this attraction or bonding varies with different oxides; in some—such as sodium oxide—it is very strong, whereas in others—such as mercury(II) oxide—it is relatively weak. Metals high in the electrochemical series form strong bonds with oxygen whilst those low in the series do not. Carbon and oxygen form quite a strong bond between them. When a metal oxide is heated with carbon, what happens depends on the relative strengths of the attraction between the metals and oxygen on the one hand and between carbon and oxygen on the other hand. If the former is the stronger, nothing will happen; if the latter, then the carbon wins the 'tug-of-war', gains the oxygen and the oxide is reduced to the metal. The experiment just carried out suggests that the carbon-oxygen bond is stronger than the copper-oxygen one. Let us now see what happens when other oxides (or compounds containing oxygen such as nitrates) are heated with carbon.

Experiment 26.5. Reduction with Charcoal.

Mix together with powdered charcoal (in separate dishes) (a) lead(II) oxide (litharge, PbO), (b) dilead(II) lead(IV) oxide (trilead tetroxide, Pb_3O_4), (c) bismuth oxide (Bi_2O_3), (d) silver nitrate ($AgNO_3$).

Take a 15 cm strip of asbestos paper and fold it lengthways down the middle to form a 'V'. Place some of mixture (a) on the strip and heat it strongly in the Bunsen flame with the airhole open. After some time (you may have to replenish the mixture if it gets blown off) you should see some grey metallic beads on the paper. They can be scraped off into a little concentrated nitric acid in a tube when brown fumes give proof that a metal has been obtained. Repeat with the other mixtures.

$$2PbO + C \longrightarrow 2Pb + CO_2$$
$$2Pb_3O_4 + 4C \longrightarrow 6Pb + 4CO_2$$
$$2Bi_2O_3 + 3C \longrightarrow 4Bi + 3CO_2$$
$$2AgNO_3 + C \longrightarrow 2Ag + CO_2 + 2NO_2$$
nitrogen dioxide

Experiment 26.6. To reduce Nitric Acid with Charcoal. Touch the surface of fuming nitric acid on a watch glass with a small bundle of red-glowing splints. The wood catches fire and brown fumes arise:

$$\overbrace{C + 4HNO_3 \longrightarrow \underbrace{CO_2 + 4NO_2}_{reduction} + 2H_2O}^{oxidation}$$

Carbon Monoxide

As well as carbon dioxide, there is another oxide of carbon in which the proportion of oxygen to carbon is only half as great; this is carbon monoxide **an extremely poisonous gas**. Although this can be made from carbon by carrrying out the combustion in a limited amount of air it is generally produced in the laboratory in other ways.

One method is to pass carbon dioxide through a combustion tube *packed* with charcoal and heated very strongly. A silica tube in a furnace should be used and the carbon dioxide dried through concentrated sulphuric acid or silica gel (see Experiment 50.7). Some of the gas is reduced by the hot carbon

$$CO_2 + C \longrightarrow 2CO$$

but a fair amount of the dioxide goes through unchanged and must be removed by bubbling it through potassium hydroxide solution.

$$2KOH + CO_2 \longrightarrow K_2CO_3 + H_2O$$

In Chapter 14 you learned that concentrated sulphuric acid is a strong dehydrating agent and will remove hydrogen and oxygen in the proportions to form water from many substances. This property is made use of in the next experiment.

Experiment 26.7. To Prepare Carbon Monoxide.

(i) Fit up the apparatus of Fig. 26.5 and let the concentrated sulphuric acid drip on to the methanoic acid (formic acid). After the air has been displaced carbon monoxide may be collected.

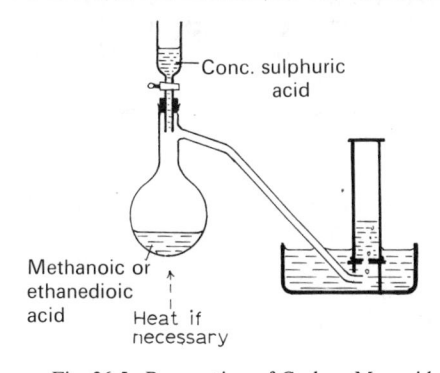

Fig. 26.5. Preparation of Carbon Monoxide.

(ii) Repeat using ethanedioic acid (oxalic acid) or sodium oxalate and pass the issuing gas through aqueous potassium hydroxide before collecting it in order to remove carbon dioxide. Collect several jars of the gas. (Heating will be necessary in this case.) **Be very careful not to breathe in any of the gas.**

Methanoic acid has the formula HCOOH (it is the acid ejected by ants when they sting their victims) and reacts with concentrated sulphuric acid according to the equation

$$HCOOH \xrightarrow{+c. H_2SO_4} CO + H_2O$$

Ethanedioic acid (a poisonous acid present in small quantities in rhubarb and spinach) is formulated $(COOH)_2$ and its reaction can be represented by

$$\begin{array}{c} COOH \\ | \\ COOH \end{array} \xrightarrow{+c. H_2SO_4} CO + CO_2 + H_2O \rightarrow$$

the broken lines showing how the molecule breaks up. Sodium ethanedioate, on the other hand, reacts chemically with sulphuric acid.

$$(COONa)_2 + 2H_2SO_4 \longrightarrow 2NaHSO_4 + H_2O + CO$$

Sodium hydrogensulphate is one of the products.

Experiment 26.8. A Look at the Properties of Carbon Monoxide.

(a) From the method of collecting carbon monoxide what conclusion can you draw about its solubility?

(b) Add some lime water to a jar of the gas and shake; is there any change?

(c) Carbon monoxide is slightly less dense than air. Hold a jar of the gas mouth downwards and insert a lighted splint well into the jar. What happens to the splint? What happens to the gas? As soon as the burning stops replace the cover and again add lime water and shake. Is there any change now?

(d) Connect the apparatus in which the carbon monoxide was produced via a Drechsel bottle of potash to a combustion tube containing some black copper oxide in a boat. Pass the gas through the tube while the oxide is being gently heated and burn the issuing gas at a jet so that none of the monoxide passes into the room. What happens to the solid? Is there any visible change?

The above tests show that carbon monoxide burns very readily. This is a reaction that takes place in a coal or coke fire. Coke is composed mainly of carbon and on burning forms carbon dioxide. (See Fig. 26.6.)

$$C + O_2 \longrightarrow CO_2$$

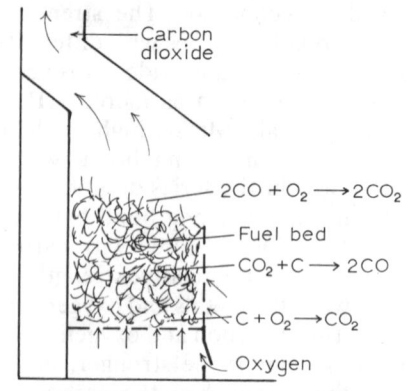

Fig. 26.6. Reactions in a coke fire.

The draught created by the chimney and the fact that hot gases are much less dense than cold air cause this carbon dioxide to rise through the hot fuel above it. Here it is reduced to carbon monoxide.

$$CO_2 + C \longrightarrow 2CO$$

At the top of the fuel bed the monoxide meets more air and burns with a blue flame.

$$2CO + O_2 \longrightarrow 2CO_2$$

A considerable amount of heat is given out when these reactions take place. This serves to maintain the temperature of the fuel bed and to warm the air in the room.

Experiment 26.8(d) will have shown you that carbon monoxide is a good reducing agent and can be used in place of hydrogen or carbon in obtaining some metals from their oxides.

$$CuO + CO \longrightarrow Cu + CO_2$$

Its reducing action can be shown in the following experiment.

Experiment 26.9. A Silver Mirror Produced by Carbon Monoxide.

Clean a test-tube with concentrated nitric acid and water. Half-fill it with a solution of silver nitrate and stand it in a bath of water at a temperature of about 60°C. Pass carbon monoxide slowly into the solution and in a short while silver should be deposited. It may come down as a grey powder, but if the cleaning has been well carried out a silver mirror should form on the inside of the tube.

Carbon monoxide is a *very poisonous gas* and deaths frequently occur through breathing coal gas (of which it is a constituent) or the exhaust fumes from cars in badly ventilated places. Oxygen is normally absorbed from the lungs by the blood to form a compound, oxyhaemoglobin. Blood containing oxygen is red in colour as in the arteries. During muscular activity carbon dioxide is produced by the oxidation of sugars by the oxygen in the blood.

The blood (now bluish in colour) carries the carbon dioxide to the heart and then to the lungs. If, however, the lungs are exposed to an atmosphere of carbon monoxide this gas is more readily absorbed than oxygen and bright pink molecules of carboxyhaemoglobin are formed. This compound is very stable and prevents the blood from performing its normal function of distributing oxygen to the different parts of the body. One part of carbon monoxide in 200 parts of air is sufficient to cause death in about two minutes and as little as one part in 500 will soon cause a collapse. Long tunnels used by cars and lorries are sometimes fitted with devices to record the concentration of carbon monoxide and automatically set fans in motion to remove it.

Fuels

Most of you could give some sort of answer to the question 'What is a fuel?'

A fuel is a substance which burns readily, but not too fast, liberating heat.

It may be a solid, a liquid or a gas. Some common fuels are listed here.

Solid	Liquid	Gaseous
Anthracite	Ethanol	Coal gas
Coal	Benzene	Producer gas
Coke	Paraffin	Water gas
Peat	Petrol	Natural gas
Wood	Diesel and other oils	(mainly methane)
Smokeless fuels such as Gloco, Cleanglow, etc.		Butane (Calor gas)

If you know anything about the composition of the above substances you will be aware that they all contain carbon and nearly all also contain hydrogen. Most fuels burn with a flame although a few—e.g. coke—only glow. The production of a flame is a sign that the fuel is vaporizing; unless it does so there is no flame. The solid fuels listed are burnt in open grates or in closed stoves, some being better for particular purposes than others. The liquid and gaseous fuels also require special burners for their combustion. The lowest temperature at which a fuel will take fire is called its **ignition temperature**, this is very low for petrol but high for coke. Coke, along with coal gas, is obtained from coal. Peat is a variety of coal which has not been so long in being formed as ordinary bituminous coal. Anthracite is similar to coal, but it requires a higher temperature to ignite it. Most liquid fuels are obtained by the refining and 'cracking' of petroleum (see p. 335). 'Cracking' involves the breaking down of large molecules into smaller ones. Coal gas, producer gas and water gas are made industrially, but natural gas (mainly methane) occurs in many parts of the world, sometimes associated with sources of oil.

Energy Values of Fuels

The choice of a fuel is often determined by the convenience of its use, but a comparison between different fuels can only be made on the basis of the amount of heat energy given out by the burning of a definite mass of fuel (its 'calorific value'). (In the case of gaseous fuels it is more convenient to use volumes rather than masses.)

The heat given out is measured in joules or kilojoules (these units have replaced the calorie and kilocalorie previously used).

The joule may be defined as the energy expended by the passage of 1 ampere of current for 1 second between two points at a potential difference of 1 volt.

In terms of *heat energy* we can write a special definition and say:

One kilojoule is the amount of heat energy needed to raise the temperature of one kilogram of water $0.24\,°C$ (or to raise the temperature of 2.4 g of water from $0\,°C$ to $100\,°C$).

In industry at the moment the unit concerned is the British Thermal Unit which is defined as

the amount of heat required to raise the temperature of one pound of water one degree Fahrenheit.

A larger unit, the therm, = 100 000 Btu and this is in general use in the gas industry at present. The calorific value of a gaseous fuel is expressed in Btu/ft^3, but as soon as practicable a change is to be made. From then onwards calorific values will be expressed in megajoules per standard cubic metre, i.e. MJ/m^3. British Thermal Units per cubic foot can be converted to the new system by multiplying by 0·0379. (A megajoule is one million joules.)

Many of the above fuels are naturally-occurring ones, but others are artificially produced, either coal or petroleum being the raw material. Coal is the remains of giant plants that lived over 200 million years ago. For a long time it has been one of the most valuable of all raw materials and many valuable substances can be obtained from it. These are lost when coal is burnt in open grates (and, in addition, much smoke is produced to pollute the atmosphere). Petroleum (Latin: *petra*, stone and *oleum*, oil) is a complex mixture of liquid and gaseous hydrocarbons, varying from the gas, methane (CH$_4$) to the black viscous liquid, pitch. It is generally thought to have originated from the decomposition of minute marine plants and animals. From it, by the process of fractional distillation, is obtained most of the petrol or gasoline, lubricating oils and fuel oils needed in the world today.

Coal gas, as the name tells us, is made from coal by heating it in iron retorts in the absence of air and purifying the gas given off. This process can be simulated in the laboratory.

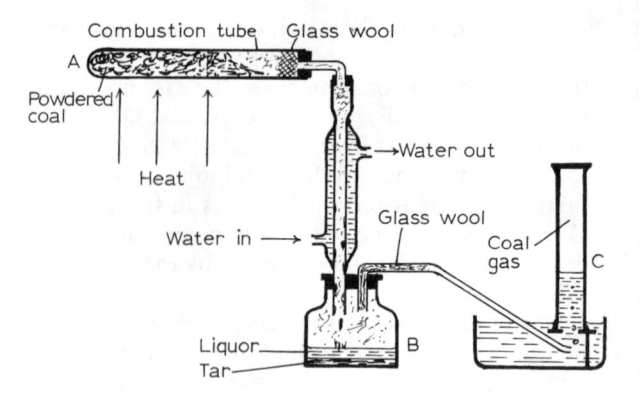

Fig. 26.7. Preparation of crude coal gas.

Example 26.10. Heating Coal in the Absence of Air.

The apparatus shown in Fig. 26.7 represents the iron retort (*A*), the tar and liquor well (*B*) and the gas holder (*C*).

When the coal is heated strongly dense fumes are evolved and these partly condense in the cooled tube to form a black tarry liquid which eventually separates into two layers, the upper one being lighter in colour. It contains dissolved ammonia and is called ammoniacal liquor. The ammonia can be recovered by heating the liquor with calcium hydroxide and is used industrially for the production of the fertilizer, ammonium sulphate. Between the tar well and the holder at a gas works, other pieces of plant are inserted in which physical and chemical reactions remove most of the impurities (such as hydrogen sulphide) from the gas. The coal gas can be collected over water and shown to burn with a luminous flame forming water (mist on the sides of the jar) and carbon dioxide (test with lime-water). Wood can be used in this experiment in place of coal and similar products will be obtained (but with charcoal formed instead of coke).

The chief constituents of coal gas are hydrogen

Calorific Value of Fuels.

Fuel (dry)	Btu/lb	MJ/kg	Btu/ft^3	MJ/m^3
Coal*	11–15 000	26–35		
Coke*	12 000	28		
Wood*	7 400	17		
Charcoal (pure)	14 555	33·8		
Petroleum*	19 800	46		
Ethanol (meths)	11 150	29·9		
Producer gas*			150–200	5·7–7·6
Hydrogen			250	9·5
Blue Water gas*			300–350	11·3–13·2
Coal gas*			450–550	17·0–20·8
Natural gas*			1000	38

*These values vary, average figure being given.

and methane with smaller amounts of carbon monoxide, hydrocarbons, oxygen, carbon dioxide and nitrogen (the last three are really impurities). Coke is left in the retorts and is used as a fuel, for the production of water gas and in preparing smokeless fuels for domestic grates.

Being mainly carbon, it is a good reducing agent and is used in the extraction of metals such as iron and zinc from their ores. Much coke is made in special coke-oven plants where the coal gas obtained is of secondary importance. From the tar, by distillation, compounds may be obtained for conversion into dyes, drugs, plastics, perfumes and explosives.

Producer gas is a much cheaper fuel used in industry and is usually made where it is required. Air is blown through red-hot coke which becomes incandescent. Carbon monoxide is formed and this with the unused nitrogen from the air makes up producer gas.

$$2C + (4N_2 + O_2) \longrightarrow (2CO + 4N_2)$$
$$\text{air}$$

About 33% of the gas will burn, giving out carbon dioxide as it does so.

Blue water gas is made at coal-gas works by blowing steam through coke burning at white heat (about 1000°). This gives a mixture of carbon monoxide and hydrogen which, theoretically, is 100% combustible.

$$CO + H_2O \longrightarrow (CO + H_2)$$

This mixture is called blue water gas. The steam used, however, cools down the coke and the reaction would stop, but the temperature of the fuel bed is maintained by blowing air and steam alternately through it. The reaction with air is an exothermic one, much heat being given out, while that with steam is endothermic. Blue water gas, so called because both its constituents burn with blue flames, is usually enriched by 'cracking' oil to add hydrocarbons and it is then known as **carburetted water gas**. This is mixed with coal gas to form **town gas.**

During the last few years a considerable increase in the price of coal suitable for gasmaking has necessitated economies in the industry in order to keep town gas at a competitive price. Many small works have been eliminated, a gas grid set up and research carried out on other raw materials to replace coal. Petroleum fractions available from oil refineries can, by catalytic treatment, be converted into a gas with a calorific value similar to that of water gas. Oil is easier to handle than coal and has a higher hydrogen-carbon ratio-, two points in favour of its use.

Modern 're-forming' gas plants use a petroleum feedstock such as naphtha which is liquid by-product of oil refining and consists of a mixture of hydrocarbons in which the number of carbon atoms is generally between 5 and 10: the compounds present vary from pentane, C_5H_{12} to decane, $C_{10}H_{22}$.

This feedstock is vaporized, mixed with steam and passed over a heated nickel catalyst to give a mixture of carbon monoxide, carbon dioxide, hydrogen and a little methane. After this has been passed through another catalyst consisting of iron and chromium and then enriched by addition of more methane (CH_4) or propane (C_3H_8) the gas has quite a high calorific value. Finally, as the gas has no smell, a small quantity of tetrahydrothiophene (a sulphur compound) is injected into the gas stream to remedy this and to give warning in case of leaks.

The first large supplies of natural gas (mainly methane) were brought to Britain in liquid form by refrigerated tanker from Algeria to be mixed with gas from more traditional sources. As the pipelines conveying natural gas from the North Sea finds creep over the countryside even these modern reforming plants will become obsolete and in a few years time coal gas—and even oil gas—will become almost as rare as it already is in the United States where natural gas is piped all over the country.

Combustion of Fuels

All fuels on burning form carbon dioxide and steam as can be shown using the apparatus of Fig. 26.8. The products of combustion, along with excess air, are drawn through (a) a tube containing a little anhydrous copper(II) sulphate, the whole being surrounded by cold water to help the condensation of any steam formed and (b) a tube containing lime-water. Can you write the equation for the change that takes place in the lime-water?

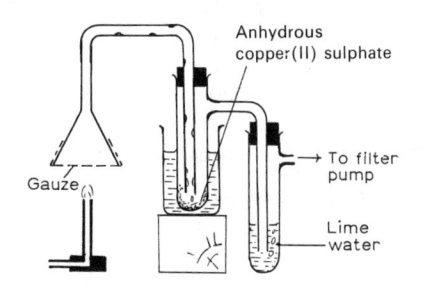

Fig. 26.8. Identifying products of combustion

If a semi-micro burner is not available an ordinary Bunsen with the barrel removed can be used; do not have the flame more than half-an-inch high. Some fuels, such as alcohol (see p. 329), provide part of the oxygen needed for their combustion. If the experiment is repeated using a small alcohol burner as shown in Fig. 26.9 the same results will be obtained.

$$C_2H_5OH + 3O_2 \rightarrow 2CO_2 + 3H_2O$$

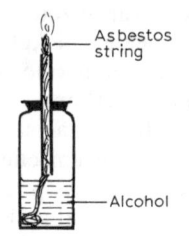

Fig. 26.9. Alcohol burner

Fig. 26.10. Luminous Bunsen flame (a); Non-luminous Bunsen flame (b)

Flame

It was once thought that fire was an actual substance, but we now know that the production of fire or flame is an indication that a chemical change is occurring between gases or vapours. It is a change which releases sufficient energy for one of the reacting substances to be raised above its ignition temperature so that it burns.

Let us consider what happens in the flames produced in the Bunsen burner. The type of flame depends on whether the air-hole at the bottom of the barrel is open or not. If it is, then air is sucked in by the rising gas stream and chemical changes can occur more readily than in the second case where the only air available to assist combination is that which diffuses in from the surrounding atmosphere.

The luminious Bunsen Flame

This is usually somewhat unsteady and flickers, but three zones can be recognized (Fig. 26.10a).

(A) a small dark zone in which the gas is unburnt and therefore cool;

(B) a large yellow region where incomplete combustion goes on. Some of the hydrocarbons in the gas, particularly methane and ethene, burn to form carbon monoxide and steam and much carbon is liberated. The amount of available oxygen is small. The main reactions are

$$C_2H_4 + O_2 \rightarrow 2C + 2H_2O$$
and $$CH_4 + O_2 \rightarrow C + 2H_2O$$

The carbon particles glow brightly and make the flame luminous. Some of the hydrogen in the coal gas also burns to form steam. If a piece of broken pot is held in this part of the flame it becomes covered with soot, particularly if the flame is small. The oxygen necessary for this partial combustion diffuses into the zone from the surrounding air, all the molecules being in a continuous state of rapid movement.

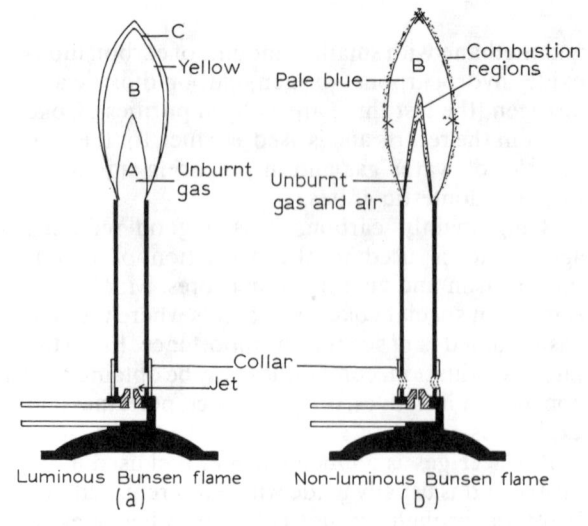

(C) a thin, almost invisible, outer mantle where combustion is complete; all the oxygen is now obtained from the air outside the flame. Here the reactions are

$$C + O_2 \rightarrow CO_2 \quad 2CO + O_2 \rightarrow 2CO_2$$
$$2H_2 + O_2 \rightarrow 2H_2O$$

The non-luminous Bunsen Flame (Fig. 26.10(b))

If the air-hole is wide open the flame roars and will become unsteady if the gas tap is gradually closed. Eventually the burner may 'light-back'. This can be corrected by turning the gas on full and striking the rubber tube sharply with the side of the hand. In normal use the air-hole should be about half-open so that the gas burns quietly without any yellow tinge to the flame.

Two main zones can be recognized if the flame is viewed against a dark background (in modern all-glass laboratories it is often extremely difficult to see if a Bunsen is actually lit).

(A) cool, unburnt gas as in the luminous flame;

(B) a pale-blue zone.

There are two regions of combustion on the inner and outer edges of zone B (remember that oxygen is needed for combustion).

On the inner edge where primary air rising through the barrel with the gas is available, carbon monoxide and steam are formed.

$$2CH_4 + 3O_2 \rightarrow 2CO + 4H_2O$$
$$C_2H_4 + 2O_2 \rightarrow 2CO + 2H_2O$$

B is a very hot zone containing molecules of coal gas, carbon monoxide and steam at a high temperature, but little if any oxygen. Further burning takes place on the outer fringe of the zone where secondary air from the surrounding atmosphere is available. Here carbon dioxide and more steam are formed as in the luminous flame.

The hottest parts of the flame are indicated by a cross. With a good bunsen the temperature may reach about 1000°C. The burners on gas cookers and in the ovens are made to give a large number of miniature non-luminous bunsen flames; the luminous type would obviously be unsuitable as soot would be deposited on the cooking vessels.

Experiment 26.11. The Bunsen Burner.
Try to find out the nature of the different zones of the flames by the following means. Try each test on both types of flames.

(*a*) Stick a pin through a match close to its head and suspend it in the barrel (Fig. 26.11(*a*)). Light the gas. Does the match take fire immediately?

(*b*) Hold a small piece of glass tubing with one end in zone *A*. Apply a light to the other end. What do you notice (Fig. 26.11(*b*))?

(*c*) Obtain a square of thick card and hold it for *a moment* across zone *A*. Take out and examine. Does this give any further clue (Fig. 26.11(*c*))?

(*d*) Hold a piece of porcelain or fire-clay in zone *B*. Is there any difference between the effects of the two flames?

(*e*) Obtain a wire-gauze without an asbestos centre and try to flatten the flame with it (Fig. 26.11(*d*)). Why does the flame eventually pass through the gauze?

(*f*) With the gas off, hold the gauze about 2·5 cm above the barrel and light the gas *above* the wire. Why does it take time for the gas below the gauze to catch fire? (The reactions given are for coal gas and will not all occur with natural gas (methane).)

The Carbon Cycle

Carbon dioxide is always present in the atmosphere, being produced by

(*a*) combustion in fires or in petrol and oil engines,
(*b*) respiration,
(*c*) 'burning' of limestone in lime-kilns.

Much of it is absorbed by growing plants in order to build up their cells. A substance called chlorophyll is present in the leaves and green parts of plants and this acts as a kind of energy trap in converting the carbon dioxide and water (taken in through the roots) into sugars and then into starch or cellulose. The energy present in sunlight brings this change about and produces oxygen to balance that taken from the air during combustion and breathing.

$$\text{carbon dioxide} + \text{water} \xrightarrow[+ \text{sunlight}]{\text{chlorophyll}} \text{sugar} + \text{oxygen}$$
$$6CO_2 + 6H_2O \rightarrow C_6H_{12}O_6 + 6O_2$$

When foods containing sugars (carbohydrates) are eaten the change is reversed. The sugar is oxidized by the oxygen absorbed in the blood stream and converted into carbon dioxide which is breathed out. Considerable energy in the form of heat is produced and this serves to keep our body temperatures above that of our surroundings and reasonably constant at 36·9°C.

$$C_6H_{12}O_6 + 6O_2 \rightarrow 6CO_2 + 6H_2O + \text{heat energy}$$

The percentage of carbon dioxide in the air breathed out is about 100 times as great as that normally present in the atmosphere, 4% instead of about 0·04%.

There are other processes helping to maintain the amount of carbon dioxide in the air approximately constant; there are, of course, local variations, the percentage being higher in industrial areas than in the country. The gas is absorbed by lakes and rivers and washed down by rain into the soil. There it may dissolve out calcium carbonate to form the hydrogencarbonate which eventually reaches the sea and is used by shell fish for building up their shells. After their death they fall to the bottom and over many centuries have laid down vast beds of chalk. Upheavals in past time have caused these beds to become part of dry land and we now use the chalk or limestone to produce quicklime and so the

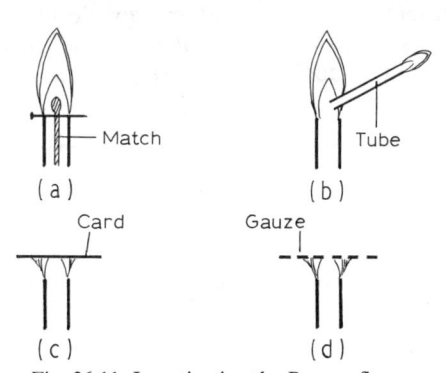

Fig. 26.11. Investigating the Bunsen flames

carbon dioxide is once more returned to the atmosphere. The sea acts as a reservoir maintaining a balance between the processes increasing the carbon dioxide and those which reduce it. So the carbon cycle continually goes on. This is represented in Fig. 26.12.

Photosynthesis (building through light), the process converting carbon dioxide into sugars, needs the energy present in sunlight for it to take place. This energy reappears when the reverse change occurs in our bodies and, as stated, helps to maintain a constant body temperature. All life on earth, animal or vegetable, ultimately depends for its maintenance on the energy of the sun. This in turn is derived from the conversion of hydrogen into helium and so it is really by atomic energy that life continues.

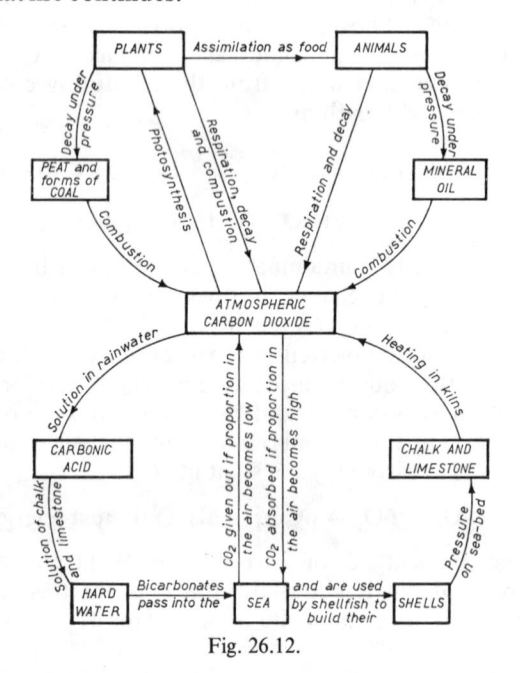

Fig. 26.12.

Sources of Energy*

All fuels are sources of energy in the form of heat, but this is only one of many different types of energy. Let us look at some of the others (see also chapter 19).

Domestic household appliances use electrical energy and convert it into some other forms e.g. light energy from the lamp or sound energy from the bell. A battery, on being charged, undergoes a chemical change and new substances having a higher energy content than the original ones are formed. The energy absorbed is chemical energy. In every chemical change there is also a change of energy content and this usually appears as heat. If it is given out we call the reaction exothermic and if it is taken in, endothermic.

Chemical actions may be brought about by the use of different forms of energy; heat energy in combustion, light energy in photography, electric energy in electrolysis, mechanical energy in explosions, solar energy in photosynthesis. We have seen that in ordinary chemical changes the nucleus of the atom is not in any way affected, only the outer electrons being concerned. But in radioactive changes disintegration of the nucleus occurs with the release of considerable heat or atomic (nuclear) energy. Similar changes can now be brought about by artificial means by bombarding the nucleus of an atom such as uranium with fast-moving neutrons. In such a reaction the law of conservation of mass no longer holds; some very small amount of mass disappears and is converted into an enormous amount of energy. (The law should really be called the law of conservation of mass-energy.) These reactions are known as nuclear fission, but light elements can also be converted into heavier ones at extremely high temperatures such as exist on the sun. It has been calculated that when four moles of hydrogen atoms are converted into one mole of helium (nuclear fusion) 0·029 g of matter is converted into 2600 gigajoules of heat.

Nuclear (fission) energy has now been harnessed for the production of electrical energy. Even so, at present the amount of energy derived from nuclear sources is only about 1% of the total used in the world, coal still supplying about 40% and oil and natural gas about 45%.

*See index for other references to the various types of energy.

Exercise 26

1 Give reasons why a motor car engine should not be allowed to run in a closed garage. (*WR*)

2 Draw a labelled diagram of a simple apparatus in which you could heat coal to show what products are formed. List four main products which are obtained. Now consider each product in turn, and say what are the main uses of it.

Where does petrol come from? Mention three other important materials obtained from the same source. (*EM*)

3 Name (*a*) two solid fuels, (*b*) two liquid fuels, and (*c*) two gaseous fuels other than coal gas.

Explain in your own words what we mean by the word 'combustion'.

Say what sort of things are produced in combustion, and the conditions necessary for combustion to occur.

What dangers arise by the incomplete combustion of fuels?

Draw two simple diagrams to show (*a*) a bunsen flame in which combustion is complete (or as complete as possible), and (*b*) a bunsen flame in which combustion is incomplete. Take care to label each, and to show the structure of each flame. (*EM*)

4 Underline the two ELEMENTS to be found in the majority of FUELS used by man. Carbon. Nitrogen. Sulphur. Chlorine. Hydrogen. Salt. Argon.

When fuels containing these elements are burnt, what compounds are most often produced? In an explosion of one of these fuels what three forms of ENERGY are likely to be released? (*SR*)

5 (*a*) Give two reactions in which carbon monoxide is a product. State the conditions in each case.

(*b*) Describe how you would prepare and collect a few jars of carbon monoxide in a fairly pure state.

(*c*) What reactions, if any, have (i) carbon monoxide, (ii) carbon dioxide, on a cold solution of sodium hydroxide and on strongly heated copper oxide? (Products and equations only.) (*C*)

[NO diagrams except in Section (*b*).]

6 (*a*) Name the FOUR chief products obtained when coal is heated strongly in large retorts.

(*b*) Explain briefly how each of these products is obtained separate from the others.

(*c*) For each product give its most important use.

(*d*) Name the TWO chief constituents of coal gas.

(*e*) Explain the reactions that take place in a coke fire. (*C*)

7 Give the approximate composition of (*a*) producer gas, (*b*) water gas. Say briefly how producer gas and water gas are manufactured. Which is the better fuel?

Draw clear, labelled diagrams to show (*a*) the construction of a Bunsen burner, (*b*) the nature of the flame when the air-hole is fully open. (*OC*)

8 Describe the preparation and collection of carbon monoxide.

How, and under what conditions, does carbon monoxide react with (*a*) oxygen, (*b*) iron(III) oxide, (*c*) steam? (*J*)

9 Neither carbon monoxide nor carbon dioxide will support animal life yet only the former is considered poisonous. Explain fully why this is so.

10 Study the reactions taking place in a coke fire and then draw a diagram of a piece of apparatus suitable for preparing carbon monoxide by a similar means in the laboratory.

11 Draw **two** clearly labelled diagrams one showing how you would prepare and collect jars of carbon dioxide, the other illustrating how you would obtain a sample of carbon monoxide, free from carbon dioxide. Illustrate each preparation with an equation.

Describe **two** tests in each case, by means of which you could identify each of these gases.

Explain why the exhaust fumes from motor cars are dangerous to health. (*C*)

27
More Laws of Chemistry; Volume Relationships

In the latter half of the eighteenth century Henry Cavendish was investigating the reaction between hydrogen and oxygen and found that in the formation of water two volumes of hydrogen always combined with one volume of oxygen. A French scientist, Gay Lussac, studied similar reactions between gases and noticed that:

1 volume of nitrogen + 1 volume of oxygen →
2 volumes of nitrogen monoxide.

1 volume of nitrogen + 3 volumes of hydrogen →
2 volumes of ammonia

2 volumes of carbon monoxide + 1 volume of oxygen → 2 volumes of carbon dioxide.

2 volumes of hydrogen + 1 volume of oxygen →
2 volumes of steam.

You will see from these figures that there is a simple relationship between the volumes concerned.

In 1808 Gay Lussac put forward what is now

known as Gay Lussac's Law and can be stated:

When gases react together they do so in volumes which bear a simple relationship to each other and to the products if gaseous, all volumes being measured under the same conditions of temperature and pressure.

This is a perfectly general law where gases are concerned, but it is not applicable to solids or liquids. In the reactions below the only *volume* relationships are between the gases.

Solid carbon + 1 volume of oxygen $\xrightarrow{\Delta}$ 1 volume of carbon dioxide.
Solid ammonium nitrate $\xrightarrow{\Delta}$ 1 volume of dinitrogen oxide + 2 volumes of steam.

Gases differ in many of their physical and chemical properties yet they obey the laws of Charles and Boyle and this one of Gay Lussac. So there must be something that gases have in common. It was about this time that Dalton was spreading his ideas about atoms, but no distinction was made between atoms of elements and 'compound atoms'. In 1811 an Italian scientist, Avogadro, put forward the idea of molecules in place of these 'compound atoms', each molecule containing at least two different atoms (although, later, molecules of elements were also recognized). He suggested that:

Equal volumes of all gases under the same conditions of temperature and pressure contain the same number of molecules.

This hypothesis was not immediately accepted; in fact it was only after Avogadro's death that another Italian chemist, Cannizzaro, turned people's attention towards his ideas which then found universal acceptance. The above statement is known as Avogadro's Hypothesis or Avogadro's Law. If we look at the reaction between hydrogen and chlorine in the light of these ideas we have:

1 volume of hydrogen + 1 volume of chlorine → 2 volumes of hydrogen chloride. i.e. *n* molecules of hydrogen + *n* molecules of chlorine → 2*n* molecules of hydrogen chloride.
∴ 1 molecule of hydrogen + 1 molecule of chlorine → 2 molecules of hydrogen chloride.
∴ each molecule of hydrogen chloride contains $\frac{1}{2}$ molecule of hydrogen and $\frac{1}{2}$ molecule of chlorine.
As the molecules of hydrogen and chlorine contain two atoms (to be proved later), the molecule of hydrogen chloride must contain 1 atom of hydrogen and 1 atom of chlorine.
∴ the formula for hydrogen chloride must be HCl.

We can illustrate this and similar reactions as below:

$$H_2 + Cl_2 \longrightarrow 2HCl$$

Similarly:

$$2H_2 + O_2 \longrightarrow 2H_2O$$

and

$$N_2 + 3H_2 \longrightarrow 2NH_3$$

The importance of this idea of Avogadro's is that it enables us to change directly from consideration of volumes (determined practically) to a consideration of molecules or vice versa.

The equation

$$CH_4 + 2O_2 \rightarrow CO_2 + 2H_2O$$

tells us that one volume of methane combines with two volumes of oxygen to form one volume of carbon dioxide and two volumes of steam. This means that in practice 1 dm³ of methane will use up 2 dm³ of oxygen when burning to form 1 dm³ of carbon dioxide and 2 dm³ of steam. If the products of the reaction cool so that steam condenses the volume of the water formed is negligible (only about 1 cm³ of water is formed from 1700 cm³ of steam) and is ignored in calculations.

It has been shown above that use is made of Avogadro's Hypothesis in determining the molecular formulae of compound gases.

EXAMPLE (a). Assuming air to contain 21% of oxygen, how much will be required for the complete combustion of 300 cm³ of methane (CH_4) and 150 cm³ of acetylene (C_2H_2)?

Write down the equations and underneath write the reacting volumes as deduced from Avogadro's Hypothesis.

$$CH_4 + 2O_2 \rightarrow CO_2 + 2H_2O$$
1 vol.   2 vols.
300 cm³   600 cm³

$$2C_2H_2 + 5O_2 \rightarrow 4CO_2 + 2H_2O$$
2 vols.   5 vols.
150 cm³   375 cm³

Total volume of oxygen needed = 975 cm³.

$$975 \times \frac{100}{21} = 4643 \text{ cm}^3 = \underline{4 \cdot 643} \text{ dm}^3$$

So far we have assumed that the molecule of hydrogen contains two atoms, but this fact must now be proved.

The Atomicity of the Hydrogen Molecule

The statements given above will, for convenience, be repeated here.

1 volume of hydrogen + 1 volume of chlorine → 2 volumes of hydrogen chloride.

∴ 1 molecule of hydrogen + 1 molecule of chlorine → 2 molecules of hydrogen chloride. (*What statement is omitted here?*).

∴ 1 molecule of hydrogen chloride contains ½ a molecule of hydrogen and ½ a molecule of chlorine.

As these ½ molecules cannot be less than 1 atom, each molecule of hydrogen (or chlorine) must contain *at least* 2 atoms.

Therefore the molecule of hydrogen (or chlorine) contains twice as many atoms of hydrogen (chlorine) as are present in each molecule of hydrogen chloride.

When an acid such as sulphuric which contains two atoms of replaceable hydrogen reacts with a base such as sodium hydroxide it is possible to form two different salts, sodium sulphate and sodium hydrogensulphate. But, under the same conditions, only one salt has ever been made using hydrochloric acid and a particular base. This shows that there can be only one atom of replaceable hydrogen in hydrochloric acid.

Therefore the molecule of hydrogen contains two atoms.

Most gaseous elements contain two atoms, i.e. they are *diatomic*, examples being hydrogen, oxygen, chlorine, nitrogen, fluorine and bromine (as vapour). The rare gases in the air, helium, neon, argon, krypton and xenon are monatomic, i.e. they contain only one atom in their molecules. Compound gases vary more in the number of atoms in the molecule. The atomicity of carbon monoxide (CO) is two, of hydrogen sulphide (H_2S) and carbon dioxide (CO_2) three, of ammonia (NH_3) four, of methane (CH_4) five and of ethene (C_2H_4) six.

Another assumption we previously made (see p.45) was that the vapour density of a gas is half the relative molecular mass. This is quite easy to prove, but it is not essential at this stage.

Remember then, that:

vapour density is half the relative molecular mass

and

relative molecular mass is twice the vapour density.

EXAMPLE (b). What volume of carbon dioxide is produced when 450 cm³ of butane (C_4H_{10}) are burned?

Remember that the only products formed when hydrocarbons are completely burned are carbon dioxide and water or steam.

The equation must be worked out:

$$C_4H_{10} \rightarrow CO_2 + H_2O$$

Obviously, 4 volumes of carbon dioxide and 5 volumes of steam will be formed from each volume of butane. The oxygen needed will be 6½ volumes. So we write:

$$C_4H_{10} + 6\tfrac{1}{2}O_2 \rightarrow 4CO_2 + 5H_2O$$

(Notice that there is no objection to using fractions in equations of this type when we are referring to *volumes*.)

1 volume of butane produces 4 volumes of carbon dioxide, so 450 cm³ of butane produce 1800 cm³ or $\underline{1 \cdot 8}$ dm³ of carbon dioxide.

Molar Volume

We have already used the term mole; one mole of hydrogen molecules has a mass of 2·016 g (we are using the *exact* figure here). Let us find the volume this would occupy at 0°C and 760 mm pressure.

1 dm³ of hydrogen at s.t.p. has a mass of 0·0899 g
2·016 g hydrogen occupy

$$\frac{2 \cdot 016}{0 \cdot 0899} = 22 \cdot 4 \text{ dm}^3$$

This volume, 22·4 dm³ is called the molar volume (M.V.), and is the same for all gases. Why should this be so?

One mole of hydrogen contains 6×10^{23} molecules (Avogadro's number).
i.e. 22·4 dm³ of hydrogen at s.t.p. contain 6×10^{23} molecules of hydrogen.

∴ 22·4 dm³ of oxygen at s.t.p. will contain 6×10^{23} molecules of oxygen.

But this is one mole of oxygen so 22·4 dm³ is also the molar volume of oxygen (and of all other gases). At room temperature and one atmosphere pressure this volume is almost exactly 24 dm³.

This suggests another way of determining the

molar mass of a gas; find the mass of it which occupies 22·4 dm³ under standard conditions of temperature and pressure (s.t.p.) or 24 dm³ at room temperature and pressure (r.t.p.).

EXAMPLE (c). *61·7 cm³ of a dry gas at r.t.p. has a mass of 0·1131 g. Find the relative molecular mass of the gas.*

61·7 cm³ of gas has a mass of 0·1131 g

∴ 1 cm³ has a mass of $\frac{0·1131}{61·7}$ g

∴ 24 000 cm³ has a mass of $\frac{0·1131 \times 24000}{61·7}$ g

$$= 44·8.$$

The relative molecular mass is there 44

EXAMPLE (d). *What volume of oxygen will be obtained by heating 8·5 g of sodium nitrate; assume the gas to be measured at room temperature and pressure?*

$$2NaNO_3 \longrightarrow 2NaNO_2 + O_2$$
170 g   24 dm³

170 g of sodium nitrate give 24 dm³ of oxygen at r.t.p.

1 g of sodium nitrate will give $\frac{24}{170}$ dm³ of oxygen

8·5 g of sodium nitrate will give $\frac{24 \times 8·5}{170}$ dm³

$$= 1·2 \text{ dm}^3.$$

EXAMPLE (e). *What volume of carbon monoxide measured at r.t.p. will reduce 50 g of lead (II) oxide?*

$$PbO + CO \longrightarrow Pb + CO_2$$
219 g   24 dm³

219 g of oxide are reduced by 24 dm³ of carbon monoxide

∴ 1 g of oxide is reduced by $\frac{24}{219}$ dm³

∴ 50 g of oxide are reduced by $\frac{24 \times 50}{219}$ dm³

$$= 5·48 \text{ dm}^3$$

Volumetric Composition of Steam

Gay Lussac's Law and Avogadro's Hypothesis can be used to determine the composition of gases. Let us see how this is done in the case of steam (*not* water, we are only dealing with gases or vapours).

10 cm³ of oxygen are mixed with 20 cm³ of hydrogen and a spark passed through the mixture which is kept at a constant temperature above 100°C (why?). The resulting volume is only 2/3 of the original, i.e. 20 cm³ (Fig. 27.1). When the apparatus is allowed to cool down all the steam condenses to water and there is no gas left. From this we can conclude that 10 cm³ (1 volume) of oxygen combine with 20 cm³ (2 volumes) of hydrogen to form 20 cm³ (2 volumes) of steam: i.e.

2 volumes hydrogen + 1 volume oxygen form 2 volumes steam.

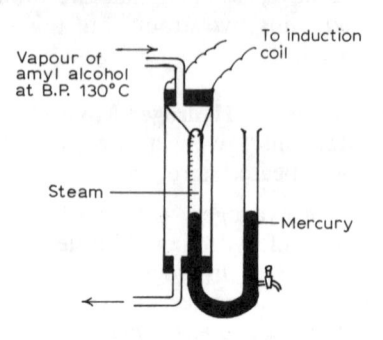

Fig. 27.1. Volume composition of steam

As equal volumes of all gases under the same conditions contain the same number of molecules we can say:

$2n$ mols. hydrogen + n mols. oxygen form $2n$ mols. steam,
or 2 mols. hydrogen + 1 mol. oxygen form 2 mols. steam,
i.e. 1 molecule of steam contains 1 molecule of hydrogen and ½ molecule of oxygen.

We have already shown that the molecule of hydrogen is diatomic so the formula for steam is H_2O_x where x has to be found.

If we determine the vapour density of steam (9) we can double this to give us the relative molecular mass of steam as 18. As the relative atomic mass of oxygen is 16 the formula must be H_2O.

However, if we have no information about the vapour density all we can say is that since one molecule of steam contains half a molecule of oxygen and cannot contain less than one atom of oxygen, then *the molecule of oxygen must contain at least two atoms.*

Finding the relative atomic mass of a Metal by Displacement

When metals are placed in solutions containing ions of other metals lower in the electrochemical series, then the first metal will go into solution and displace the other. Zinc, for example, will displace copper from a solution of copper(II) sulphate and form one of zinc sulphate, heat being given out. This can be simply expressed:

$$Zn(s) + Cu^{2+}(aq) \rightarrow Cu(s) + Zn^{2+}(aq)$$

The action of a metal on a dilute acid to liberate hydrogen is one example of this, as acids can be regarded as solutions containing hydrogen ions, and only metals higher in the series, i.e. more electro-positive, will displace the hydrogen.

If the metals have the same valency, then the masses of the two concerned will be in the same ratio as their atomic masses and this method can be used to find the atomic mass of one of them if that of the other is known.

EXAMPLE (f). If $0.63\,g$ of iron (*relative atomic mass = 56*) is placed in a solution of copper(II) sulphate until all action ceases, $0.715\,g$ of copper is precipitated. Find the relative atomic mass of copper.

As both metals have a valency of 2 in this reaction we can make use of the relationship

$$\frac{\text{mass of copper}}{\text{mass of iron}} = \frac{\text{relative atomic mass of copper}}{\text{relative atomic mass of iron}}$$

Substituting;

$$\frac{0.715}{0.63} = \frac{\text{relative atomic mass of copper}}{56}$$

$$\frac{\text{relative atomic mass}}{\text{of copper}} = \frac{0.715 \times 56}{0.63} = \underline{63.5}$$

Experiment 27.1. To Determine the relative atomic mass of Iron by Displacement of Copper (*relative atomic mass = 63.5*).

Weigh out about 1 g of 'reduced' iron and add it to a beaker containing about 10 g of copper(II) sulphate dissolved in 100–150 cm³ of water. (A saturated solution diluted with an equal volume of water is suitable.) Leave for about twenty minutes, stirring occasionally. The grey iron should by now have been replaced by red copper. Let this settle and then filter. Wash the copper on the paper with distilled water followed by alcohol (methylated spirits) and then dry it (along with another similar filter paper) in an oven at a temperature of about 90 °C or under an infra-red lamp. When it is dry, place the paper with the copper on the left-hand pan of a balance and the other paper on the right-hand pan (why?). Now find the mass of copper. If using a top-pan balance, find the mass of the filter paper first.

Use the relationship given in the example above to find the atomic mass of the iron.

Exercise 27

Try these mentally, writing down only the equation:
A. How much sulphur dioxide will be formed when the theoretical amount of sulphur is burnt in 2·5 dm³ of oxygen?
B. What volume of steam will theoretically be obtained by passing 3 dm³ of hydrogen over heated tri-iron tetroxide, all measurements being made at 100 °C?
C. How many dm³ of nitrogen will combine with 6·6 dm³ of hydrogen, and how much ammonia will be formed?
D. In the production of dinitrogen oxide from ammonium nitrate, how much steam will be formed if the volume of the oxide is 1·8 dm³?

1 Two chlorides, *A* and *B*, of the same metal contain respectively 84·93 and 73·8 per cent of the metal. Calculate the mass of the metal combined with 35·5 g of chlorine in each chloride and state the law which is illustrated.

If the relative atomic mass of the metal is 200, what are the two valencies of the metal and what are the simplest formulae of the chlorides? *(O)*

2 Using the fact that magnesium liberates hydrogen from dilute hydrochloric acid, describe how you would determine the relative atomic mass of this metal. Indicate how the required answer would be deducted from the experimental observations.

What is the percentage of magnesium in Epsom salts (hydrated magnesium sulphate of formula $MgSO_4.7H_2O$)?

State and explain *two* uses made of magnesium in everyday life. *(OC)*

3 State Avogadro's law.

When a certain gaseous oxide of nitrogen *X* is heated with iron, the gaseous residue is found to consist of nitrogen, the volume of which is half the volume of the original *X* (the volumes being measured at the same temperature and pressure). What conclusions can be drawn from this information about the molecule of nitrogen? Explain your reasoning.

The relative atomic mass of nitrogen is 14. The vapour density of *X* is 15. Say clearly and concisely what these two statements mean, and use them to determine the formula of *X*. *(OC)*

4 What is the cause of the pressure exerted by a gas on the walls of the containing vessel, and why does the pressure change as (*a*) the volume, (*b*) the temperature of the vessel is altered?

State Avogadro's rule. Why do you believe that the hydrogen molecule contains two atoms?

What is the relative molecular mass of a gas if one dm³ of it at s.t.p. has a mass of 0·76 g? *(O)*

5 Calculate the percentage of water of crystallisation in Epsom salt ($MgSO_4.7H_2O$).

(*b*) A compound contains 80% carbon and 20%

hydrogen. What is the simplest formula it can have?
If its relative molecular mass is 30, what is its true molecular formula?

(c) What volume of chlorine, measured at standard temperature and pressure, would be used in converting 8 g iron into iron(III) chloride ($2Fe + 3Cl_2 = 2FeCl_3$)? (C)

6 The gas propane burns in oxygen according to the equation: $C_3H_8 + 5O_2 = 3CO_2 + 4H_2O$.
Calculate, explaining carefully how you obtain your answers,
(a) the volume of oxygen needed for the complete combustion of 1·5 dm³ of propane, assuming that the temperature and pressure remain constant throughout the experiment;
(b) the mass of carbon dioxide produced by the complete combustion of 1100 g of propane. (S)

7 What volume of hydrogen measured at (a) s.t.p., (b) r.t.p. will completely reduce 1·59 g of copper(II) oxide to the metal? (c) What will be the mass of hydrogen in each case? (d) If the same mass of copper(I) oxide is used instead, what will be the two volumes of hydrogen needed under similar conditions and (e) what will be the mass of hydrogen in this case?

8 (a) What do you understand by the *relative molecular mass* of a compound?
(b) 3·3 g of a certain compound contain 1·8 g carbon, 0·3 g hydrogen, and 1·2 g oxygen. What is the simplest formula it can have? If its relative molecular mass is 88, what is its molecular formula?
(c) What volume of carbon monoxide, measured at s.t.p., should be obtainable from 2·3 g of methanoic acid (H_2CO_2)?

$$H_2CO_2 = H_2O + CO.$$

(d) What is the maximum mass of iron(III) oxide that can be reduced by three-tenths of a mole of carbon monoxide?

$$Fe_2O_3 + 3CO = 2Fe + 3CO_2. \quad (C)$$

9 How much potassium nitrate would have to be heated to give one dm³ of oxygen measured at r.t.p.? If the same mass of potassium chlorate(V) was used instead, what volume of oxygen, again measured at r.t.p., would be obtained? If the chlorate(V) is twice as expensive as the nitrate which would be the more economical to use?

10 Define molar volume.
(a) What volume of hydrogen at s.t.p. would be obtained by the action of water on 2·3 g sodium?
(b) What volume of nitrogen is left when 100 cm³ of dinitrogen oxide are passed over heated copper, both volumes being measured at the same temperature and pressure?
(c) Calculate the percentage of copper in copper(II) sulphate crystals, $CuSO_4.5H_2O$.

11 (a) A sample of water was found to have a little calcium hydrogencarbonate dissolved in it. On boiling 250 cm³ of solid was deposited, and the water was then 'soft'.

$$Ca(HCO_3)_2 = CaCO_3 + CO_2 + H_2O.$$

(i) What was the mass of calcium hydrogencarbonate originally present in one dm³ of the water?
(ii) What volume of carbon dioxide, measured at room temperature and pressure, came off during the boiling?
(b) (i) Explain the meaning of the *valency* of an element.
(ii) 0·108 g of a certain divalent metal M displaced 100 cm³ hydrogen, measured at standard temperature and pressure, from dilute sulphuric acid. Calculate the relative atomic mass of the metal M.
(iii) Give the formula of its oxide, and find the relative molecular mass of its chloride.

12 (a) Calculate the volume at s.t.p. of hydrogen given off when 1·38 g of sodium are carefully added to cold water.
(b) Calculate the percentage decrease in volume when carbon monoxide is completely burned, assuming that the conditions before and after combustion are the same. (S)

13 (a) Describe, with a sketch of the apparatus, how you would determine the volume composition of steam.
(b) State Avogadro's law. What information can be obtained about the molecule of oxygen by applying this law to the result of the experiment described in (a)? (OC)

14 (a) Calculate the percentage of water of crystallization in Epsom Salts, $MgSO_4.7H_2O$.
(b) 1·12 g of a metal X dissolved in excess dilute sulphuric acid displaces 448 cm³ of hydrogen at s.t.p. The metal has a valency of 2. Find, showing clearly how you arrive at your answer in each case:
(i) the mass of hydrogen produced,
(ii) the relative atomic mass of the metal,
(iii) the relative molecular mass of the metallic chloride. (S)

15 It is found by experiment that 100 cm³ of hydrogen and 100 cm³ of chlorine react to give 200 cm³ of hydrogen chloride, all volumes being measured at the same temperature and pressure.
(a) State the law which this result illustrates.
(b) Show what conclusions can be drawn from it about the number of atoms in the molecules of hydrogen and chlorine.
Describe with the aid of a sketch how you would prepare a solution of hydrochloric acid, starting from common salt. (OC)

16 (a) State Gay Lussac's Law. Write the equation representing the formation of hydrogen chloride from its elements and show how this reaction illustrates the Law.
(b) 4·35 g of manganese(IV) oxide are warmed with excess concentrated hydrochloric acid. Calculate the volume at r.t.p. of chlorine evolved.
(c) Calculate the percentage of water contained in $CaSO_4.2H_2O$. If the percentage of water in another hydrate of calcium sulphate is 6·2, find the formula of the hydrate. (S)

28
Light Rays

When we 'see' an object it is clear that something passes from the object to our eyes. That which passes we call light and a study of the properties and nature of light has occupied scientists for many centuries. In the next few chapters we shall consider some of these properties and how an understanding of them has led to the development of a wide range of optical instruments.

'Light travels in straight Lines'

You will have seen a shaft of sunlight passing through a gap in clouds or the beam of light from a ciné projector and observed how the edge of the beam is always straight. It certainly looks as though light travels in straight lines and a simple experiment will confirm this.

Experiment 28.1. Make a hole rather large than a pinhole in the middle of each of three cards. Mount two of them (Fig. 28.1) so that the light from a lamp may be seen through the holes. Now place the third card between the other two and move it about until the light can once again be seen. This time it passes through all three holes. Clamp the third card in this position. Now thread a length of string or cotton through the three holes and pull it taut. You will observe that the three holes are in a straight line.

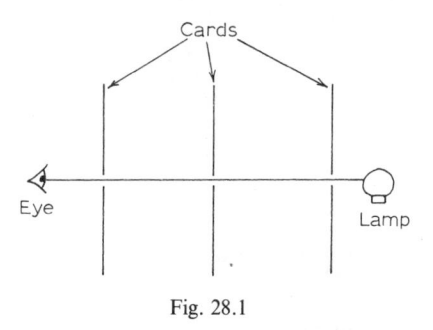

Fig. 28.1

It is as well to note here that more careful experiments than this do show that light does, to a very small extent, bend around corners. Although these experiments give very important information about the nature of light, they are beyond our scope at present and we shall go on saying that light travels in straight lines. In fact it is convenient to talk about light rays which we can represent in diagrams by straight lines. A light ray is in fact the path taken by light in going from one point to another. A 'bundle' of rays we refer to as a beam of light.

Shadows and Eclipses

The fact that light does not to any extent bend around corners results in the formation of the sharp shadows with which we are familiar. However if you examine shadows carefully you may sometimes find that it is not a simple case of darkness inside the shadow and light outside.

Experiment 28.2. Fix an opaque card (*C*) with a small circular hole close to a large pearl electric lamp bulb (Fig. 28.2(*a*)). Place an opaque obstacle in the path of the light emerging from the hole and observe the shadow cast on a white screen. It should be quite sharp. The light is coming from something like a point light source.

Now remove the card (*C*) and observe the shadow on the screen. The really black portion of the shadow has become much smaller but it is sur-

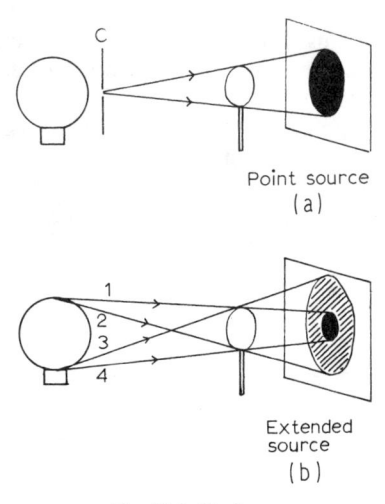

Fig. 28.2. Shadows

159

rounded by a region of partial darkness. The reason for the shadow being formed is shown in Fig. 28.2(b). The diagram shows two rays of light from the top of the lamp passing either side of the obstacle (1 and 2) and two from the bottom of the lamp (3 and 4). No light at all will reach the screen between rays 1 and 4. In the larger area bounded by rays 2 and 3, only some of the rays that would fall on the screen if the obstacle were missing will be blocked.

Examples of shadows on a very large scale are eclipses of the sun and moon. In each case the light source is the sun and it is an extended light source as in Fig. 28.2(b). So regions of total darkness (umbra) and partial darkness (penumbra) are sometimes obtained.

Fig. 28.3 shows the state of affairs leading to an eclipse of the sun. The moon on its more or less circular path around the earth is occasionally situated between the earth and the sun and it casts a shadow on the earth's surface.

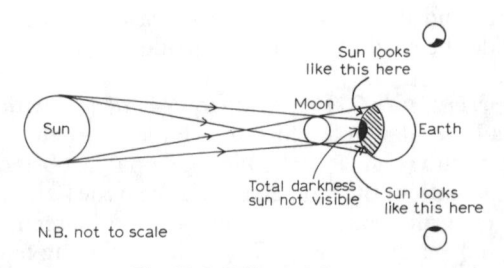

Fig. 28.3. Eclipse of the sun

Total eclipses of the sun are occasions of very great interest to astronomers because they provide an opportunity (lasting only a few minutes) to study the region around the sun. Normally the effects of this region, called the corona, are largely swamped by the enormous quantity of radiation from the sun itself. During a total eclipse when the disc of the sun is obscured, huge arches and columns of incandescent gases can be seen rising from the sun's surface.

Fig. 28.4 shows the conditions leading to an eclipse of the moon. The moon is only visible because it reflects sunlight and so when it passes into the shadow cast by the earth in light from the sun it is eclipsed.

It should be noted that an eclipse of the moon is a comparatively rare occurrence and quite different from the regular 'phases of the moon', the obvious difference being that an eclipse of the moon is all over in a matter of two or three hours while the moon passes quite regularly through its phases over a period of a lunar month (28 days).

The Pinhole Camera

A device which in itself is of no real practical importance, but which shows well this property of light not bending round corners, is the pinhole camera. This is something which you can make quite easily.

Experiment 28.3. The materials needed are a fairly rigid cardboard box (about 15 cm cubed is ideal but the dimensions are not vitally important), a sheet of greaseproof paper and some glue or paste. Cut out one face of the cardboard box and replace it by stretching the greaseproof paper over the open end and sticking it in position. Now make a fairly small pinhole in the middle of the box face opposite the greaseproof screen. If you now point the pinhole at a brightly illuminated object, for example the picture on the screen of a television set in a darkened room, you will see a faint inverted image of the object on the greaseproof paper screen.

If you gradually increase the size of the pinhole you will find that the image becomes brighter but less sharp. When the hole is considerably larger than a pinhole, only a patch of light is obtained on the screen. The formation of the image is illustrated in Fig. 28.5.

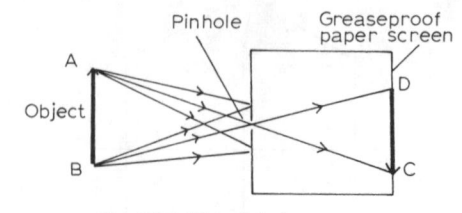

Fig. 28.5. The pinhole camera

Rays of light are spreading out from point A at the top of the illuminated object and the diagram shows a beam of these rays falling on the front of the box. Since the hole in the box is very small only a few of these rays will get into it and they will form a spot of light on the screen at C. A similar bundle of rays is shown coming from B at the bottom of the

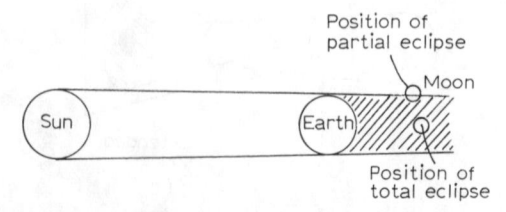

Fig. 28.4. Eclipse of the moon

object and again only a few of these rays will get in giving rise to a spot of light at *D*. Thus each point on the object produces a small spot of light on the screen and we obtain quite a sharp inverted image.

Fig. 28.6 shows what happens if we enlarge the hole in the box. Quite a large bundle of rays from *A* reaches the screen and forms there a patch of light. Similarly each point on the object produces a patch of light on the screen and the net effect is of one large patch of light.

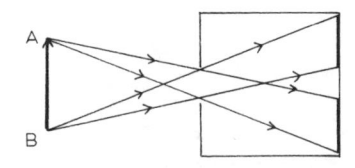

Fig. 28.6. Pinhole camera (aperture enlarged)

A pinhole camera may be used for taking photographs if the screen is replaced by a light sensitive photographic plate or film. It is worth considering the advantages and disadvantages of the pinhole camera compared to a normal camera using a lens. The big disadvantage of the pinhole camera is that so little light passes through the pinhole that it takes a long time for enough light to fall on the plate to produce a picture. This means that only absolutely still objects like buildings may be photographed. On the other hand most cameras need to be focussed for objects at any distance from it. Furthermore even the best lenses inevitably produce certain distortions of the image which do not arise with a pinhole camera.

Can we see Light?

The answer to this question is not quite as obvious as it might seem. Certainly we see objects because our eyes are sensitive to light which enters them. But can we see light which passes in front of our eyes without entering them. The answer to this can be seen by considering Fig. 28.7 which shows someone standing on the side of the earth which is in darkness, looking out into millions of kilometres of space filled with light from the sun, but apart from the moon and stars it seems quite dark. So it would seem that light which does not enter the eye is invisible. But this leaves a number of questions to be answered. Why is it that we do see the beam from a ciné projector? Why is it that when we look into the sky in day time, the whole sky is bright even if the sun is obscured by cloud?

An answer to both these questions is obtained by considering the projector beam. It becomes more

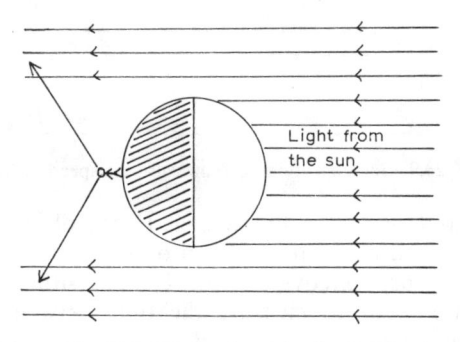

Fig. 28.7. Why is the night sky dark?

clearly seen if, for example, a lot of cigarette smoke or chalk dust is introduced into the atmosphere of the room. What we are doing is providing a large number of tiny particles which will scatter some of the light (Fig. 28.8(*b*)). It is this scattered light which is received by the observer. In a completely dust and smoke free room the beam of the projector would be invisible as shown in Fig. 28.8(*a*).

The brightness of the sky may be explained also in terms of the scattering of light by millions of particles in the earth's atmosphere. In fact to the astronaut outside the atmosphere although the sun may be brilliantly visible the rest of the sky will be quite black.

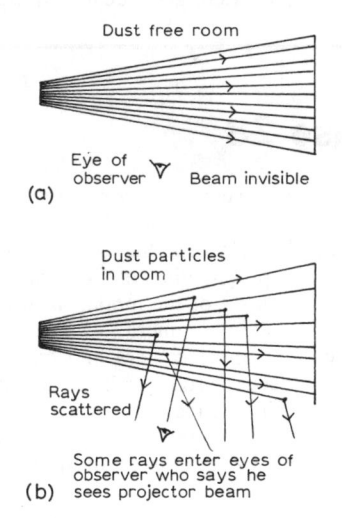

Fig. 28.8. How we see a projector beam

How Fast Does Light Travel?

The first logical attempt to find an answer to this question was made by the great Italian scientist Galileo in about 1600. He had two observers, *A* and *B* (Fig. 28.9), a considerable distance apart and each equipped with a lantern which was

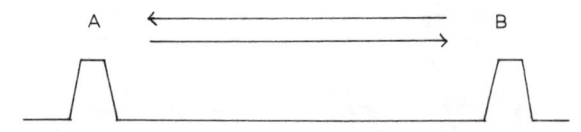

Fig. 28.9. Galileo's attempt to measure the speed of light

covered up. At a pre-arranged time observer A uncovered his lantern. When observer B saw this he in turn uncovered his. Galileo situated at A attempted to measure the time between the uncovering of A's lamp and seeing B's. This time he thought would be the time taken by light going from A to B and back.

We know now that the velocity of light is so great that such an experiment could have no hope of giving a useful answer. But it was a necessary first step. The first satisfactory value for the velocity of light was obtained by the Danish astronomer, Römer, in 1676. He studied the moons of the planet Jupiter and observed that one of them was eclipsed fairly regularly. However, careful study showed variations in the time intervals between successive eclipses of this particular moon. He found that these variations depended on the relative positions of Jupiter and Earth in their orbits. Fig. 28.10 shows the orbits of Jupiter and Earth. E_1 and J_1 show the positions of the planets when they are closest together. Since Jupiter takes nearly 12 years to go once round the sun while the Earth does it in one year, after some months Jupiter will be at J_2 and Earth at E_2. Römer argued that, although Jupiter's moon went on orbiting at quite a steady rate, the appearance of its eclipses seemed irregular on earth due to the different distances the light had to travel in reaching us. The details of Römer's calculations are beyond our scope, at present, but knowing the diameter of the earth's orbit and by studying the intervals between eclipses over a year he was able to calculate a good value for the velocity of light.

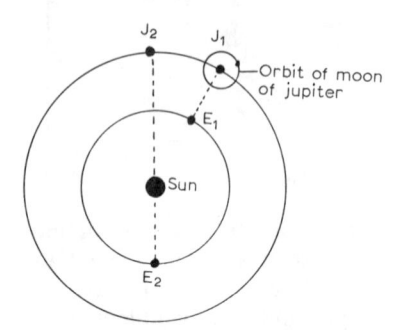

Fig. 28.10

Many other experiments have been carried out to measure the velocity of light since Römer, some astronomical and some terrestrial. The value usually quoted is 3×10^8 m/s.

Exercise 28

1 Describe an experiment you could perform to show that light travels in straight lines.
  Show, with the help of a diagram in each case, how the straight line propagation of light can be used to explain (a) an eclipse of the moon, (b) the formation of an image in a pinhole camera. (C)

2 Describe a pinhole camera and explain how an image of an object is produced by it. What would be the effect of (i) increasing the length of the camera, (ii) increasing the size of the hole? (L)

3 Draw labelled diagrams to show the differences between the shadows of an object formed by (i) a point source, (ii) an extended source. Give reasons for the differences. (C)

4 (a) Describe briefly a pinhole camera. State what it demonstrates and why it is rarely used for taking photographs.
  The image of an object is produced in a pinhole camera. Explain the effect on the image of (i) reducing the size of the hole, (ii) increasing the object distance by 50%, and (iii) increasing the camera length by 50%. (L)

29 Reflection

In the last chapter we stated that light travels in straight lines. This may not be quite true however if some material is placed in the path of the light rays. Three things may happen in this case: (1) the light may pass through the material. We say the material is transparent or translucent. The light may or may not continue in the same straight line as before striking the surface. (2) Light may be completely absorbed in the material. We say it is opaque. (3) The light may be reflected back from the surface. Materials which do this most effectively have light, smooth surfaces.

In most cases, two or all three of these effects will occur at once. We are going to consider now the last of them. The most convenient kind of surface for studying reflection is a mirror, which may consist of polished metal or a layer of shiny metal deposited on glass. For most experiments a mirror mounted on a wooden block so that it will stand up vertically on the bench is most convenient.

Experiment 29.1. The Laws of Reflection.

The ray box in Fig. 29.1 may consist simply of an electric lamp bulb (usually a 12-volt car bulb) mounted in a box fitted with a narrow slit. It is placed on a sheet of drawing paper together with the mirror and the ray of light produced is allowed to fall on the mirror surface. A reflected ray BC will be observed. Mark in pencil the position of the mirror surface, the path of the ray AB (known as the incident ray) and the path of the reflected ray BC. It is convenient to mark the rays by two or three dots on each, with the ray box and mirrors in position, and join them using a ruler when the apparatus has been removed. Construct a line BD making a right angle with the mirror at B. (We call this the *normal* at the point of incidence.) Now use a protractor to measure the angles i (angle of incidence) and r (angle of reflection). Repeat this experiment with the ray box in various positions producing a range of values of the angle of incidence i. Enter your results in a table as shown below:

Angle of incidence (i)	Angle of reflection (r)

Note that these are the angles between the rays and the *normal*.

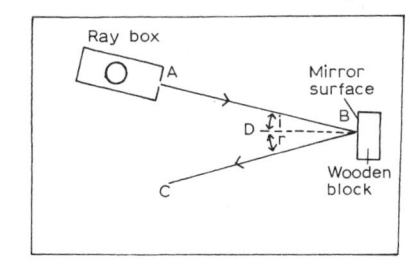

Fig. 29.1. Experiment to verify the laws of reflection

When your table is complete you will observe that in each case the angle of incidence is very close or equal to the angle of reflection. Small experimental errors will of course appear. One source of these will be the thickness of the ray, and as thin a ray as possible should be used. This experiment verifies one of the two laws of reflection which states:

Angle of Incidence = Angle of Reflection.

The other law, which is rather more difficult to verify exactly, but which our experiment shows reasonably well, states:

The reflected ray is in the same plane as the incident ray and the normal at the point of incidence.

Clearly if the ray reflected at B were going upwards or downwards from the plane of our drawing paper it would not be visible on it.

Images in Plane Mirrors

If we look at ourselves or anything else in a mirror, our eyes receive light which has been reflected at the mirror. However we tend to assume that light reaching our eyes has travelled in straight lines and we see the reflected rays as coming from somewhere behind the mirror. We say that we see an image of the object and the image is situated behind the mirror. The following experiment will enable us to find where in fact such an image does appear to be.

Experiment 29.2. Place the mirror, mounted on a wooden block, on a sheet of drawing paper which is fixed if possible on a drawing board and mark the position of the mirror surface. Stick a large pin O in front of the mirror (Fig. 29.2) and view the image of O in the mirror with the eye in some position E_1. Stick a pin A in a position such that it covers up the image of O and now very carefully stick in a second pin B so that it covers up both A and the image of O. B, A and the image of O are all in the same straight line. Remove pins A and B and repeat the process with the eye in a new position (E_2). Now remove the mirror and all the pins. We know that the image of O in the mirror was in line with A and B and also in line with C and D. Thus if we produce BA and DC until they meet we have found the position of the image (I). Join I and O with a straight line and measure along it the distance of the image from the mirror and of the object from the mirror. Note also the angle which line IO makes with the mirror. Repeat the experiment with the object at different distances from the mirror and enter your results in a table as follows.

OP	IP

The following general conclusions should be clear from your results.

1. The image is as far behind the mirror as the object is in front.
2. The line joining object and image makes a right angle with the mirror.

These are NOT laws of reflection, but with a little simple geometry you may be able to prove them from the two laws of reflection we verified in the previous experiment.

Fig. 29.3 shows how the eye can receive rays reflected from a plane mirror. The two rays OA and OB are diverging from the object (O) and are reflected at A and B along directions AC and BD determined by our two laws of reflection. The eye (E) receives these rays and sees them as diverging from I. It should be emphasized here that since the mirror is opaque, no rays of light from O pass through I and if we put a screen at I we should obtain no image on it. The image at I is thus called a virtual image.

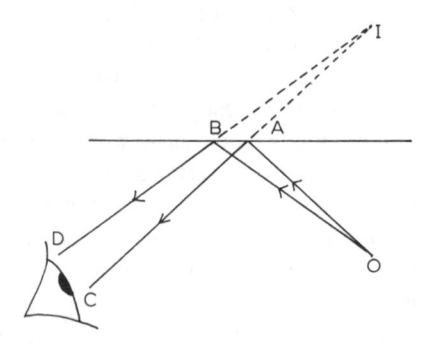

Fig. 29.3. How the eye sees an image in a plane mirror

One well known property of images in plane mirrors is that they are 'laterally inverted', which means for example that the left hand side of one's face seen in a mirror appears to be on one's right hand side. Fictional detectives have long been aware that if they wish to read something from blotting paper they should look at it in a mirror. The effect is illustrated in Fig. 29.4.

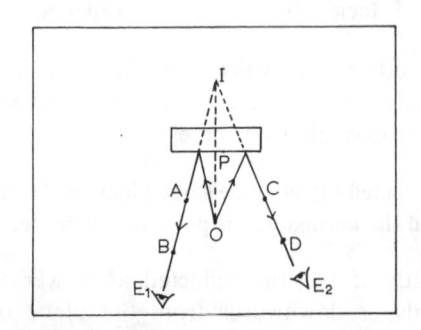

Fig. 29.2. Experiment to locate an image in a plane mirror

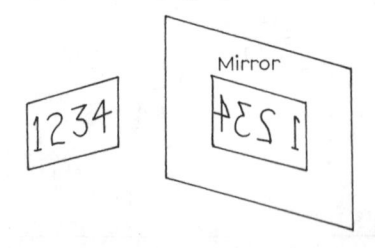

Fig. 29.4. Lateral inversion

Uses of Plane Mirrors

(a) Elimination of parallax error

You may have misread the time by several minutes by trying to read a clock at an oblique angle instead of standing directly in front of it. This error, known as a parallax error (Fig. 29.5), may arise in reading any instrument which has a pointer a distance away from the scale. Plane mirrors are often incorporated in the scale of instruments such

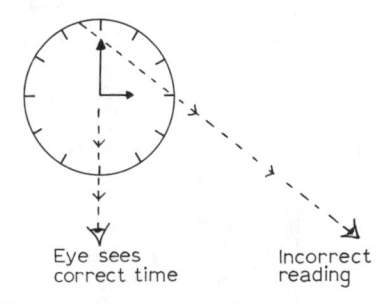

Fig. 29.5. Parallax error in reading a clock

as ammeters to enable this parallax error to be eliminated. If the person reading the instrument ensures that the pointer obscures the image of it in the mirror, he can be certain that he is viewing the scale at right angles to it, and there will be no parallax error (Fig. 29.6).

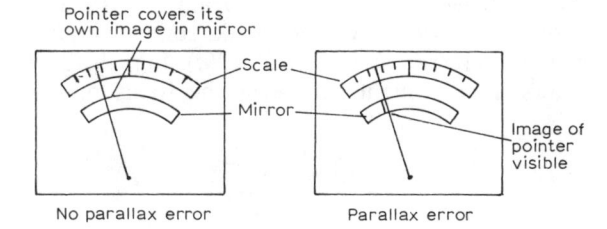

Fig. 29.6. Elimination of parallax errors with a plane mirror

(b) The Mirror Periscope

The simple mirror periscope shown in Fig. 29.7 is a device for looking over the top of obstacles. In various forms it has had many military uses and is much in evidence at the back of crowds trying to watch ceremonial parades and similar activities.

Reflection at non Mirror Surfaces

We must now try to see why it is that mirrors behave differently from other surfaces even though they reflect quite well. It must be realized that the laws of reflection apply to all reflecting surfaces. In the case of a mirror this leads to the formation of an image of an object as shown in Fig. 29.8(a). Since the surface is quite smooth, all the normals

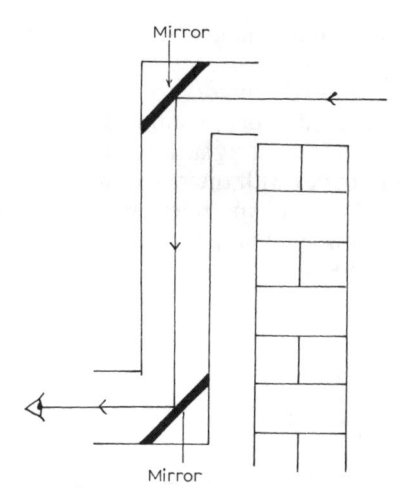

Fig. 29.7. Mirror periscope

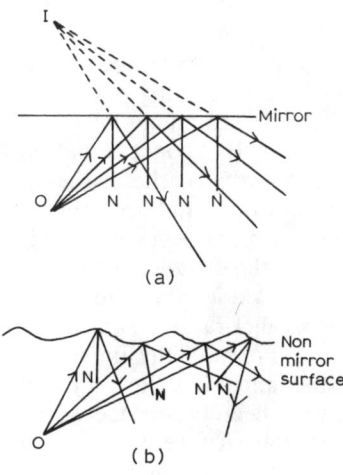

Fig. 29.8. Reflection at mirror and non-mirror surfaces

(N) are parallel and all rays from the object after reflection appear to be diverging from a point I. In the case of the non-mirror surface, for example a white painted surface, which if examined carefully enough will be seen to be irregular, the normals will not be parallel. Hence the bundle of rays diverging from O will after reflection be scattered in all directions as in Fig. 29.8(b) and no image will be formed.

Curved Mirrors

For many applications curved mirrors are more useful than flat ones. The dentist uses a curved mirror to obtain an image of his patient's teeth, many car driving mirrors are curved and there are a number of other uses which we shall mention.

We shall consider mainly 'spherical' mirrors, that is, mirrors whose surfaces are parts of spheres although curved mirrors of other shapes are sometimes used. Curved mirrors are either concave, with the reflecting surface on the inside of the curve, or convex with the reflecting surface on the outside. The next experiment shows the basic difference between the properties of curved and plane mirrors.

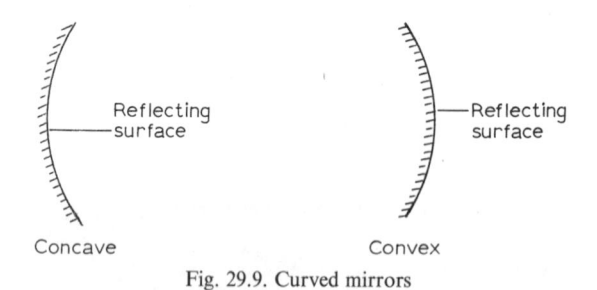

Fig. 29.9. Curved mirrors

Experiment 29.3. A cylindrical mirror is most convenient for this experiment. It may consist of a strip of metal forming part of the surface of a cylinder and if chromium plated on both faces it may be used as both a concave and convex mirror. The ray box described earlier in the chapter should be fitted now with a multiple slit instead of the single one and a biconvex cylindrical lens as shown in Fig. 29.10. If the distance between the lamp and this lens is adjusted it will be possible to produce a number of parallel rays of light from the ray box. Place the ray box and the mirror on a sheet of drawing paper and allow the parallel rays to fall on the concave reflecting surface as shown in Fig. 29.10. Mark the surface of the mirror and the paths of the rays falling on and reflected from the mirror. You will probably observe that the reflected rays pass through a single point although those coming from near the edge of the mirror may miss this point by a small distance. This point is known as the *principal focus* and its distance from the centre of the mirror (known as the pole) is the *focal length*. You can measure the focal length of your mirror from your diagram.

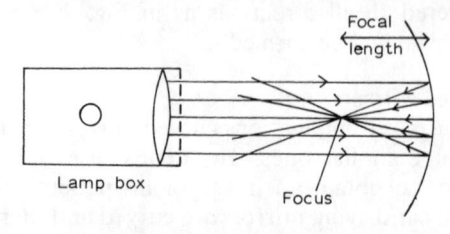

Fig. 29.10. Focus of a concave mirror

Repeat the experiment using the convex face of the mirror. You will observe that the parallel rays are reflected as shown in Fig. 29.11. In this case the parallel rays are made to diverge, whereas with with the concave face they converged after reflection.

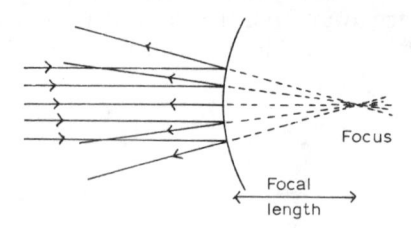

Fig. 29.11. Focus of a convex mirror

Having drawn in the mirror surface and plotted the incident and reflected rays for a convex mirror, remove the ray box and mirror. Now produce the reflected rays back behind the mirror. You will observe that the reflected rays diverge as though they were coming from a single point. This is the principal focus of the convex mirror. Once again you may measure the focal length from your diagram.

Converging and Diverging Mirrors

A concave mirror is known as a converging mirror. It will make a parallel beam of light converge to its focus. It will also make some diverging beams converge as in Fig. 29.12(*a*). Fig. 29.12(*b*) shows a point light source at the focus of the

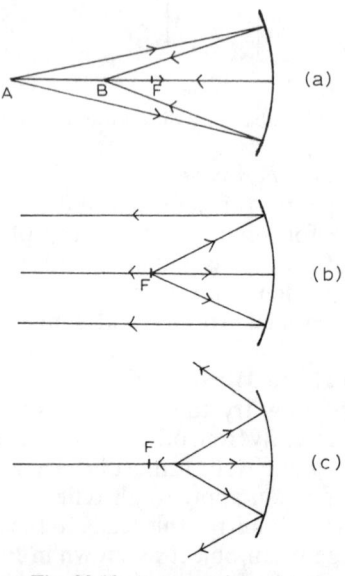

Fig. 29.12. A converging mirror

mirror. The diverging beam from this becomes parallel after reflection. If the incident light is diverging even more sharply than this, i.e. from a light source inside the focus (Fig. 29.12(c)) it will not converge after reflection but will diverge less sharply than before.

Fig. 29.12(b) shows the use of a concave mirror in a searchlight or spotlight. A lamp is placed at the focus of the mirror and a parallel beam of light is obtained. In practice the reflectors used are neither spherical nor cylindrical in shape. In section they appear as in Fig. 29.13. Such mirrors are called paraboloidal. The shape shown in Fig. 29.13 is a parabola.

Convex mirrors are called diverging mirrors. All parallel and diverging beams will diverge after reflection at a convex reflecting surface.

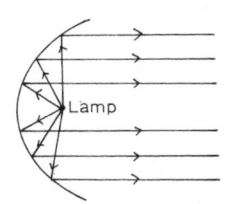

Fig. 29.13. A paraboloidal mirror used in a spotlight

Images in Curved Mirrors

In discussing the formation of an image by a pinhole camera (page 160) it was noted that a point on the object produced only a point on the screen and hence a sharp image was obtained. Now Fig. 29.12(a) shows a beam of light diverging from point A and being brought back to a point at B by the mirror. It would therefore seem that an object at A should produce an image at B. We shall now describe an experiment to investigate the formation of images by curved mirrors.

Experiment 29.4. A convenient object for this experiment is a piece of wire gauze or a piece of ground glass with a thick line marked in pencil on it. It should be illuminated by a lamp box and placed in front of a concave spherical mirror

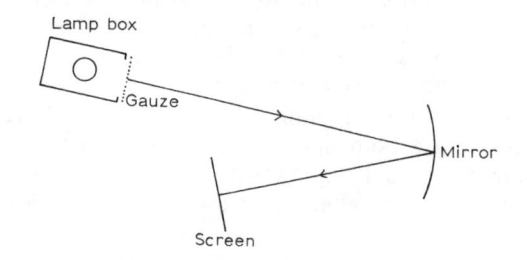

Fig. 29.14. Experiment on images in a concave mirror

mounted in a mirror holder. The image may be focused on a screen placed as in Fig. 29.14. Start with the object about 100 cm from the mirror and move the screen until a sharp image of the object is obtained on it. It will be noted that the image is inverted. Now measure carefully the distances of the object and the image from the mirror and also the sizes of the object and image. The image size is easily measured if a piece of mm graph paper is pinned on the screen. Repeat the experiment with the object to mirror distance 90, 80, 70 cm, etc. Tabulate your results as shown below.

Some typical approximate results are entered in the table but they apply to a particular mirror and your results will certainly differ from them. The last column in the table gives the magnification which is defined as:

$$\frac{\text{Size of image}}{\text{Size of object}}$$

The following points may be noted:

1. As the object gets closer to the mirror the image gets further away and bigger.
2. No image can be obtained on the screen for object to mirror distances of 20 cm or less. We mentioned above that light diverging from a point between the mirror and its principal focus would not converge after reflection. This suggests that the focal length of the mirror in this case is about 20 cm and for objects inside the focus we can obtain no real image.

Object to mirror distance (cm)	Image to mirror distance (cm)	Size of object, O (cm)	Size of image, I (cm)	Magnification $= I/O$
100	25	4	1	0·25
60	30	4	2	0·50
40	40	4	4	1·0
30	60	4	8	2·0
20	No image obtained on screen			

3. A position for the object may be found which produces an image at the same distance from the mirror as the object. Under these circumstances the object and image are the same size. If we consider a point object (O) (Fig. 29.15) producing a point image coinciding with O, it is clear that rays diverging from O must be reflected from the mirror along their own path. For this to happen the rays must make right angles with the mirror and the lines OA, OB, etc. must be radii of the surface of the mirror. O, then, is the centre of the sphere of which the mirror is a part, and is called the centre of curvature of the mirror, and OC is a radius of the curvature. From the figures in the table it appears that the radius of curvature for this mirror is 40 cm. This is about twice the focal length. In fact it can be shown that for both concave and convex mirrors:

Radius of curvature = 2 × focal length

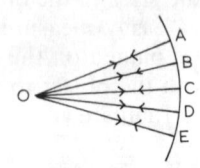

Fig. 29.15. When object and image coincide they are at the centre of curvature of the mirror

Real and Virtual Images

The images we have been dealing with, like those obtained in a pinhole camera, are formed by rays of light falling on a screen. The rays actually pass through the position of the image. Such images are called *real*. In the case of a plane mirror, the image is formed behind the reflecting surface, which is opaque. Thus no rays of light actually pass through the apparent position of the image and nothing would be seen on a screen if it were placed in this position. Images of this kind are known as *virtual*. If Experiment 29.4 is attempted using a convex mirror no real images will be obtained. However virtual images may be obtained using a convex mirror. These may be observed simply by viewing any object reflected in the mirror (Fig. 29.16). The images will be seen to be always upright and diminished in size.

A virtual image is also obtained in a concave mirror *if the object is inside the focus*. In this case a magnified upright image is obtained. A concave mirror used in this way acts as a sort of magnifying glass. It is used like this by the dentist to produce an image of the patient's teeth. Quite large, long focus concave mirrors are also used as shaving mirrors in the same way.

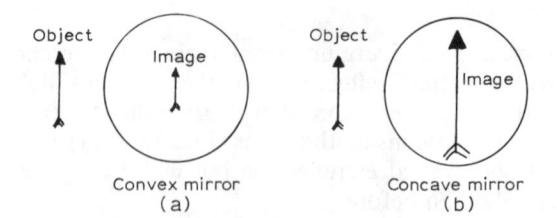

Fig. 29.16. Virtual images

Other uses of Curved Mirrors

(a) Driving mirrors

Convex mirrors are sometimes preferred to plane ones in cars as rear-vision mirrors. Their advantage is that they give a large field of view as may be seen in Fig. 29.17.

There is however the accompanying disadvantage that the image will be diminished and drivers may get a wrong impression of the distance of vehicles behind them.

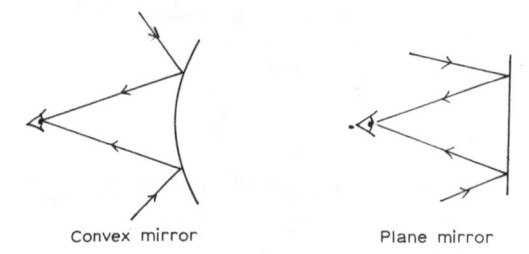

Fig. 29.17. A convex mirror gives a wider field of view than a plane mirror

(b) Reflecting telescopes

In building telescopes for astronomical use one important consideration is collecting and focussing as large an amount of light as possible from the object being viewed. This is particularly true since astronomers are now most interested in stars and galaxies which are very distant and very faint. If a lens is used for this purpose it must be of large area and therefore consists of a very heavy slab of glass. When mounted in the telescope it can only be supported around its edge and hence it will tend to sag under its own weight (Fig. 29.18), and the surfaces will be distorted. This is one reason why most large astronomical telescopes are made with concave mirrors instead of lenses. The mirror may be supported behind the reflecting

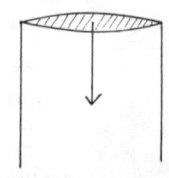

Fig. 29.18. A large lens tends to sag under its own weight

surface over its whole area. One possible arrangement of a reflecting telescope is shown in Fig. 29.19. Several reflecting telescopes with mirrors 3 metres or more in diameter are in use in the U.S.A. and U.S.S.R.

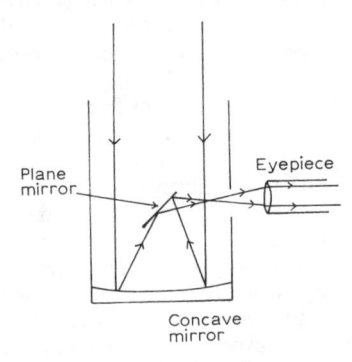

Fig. 29.19. A reflecting telescope

Scale-drawing Method of Finding Position of Images

If the focal length of a curved mirror and the position of an object with respect to it are given, it is possible to find the position of the image which will be formed and the magnification by means of a scale drawing. As in the case of a plane mirror, if we wish to find the position of the image of a point object we must plot the paths of at least two reflected rays. The method is probably best explained by taking some simple examples. Fig. 29.20 shows an object AB placed at a position outside the centre of curvature (C) of a concave mirror. The line CP (where P is the pole of the mirror) is called the axis and the point A lies on this. F is the focus and is half way between C and P. The mirror is most conveniently represented by a straight line with oblique lines top and bottom to indicate the type of mirror we are dealing with. The image of A must fall on the axis since a ray AP would fall on the mirror normally and be reflected back along its own path. The image of B may be located by plotting at least two rays from this point. A ray BD parallel to the axis will, after reflection, pass through the focus (see Fig. 29.10). A ray BF produced will go, after reflection, parallel to the axis (see Fig. 29.12(b)). A ray BP makes an angle of incidence $B\hat{P}A$ (AP is a normal) and it will thus make an *equal* angle of reflection $E\hat{P}A$. We have drawn three reflected rays which converge to the point E. Any two of them will be sufficient to establish that E is the position of the image of B and EG is the image of BA. The following points should be noted in using this method.

1. If the distances of AB, C and F from the mirror are marked accurately to scale, GP will give the distance of the image from the mirror.

2. If the size of the object is given and AB is accordingly drawn to scale, GE will give the image size. Alternatively $\dfrac{GE}{AB}$ is the magnification.

N.B. The scale for object and image sizes *need not* be the same as for object and image distances.

3. The diagram shows that in this particular case the image is inverted and real.

Fig. 29.21 is a ray diagram showing the formation by a concave mirror of a virtual image of an object inside the focus. Ray BD after reflection passes through the focus. Ray BP is reflected in direction PH so that angle of incidence $(B\hat{P}A)$ = angle of reflection $(A\hat{P}H)$. The eye receiving these two rays sees them as diverging from E rather than from B. Hence we have a virtual image EG which is upright and larger than the object.

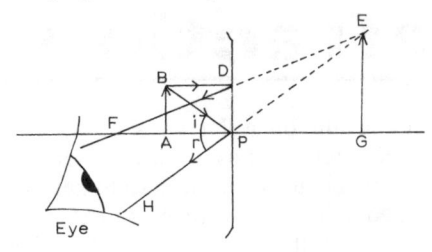

Fig. 29.21. A virtual image formed in a concave mirror

Fig. 29.22 deals with a convex mirror. In this case ray BD is reflected in direction DJ as though coming from F. Ray BP is reflected so as to make $< i = < r$. Now an eye receiving these reflected rays sees them as coming from E and we have a virtual image EG which is upright but smaller than the object.

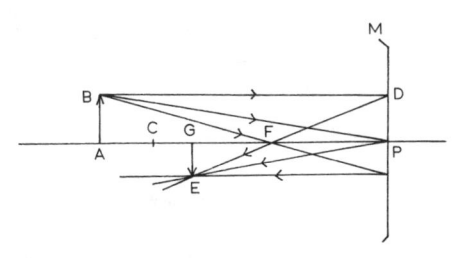

Fig. 29.20. Finding an image in a concave mirror by drawing

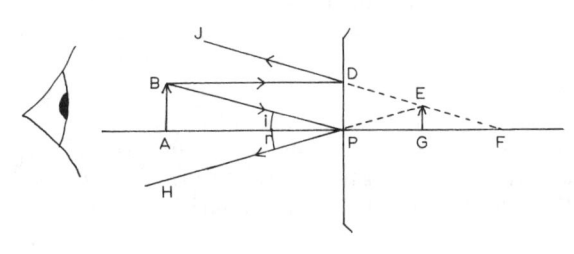

Fig. 29.22. A virtual image formed in a convex mirror

169

Exercise 29

1 Describe how you would find experimentally the relation between the angle of incidence and the angle of reflection of light at the surface of a plane mirror. *(C)*

2 How would you obtain, by a series of experiments, the position of the image of a pin placed upright in front of a plane mirror? State the result you would expect to obtain. *(L)*

3 The image formed by a plane mirror is said to be (i) a *virtual* image, (ii) *laterally inverted*. Explain the words in italics. *(S)*

4 A concave mirror is used to form an image of an object pin. Where must the object be placed to obtain: (a) an upright, enlarged image, (b) an image the same size as the object? *(J)*

5 What is (i) a *real* image, (ii) a *virtual* image?
Show, by means of a diagram, how an erect magnified image of an object may be produced by reflection at the surface of a concave mirror. *(L)*

6 A small object, 1 cm high, is placed 15 cm in front of a concave mirror of focal length 5 cm and along its principal axis. Find by drawing or otherwise, the nature, position and size of the image produced. *(O)*

7 An object is placed 5 cm from a concave mirror, on and perpendicular to the axis of the mirror. The real image produced is three times as long as the object. When the object is similarly placed 5 cm from another concave mirror the image produced is also three times as long as the object but is virtual. Illustrate each of these arrangements by a ray diagram and determine by scale diagrams or by calculation the focal length of the mirror used in each case.

Describe and explain (a) ONE common use of a concave mirror, (b) ONE common use of a convex mirror. *(J)*

30
Refraction

When light falls on a flat glass or water surface some of it is reflected as we have described in the last chapter. In fact such surfaces sometimes make quite good mirrors. However most of the light passes through the surface, and it is this effect that we are going to investigate now.

Experiment 30.1. Fill a large glass trough (preferably deep and rectangular in shape) with water and add a little fluorescein. This is a fine powder that will remain in suspension in water and make a beam of light visible, as smoke or chalk dust will make it visible in air. Mount a ray box fitted with a broad slit above the surface of the water and allow the light from it to fall, first at right angles to the surface (Fig. 30.1(a)). The path of the beam through the air may be seen by placing a white screen behind the tank. The light will be seen to travel straight on past the water surface as shown in the diagram. Now move the light box so that the beam falls obliquely on the water (Fig. 30.1(b)). In this case the beam is 'bent' downwards at the surface.

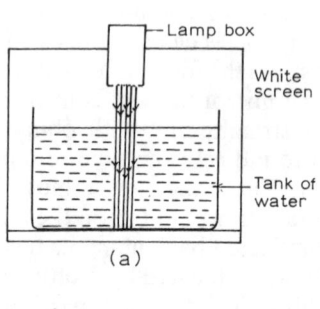

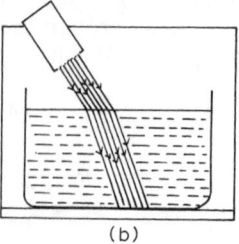

Fig. 30.1. Refraction in water

This bending effect which occurs when light passes from one material to another is known as refraction. On it depends the action of lenses which are the basis of almost all optical instruments. It also produces a number of well known effects. For example you may have noticed that a swimming pool sometimes appears less deep than it really is or that a stick dipping into water seems bent at the surface. Both of these illusions are readily explained in terms of refraction of light. We shall now try to investigate the effect more closely.

Experiment 30.2. Place a rectangular glass block on a sheet of drawing paper on a drawing board and mark in its outline. Place two pins A and B (Fig. 30.2) to define a ray of light falling on the block and view them through it. Move the eye until the two pins *seen through the block* seem in line and stick in two more pins, C and D so that all four appear to be in a straight line. C and D now define the path of ray AB as it emerges from the block. Remove the block and pins. Join AB and CD and produce each to the edge of the block at X and Y respectively. The line XY is thus the path of the ray through the block. Draw

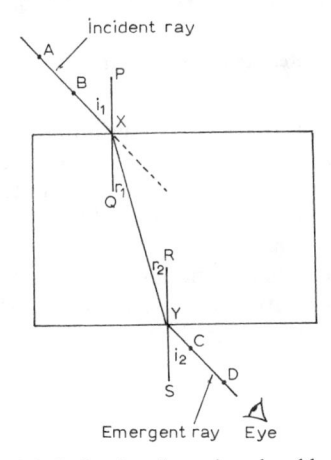

Fig. 30.2. Refraction through a glass block

normals to the surface PQ at X and RS at Y. Measure the angles i_1, i_2, r_1 and r_2. Repeat the experiment with different positions of A and B giving a wide range of values of i but avoiding values of 10° or less. Tabulate your results as follows:

The following points should emerge from your results:

1. When light passes from air into glass it is bent *towards the normal*. (The same sort of bending occurred in Experiment 30.1 with light passing from air to water.)

2. The bending was away from the normal when the light passed back from glass into air. In general light bends towards the normal when passing from a rare medium to a dense medium and away from the normal when travelling in the opposite direction.

3. $i_1 = i_2$. Hence the amount of deviation was the same at both faces and *for a parallel sided block* the emergent ray is parallel to the incident ray.

4. $\frac{\sin i}{\sin r}$ is a constant quantity for this glass block. It is known as the refractive index of the material and will probably be about 1·5 or 1·6 depending on the kind of glass used. If the angle of incidence and the refractive index are known the angle of refraction (r) may be calculated.

e.g. Refractive index = 1·5
Angle of incidence = 40°

$$\therefore \frac{\sin 40}{\sin r} = 1·5$$

$$\sin r = \frac{\sin 40}{1·5}$$

$$= \frac{·643}{1·5}$$

$$= ·429$$

$$r = 25·4°$$

Strictly speaking the refractive index of glass is $\frac{\sin i}{\sin r}$ when light passes from a *vacuum* into glass. We have found experimentally the refractive index for light passing from air to glass ($_{air}n_{glass}$). However the two quantities differ by an amount which could not be detected by a simple experiment of the kind described here.

Laws of Refraction

1. $\frac{\sin i}{\sin r}$ = constant. (This is known as Snell's Law.)

2. The refracted ray lies in the same plane as the incident ray and the normal.

i_1	r_1	i_2	r_2	i (Mean of i_1 & i_2)	r (Mean r_1 & r_2)	$\frac{\sin i}{\sin r}$

Graphical Construction of Refracted Ray

It is sometimes convenient to plot the path of a refracted ray in a particular case without the use of sine tables. The light is passing from air to water ($_{air}n_{water} = 1\cdot3$) in the example shown in Fig. 30.3 and the angle of incidence is 40°. The refracted ray may be drawn by the following construction:

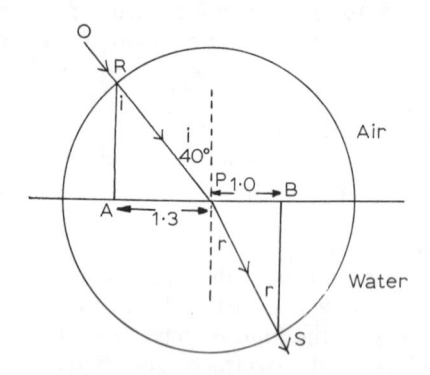

Fig. 30.3. Construction of a refracted ray $i = 40°$ $n = 1\cdot3$

Draw the normal at point P on the surface and the incident ray OP making an angle of 40° with it. Mark points on the surface A and B $1\cdot3$ and $1\cdot0$ units respectively from P ($\frac{1\cdot3}{1\cdot0} = {_{air}n_{water}}$). Construct normals to the surface at A and B. The normal at A meets OP at R.

With radius PR, and centre P construct a circle. The circle cuts the normal from B at S. The line PS is the refracted ray.

You will see from the diagram that $\sin i = \frac{1\cdot3}{a}$ and $\sin r = \frac{1\cdot0}{a}$ where a is the radius of the circle.

$\therefore \sin i / \sin r = 1\cdot3$ as specified.

Real and Apparent Depth

We mentioned earlier some well known optical illusions which may be explained in terms of refraction. Fig. 30.4 represents a straight stick AB dipping into water which is viewed from a position E. The diagram shows two rays diverging from B and incident on the water surface at X and Y. Since they are passing from water to air they will be bent away from the normal and the eye will see them as diverging from C rather than B. Thus the part of the stick below the surface seems to occupy the position indicated by the broken line.

A similar diagram, Fig. 30.5, shows why it is that a pool of water, viewed this time from vertically above, seems shallower than it is. The real depth of the pool is AC and the apparent depth AB.

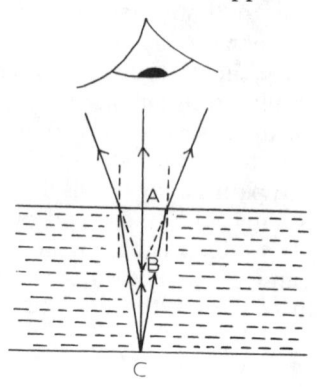

Fig. 30.5. Real and apparent depth of a pool of water

A consideration of this diagram in terms of Snell's Law $\left(\frac{\sin i}{\sin r} = n\right)$ shows that $\frac{\text{real depth}}{\text{apparent depth}} = n$.

The proof is not difficult but we shall not go through it here.

This expression applies of course to all liquids and it is sometimes convenient to find the refractive index of a liquid by a method based on it.

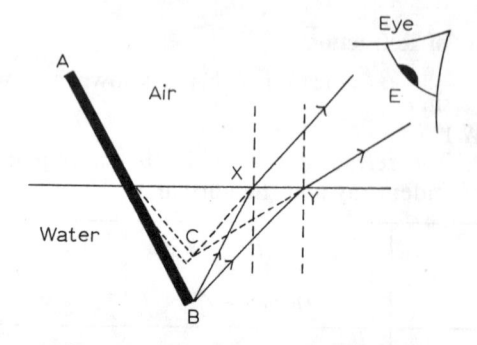

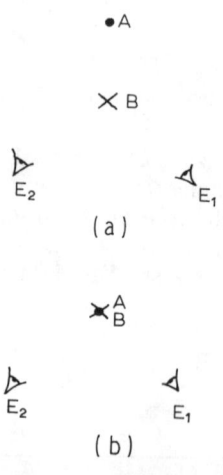

Fig. 30.4. Why a stick appears bent in water

Fig. 30.6. The method of no parallax

Before we describe such an experiment it is necessary to discuss a technique which is used in it.

This is known as the method of no parallax. In Fig. 30.6(a) when viewed from position E_1, the dot A appears to the right of cross B while from E_2 A seems to be to the left of B. Thus when the eye moves from E_1 to E_2, A seems to move relative to B. Only if, as in Fig. 30.6(b), A and B coincide, is there no motion of one relative to the other when the viewer's eye is moved. We shall now use this effect to find the apparent depth of a liquid.

Experiment 30.3. Place a pin (A) on the bottom of a large measuring cylinder (Fig. 30.7) and partially fill it with the liquid whose refractive index is to be measured. Clamp a second pin (B) horizontally and move it so that its head is close to the side of the cylinder. Now move pin B up and down until, when the eye is moved across the top of the cylinder there is no movement of B relative to A viewed through the liquid. Pin B is now at the same level as the apparent position of A. We can now measure the real and apparent depths by measuring the distances of A and B below the liquid surface. Repeat the experiment with a number of different liquid levels in the cylinder and tabulate your results as follows:

Real depth (cm)	Apparent depth (cm)	$\dfrac{\text{Real depth}}{\text{Apparent depth}} = n$

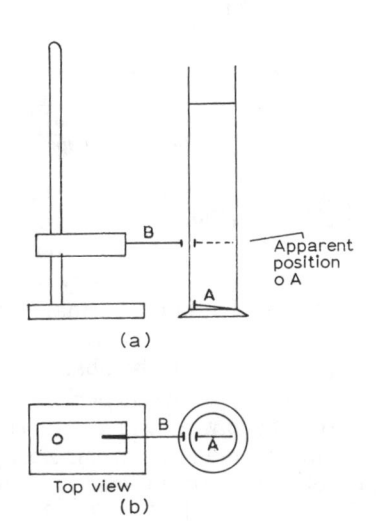

Fig. 30.7. Finding the apparent depth of a liquid

Total Internal Reflection

Experiment 30.4. Place a semi-circular glass block and a ray box fitted with a single narrow slit on a piece of drawing paper. Arrange the ray box as in Fig. 30.8(a) so that the ray falling on the curved face is directed towards O, the mid point of the flat face. Under these circumstances the ray will always make a right angle with the curved face and pass through without any refraction (undeviated). In Fig. 30.8(a) the ray also makes a right angle with the flat surface and passes through that undeviated. Now move the ray box round the curved face, ensuring still that the ray is directed towards O. You will observe that of the light falling at O some is reflected back through the glass and the rest passes through the surface and is refracted away from the normal (Fig. 30.8(b)). If you further increase the angle of incidence at O you will reach a condition where the refracted ray travels almost along the surface. The angle of refraction (r) is nearly 90°. This is the condition shown in Fig. 30.8(c). Move the ray box just a little further and the refracted ray disappears completely. The light is *totally internally reflected* (Fig. 30.8(d)). Its direction after striking the surface is determined by the laws of reflection. Mark the incident ray and pencil in the outline of the block for the case shown in Fig. 30.8(c). Now measure the angle of incidence, i, in your drawing. This is known as the **critical angle** for the glass. It will probably be about 41° or 42°.

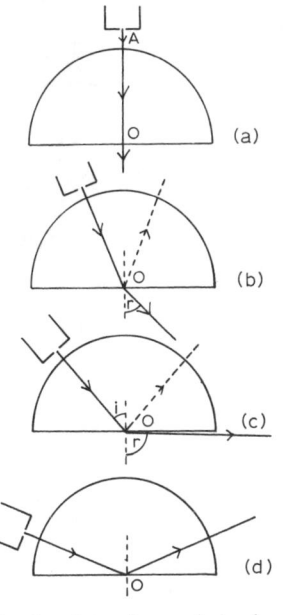

Fig. 30.8. Refraction through a semi-circular glass block

173

This experiment implies that when light, travelling in a dense material, meets a surface with a less dense material at an angle of incidence greater than the critical angle it will not pass through the surface but will be totally internally reflected. The critical angle varies with the refractive indices of the materials. For glass and air as we have mentioned it is 41° or 42°, for water and air it is about 48°.

Some Applications of Total Internal Reflection

Experiment 30.5. Allow a narrow beam of light from a ray box to fall normally on one of the shorter sides of a right angled isoceles glass prism (Fig. 30.9). You will observe that the emerging ray is at right angles to the incident ray. The ray passes through the first surface undeviated and makes an angle of incidence of 45° with the hypotenuse. This is a glass-air surface for which the critical angle is about 42°. Thus the ray will be totally internally reflected, the angle of reflection being 45°. Finally the ray strikes the third face normally and goes through without being bent. Altogether the light has been deviated through 90°.

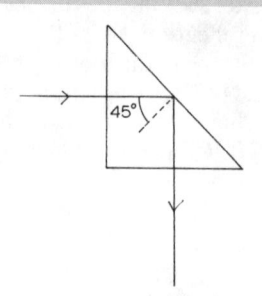

Fig. 30.9. Deviation through 90°

Now allow the beam to fall at right angles to the hypotenuse (Fig. 30.10). You will see that it comes out parallel to the incident ray having been deviated through 180°. Once again, there is no deviation at *A*, but the angle of incidence at *B* is 45° so total internal reflection occurs. A second reflection occurs at *C* and the ray emerges at *D* without deviation.

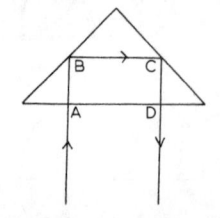

Fig. 30.10. Deviation through 180°

In the first of these experiments the prism behaves just as a plane mirror would if placed in the position of the longer face of the prism. Prisms are in fact preferred to mirrors in a number of optical instruments. Prism periscopes (Fig. 30.11) are used if precision is important as in submarine periscopes.

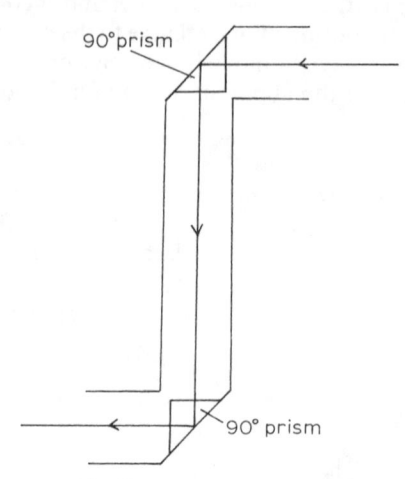

Fig. 30.11. A prism periscope

Right angled prisms are also used to deviate light through 180° (Fig. 30.10) in certain optical instruments, notably prismatic binoculars. A simple telescope may consist simply of two convex lenses at each end of a tube. Two major snags arise if such an arrangement is to be used in binoculars. If good magnification is to be obtained the lenses must be rather far apart and hence the tube will be long. Also the image will be inverted. The famous optical instrument manufacturers, Zeiss, in 1895 overcame both of these difficulties by using two right angled prisms as in Fig. 39.12. By making the light travel three times backwards and forwards along the tube they obtained the effect of a long

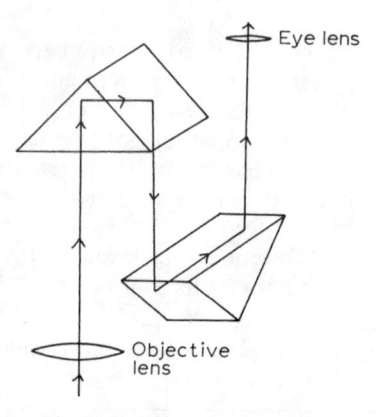

Fig. 30.12. Prismatic binoculars

tube. The first prism laterally inverted the image while the second inverted it vertically, hence the final image seen through the eyepiece was the right way up.

Exercise 30

1 What is meant by *refraction* of light? State Snell's law of refraction and describe how you could verify the law, using a rectangular block of glass.

A cube of glass of 10 cm side lies on a table. A ray of light is shone along the table and strikes one face of the block at an angle of 45°. Calculate the angle of refraction. The light passes to the opposite face and then out of the block. Draw a full size diagram showing the passage of the ray through the block and from the diagram find the perpendicular distance between the directions of the incident and emergent rays.

[Refractive index of glass = 1·5.]   (C)

2 Under what conditions does total internal reflection occur? Show, by means of a diagram, how you would use a glass prism to turn a ray of light through 180°. Mark the angles of the prism and explain why the ray takes the path shown. (C)

3 State Snell's law of refraction. Describe how you would verify the law by tracing a ray of light through a rectangular block of glass.

The depth of water in a swimming bath varies from 1 metre at the shallow end to 2 metres at the deep end. To an observer standing on the edge of the bath at the shallow end, the depth of water appears to vary little from the shallow end to the deep end. Give a clear explanation of the observation and draw a diagram to help your answer. (C)

4 Draw a diagram to show how you would arrange two right-angled isosceles glass prisms as a periscope in order to see over the heads of a crowd. Draw the path of one ray of light from the object to your eye and explain why the ray follows the path shown.

5 What is meant by the terms *total internal reflection* and *critical angle*?

Describe in detail how you would measure the critical angle for light travelling from glass to air.

How may total internal reflection be used in a glass prism to turn a ray of light through 90°? Name **one** optical instrument which makes use of this technique.

(C)

31
Lenses; Optical Instruments and the Eye

Lenses were widely used in spectacles as long ago as 1300 and in the seventeenth and eighteenth centuries very big advances were made in biology using mostly single lenses as magnifying glasses. Dutch scientists discovered in about 1600 how to put two lenses together to form what we call telescopes and compound microscopes. Galileo and other scientists quickly made many new discoveries in astronomy using telescopes, but the development of the compound microscope was very much slower. However, the development of high power microscopes in the nineteenth century was all important to the progress in biology at that time. Even during recent years with the discovery of the electron microscope, improvements in optical instruments have made an essential contribution to biological research. Other important instruments using lenses are cameras and projectors. In order to understand these devices it is necessary first that we investigate the properties of lenses. A simple lens is a piece of glass whose surfaces are usually spherical, although we shall sometimes use cylindrical lenses and some, made for special purposes, have much more complicated shapes. We shall for the most part confine ourselves to dealing with biconvex and biconcave lenses (Fig. 31.1).

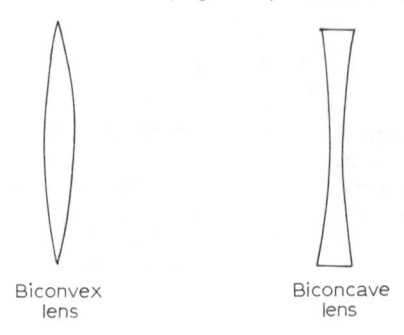

Fig. 31.1.

Experiment 31.1. Set up a ray box with a multiple slit so that it produces a number of parallel rays of light. Allow them to fall on a pair of equilateral prisms arranged as shown in Fig. 31.2(a). Plot on a sheet of drawing paper the outline of the prisms and the paths of the incident and emerging rays. Your drawing will probably look like Fig. 31.2(a), although if a ray after refraction at the first surface, next strikes the face in contact with the other prism, a total internal reflection may occur. The paths of the rays shown may be simply explained since at the first surface the refraction is towards the normal and at the second away from it.

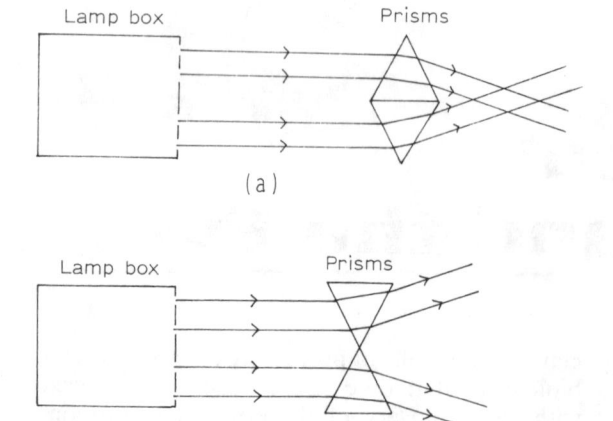

Fig. 31.2. (a) Prisms arranged to show the action of a converging lens
(b) Prisms arranged to show the action of a diverging lens

Repeat the experiment with the prisms arranged as in Fig. 31.2(b). You will see that the first arrangement is a converging system and the second produces divergence. However, in both cases all the rays passing through one of the prisms emerge, just as they went in, parallel to each other. It is only the two sets of rays which converge or diverge.

Now carry out the same experiments with cylindrical, biconvex and biconcave lenses.

The similarities and differences between the lens and prism experiments will be clear. The biconvex lens, like the prism in Fig. 31.2(a) is converging, but in the case of the lens *ALL* the rays converge to a point (F)—(as long as none of them has passed through the lens near its edge). The biconcave lens is diverging as are the prisms in Fig. 31.2(b). If the diverging rays emerging from the lens are produced back, they will be seen to be diverging as though from a point (F)—(once again as long as only the central portion of the lens is used). The points (F) in Fig. 31.3 are known as the principal foci of the lenses, and the distances from these points to the centres of the lenses are known as the focal lengths.

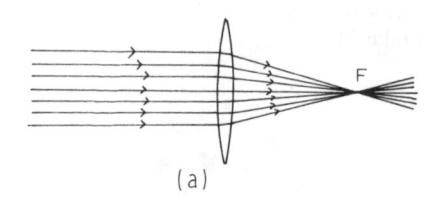

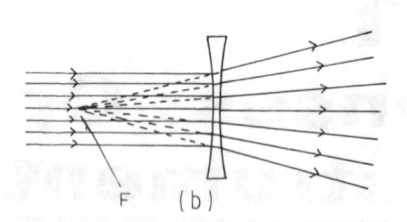

Fig. 31.3. (a) Focus of a converging lens
(b) Focus of a diverging lens

Lenses (a) and (b) (Fig. 31.4) are converging like biconvex. Types (c) and (d) are diverging. A meniscus lens is convex or converging if the radius of curvature of the convex surface is *LESS* than that of the concave surfaces.

It should be noted that although the radii of curvature of the surfaces of a lens are related to the focal length, it is not a simple relationship and we shall not deal with it in this book. (For a spherical *mirror* Radius of curvature = 2 × focal length.)

Plano-
convex
(a)

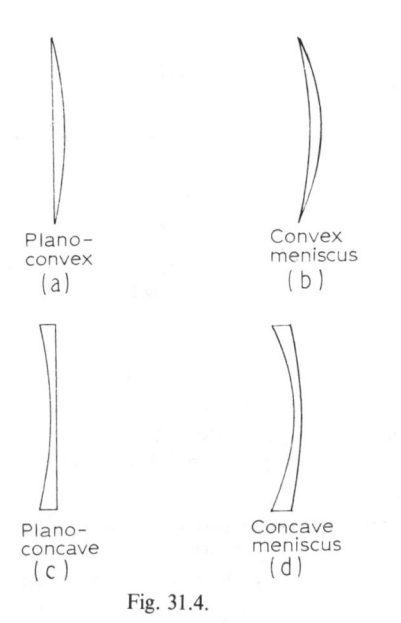

Convex
meniscus
(b)

Plano-
concave
(c)

Concave
meniscus
(d)

Fig. 31.4.

Images formed by Lenses

Experiment 31.2. Set up an illuminated object (see Experiment 29.4), a biconvex spherical lens in a lens holder and a screen as in Fig. 31.5.

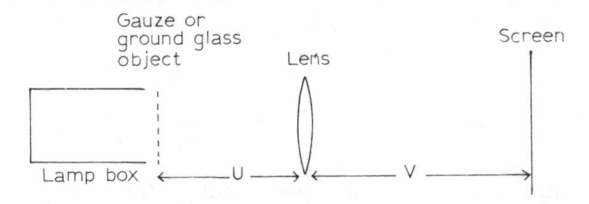

Fig. 31.5. Experiment to show real images obtained with a converging lens

Start with the distance between the object and lens about 100 cm, and adjust the position of the screen until a sharp image of the object is obtained on it. You will see that the image is inverted. Measure accurately the distances of the object and screen from the lens. Note also the sizes of the object and image (once again the image size may be easily measured by pinning a piece of mm. graph paper to the screen). Repeat the experiment with object to lens distances of 90 cm, 80 cm, 70 cm, etc., and tabulate your results as below.

The conclusions that will emerge from your table will be very similar to those derived in the case of a concave mirror.

(1) As the object is brought closer to the lens the image becomes further away and larger.

In fact: $\dfrac{\text{Image distance}}{\text{Object distance}} = \dfrac{\text{Size of image}}{\text{Size of object}}$
$= \text{Magnification}$

(2) There is a minimum object distance for which an image may be obtained on the screen. This minimum distance is a little more than the focal length of the lens.

Virtual Images

Experiment 31.2 deals only with real images which always arise with a converging lens as long as the object is outside the focus. If it is attempted with a diverging (biconcave) lens, no real image will be obtained. If an object is viewed through a biconcave lens an upright, diminished, virtual image will be seen as in Fig. 31.6(a).

Virtual image in a diverging lens
(a)

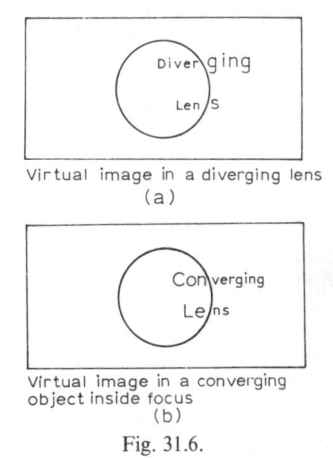

Virtual image in a converging object inside focus
(b)

Fig. 31.6.

Fig. 31.6(b) shows the very familiar use of a converging lens as a magnifying glass. Here a magnified upright virtual image is obtained if the object is placed inside the focus.

Finding the Focal Length of a Converging Lens

Experiment 31.3. A simple convenient method of finding the focal length of a converging lens uses a

Object to lens distance (u) cm	Image to lens distance (v) cm	Size of object (o) cm	Size of image (i) cm	Magnification $\dfrac{(i)}{(o)}$	Image distance (v) $\dfrac{}{\text{Object }(u)}$ distance

plane mirror and a pin clamped in a stand as well as the lens. Place the lens on the plane mirror which is resting on the bench, and clamp the pin (P) vertically above the centre of the lens (Fig. 31.7). View the image of the pin seen in the mirror through the lens from above P. Now slide the clamp, holding P, vertically up and down until the pin coincides with its own image, in the lens-mirror system. This position will be found by the no parallax method described in experiment 30.3, page 172, and it is the focus of the lens. Measure the distance from P to the lens which is the focal length.

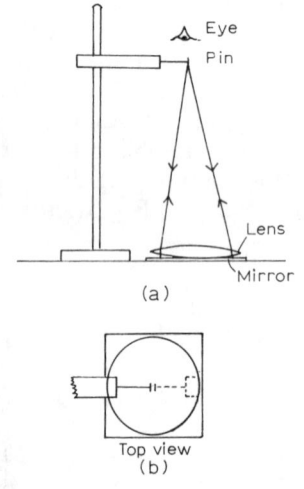

Fig. 31.7. Finding the focal length of a biconvex lens

In an experiment of this kind you should never be satisfied with one, unverified value, so the experiment should be repeated several times and the mean of your values taken.

If P is at the focus of the lens, then light rays diverging from this point and falling on the lens will emerge from it parallel. They will fall on the mirror normally, be reflected back along their own path, and fall on the lens again still parallel. Finally, after a second refraction they will converge to the principal focus, i.e. P. Thus if the pin and its image coincide they must be at the principal focus.

Graphical Location of Images in Lenses

As in the case of mirrors there is a simple graphical construction which enables us to find the position and nature of an image and its magnification if we know the object position and focal length of the lens. Fig. 31.8 shows a simple example of this method. An object AB is placed outside the focus (F) of a biconcave lens. (N.B. a lens has a focus on each side and F_1 and F_2 are equidistant from it.)

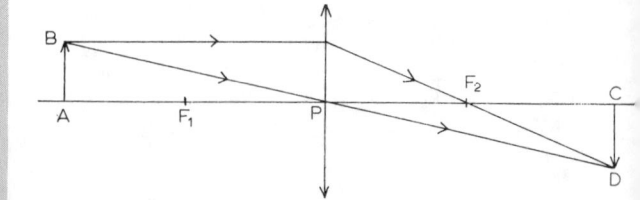

Fig. 31.8. Construction to find image positions with a converging lens

A ray from B parallel to the axis of the lens after refraction will pass through the focus (F_2). A ray from B falling exactly on the middle of the lens will pass through undeviated. The reason for this may be seen by considering Fig. 31.9. The shaded portions of both of the lenses shown are close approximations to parallel sided blocks of glass and if the lenses are thin, light will pass through these portions as through a fairly thin sheet of glass. Thus the image of B will occur at D, the point of intersection of these two refracted rays.

Fig. 31.9. The central part of a lens is almost parallel sided

Since a ray AP would pass through the lens undeviated, the image of A must lie on the axis. Hence the image of AB is CD. If the object distance and focal lengths are drawn accurately to scale, CP will give the image distance and $\frac{CD}{AB}$ the magnification. The image in this case is real (the rays from B actually pass through D), inverted and magnified—Fig. 31.10 shows a biconcave lens with an object in a similar position with respect to the lens to Fig. 31.8.

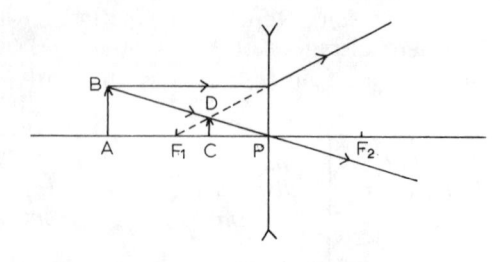

Fig. 31.10. Construction to find the image positions with a diverging lens.

If *BA* is viewed through the lens, the rays drawn diverging from *B* will be seen as coming from *D*. A virtual, upright, diminished image of *BA* is seen at *DC*.

The simple Magnifying Glass

As we have already seen, if a converging lens is used, with an object inside the focus, a magnified upright virtual image is obtained. The lens is then being used as a magnifying glass and the formation of the image is illustrated in Fig. 31.11.

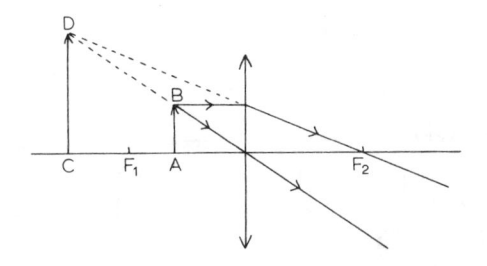

Fig. 31.11. A converging lens used as a magnifying glass.

The rays diverging from *B* after being refracted by the lens will be seen by the eye as diverging from *D* and a magnified image is obtained at *CD*. This use of a converging lens is widely applied these days in the simple 'viewer' for colour-transparencies shown in Fig. 31.12.

The transparency is illuminated by a battery-powered lamp placed behind a ground glass screen so that the illumination is uniform. The converging lens forms a magnified virtual image of the slide.

The eyepieces of most telescopes and microscopes, although in practice they usually consist of more than one lens, are basically converging lenses used in this way.

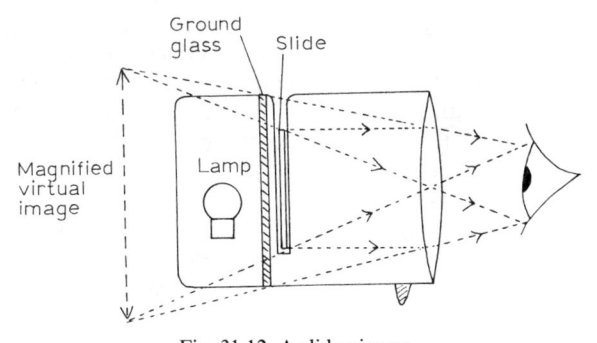

Fig. 31.12. A slide viewer.

Other Optical Instruments

All the other optical instruments we have now to consider use converging lenses to produce real images, i.e. images that can be produced on some sort of screen.

(a) The camera and photographic enlarger

The production of a photographic negative depends on changes which occur to certain chemicals when light falls on them. Almost always these changes are the conversion of silver bromide or silver iodide to silver (which appears black on the negative). The film which is put in a camera consists of a suspension of tiny crystals of these silver salts suspended in gelatine (the emulsion) coated on a transparent plastic material—cellulose acetate. The function of the camera is to produce a sharp real image of an object on this film and to leave it there for just long enough to produce a satisfactory 'latent' image on the film. This can then be 'developed' in certain chemicals to give a usable negative, and 'fixed' in other chemicals so that further exposure to light produces no more chemical change. The details of the mechanisms involved in these processes are complicated and some of them have only recently been really understood. However, our main concern now is with the optics of the camera and we shall say no more of the photographic process.

Fig. 31.13 shows the essential features of a simple roll film camera. The converging lens, which in cheap cameras may be a meniscus lens but in more expensive cameras will be an arrangement of a number of lenses, has to produce a real image on the film. In order to cope with different object distances there may be a mechanism to vary the distance between the lens and the film. However, box-cameras are often 'fixed focus' and a satisfactory image is obtained for an object, say, 3 metres or more from the lens.

There are two other factors which the photographer can control by adjusting his camera. They are the 'aperture' and the time for which the light falls on the film. We have previously mentioned that if the edge of the lens is used it will not behave

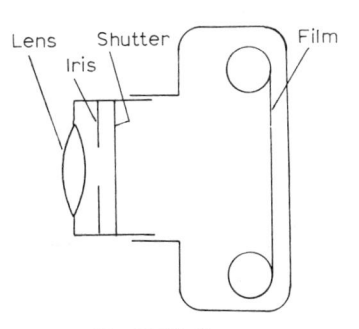

Fig. 31.13. Camera

in the simple way we have described in this chapter. It will in fact produce distorted images. It is therefore necessary to incorporate an 'iris' which prevents light passing through the edge of the lens. The hole in the iris through which the light actually passes may be adjustable in diameter. If photographs are being taken in very dim light conditions it will be quite large, if the light is bright the lens will be 'stopped down'.

With modern films it is usually sufficient to allow the image to fall on the emulsion for only 1/50 or 1/100 s or even less. However, once again, the time will be determined by the light conditions. In order to expose the film for very short periods of time like this, various forms of mechanical shutters are made in cameras. In the inexpensive box cameras the shutter is set for 1/50 s only. In more expensive instruments it may be possible to select any one of perhaps ten different times, ranging from 1/1000 s up to 1 s.

When a film has been exposed in a camera, developed and fixed, it will be blackest where most light has fallen, and the black parts of the object will be represented by a completely transparent area on the film. This we call a negative.

In order to produce an ordinary photograph, or positive, on paper we may use an enlarger (Fig. 31.14). With this instrument one uses a converging lens to focus on a piece of paper a real magnified image of the negative. The paper is sensitive to light in a similar way to the film so that it also interchanges blacks and whites. The negative must be illuminated from behind by a lamp and uniform illumination may be achieved by placing a ground glass plate between lamp and negative. The paper also has to be developed and fixed before a final image is obtained.

(b) The slide projector

The principle of the slide projector (Fig. 31.15) is basically the same as that of the enlarger mentioned. A converging lens is used to project a magnified image of an illuminated object on a screen. The object in this case is usually a 'positive' slide or transparency. The following points are important in the design of projectors:

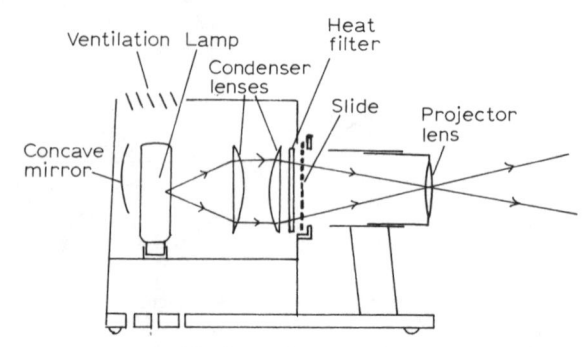

Fig. 31.15. Slide projector

(1) The slide must be as brightly and evenly illuminated as possible. The lamp must therefore be of high light output, and a concave mirror is placed behind it so that light from it going in the 'wrong' direction is not wasted. The two plano-convex lenses constitute what is called a condenser system. They concentrate the light on the slide and ensure even illumination.

(2) The distance between the slide and the projector lens must be a little more than its focal length in order that a large magnified image may be formed on a screen at a longer distance from the projector. The lens will be mounted so that its position may be varied a little in order to focus the image on screens at different distances.

(3) A great deal of heat will be produced by the lamp and the slide may be damaged if it gets too hot. The case of the instrument may be slotted to allow cooling by convection, or a fan may be fitted for cooling purposes. Also a plate of glass specially made to absorb radiant heat may be placed between the lamp and slide.

(4) Since an inverted image is formed the slide will be put in upside-down.

The Eye

The eye may be considered as an optical instrument similar in its action to the camera. It has a lens, or strictly speaking a system of lenses, which

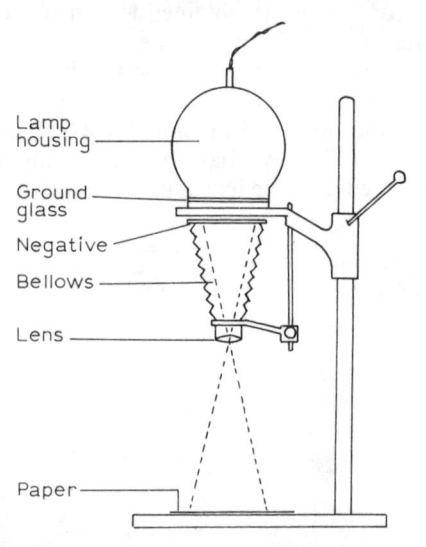

Fig. 31.14. Photographic enlarger

back of the eye. The eye is almost spherical and light enters through a circular aperture a few millimetres in diameter in the iris (*I*). The iris is that part of the eye which varies in colour from person to person. Having passed through the iris light falls on the lens (*L*) which focuses it on the back surface of the eye, which is called the retina (*R*). The image falling on this sensitive surface is transmitted via the optic nerve (*O*) to the brain (Fig. 31.16).

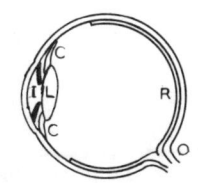

Fig. 31.16. The human eye

We are principally concerned with the optics of the eye here and its most remarkable feature in this respect is its ability to focus on objects from a few centimetres away to infinity.

In a camera this may be achieved by varying the distance between the lens and the film. In the eye, although some small movement of the lens does occur, it is not the main factor. The lens is attached to the walls of the eye by the ciliary muscles (*C*) which are able to alter the radii of curvature and hence the focal length of the eye. In a 'normal' eye these muscles are relaxed when distant objects are in focus on the retina. When closer objects are viewed the muscles decrease the radii of curvature and focal length of the lens. This process we call accommodation. The tension in the muscles becomes excessive and the eye subject to strain if an object closer than about 25 cm is focused. We say that a normal eye has a least distance of distinct vision of 25 cm.

The other remarkable feature of the eye is its ability to adjust for light conditions. The aperture in the iris decreases in size in bright light and increases in poor light. The response is not instantaneous and so we are sometimes dazzled when we come out of the dark into a bright room, or we see nothing at all on entering a cinema from the daylight outside.

Defects of Vision

(1) *Presbyopia or 'loss of accommodation'*

The material of the eye lens often becomes harder with increasing age, with the result that near objects (Fig. 31.17(*a*)).

They are unable to make their eye lens 'sufficiently converging'. The remedy is to help the eye lens by providing a converging spectacle lens which will be required for close work only (Fig. 31.17(*b*)).

Unaided eye, image out of focus
(a)

Converging spectacle lens, image in focus
(b)

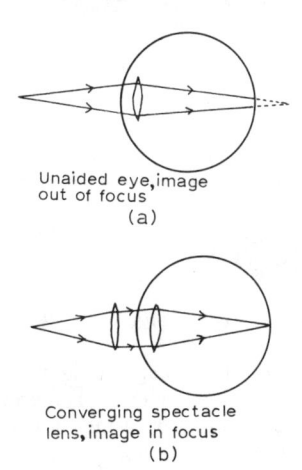

Fig. 31.17. Presbyopia (loss of accommodation)

(2) *Myopia or short sight*

People suffering from this defect are unable to focus on distant objects. When the ciliary muscles are relaxed the lens is still too converging to focus distant objects on the retina (Fig. 31.18(*a*)).

A diverging spectacle lens will be used in this case (Fig. 31.18(*b*)).

Light from distant object

(a)

Unaided eye Image in front of retina

Diverging spectacles Image on the retina

(b)

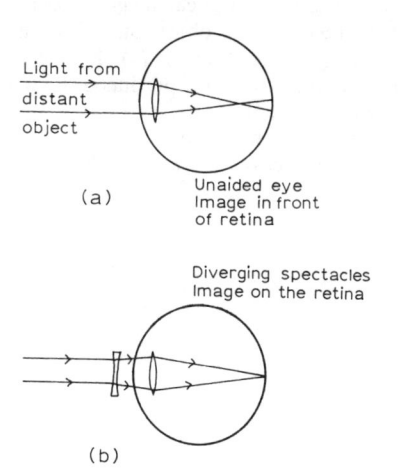

Fig. 31.18. Myopia (short sight)

(3) *Hypermetropia or long sight*

This defect arises as a result of the eye lens, when relaxed, having too long a focal length. Thus as in Fig. 31.19(*a*), distant objects are not in focus.

The eye may accommodate for this but the accommodation is not sufficient to make near objects clearly visible (Fig. 31.19(b)).

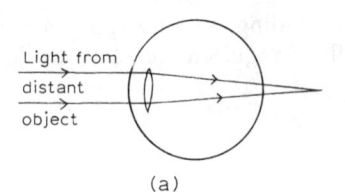

A converging spectacle lens for both close and distant work will correct this defect.

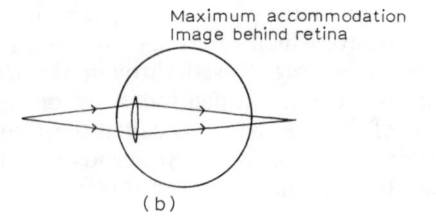

Fig. 31.19. Hypermetropia (long sight)

Exercise 31

1 What is (a) a *converging lens*, and (b) a *diverging lens*?

Show by means of ray diagrams how:
(i) an erect magnified image, (ii) an erect diminished image, may be obtained by using, in each instance, a suitable lens. For what purpose is (i) commonly used? (*L*)

2 Define *focal length* of a converging lens.

Describe an accurate method for finding the focal length of such a lens and show how you would obtain the result from your observations.

A converging lens of focal length 10 cm is used to view an object placed 8 cm from the lens. Find, by drawing an accurate scale diagram, the position of the image and the magnification produced. (*C*)

3 Distinguish between a real and a virtual image.

A converging lens of focal length 8 cm is used to produce an image of an object placed 4 cm from the lens. Find by a scale drawing (a) the position of the image, (b) the magnification produced, (c) whether the image is real or virtual. (*C*)

4 Describe, with the aid of a diagram, the construction of any projection lantern, such as a film-strip projector.

In such a projector, the projecting lens has a focal length of 10 cm and is 12 cm from the slide which is 4 cm square. Find, by drawing or calculation, how far the screen must be placed from the lens and how large the image is. (*C*)

5 Draw a labelled diagram of a camera with a single movable lens and state the function of each NAMED part. Compare the focusing arrangement of the eye with that of the camera.

A converging (convex) lens is used to form an enlarged image of a slide on a screen. The slide is 3·0 cm high and is placed 12·0 cm from the lens which has a focal length of 9·0 cm. Find by a scale drawing, the position of the screen and the size of the image. (*C*)

6 The accommodation of a short-sighted eye extends over a range 20 cm to 8 m. Explain what this statement means. Name the type of spectacle lens required in order that the eye may see a distant object, and draw a diagram to show how the image is formed. (*C*)

7 Two persons, *A* and *B*, are wearing spectacles, *A* having converging lenses and *B* diverging lenses. State the defect of vision from which each suffers and explain how the defect is corrected. (*C*)

32
The Spectrum and Colour

In carrying out experiments with prisms you may have noticed that objects viewed through prisms appear edged with narrow bands of colour. Images formed in lenses may also show this effect. It is often said that these effects were first explained by Newton in 1666, but a scientist named Marci (1595–1667) carried out important experiments on the subject earlier than this.

Newton allowed a narrow beam of sunlight coming through a circular hole in a shutter to fall on a prism, and after refraction, on to a white screen. He observed that instead of a circular patch of white light, he obtained an oval made up of a number of different colours which he called a spectrum. He named seven colours which he saw: red, orange, yellow, green, blue, indigo and violet.

Using a filament lamp instead of the sun, and a slit instead of a circular aperture, you can quite easily show this effect.

Experiment 32.1. Allow a narrow beam of white light from a ray box to fall on a prism, preferably made of flint glass, as in Fig. 32.1. Move the white screen until the light emerging from the prism is seen on it. Now examine very carefully the appearance of the patch of light. You may find that you cannot see all the colours that Newton named. Certainly it is difficult to distinguish between blue and indigo. In fact Newton may have named seven colours simply because he was comparing colour with musical pitch, and he was thinking in terms of musical scale. However, we shall see later how we can produce a spectrum in which the colours are more clearly defined.

Now introduce a second prism arranged as in Fig. 32.2, and move the screen again to receive the emerging light. You will see that the light on the screen is almost white.

Finally turn the second prism to a position as in Fig. 32.3, and examine the patch of light on the screen. The colours have returned and are more widely spread out than in the first instance.

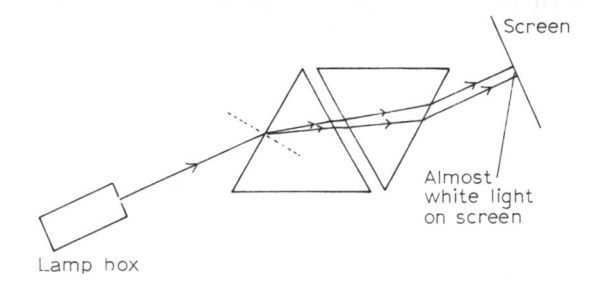

Fig. 32.2. Recombining the colours of the spectrum

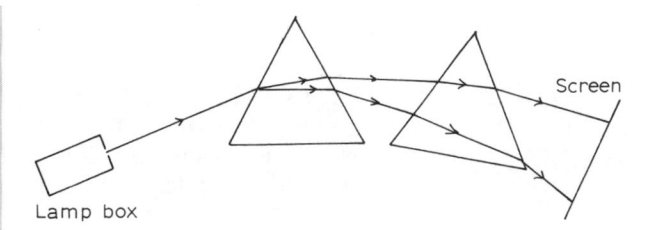

Fig. 32.3. A second prism splitting up the colours still further

Newton's interpretation of these experimental results was as follows: White light is a mixture of light of all the colours in the spectrum. When the white light falls on the first surface refraction occurs, but the refractive index for violet light is greater than that for red. Hence the violet light is bent more than the red. The refractive indices for the other colours lie between these extremes and they are bent by different amounts. Thus at the first surface *dispersion* of the white light into its component colours occurs. At the second surface of the prism a further refraction occurs which results in the colours diverging still more sharply. In the case illustrated in Fig. 32.2, the second prism recombines the colours produced

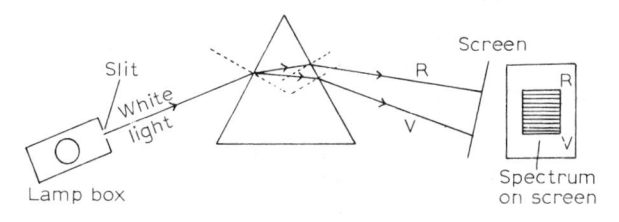

Fig. 32.1. Splitting up of white light

183

by the first, so that white light emerges. The case is comparable to a parallel-sided glass block, which does not produce dispersion.

Figs. 32.1, 32.2 and 32.3 show the paths of single rays of white light through the prism. In practice we shall always have a beam consisting of a number of rays, each of which will be dispersed to form its own spectrum (Fig. 32.4). These spectra overlap with each other and hence we obtain an ill-defined 'impure spectrum' by this simple method. In fact, if the beam is too broad we shall obtain a white patch on the screen which is just edged with red and orange at one end and blue and violet at the other.

The arrangement used in the following experiment was first described by the German scientist Fraunhofer. It enables us to obtain a pure spectrum of white light in which there is no overlapping of the colours.

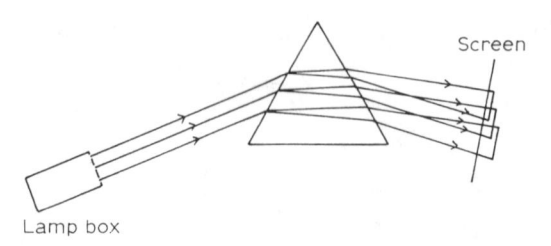

Fig. 32.4. An 'impure' spectrum

Experiment 32.2. Arrange the apparatus as in Fig. 32.5.

The distances from the slit to the first lens and from the second lens to the screen will be the focal lengths of these lenses. The prism, as before, should be of flint glass, if possible, to give a maximum dispersion. A good continuous spectrum should be obtained on the screen.

Parallel white light falling on the prism is split up into its component colours. In the light which emerges, all the rays of a particular shade will be parallel, since the prism will deviate them all by the same amount. The second lens will then focus all rays of this shade to a point on the screen.

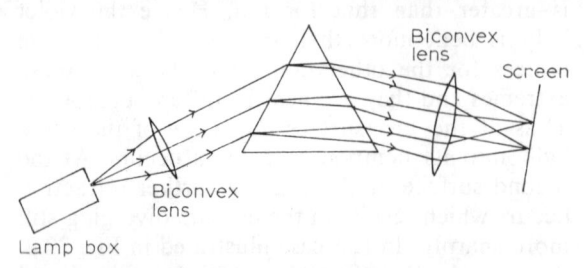

Fig. 32.5. A 'pure' spectrum.

Fig. 32.5 shows the paths of the extreme red and violet rays.

You may be able to identify six or seven colours in the pure spectrum obtained in this experiment. However, you should note that in fact what was obtained was a continuous range of hues which we group together under the headings: red, orange, yellow, etc.

A colour-blind person groups them quite differently. People with normal sight have different ideas about where one colour ends and another begins. Animals have a very limited sense of colour indeed.

We have mentioned that Newton compared colour to pitch in musical sounds. As we shall see in Chapter 34, the pitch of a note is determined by the frequency of the vibrating source or the wavelength of the sound. Light also may be shown to be a sort of wave, and colour is determined by the wavelength. The white light spectrum contains a continuous range of wavelengths.

Coloured Lights and Filters

Pieces of coloured glass and coloured transparent plastics are known as filters and are widely used in photography and in stage lighting. The effects which they produce may be studied using the pure spectrum apparatus shown in Fig. 32.5.

Experiment 32.3. Observe the effects of placing red and blue filters in turn over the ray box slit of Fig. 32.5. You will see that first only the red portion of the spectrum appears on the screen, and then only the blue part. Now place both blue and red filters over the slit. Practically no light reaches the screen at all. A coloured glass is one which allows through only one colour or strictly a small range of colours and absorbs any others falling on it. Now dispense with the slit, lens and prism arrangement, and place the red filter over the lamp-box opening. *In a blacked-out room*, allow the red light produced to fall on a piece of white paper. It seems to be red. Replace the red filter with a blue one, and the white paper appears blue.

We call the paper 'white' because in white light it reflects all the component colours about equally strongly. In red light it can only reflect red and in blue light blue. If now you use two lamp boxes (Fig. 32.6) one fitted with the red filter and the other with the blue one, and allow red and blue light to fall on the same piece of white paper, it may give the appearance of being almost white. In fact if the red and blue are of just the right wavelengths,

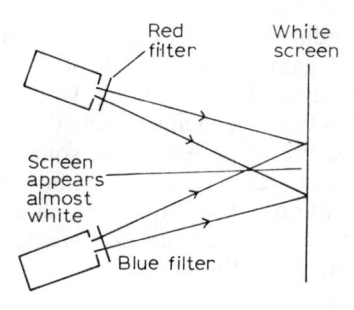

Fig. 32.6.

they will give pure white. Pairs of colours which like this together give white are called *complementary*. Pairs of complementary colours are:

Red and Greenish-blue,
Orange and Blue,
Greenish-yellow and Violet.

If three colours are correctly chosen, it is possible by varying their intensities and mixing them, to produce any shade. Three such colours, red, green and blue, are known as primary colours. It must be explained that until now we have been talking about the mixing of coloured lights. The artist would say that the primary colours are red, yellow and blue, because the rules of pigment mixing are very different.

Pigments

An object which gives off no light of its own is visible only because of reflected daylight or artificial light. We have said that a white object reflects all wavelengths more or less equally. A black object reflects very little light of any colour. Coloured objects reflect a certain range of shades and absorb all the other light falling on them. For example: an object may reflect orange, yellow and green, and absorb red, blue and violet. Such an object in white light will appear yellow. In green light it will appear green, but in violet light no light will be reflected and it will seem black. This experiment with the colour of objects may be conveniently verified by examining scraps of wool of various colours in the light produced by a lamp box fitted in turn with various filters.

Suppose we colour an object by mixing the pigment mentioned above (which reflects orange, yellow and green, and absorbs red, blue and violet) with a blue pigment which reflects green, blue and violet and absorbs red, orange and yellow, the resulting pigment will be green in colour. It will absorb all the colours absorbed by either of the constituents, leaving only green to be reflected (Fig. 32.7).

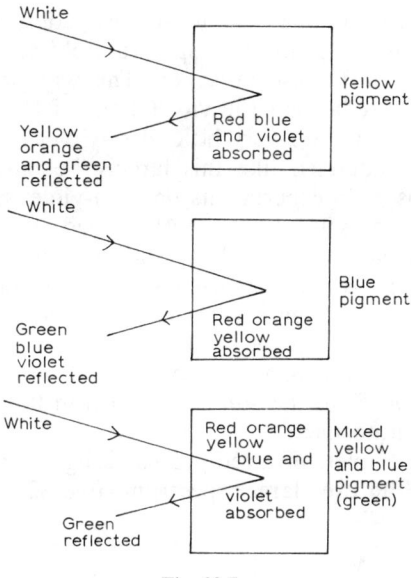

Fig. 32.7.

Since all pigments produce some absorption it is not possible to produce white by mixing, although black can be obtained.

In the case of coloured lights, as we have seen, we may obtain white but not, of course, black.

Invisible Radiations

If the screen in our apparatus used to produce a pure white light spectrum (Fig. 32.5) was replaced by a photographic plate or film, after developing and fixing we might observe a little blackening of the plate beyong the red and violet ends of the spectrum. This would be due to two types of radiation which always arise from light sources called infra-red and ultra-violet.

Infra-red is radiant heat. It is radiated strongly by the radiant type of electric heater (Fig. 37.1, p. 217) in which it is usually reflected by a polished metal concave surface. Infra-red may be most easily detected by a phototransistor. This is a device which becomes a better conductor of electricity when radiation is falling on it. If it is connected in a circuit with a battery and a sensitive galvanometer, the galvanometer will show a reading which will increase if light falls on the phototransistor. Thus if the phototransistor is situated in the visible light spectrum of Fig. 32.5, an increased reading will be obtained. This increased reading will be maintained even when the phototransistor is held beyond the red end of the spectrum, i.e. in the infra-red region. Enormous quantities of infra-red radiation reach us from the sun and you will probably have observed how heat from the sun can be focused using a converging lens.

Ultra-violet may be most easily detected by receiving the continuous spectrum of Experiment 32.2, on a fluorescent screen. This will glow in a region just beyond the violet limit of the visible spectrum. Only a very little ultra-violet radiation will be obtained like this largely because glass absorbs it. In experiments on ultra-violet spectra, quartz lenses and prisms must be used. A mercury vapour lamp (the blue lamp widely used for street lighting) is a much stronger source of ultra-violet than a filament lamp. Ultra-violet is also part of the radiation we receive from the sun and it causes tanning of the skin. However, much ultra-violet radiation from the sun is absorbed in the earth's upper atmosphere.

Ultra-violet, infra-red and visible light constitute part of an even larger spectrum (Fig. 32.8) which contains radio waves, X-rays and gamma rays. X-rays and gamma rays are dealt with in some detail in Chapter 49 of this book. Under the heading 'radio waves' we include many different kinds of radiation including television, sound radio and the very short waves used for communications of all sorts. Although the ways in which these radiations are produced are widely different, they do to some extent share the properties of light such as reflection and refraction. Most strikingly they all travel with the same velocity as light and all exhibit wave properties. The basic difference between them all is their wavelength. We shall have more to say about this quantity in Chapter 33, but Fig. 32.8 shows the wavelengths of radiations in the electromagnetic spectrum.

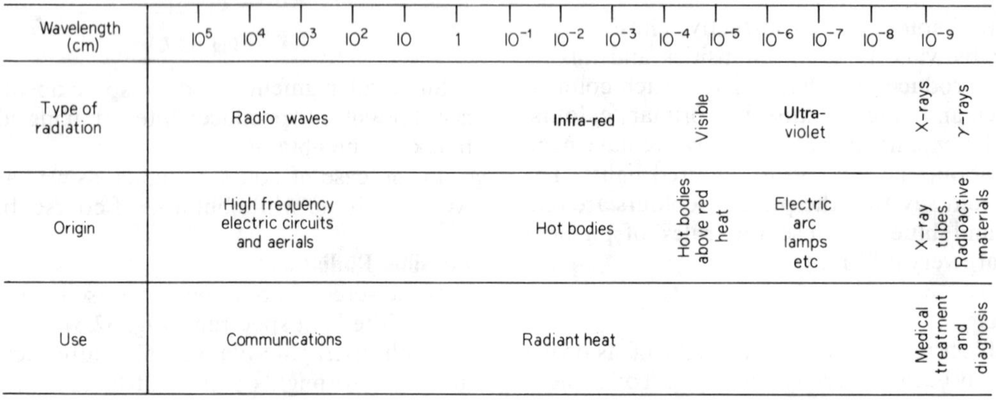

Wavelength (cm)	10^5 10^4 10^3 10^2 10 1 10^{-1} 10^{-2} 10^{-3} 10^{-4} 10^{-5} 10^{-6} 10^{-7} 10^{-8} 10^{-9}						
Type of radiation	Radio waves	Infra-red	Visible	Ultra-violet	X-rays	γ-rays
Origin	High frequency electric circuits and aerials	Hot bodies	Hot bodies above red heat	Electric arc lamps etc	X-ray tubes.	Radioactive materials
Use	Communications	Radiant heat			Medical treatment and diagnosis	

Fig. 32.8.

Exercise 32

1 A narrow parallel beam of white light falls on a prism and after passing through the prism the light is received on a white screen. State what happens to the light as it passes through the prism and what you would see on the screen. What would be the effect of (i) removing the prism, (ii) retaining the prism but replacing the white light with yellow (monochromatic) light? Illustrate your answer with suitable diagrams. (C)

2 (a) State briefly what happens when a ray of white light is passed through a triangular glass prism. Illustrate your answer by means of a diagram.
(b) What will be the appearance of a red flower when it is illuminated by blue light? Explain this briefly. (S)

3 Describe fully, with the aid of a diagram, how you would project on a screen a reasonably pure spectrum of white light. Draw a labelled diagram to show the appearance of this spectrum. (O)

4 A white card on which are painted a blue square and a red triangle is illuminated by red light. Describe and account for the appearance of the card. (J)

5 (a) Draw a diagram to show what happens when a ray of white light is shone through a triangular glass prism. What does this experiment tell us about white light?
(b) The refractive index of glass is different for different colours of light. How do we know this? For which colour is the refractive index of glass greatest?
(c) A red lamp is placed in a dark room. What will be the colour of the following bodies when viewed by the light of this lamp:
   (i) a body which appears red in white light?
   (ii) a body which appears blue in white light?
   Give reasons for your answers. (C)

33 Waves

In chapter 31 we mentioned briefly that light together with X-rays, radio, ultra-violet and infra-red radiation exhibit certain properties which can only be explained by assuming that these radiations behave like waves. We shall not in this book describe this evidence which was obtained by a number of scientists during the nineteenth century. However it is useful for us to look quite carefully at water waves for we shall be able to see how those effects of light with which we are familiar, like reflection and refraction, can be described in terms of a wave theory of light. Furthermore in the next chapter we shall be thinking about 'sound' and sound also is known to be a kind of wave motion. A convenient piece of apparatus for our study of water waves is a simple ripple tank. It consists (Fig. 33.1) of a shallow glass bottomed tank containing water to a depth of about 1 cm supported on four legs about a metre long. The tank should have sloping 'beaches' of wire gauze round the edge to help to eliminate the reflection of waves from the sides of the tank. A low voltage lamp mounted above the tank as shown in the diagram casts an image of the wave pattern produced on to a piece of white paper placed below the tank. We shall first look at some simple wave patterns.

Experiment 33.1. Use a teat pipette to drop water, one drop at a time, in the middle of the tank and observe the wave pattern. Fig. 33.2 shows in a sequence of pictures the sort of effect you are likely to observe. Now take a piece of wooden dowel rod of rather more than 1 cm. in diameter and perhaps 15 cm long and rest it cross the tank. Rock the rod gently to produce a good straight wave travelling along the tank. Fig. 33.3 shows the sort of wave you may obtain in this case.

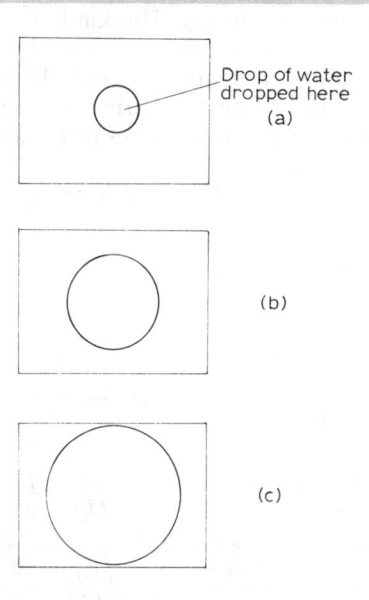

Fig. 33.2. A circular wave

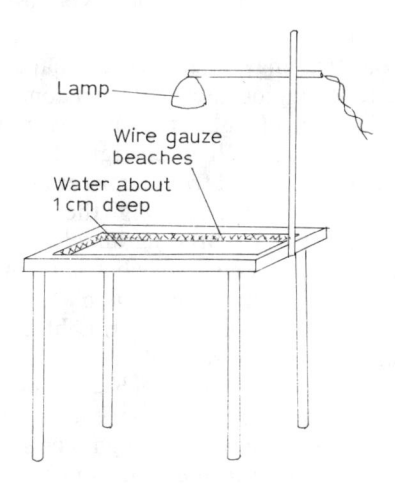

Fig. 33.1. Simple ripple tank

Although these wave patterns seem very simple and familiar to us it is worth while our thinking about them rather carefully. Perhaps the most important and obvious thing for us to note is the fact that while the wave moves steadily across the tank there is no substantial movement of the water itself in the direction of the wave. If you are not too sure of this, it is an easy matter to check by floating perhaps a small piece of cork on the water surface and observing its behaviour as a wave passes. In fact the movement of the water surface is mostly up and down, or at right angles to the direction in

187

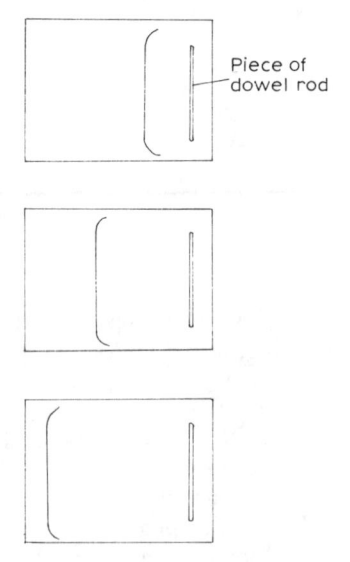

Fig. 33.3. A straight wave

which the wave is moving. This kind of wave (Fig. 33.4) is known as a transverse wave. We shall see in the chapter on sound that not all waves are transverse. In the next experiment we shall see what happens when water waves meet an obstacle.

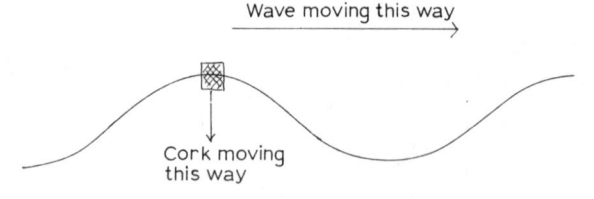

Fig. 33.4. A transverse wave

Experiment 33.2. Place a straight barrier across the tank as shown in Fig. 33.5(a). Now use the teat pipette again to produce single circular waves and note how they behave. The sequence of patterns shown in Fig. 33.5(b) is what you are likely to see. Clearly the wave is being reflected at the barrier. The wave after reflection spreads out as if it is

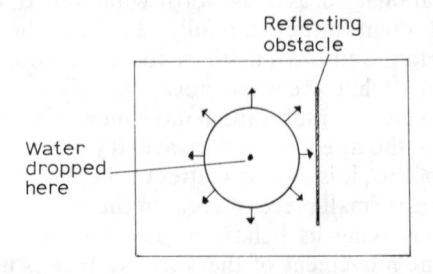

Fig. 33.5.(a) Circular wave spreading out

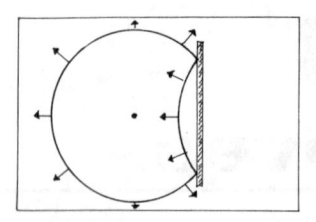

Fig. 33.5. (*b*) Wave being reflected from obstacle

coming from some point behind the obstacle. It is interesting to try to decide exactly where this point is. Make a whole series of circular waves (Fig. 33.6) by moving a finger up and down on the water surface. Now using a finger of the other hand, try to produce waves on the other side of the barrier which complete the circle of the reflected waves (Fig. 33.7). It will probably be clear to you that to

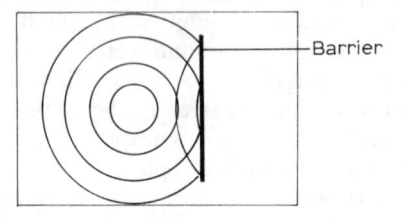

Fig. 33.6. Continuous circular waves reflected at a barrier

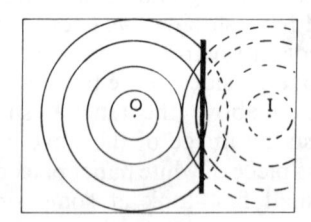

Fig. 33.7. Circular waves on both sides of a barrier

do this your two fingers need to be equal distances from the reflecting obstacle. We can compare this to the reflection of light from a mirror surface. Fig. 33.8 shows rays of light diverging from a point object O being reflected from a mirror after which they continue to diverge as though they are spreading from the image I. This diagram also shows the waves in this case. See also what happens when straight waves produced with a piece of dowel rod, as previously, fall on the straight barrier. It is possibly most interesting if the obstacle is set obliquely to the waves in which case the pattern is likely to be as in Fig. 33.9. Once again we could relate this to the behaviour of light. This time the effect is comparable to the reflection of a parallel beam of light from a flat mirror.

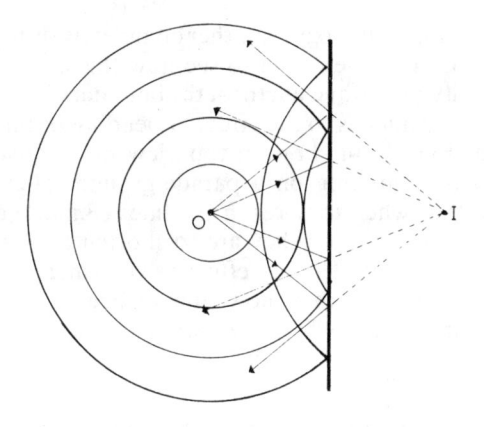

Fig. 33.8. Reflection by a plane mirror (rays and waves)

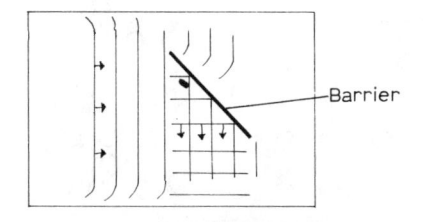

Fig. 33.9. Reflection of straight waves from a flat obstacle

For the next experiment it will be convenient to use a device for producing a continuous stream of waves, either circular or straight. This consists of a wooden bar hanging by two rubber bands from a frame fixed across the tank (Fig. 33.10). A small low voltage electric motor with a weight fixed to its axle, is mounted on the bar. When the motor axle turns, the weight which is a little off axle causes a vibration. If straight waves are needed, the bar is allowed to hang so that it just rests on the water surface. If circular waves are required, the bar is raised clear of the water and a ball fixed to and a little below the bar rests on the surface.

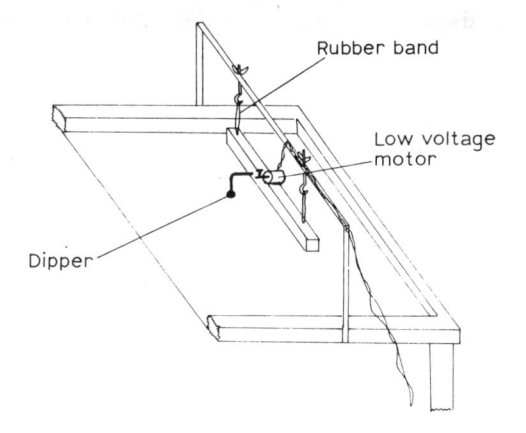

Fig. 33.10. Ripple tank vibrator arrangement

Before using this arrangement for further experiments we will consider in a little detail the continuous waves which it produces. The distance between two consecutive wave crests we call the wave length (λ)(Fig. 33.11). If the electric motor makes f rotations each second then there will be produced f wave crests per second. This tells us that one second after a certain wave crest has been formed by the vibrating bar, it will have travelled a distance $f \times \lambda$ from its source. This means that the wave velocity is $f \times \lambda$. The quantity f is known as the frequency of the wave and is measured in hertz (Hz). A wave has a frequency of f Hz if its source makes f vibrations per second. If the wavelength is measured in metres, then the wave velocity is given in metres per second by:

$$v = f \times \lambda$$

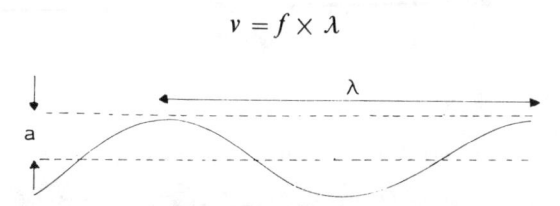

Fig. 33.11. Amplitude a Wavelength λ

There is just one more term used in connection with waves which we need to explain. This is the amplitude of a wave, a in Fig. 33.11. It is the maximum displacement of the water surface from its normal level. If you like it is the height of a crest above the level of the undisturbed surface.

We have already described experiments which show the reflection of water waves and we have discussed the way in which this may be compared to the reflection of light from a mirror surface. We shall now go on to consider the other main property of light described earlier in the book, namely refraction. The following experiments should show a comparable effect with water waves.

Experiment 33.3. Adjust the vibrator so that it produces good straight waves. Now slope the tank from end to end so that the glass bottom at the end farthest from the vibrator is not covered with water, in other words so that the water gets shallower as the waves move away from their source. You should be able to see that the wavelength gets shorter as the waves move into shallower water, (Fig. 33.12). Now restore the tank to a level position and provide some shallow water in this instance by placing in the water a rectangular piece of glass or perspex whose thickness is such as to reduce the depth of the water over it to 2 or 3 mm.
Set the edge of the rectangle obliquely to the

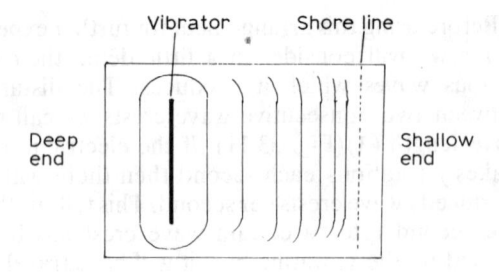

Fig. 33.12. Straight waves moving into shallower water

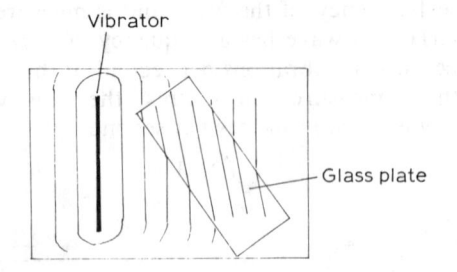

Fig. 33.13. Refraction of water waves

straight waves produced by the vibrator and observe the effect on the waves. Perhaps you will see something like Fig. 33.13.

Some consideration of these effects and how they relate to the refraction of light is clearly called for. Although the wavelength gets less in shallow water, there is no reason why the frequency should change. Hence using our equation $v = f \times \lambda$ we can see that decreased wavelength in this case means decreased wave velocity. In simple terms the waves travel more slowly in shallower water. The second part of this experiment illustrated in Fig. 33.13 shows how this reduced velocity over shallow water leads to a bending of the waves. This clearly suggests a comparison with the bending of light rays at the boundary between two media. Could it be that light, when it passes from air to glass say, travels more slowly, and that this reduction in velocity is responsible for the bending? This idea was suggested by a Dutchman, Huygens, towards the end of the seventeenth century. Although Romer (Chapter 28) had by that time obtained a value of the velocity of light in space (a vacuum), it was not possible to measure the speed of light in glass or any other medium. Huygens' idea therefore had to wait for confirmation until the nineteenth century, when a French scientist Foucault showed that light did indeed travel more slowly in a medium than in a vacuum.

In using water waves to give a picture which helps us to understand the behaviour of light, we are using an analogy. Another completely different analogy may help you to see how this change of velocity (and wavelength) at the boundary between two media leads to the effect of bending or refraction. Fig. 33.14 shows a top view of a squad of soldiers marching on a parade ground. They are told that when they reach the line AB marked on the parade ground they are to shorten the length of their paces. The net effect as you can see from the diagram, is that the soldiers wheel or change the direction in which they are marching.

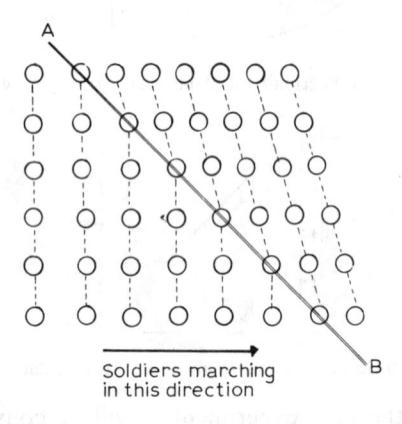

Fig. 33.14.

We may carry our analogies for refraction just a little further to get some idea of the significance of the quantity refractive index which we introduced in Chapter 30. Fig. 33.15(a) shows parallel rays of light falling on a boundary between air and glass. The refractive index $_{air}n_{glass}$ is $\dfrac{\sin i}{\sin r}$.

Fig. 33.15(b) shows the same effect considered in terms of waves and fig. 33.15(c) is a combination of a part of each of the previous two diagrams. The line PQ in Fig. 33.15(c) represents a wave crest at a certain instant in time and RS is the same wave

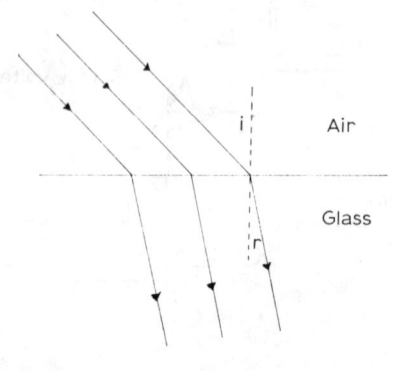

Fig. 33.15. (a) Refraction (rays)

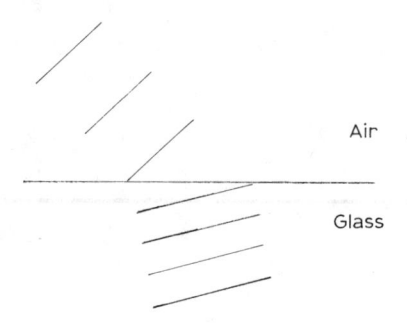

Fig. 33.15. (b) Refraction (waves)

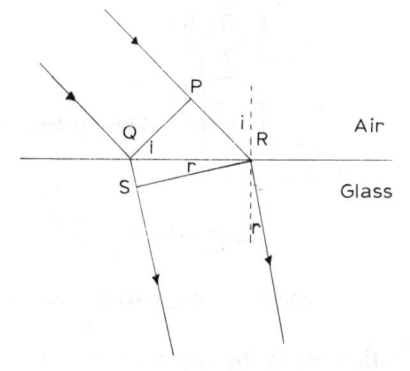

Fig. 33.15. (c) Refraction (rays and waves)

crest a short time (t seconds) later. One end of the wave has travelled the distance PR during this time in air. The other end in the same time has travelled the shorter distance QS since it has travelled in glass and therefore moves more slowly.

Now PR = $c \times t$ m —————————— (1)
(c is the velocity of light in air in m/s)
and QS = $v \times t$ m —————————— (2)
(v is the velocity of light in glass in m/s)

By simple geometry it is possible to show that the angle PQ̂R is the angle of incidence $î$ and QR̂S is the angle of refraction, $r̂$. Hence from the two right angled triangles PQR and QRS

$$\sin i = \frac{PR}{RQ} \text{ and } \sin r = \frac{QS}{RQ}$$

and $\dfrac{\sin i}{\sin r} = \dfrac{PR}{RQ} \times \dfrac{RQ}{QS}$

$= \dfrac{PR}{QS}.$

Thus from equations (1) and (2) above

$$\frac{\sin i}{\sin r} = {}_{air}n_{glass} = \frac{c \times t}{v \times t}$$

$$= \frac{c}{v}$$

We have now shown what is perhaps a more significant meaning to the quantity refractive index. It is the ratio of the velocities of light in the two media.

There are many more experiments which we might carry out with a ripple tank to reveal further properties of water waves, and these properties can generally be related to light and other forms of electromagnetic radiation and to sound. However it would be inappropriate to proceed further with the subject here. Perhaps a word of caution is needed though before we leave it. There is good reason to believe that light has some kind of wave nature. It is beyond the scope of this book to be more precise about what kind of wave it is. It is certainly not anything directly to do with water waves. Our discussion of water waves has been simply to give us a useful picture, and incidentally to enable us to make some predictions like the one made by Huygens about the change in the velocity of light when it goes from one medium to another.

Exercise 33

1 Draw diagrams to show the reflection of water waves in a ripple tank in the following circumstances:
  (a) Straight waves approaching a concave reflecting obstacle.
  (b) Circular waves approaching a concave obstacle.
  (c) Straight waves approaching a convex obstacle.
  (d) Circular waves approaching a convex obstacle.
2 BBC Radio 1 broadcasts are transmitted on a wavelength of 247 metres. The velocity of radio waves is 3×10^8 m/s. What is the frequency of this transmission?
3 The vibrator of a ripple tank vibrates 5 times per second and produces waves of length 4 cm. How long will a wave take to go from the vibrator to the end of the tank 30 cm away?
4 The refractive index of glass is 1·5 and the velocity of light in air is 3×10^8 m/s. What is the velocity of light in glass?

34 Sound

We have mentioned in Chapter 32 the vast quantities of light, and of ultra-violet and infra-red radiation given off by the sun. It is thought that this energy arises from something like an enormous hydrogen bomb explosion going on continuously. You may wonder then why we *hear* nothing of this. Is it because there is a basic difference between light and sound? The visible radiation, in order to reach us, has to travel through 148 million kilometres, most of which is empty space. Is sound incapable of travelling through a vacuum? The answer is provided by a simple laboratory experiment.

Experiment 34.1. Suspend an electric bell by means of a piece of sponge rubber over the base plate of a vacuum pump (Fig. 34.1). The electrical connection between the bell and battery may be made by way of electrodes which pass through the base plate but are insulated from it. Place a bell-jar over the electric bell. It will sound less loud than without the bell-jar in position, but will still be quite audible. Now switch on the vacuum pump. As the air is removed from the jar, the sound of the bell will become more and more faint until, if a good vacuum is obtained, the hammer can still be seen vibrating although no sound at all is heard. As the air is readmitted to the bell-jar the sound gradually becomes audible again. Clearly sound will travel in air but not in a vacuum. If the bell is simply placed on the pump base plate instead of being suspended, sound will be heard even when the bell-jar is evacuated. Sound will travel through solids and, in fact, through liquids as well as gases.

How are Sounds Produced?

Many sounds, both musical and nonmusical, are produced by solid objects being struck, plucked or bowed. The gong of an electric bell, the stretched skin of a drum and the strings of a piano are struck. The strings of a guitar are plucked, and those of a violin bowed. In all these cases a solid object is made to vibrate. You can see the vibration, for example, of a guitar string after it has been plucked, and if you lightly touch the gong of an electric bell which is ringing you may be able to feel it vibrating.

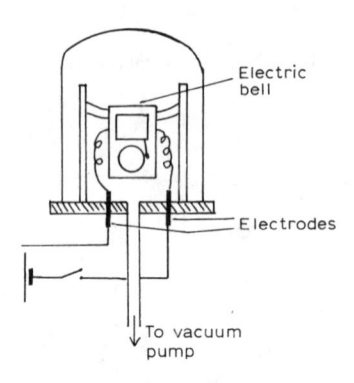

Fig. 34.1. Sound does not travel in a vacuum

The vibrations of the prongs of a tuning fork which has been struck can be shown most clearly by gently touching a suspended pith ball with the fork, (Fig. 34.2). The ball will bounce away from the vibrating prong through a considerable distance.

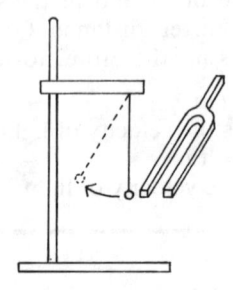

Fig. 34.2 The prongs of a sounding tuning fork are vibrating

Sounds then are produced by things vibrating and need a medium (solid, liquid or gas) to travel from one place to another. But what is sound? Fig. 34.3 shows what happens to the air in the neighbourhood of a vibrating tuning fork. In Fig. 34.3 (a) the prongs of the fork are shown moving outwards compressing the air close to them. A fraction of a second later (Fig. 34.3(b)) the prongs are moving in the opposite direction allowing this air to expand again. However, the original compression has moved into air a little further from the fork. The next outward movement of the prongs produces another compression (Fig. 34.3(c)) which

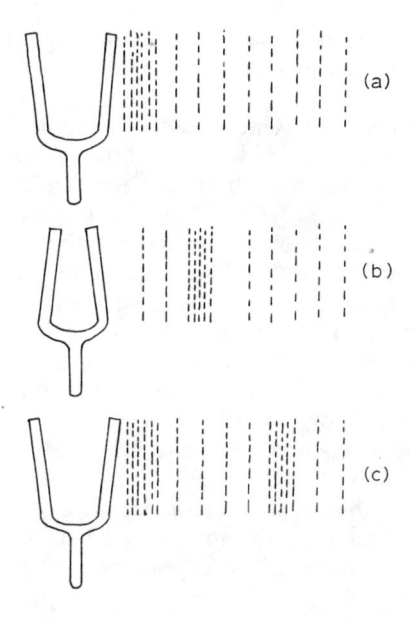

Fig. 34.3. Compressions and rarefactions

in turn moves away, and is followed by an expansion or 'rarefaction'. This series of compressions and rarefactions in air or any other material is what we mean by sound.

It should be emphasized that the air as a whole does not move any distance in the direction in which the sound is moving. As in the case of water waves, it is only the disturbances which move.

The comparison between sound and water waves must not be carried too far though. In the case of ripples on water, the surface moves at right angles to the direction in which the waves are travelling (Fig. 34.4(*a*)). Such disturbances are known as transverse waves. The vibrations of the molecules of the material through which sound waves are passing are in the same direction as that taken by the sound (Fig. 34.4(*b*)). Sound waves are said to be longitudinal.

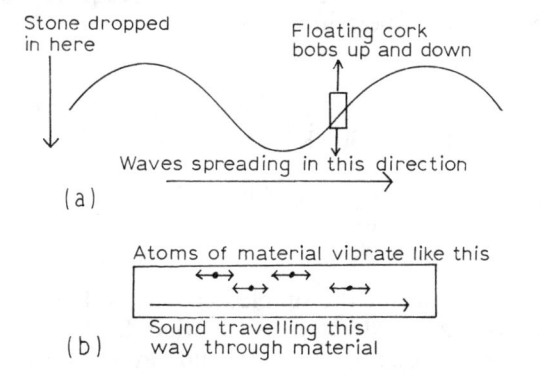

Fig. 34.4. (*a*) Transverse water waves
(*b*) Longitudinal waves, e.g. sound

A good idea of how a longitudinal wave progresses may be obtained with the aid of a spiral spring of the type sold in toy shops known as a 'Slinky'. If such a spring is held, slightly stretched on a smooth bench, and one end is given a sharp push, the disturbance will travel along the spring. A number of such pushes in quick succession will produce a number of compressions travelling along the spring (Fig. 34.5).

In the case of sound the wavelength is the distance between two compressions.

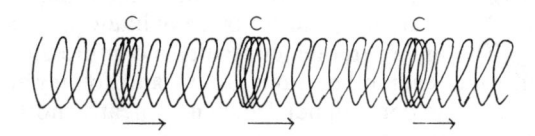

Fig. 34.5. Compressions travelling along a 'slinky' spring

Loudness and Pitch

We shall now try to investigate how the kind of sound we get depends on the vibrations producing it.

Experiment 34.2. Clamp a strip of springy metal (e.g. a 30 cm metal ruler or hacksaw-blade) so that it projects over the edge of the bench (Fig. 34.6). Push it down at the projecting end and let it go. It will vibrate and produce a low-pitched sound. Now stop it and push it down further before releasing it. The strip of metal will make larger vibrations and a louder noise of the same low pitch. The loudness of the sound is seen to depend on the size of the vibrations. The distance between an extreme position and the rest position of a vibrating body (as in Fig. 34.6), is known as the *amplitude* of vibration (*a*). The loudness or intensity of a sound is bigger, the bigger the amplitude of the vibrating source producing it. Reduce the length (*l* in Fig. 34.6), projecting from the edge of the bench and set the piece of metal vibrating again. You will be able to see that the metal is vibrating more rapidly, and the pitch of the sound produced has been raised. The pitch of a sound is determined by the

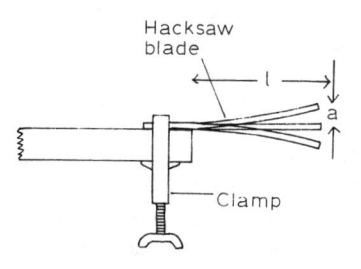

Fig. 34.6. Pitch and loudness

number of vibrations per second of the source producing it. This quantity is called the *frequency*. The higher the frequency of a vibrating object, the higher the pitch of the sound produced.

Fig. 34.7 shows the effect on the air around the vibrating hacksaw-blade of differences of amplitude and frequency.

Fig. 34.7(a) shows the compressions and rarefactions arising from the vibrating piece of metal. Fig. 34.7(b) has a larger amplitude and the compressions are more intense, i.e. the sound is louder. Fig. 34.7(c) shows a shorter piece of blade projecting. The higher frequency results in compressions following each other more rapidly and so they are closer together. Higher frequency means shorter wavelength.

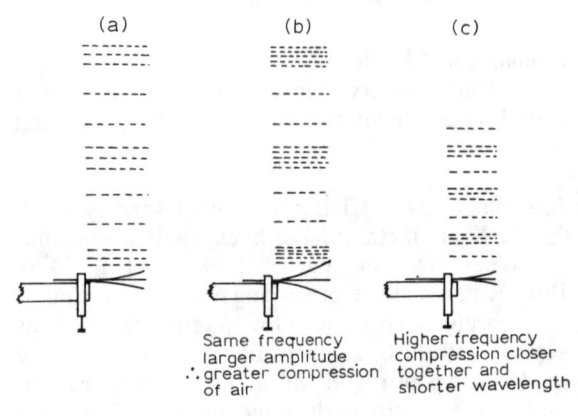

Fig. 34.7.

Velocity of Sound

You will be familiar with many examples which show that sound does take time to travel from one place to another. If you have watched a cricket match from a considerable distance away, you may have noticed the time interval between seeing the batsman making his shot and hearing the sound of bat against ball. A considerably longer time lag usually exists between seeing lightning and hearing the thunder. Only if the storm is very close indeed does the thunder come at almost the same time as the lightning. In each of these cases the sound takes considerably longer to travel over a certain distance than does light. In fact over distances of even many kilometres the time taken by light is quite negligible. This is why the time-keeper at a race will attempt to start his stopwatch when he sees the smoke from the starter's pistol, and not when he hears the bang. It is possible to get a rough idea of how fast sound does travel in air using a starting pistol and stopwatch if you can use a sufficiently large piece of open fairly flat ground.

Experiment 34.3. One observer equipped with a starting pistol loaded with blank rounds should be situated if possible about a kilometre from the other observers who should have stopwatches which measure to one-tenth or one-fifth of a second. A distance of less than half a kilometre is probably too short for satisfactory results. After a pre-arranged signal from the experimenter with the gun, those with stopwatches should start them when they see the smoke from the gun, and stop them when they hear the report.

If this is repeated several times by a number of different observers a good average value for the time taken by the sound to cover the distance may be obtained. The distance may be measured by means of a surveyor's tape measure and the velocity found.

Velocity of sound $= \dfrac{\text{Distance}}{\text{Time}}$

Typical values might be:
 Distance = 600 m
 Time   = 2·0 second
 Velocity of sound in air = 300 m/s

It is clear that in measuring a time interval as short as two seconds, the time taken for the observer to react to the signals by starting or stopping his watch will be relatively large. Consequently we cannot really expect an accurate value from simple experiments of this kind. More accurate experiments, which are beyond the scope of this book, give values close to 330 m/s. The exact figure will depend on the temperature of the air, although the variation will be less than 30 m/s over the whole range of air temperatures we are likely to encounter. In fact it may be shown that the velocity of sound is proportional to the square root of the air temperature in kelvins.

$$V \propto \sqrt{T}$$

Thus if the velocity is 330 m/s at $-3\,°\text{C}$ (270 K) then V at $27\,°\text{C}$ (300 K) is given by:

$$\frac{V}{330} = \sqrt{\frac{300}{270}} = \sqrt{\frac{10}{9}}$$

or $V = 330 \times 1\cdot05 = 346$ m/s.

The results of the simple experiment described above may be affected also by wind. If the wind is in the same direction as the sound travels, the value obtained will be high by comparison with the value for still air. If wind and sound are moving in opposite directions the result will be low. This effect may be corrected by making a second set of timings with the two sets of experimenters having exchanged positions. An average for the whole experiment will refer to still air conditions.

The speed of sound in solids and liquids is generally much higher than in gases. The velocity of sound in water for example is about four times that in air. In steel sound travels at about 5200 m/s. A knowledge of these values is useful in many sets of circumstances. For example, most ships, and in these days even quite small boats, are fitted with water-depth measuring equipment. This usually consists of a device which produces a sound which goes down to the sea bed and is there reflected back to the ship. The time taken by the sound in travelling this distance is measured electronically. If the velocity of sound in water is known the depth of the sea is readily obtained.

Velocity, Wavelength and Frequency

We saw in Chapter 33 that for water waves the velocity (v), wavelength (λ) and the frequency (f) are related by the formula $v = f \times \lambda$. This relationship applies equally to sound waves. Let us consider a tuning fork of frequency 256 Hz. This means that the fork produces a compression in the air close to it every $\frac{1}{256}$ s. The distance between compressions is the wavelength, λ, (Fig. 34.8). So in $\frac{1}{256}$ s the compression moves a distance λ, or in one second, it moves 256λ. But the distance moved in one second is the velocity. Thus in general:

$$v = f \times \lambda$$

Since the velocity of sound in air is 330 m/s, the wavelength in air of a sound of frequency 256 Hz (the note middle C) is given by:

$$330 = 256 \times \lambda$$
$$\text{or } \lambda = \frac{330}{256}$$
$$= 1 \cdot 3 \text{ m}$$

Reflection of Sound

We have mentioned the method of finding the depth of water under a boat by finding the time taken for a pulse of sound to return to the underside of a boat having been reflected by the sea bed. The method is known as 'echo-sounding'. Echoes in air are a familiar experience. They will be obtained most noticeably in large empty buildings such as aircraft hangars, or in the open when a big reflecting surface such as a cliff-face is nearby. Experiment shows that the laws of reflection obtained for light apply also to sound. Thus sound is reflected as in Fig. 34.9, the angle of incidence being equal to the angle of reflection. However, it is difficult to obtain anything like a narrow beam of sound, so verification of this with any accuracy is not easy.

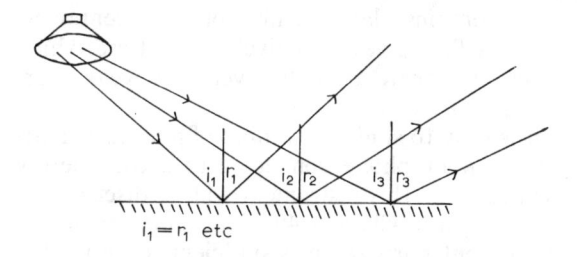

Fig. 34.9. Reflection of sound

Room Acoustics

Hard smooth surfaces like glass, metal and stone reflect sound most strongly. Soft materials like curtains and carpets absorb sounds and reflect very little. Much attention is given these days in designing buildings to the amount of echo that will occur in them. For example, in restaurants there is likely to be a great deal of clatter which the customers will find objectionable. The noise can be very much reduced by the use of curtains and carpets, or less expensively by covering the ceiling and walls with soft plastic tiles with strongly absorb sound. If this treatment is given to a hall which will be used for musical performances, the music will sound 'dead'. A certain amount of echo or reverberation (but not too much) is needed in a concert hall. The study of what is the ideal amount of echo for a particular building and how this may be achieved by choosing the right shape and materials is the work of the acoustics engineer.

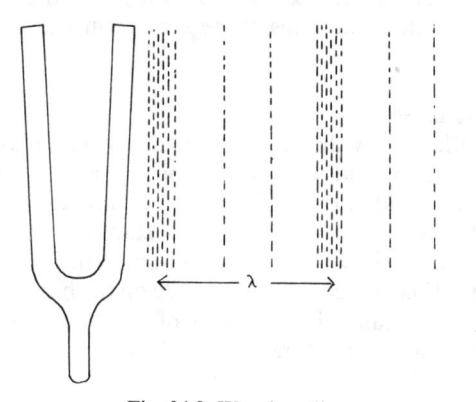

Fig. 34.8. Wavelength

Musical Sounds

Earlier in this chapter we discussed two properties of musical sounds, pitch and loudness. It was noted that pitch depends on the frequency of the vibration causing the sound, and loudness on the size or amplitude of these vibrations. But these two properties alone do not completely describe a musical sound. You would have no difficulty in distinguishing between the sounds produced by a piano and a clarinet, even if they were of exactly the same pitch and loudness. There is a third characteristic known as the 'quality' or 'timbre' of a musical sound. If Middle C is sounded on the piano the sound waves produced are principally of one frequency, called the 'fundamental'. For Middle C this is about 260 Hz, and is usually quoted as 256 for scientific purposes. However, the string may also vibrate to produce sounds of frequencies about 2×260, 3×260, 4×260, etc. The pitch of these higher frequencies called 'overtones', together with the proportions in which they occur in the whole sound, determine its quality. A 'rich sound like that produced by a piano or violin contains a large number of prominent overtones. A flute gives a relatively 'pure' tone, which means that there are only a very few weak overtones.

It is easy to study the timbre of musical sounds using a microphone connected to a cathode ray oscillograph. This instrument will be discussed in more detail in Chapter 49 of this book. However for present purposes it is sufficient to know that it will show on its screen the wave form of sounds produced by instruments which are sounded near the microphone. Fig. 34.10 shows some such wave forms. The 'spikey' pattern obtained with a violin shows the presence of many overtones, while the smooth curve produced by a recorder indicates that it is a relatively pure tone.

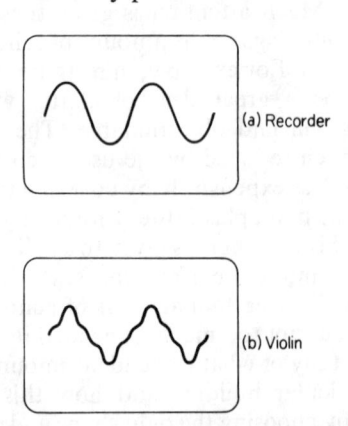

Fig. 34.10. Cathode ray oscilloscope wave patterns

Stringed Instruments

A vibrating stretched string alone produces little sound. In all stringed instruments the strings are situated near to a sounding board, or sounding box. This contributes to the quality of the sound produced as well as to its loudness. The quality is also very much affected by the way in which the strings are made to vibrate. In a piano they are struck by hammers with felt heads, in a guitar or harp they are plucked, and on a violin they are usually bowed. The pitch of the sound produced by a stringed instrument is determined, as we shall see in the next section, by the length, material and tension of the string. On an instrument of the guitar or violin type (Fig. 34.11) a few strings (four on a violin) are used. They are all of about the same length but of different thicknesses and sometimes different materials. The instrument is tuned before use by adjusting the tension by means of the pegs. In use, different pitches are obtained by varying the sounding length of the string by pressing down on it on the finger board.

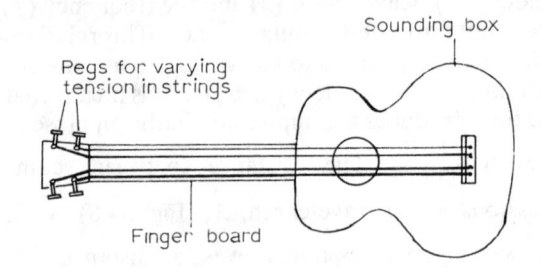

Fig. 34.11. A guitar

Pianos and harps are also tuned by varying the tension in the strings. In the former case only from time to time, but with a harp always before use. In these instruments the whole range of pitches may be obtained by having a large number of strings whose lengths may not be varied. Thus for each note on the keyboard of a piano there will be one or sometimes three strings in the instrument.

Musical Scales

Although with many musical instruments it is possible to produce a continuous range of pitches, it was found in the earliest days of musical history that it was necessary to pick certain fixed pitches to make satisfactory musical sounds. In European music this choice has been made on the basis of the 'diatonic scale'. This consists of eight notes whose pitches are in fixed ratios as follows:

C′	D	E	F	G	A	B	C″
1	$1\frac{1}{8}$	$1\frac{1}{4}$	$1\frac{1}{3}$	$1\frac{1}{2}$	$1\frac{2}{3}$	$1\frac{7}{8}$	2

These eight notes are known as an octave. It is not really important what pitch is chosen for C' as long as the ratios are correct. However, for scientific purposes, C' is normally 256 Hz. The other notes of the scale are therefore:

C'	D	E	F	G	A	B	C"	
256	288	320	341	384	426	480	512	Hz

The Factors Influencing the Pitch Produced by a Stretched String

With the aid of a stringed instrument such as a guitar or violin it is an easy matter to verify the following points:
(1) Shortening a string will raise its pitch.
(2) Increasing the tension of a string will raise its pitch.
(3) A thin string will produce a higher pitch than a thick string of the same material and about the same length and tension.

To consider these results more carefully it is convenient to use a laboratory 'stringed instrument' which is known as a Sonometer (Fig. 34.12).

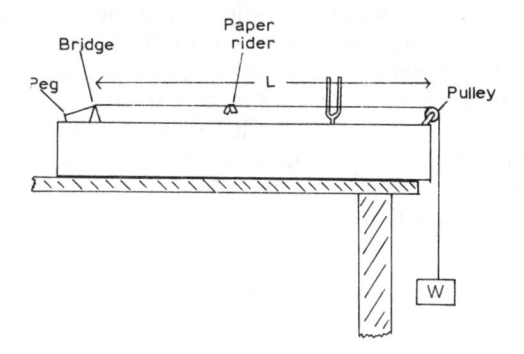

Fig. 34.12. The Sonometer

A string or wire is fixed on a peg at one end of a long wooden box and passes over a pulley at the other end. A tension is maintained in the wire by attaching weights (W) to its free end.

Experiment 34.4. Frequency and Length.

With the suspended weight (W) remaining constant, tune the wire in turn to the pitch of a number of tuning forks by varying its sounding length (L). Since the frequencies of the tuning forks (f) will generally be marked on them, a table of values of (f) and (L) may be produced.

The tuning of the wire to the forks may be done entirely by ear or with the aid of a paper rider. First strike one of the forks on a cork or rubber pad, and listen to it while you pluck the sonometer wire near its mid-point. If the sound produced by the wire is lower in pitch than that produced by the fork, move one of the bridges to shorten the wire and pluck it again. By sounding the wire and fork simultaneously you should be able to get the two almost exactly in tune by adjusting the length of the wire only.

The final exact tuning may be done by means of the method of 'beats'. If you listen very closely to two sound sources which are nearly but not quite in tune, you will hear a series of rapid variations in the loudness of the note. A sort of 'wowing' or throbbing sound is produced. This effect of beats is usually used by stringed instrument players for tuning. Alter the sounding length of the sonometer wire very carefully so as to make the beats slower. When the lowest possible beat frequency is obtained the wire and fork are in tune.

If you find difficulty in tuning the wire entirely by ear, cut a narrow strip of thin paper, fold it in half, and rest it on the middle of the wire after it has been tuned as well as possible by ear. Strike the fork and place it in contact with the sound box of the sonometer, as shown in Fig. 34.12. Now carefully adjust the position of one of the bridges until the paper rider is seen to vibrate and perhaps fall off the wire. This occurs because when the fork and wire are in tune the latter will vibrate, without being plucked, in 'sympathy' with the vibrating fork.

Table 34.1 shows a typical set of results for this experiment, the tension being constant throughout.

Table 34.1

Pitch and Frequency (Hz)		Length (L) cm	$\frac{1}{length}$
C	256	60.4	0.016
D	288	55.6	0.018
E	320	50.1	0.020
F	341	48.0	0.021
G	384	41.7	0.024
A	426	37.0	0.027
B	480	32.7	0.030
C	512	30.4	0.032

It will be observed from the first and last set of values in the table that in order to double the frequency of the wire we have had to halve its length. This suggests that the frequency is inversely proportional to the length $f \propto 1/L$.

This may be verified by calculating values for $1/L$ and plotting these against the frequency (f).

A graph of the figures in Table 34.1, is shown in Fig. 34.13.

As it is a straight line passing through the origin it shows a relationship of the form:

$$f = \text{constant} \times 1/L.$$
$$\text{or } f \propto 1/L.$$
$$\text{or } f \times L = \text{constant}.$$

Thus, for a certain wire at constant tension, if we know one pair of values of frequency and length, we may find the frequency of any other given length, e.g. if, as in the experiment above, a wire tunes to Middle C (256) at a length of 60·4 cm, we can obtain the length of wire (L) which would tune to the G below Middle C (192):

$$256 \times 60\cdot 4 = 192 \times L$$
$$L = \frac{256 \times 60\cdot 4}{192}$$
$$= 80\cdot 4 \text{ cm}.$$

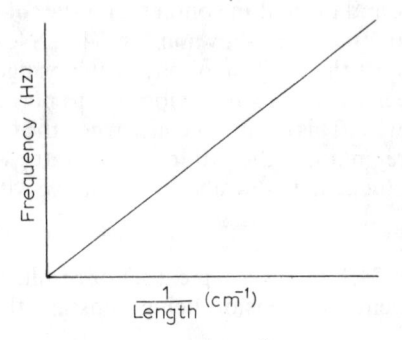

Fig. 34.13. Graph of frequency $\times \frac{1}{\text{length}}$ for a stretched string

Experiment 34.5. Frequency and Tension.
Set the bridge of the sonometer to give a convenient value of L and tune the wire to the range of forks by varying the weight attached to it and hence the tension (T). Table 34.2 shows a typical set of values for frequency and tension.

Here the frequency increases with the tension but it is not a case of simple proportionality. Doubling the tension does not double the pitch. The form of relationship may be seen by plotting f against $\sqrt{T}$. This will give a straight line graph passing through the origin showing that for a given wire of constant length:

$$f \propto \sqrt{T}$$
$$\text{or } f = \text{constant} \times \sqrt{T}$$
$$\text{or } \frac{f}{\sqrt{T}} = \text{constant}.$$

Table 34.2

Frequency (f Hz)	Tension (TN)	√Tension
256	50	7·1
320	73	8·5
384	109	10·4
480	168	13·0
512	194	13·9

Thus if a certain length of wire tunes to Middle C (256 Hz) at a tension of 20 N, it will have a frequency f Hz at 40 N tension where f is given by:

$$\frac{256}{\sqrt{20}} = \frac{f}{\sqrt{40}}$$
$$f = \sqrt{\frac{40}{20}} \times 256$$
$$= 361 \text{ Hz}.$$

The Material and Thickness of the Wire

An experiment to investigate how the frequency of a stretched string depends on its material and thickness would be rather more complicated than the two above and would involve the use of a considerable number of different strings. Such experiments show that the frequency depends on the mass per unit length (M) of the wire. In fact the relationship is:

$$f \propto \frac{1}{\sqrt{M}}$$

or

$$f \times \sqrt{M} = \text{constant}.$$

If two wires of equal length and tension are producing Middle C (256 Hz) and the C above this (2 × 256 = 512 Hz), then the mass per unit length of the lower pitched string will be four times that of the other.

Thus in a piano where the strings are all of metal wire, those producing the low notes are not only longer but also thicker than the higher pitched ones. On a violin the strings may be of wire or gut. If a certain wire string is replaced by one of gut, the replacement will be much thicker than the original in order that the mass per unit length shall remain about the same.

Wind Instruments

A wide variety of musical instruments from saxophones to pipe organs comes in the general

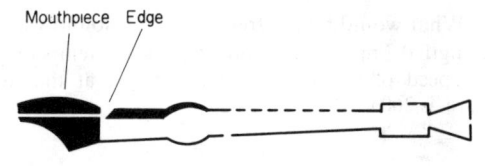

Fig. 34.14. A recorder

category of wind instruments. In all cases the sound is produced by setting up vibrations in a column of air contained in a tube. A vibrating column of air, like a stretched string, has a natural frequency or pitch which depends on its length. Fig. 34.14 shows a recorder which is one of the simplest of wind instruments. When all the holes are covered with the player's fingers the length of the vibrating air column is the full length of the tube, and the note produced is the lowest the instrument is capable of sounding. As the holes are opened the effective length of the tube decreases and the pitch of the sound is raised.

Most wind instruments enable the pitch to be varied in this way although they differ widely in the exact way in which the length of the column of air is varied. Wind players also vary pitch by 'overblowing' their instruments, i.e.: by blowing rather harder than usual which produces the first overtone of the fundamental being sounded.

In the case of a recorder, a pipe organ, and a flute, the air in the tube is set in vibration by the effect of a jet of air falling on an edge. In other wood-winds, e.g.: clarinet and oboe, the vibrations are first caused by a reed, while brass players set up the vibrations in the first instance with their lips and tongues.

Exercise 34

1 How would you show by experiment that (a) a source of sound is vibrating, and (b) sound cannot travel through a vacuum?

How would the sound produced by a vibrating string be affected by shortening the vibrating portion of it? Give the reason for your answer. (L)

2 Give a brief account of the wave motion by which sound is transmitted, and explain the terms *frequency* and *wave-length*.

An observer sees the flash of a gun being fired and hears the sound 2·4 s later. He then hears the echo from a wall 50 m immediately behind him after a further 0·3 s. Calculate (a) the velocity of sound, (b) the distance of the gun from the observer. (O)

3 Define *frequency*, *amplitude*, as applied to a sound wave.

How will the sound produced by a vibrating string be affected by increasing (i) the frequency, (ii) the amplitude of its vibration? State how the frequency and amplitude of a vibrating string may be increased. (L)

4 Describe an experiment to show that a material medium is necessary for the transmission of sound waves.

What is the wavelength in air of a note emitted by a tuning fork of frequency 320? Assume that the velocity of sound in air is 330 m/s.

What would be the effect on the frequency of a tuning fork of loading one of its prongs with a small piece of wax?

A taut piece of wire is plucked. What would be the effect on the frequency of the note emitted of (a) increasing the tension of the wire, (b) reducing the length of the wire? (J)

5 What is the relation between the length of a stretched string and the frequency of the note produced by its vibration?

A sonometer wire, the length of which may be varied by means of a bridge, is kept at constant tension and is found to have lengths of 55·7 cm, 37·3 cm and 27·9 cm when tuned to unison with forks giving notes C, G and C^1 respectively. Explain how these values bear out the law you have stated. The frequency of note C is 256 Hz and of the note G 384 Hz.

What length of the above wire on being plucked would emit the note one octave below C?

What method, other than by judging the pitch of the note by ear, could you use to show that the wire is tuned to the pitch of the fork? (L)

6 What is an *echo*? State the conditions required for an echo of a sound to be heard.

Describe an experiment to determine the velocity of sound in air.

A ship sounds its siren when it is some distance in front of a vertical cliff. An echo is heard after 3·7s. How far is the ship from the cliff? (The velocity of sound over water is 330 m/s.) Give your answer to 2 significant figures. (L)

7 Explain how molecules of air carry sound from one place to another. Give the meaning of the following terms in connection with sound waves: frequency and wavelength.

Write down the range of frequencies of audible sound waves.

A sound takes 0·9 s to travel vertically from a point 240 m under water to a point 240 m in air above the surface of the water. The speed of the sound in air is 320 m/s. From this information calculate the speed of this sound in water.

The speed of sound in a certain medium is 1200 m/s. (L)

What is the wavelength of sound of frequency 300 Hz in

this medium? (C)

8 (a) A tuning fork is made to vibrate in air and the note produced is observed. What change in frequency would be observed if some plasticine were attached to the prongs of the fork?

(b) What would be the frequency of a sound wave of wavelength 0·2 m in carbon dioxide, at room temperature, if the speed of sound in carbon dioxide, at that temperature, is 260 m/s? (C)

35 Solution

Solubility is a topic with which you should be quite familiar. You come across it every day; soap dissolves in water and sugar dissolves in milk. Air dissolves in water and is extracted by fishes in breathing. Although water is often regarded as an almost 'universal solvent' there are many other liquids which are far better in certain cases. Substances such as rubber, sulphur and celluloid, are completely insoluble in water, but solutions of them can be made. Rubber dissolves in benzene, sulphur in methylbenzene or in carbon disulphide and celluloid in propanone (acetone).

A *solution* consists of two parts; the liquid which does the dissolving is called the *solvent* and the substance dissolved is known as the *solute*. A true solution is perfectly homogeneous, i.e. the same all the way through as regards colour, density and other properties. The solvent can be recovered from the solution by distillation and the solute by evaporation.

One effect of dissolving a substance in water is to raise the boiling point slightly. A solution of salt will boil at a temperature somewhat higher than 100°C and freeze at a temperature lower than 0°C. Salt sprinkled on newly-formed ice at 0°C causes it to go into solution (or 'melt') because a mixture of the two substances is liquid at this temperature.

Experiment 35.1. Solubility in Various Solvents.

The solvents to be used should include benzene, ethanol, tetrachloromethane (carbon tetrachloride), ether, water and carbon disulphide. For solutes use iodine, rubber, paraffin wax and camphor.

To each solvent in a test tube add a little iodine. Cork the tubes and shake. Notice the different colours produced. How does the solubility of iodine in the organic solvents compare with that in water? Repeat this with the wax and camphor and leave for a few days, shaking when the opportunity allows. Cut small pieces of black rubber tubing and place one in each solution. Again cork and leave for a few days before examining the tubes. Does the wax or camphor appear to have dissolved in any of the liquids? Have the pieces of rubber swollen at all? Take them out and stretch them; what do you notice? Record your results.

solvent \ solute	iodine	rubber	wax	camphor
benzene				
ethanol				
tetrachloro-methane				
carbon disulphide				
ether				
water				

When a solid has dissolved in a liquid the particles are invisible under the most powerful optical microscope. But if a substance such as sand or chalk is shaken with water it does not dissolve; it forms a cloudy suspension.* The coarse

*Particles of a suspension range in size from about one-thousandth of a millimetre upwards; those in a solution are less than one-millionth of a millimetre.

heavier grains of sand soon settle down at the bottom, while those of the chalk, being finer, take much longer. The solid can be separated from the liquid by filtration. If a little of a solid such as saltpetre is added to water and stirred it may all dissolve; if more is added and the shaking or stirring continued this too may dissolve. However, a point will eventually be reached when no more will go into solution and some remains undissolved at the bottom. Such a solution is said to be saturated. We may define a *saturated solution* in this way:

A solution is said to be saturated with respect to a solid when it contains the maximum amount of that solid which can remain dissolved at a given temperature in the presence of some undissolved solid.

Notice particularly two points here. The last part of the definition is included because without some undissolved solid at the bottom of the vessel we cannot be certain that the solution is saturated. Why must the temperature be stated? You know that most solids are more soluble in hot water than in cold and if a saturated solution of, say, sodium nitrate were made at 20°C and the temperature then rose to 25°C we could not be certain the solution was still a saturated one unless some undissolved solid remained.

Most saturated solutions are made by dissolving as much of the solid as possible in hot water and letting the saturated solution cool down to the required temperature. The excess solid then settles out. A few substances show a retrograde solubility, i.e. they are less soluble in hot water than in cold; one such solid is calcium hydroxide and all gases are more soluble in cold water (why?).

We all understand what is meant by 'solubility' in a general sense, but in chemistry it has a particular meaning as well.

Solubility is the number of grams of a substance which will dissolve in 100 g (100 cm^3) of water at a given temperature to form a saturated solution.

Notice that solubility is a *number*, not a mass. When we say that the solubility of common salt at 100°C is 40 we mean that 40 g of it will dissolve in 100 g of water at this temperature. 100 g of water will be saturated by different amounts of different solids at a given temperature and these differences may be considerable as can be seen from the examples below.

The way in which solubility varies with temperature can more easily be seen from a solubility curve than from a set of figures. Such a curve is a graph of the values of the solubility at various temperatures plotted against the temperatures. The latter figures are plotted along the horizontal axis and the solubilities vertically. Look at the diagram (Fig. 35.1) and you will notice that:
(a) the solubility of salt varies little with temperature
(b) potassium nitrate is very soluble in hot water, but much less so in cold water
(c) the two substances have the same solubility at about 24°C.

Experiment 35.2. To Determine the Solubility of a Solid at a given Temperature.

Let us suppose that the solubility of lead(II) nitrate at (say) 70°C is to be determined. The

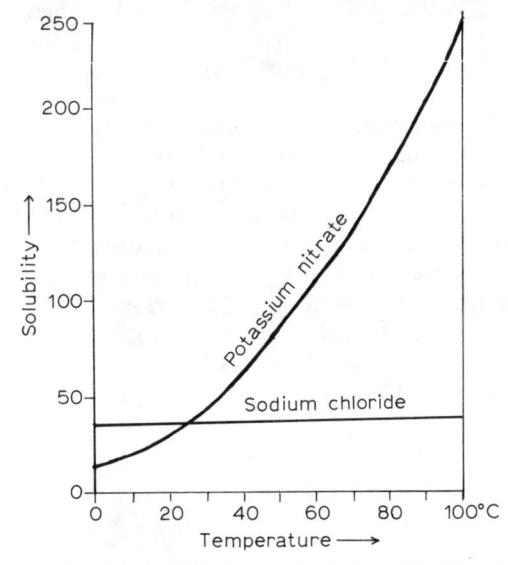

Fig. 35.1. Solubility curves for sodium chloride and potassium nitrate

Substance	Grams of solute dissolving in 100 g water		
	20°C	60°C	100°C
Sodium chloride	36	37	40
Ammonium chloride	37.2	55.2	77.5
Potassium nitrate	31.6	110	246
Copper(II) sulphate	20.7	40	75

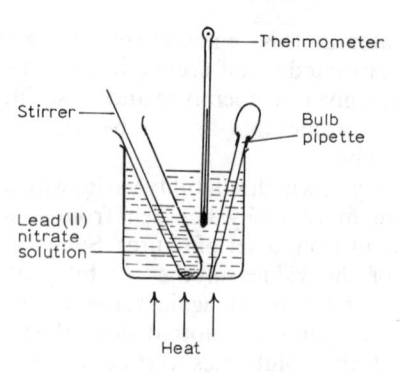

Fig. 35.2. Determining solubility

apparatus used is shown in Fig. 35.2. The water in the beaker is heated to about 75°C, the tube itself containing about 30 cm³ of distilled water. Powdered lead nitrate is now added to the tube and stirred; if it all dissolves more should be added with stirring until some remains undissolved at the bottom of the tube. The burner is then removed and the water allowed to cool down. When the temperature has fallen to about 71°C the thermometer is taken out, shaken dry and placed in the tube. When 70°C is reached the bulb pipette is removed, squeezed empty, shaken dry and quickly inserted into the lead nitrate solution. The pressure on the bulb is then released and some of the saturated solution withdrawn and emptied into a small weighed evaporating dish. (All this must of course be done quickly.) Now the dish has to be weighed again and then the solution evaporated to dryness over a water bath and once more weighed.

Let these masses be x, y and z grams respectively;
then, mass of solution $= (y - x)$ g
and mass of solid $= (z - x)$ g
This means that the mass of water
$= (y - x) - (z - x) = (y - z)$ g
So we have
$(y - z)$ g of water dissolve $(z - x)$ g lead(II) nitrate

100 g of water dissolve $\dfrac{100(z - x)}{(y - z)}$ g

This gives a figure for the solubility of lead nitrate at 70°C. If several dishes are previously weighed the experiment can be repeated at temperatures of (say) 60°, 50°, 40° and 30°C. If two pupils work together one can look after the weighings while the other carries out the rest of the experiment. In order to get a value for the solubility at room temperature the solid is shaken with water at a slightly higher temperature to make a saturated solution of it, cooled to room temperature, part withdrawn and weighings carried out as above.

From all the points obtained a solubility curve should be drawn.

Purification by Recrystallization

Crystalline substances may be contaminated with some impurity. If the two substances differ considerably in solubility purification can be carried out fairly easily. The solubility curves for potassium chlorate(V) and potassium chloride are shown in Fig. 35.3. Suppose we have 50 g of the chloride and we wish to obtain a pure sample of the former. Let us look at the solubilities.

	20°C	100°C
Potassium chlorate(v)	7	57
Potassium chloride	34	57

It should be apparent that a solution of the chlorate saturated at 100°C on cooling to room temperature (taken as 20°C) would deposit most of the solid and we can make use of this fact.

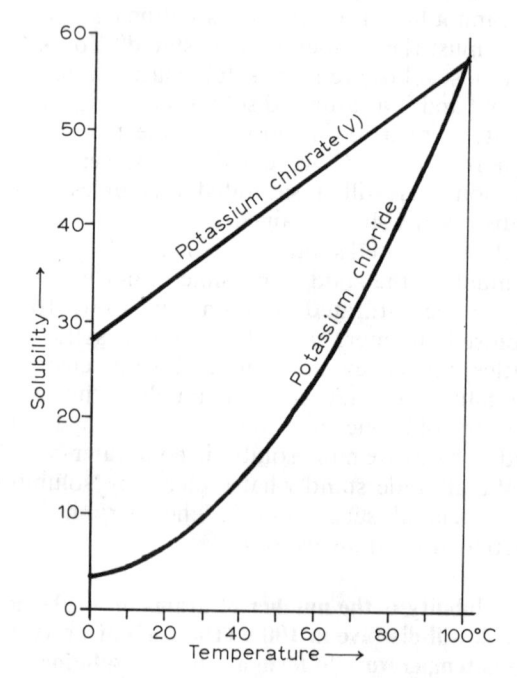

Fig. 35.3. Solubility curves for potassium chloride and potassium chlorate(V)

If the impure mixture is stirred with 100 cm³ of distilled water at 100°C it will all dissolve. On cooling down to 20°C, 43 g (50 − 7) of the chlorate will crystallize out, the remaining 7 g along with the one gram of the chloride staying in the solution. After filtration the chlorate(V) is left on the filter paper and should be washed with a little cold water to remove any trace of chloride and then left to dry.

EXAMPLE (a). 60 g of a saturated solution of a salt were evaporated and 36 g of solid were left. Calculate

(i) the solubility of the solid at the temperature of the experiment; (ii) the percentage by mass of the solid in the saturated solution.

(i) 60 g of solution contains 36 g of the solute
∴ 24 g of water dissolve 36 g of solid
100 g of water dissolve

$$100 \times \frac{36}{24} \text{ g} = 150 \text{ g}$$

∴ solubility of solid = <u>150</u>

(ii) 60 g of saturated solution contain 36 g of solid
∴ 100 g of saturated solution contain

$$100 \times \frac{36}{60} \text{ g} = 60 \text{ g}$$

∴ percentage of solid in solution = <u>60%</u>

Solubilities of Compounds

Try to remember the following:
All nitrates are soluble in water.
All hydrogencarbonates (bicarbonates) are soluble in water.
All common sodium, potassium, lithium and ammonium compounds are soluble in water.
All common metal chlorides are soluble in water excepting those of lead and silver.
All common sulphates are soluble in water excepting those of barium and lead; (calcium sulphate is slightly soluble).
The only common carbonates soluble in water are those of sodium, potassium, lithium and ammonium.
The only common hydroxides soluble in water are those of sodium, potassium, lithium, calcium, barium, and ammonium.

Conductivity of Solutions

One property of solutions we have not yet referred to is their ability (or otherwise) to conduct an electric current. This will be considered in detail in Chapter 38.

Hard and Soft Waters

There is no natural source of pure water. Reference has been made above to water being a 'universal solvent'; wherever it is found it has something dissolved in it. The purest natural form of water is rain water, but even here some carbon dioxide in the air will be dissolved in it to give a very dilute solution of carbonic acid. It was mentioned in Chapter 25 (page 139) that during thunderstorms some nitrogen oxide may be produced and this forms nitrogen dioxide which is washed down in the rain as nitric acid. In town atmospheres industrial gases—usually acidic— are often present along with soot and smoke and these cause much of the 'eating away' and blackening of buildings.

Evaporation by the sun of water from rivers and lakes causes the water vapour to rise into the air and to be held there until with a fall in temperature it drops as rain. Some of the rain that falls on the land runs along the surface (especially in towns) and via drains eventually gets back to the rivers, but most of it has to soak through the soil before this happens. In this way it dissolves many minerals, some of which may be beneficial to those who drink it.

Calcium carbonate in the form of limestone or chalk is found in many countries. Although this compound is insoluble in pure water, it is dissolved by rain containing carbon dioxide in solution, calcium hydrogencarbonate being formed.

$$CaCO_3 + H_2O + CO_2 \rightarrow Ca(HCO_3)_2$$

(Where magnesium carbonate is present as in the Dolomites in Central Europe it dissolves similarly to form magnesium hydrogencarbonate.) Another mineral, calcium sulphate or gypsum, is also dissolved slightly by rain water and so water percolating through the ground picks up many minerals on its way.

Water containing calcium compounds does not easily form a lather with soap and is said to be *hard*. The reason for this is fairly easy to understand.

There are various kinds of soap, a common one being sodium stearate (made from sodium hydroxide and glyceryl stearate, see p. 331). When this is put into water containing calcium compounds (i.e. calcium ions, Ca^{2+}) double decomposition occurs and calcium stearate is formed. This is insoluble in water and we get a white 'scum' on the liquid. If we represent the stearate radical by '$\overline{St}$' we can write an equation for this action.

$$CaSO_4 + 2Na\overline{St} \rightarrow Ca\overline{St}_2 + Na_2SO_4$$
$$\text{or } Ca^{2+}(aq) + 2Na\overline{St}(aq) \rightarrow$$
$$Ca\overline{St}_2(s) + 2Na^+(aq)$$

As long as there are any calcium ions left in the solution the soap continues to form calcium stearate and none of it is available for producing a lather.

You will all know quite well that it is easier to get a lather with hot water than with cold. Why should this be so? Your earlier study of calcium carbonate will have told you that the calcium hydrogencarbonate decomposes on heating and the carbonate is re-formed.

$$Ca(HCO_3)_2(aq) \xrightarrow{\Delta} CaCO_3(s) + H_2O(l) + CO_2(g)$$

As this is insoluble in water (no carbon dioxide is now present to keep it in solution) it is deposited. There are then no calcium ions left in the solution to cause hardness and so the water is softened.

Hardness which is caused by calcium hydrogencarbonate and which is destroyed by boiling is called *temporary hardness* (or carbonate hardness).

Calcium sulphate and chloride are not affected by heat and hardness caused by these compounds (or the corresponding magnesium ones) is called *permanent hardness* (or noncarbonate hardness).

Do not be confused by this name; it does not mean that the hardness cannot be removed at all; there are many ways of doing this. Most hard water contains both calcium hydrogencarbonate and the sulphate and we can determine the relative amounts of each by a simple experiment.

Experiment 35.3. To Compare the Hardness of a Sample of Tap Water with that of Distilled Water.
(a) Take 25 cm³ of distilled water in a conical flask and run in from a burette some pure soap solution. After each 0·2 cm³ cork the flask and shake vigorously to see if a lather (lasting for about one minute) is formed. When this happens note the burette reading.
(b) Repeat using 25 cm³ of tap water. Notice here how the 'tinny' sound heard on shaking the flask is gradually muffled as a lather is formed.
(c) Boil some tap water and allow it to cool. Put 25 cm³ of it in a flask and again run in the soap solution until a lather is produced.

Let the volumes of soap solution used in (a), (b) and (c) be x, y and z cm³ respectively.

The relative hardness of the tap water to that of the distilled water is y/x.

The amount of permanent hardness is represented by z and that of the temporary hardness ($y - z$).

The experiment can be carried out more simply by using about 10 cm³ of each sample in a test-tube and adding the soap solution from a dropping pipette, counting the number of drops in each case.

In some districts the water is naturally very soft and it may need to be artificially hardened by the addition of calcium sulphate and calcium hydrogencarbonate solutions.

Methods of Softening Hard Water

Hard water has many disadvantages, especially in industry. You have seen that when water possessing temporary hardness is boiled, calcium carbonate is deposited. This can result in boilers used for steam raising becoming lined with scale or steam pipes getting badly furred. This reduces the efficiency of the plant considerably and, due to the uneven heating which ensues and the high pressures set up, may be dangerous. Periodically the scale must be removed and the pipes replaced. Disadvantages at home include the furring of kettles and the deposition of calcium stearate on garments that have been washed with soap. This gets ironed in and white clothes gradually become grey.

Methods used in softening hard water include the following:
(i) Temporary Hardness alone.
(a) *Boiling*. This is only suitable on a small scale and where the deposition of the carbonate is not important, e.g. in the home.
(b) *Addition of Slaked Lime*. This may appear a strange method as we have seen that calcium ions are the cause of hardness and in this case more of them are being added. The actual amount must be controlled so that all calcium ions are removed in the form of a precipitate of calcium carbonate.

$$Ca(OH)_2(aq) + Ca(HCO_3)_2(aq) \rightarrow$$
$$2CaCO_3(s) + 2H_2O(l)$$

(see also method (*f*)).

This formation of a precipitate occurs in most methods of softening water; ions are withdrawn from the solution and the substance precipitated, being insoluble in water, has no further effect.
(c) *Addition of Ammonia*. Household ammonia is often used to soften water; it also helps to remove grease. Again, calcium carbonate is precipitated.

$$Ca(HCO_3)_2(aq) + 2NH_4OH(aq) \rightarrow$$
$$CaCO_3(s) + (NH_4)_2CO_3(aq) + 2H_2O(l)$$

(ii) Temporary and Permanent Hardness.
(d) *Distillation*. This removes all the mineral deposits, but is much too expensive to use on a large scale.
(e) *Addition of Soap*. This is another expensive method and is very wasteful of soap.

Here the insoluble compound formed is calcium stearate.
(f) *Addition of Soda*. The use of washing soda is very common in most households, but it is also an industrial method on a large scale. Calcium ions are removed as the carbonate.

$$CaSO_4 + Na_2CO_3 \rightarrow CaCO_3 + Na_2SO_4$$
$$\text{or } Ca^{2+}(aq) + CO_3^{2-}(aq) \rightarrow CaCO_3(s)$$

Industrial undertakings which use calcium hydroxide for softening water usually add sodium carbonate also so that both temporary and permanent hardness are removed.

(g) *Addition of Permutit*. There are some rocks, classified as zeolites, which are compounds of sodium, aluminium, silicon and oxygen in the form of sodium aluminium silicate. Hard water, percolating through these rocks is softened. The calcium ions are replaced by an equivalent amount of sodium ions. Artificial zeolites, called permutits, have been prepared and are used in domestic and industrial water softeners. A section through a domestic one is shown in Fig. 35.4. Hard water runs through a bed of sodium permutit and is converted into calcium permutit. If we represent the permutit radical by $\overline{P}$ the equation is

$$CaSO_4 + Na_2\overline{P} \rightarrow Na_2SO_4 + Ca\overline{P}$$
$$\text{or } Ca^{2+} + \overline{P}^{2-} \rightarrow Ca\overline{P}$$

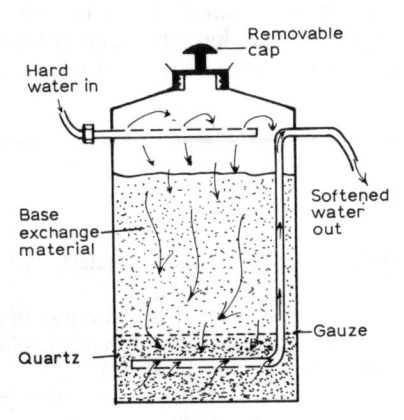

Fig. 35.4. Domestic water softener

This method is called an ion exchange method because sodium ions take the place of calcium ones in the water and the formation of calcium stearate on addition of soap is no longer possible. Eventually in the softener all the permutit layer has been changed into calcium permutit and water going through is not softened. The permutit has to be 'regenerated'. A fair amount of common salt (Na^+Cl^-) is placed in the top of the container and washed through. The high concentration of sodium ions allows the action

$$Ca\overline{P} + 2Na^+ \rightarrow Na_2\overline{P} + Ca^{2+}$$

to go on. Sodium permutit is reformed and the calcium ions are washed away. When the effluent no longer tastes salty the regeneration is complete.

When water is softened by soda or permutit the calcium ions are removed, but the water now contains an equivalent amount of sodium ions. For domestic and many industrial purposes this is unimportant, but in some industries it is essential that these should not be present. Although distillation would be the ideal method to use, the cost makes it prohibitive. However, other ion exchange methods have been devised for replacing *all* the metallic ions (the cations) by hydrogen ions and the non-metallic ions (the anions) by hydroxyl ones. As the only ions then present are hydrogen (H^+) and hydroxide (OH^-) the water is chemically pure.

Detergents

In districts where water is naturally soft the use of soap for domestic cleansing is quite sufficient. Such districts in Britain are those which obtain their water supplies from regions of hard impervious rocks containing no calcium minerals. In Britain Birmingham gets its water from central Wales, Manchester from the Lake District and Glasgow from the mountains to the north. Both soap and soda make water alkaline and this helps to dissolve dirt.

In other regions where the water is hard, soap (except for personal washing) has been replaced to a large extent by one of the modern synthetic detergents. These are generally complex compounds, such as sodium lauryl sulphate, $C_{12}H_{25}SO_4Na$. They do not form insoluble calcium compounds and consequently no scum is produced when they are added to water. They are often synthesized from petroleum products. Washing powders usually contain anhydrous sodium carbonate, Na_2CO_3, powdered soap and some whitening agent.

Advantages of Hard Water

Although much has been said about the disadvantages of hard water do not imagine that there are no advantages, there are several. Water softened by lime and soda is not suitable for use in the tanning of leather; calcium compounds are beneficial in building strong teeth and bones; the brewery industry at Burton-on-Trent is located there because hard water is preferable to soft in brewing.

Very soft water actually dissolves some of the lead of the pipes through which it runs and lead ions may be present in it and eventually give symptoms of lead poisoning. Such water is made artificially hard and then any lead(II) compounds produced are either the sulphate or carbonate.

These are both insoluble in water and, forming a protective layer on the inside of the pipe, stop any further solution.

Stalactites and Stalagmites

Those of you who are familiar with limestone districts will know that many caves occur in such areas. Peak Cavern (at Castleton in the Peak District of Derbyshire), Wookey Hole (in the Mendips) and Kent's Cavern (at Torquay) are well known and the hobby of 'pot-holing' (speleology) or of descending into underground caverns and exploring the many passages connecting them has many followers in northern England and in many regions of Europe.

How have caves been formed? As explained earlier in the chapter, rain water dissolves carbon dioxide from the atmosphere and from the air present in the soil. It then has a solvent action on calcium carbonate. It may seep through limestone or chalk rocks and gradually dissolve much of the solid earth to form a cave. Underground rivers help in this action; there are quite a number in the Pyrenees district of southern France and a few in England; they are found in limestone districts all over the world. The calcium carbonate dissolved remains in the water as calcium hydrogencarbonate.

$$CaCO_3 + H_2O + CO_2 \rightarrow Ca(HCO_3)_2$$

If drops of the solution hang from the roof of a cave decomposition and evaporation gradually occur.

$$Ca(HCO_3)_2(aq) \rightarrow CaCO_3(s) + CO_2(g) + H_2O(l)$$

This leaves a deposit of calcium carbonate and over many hundreds of years this action has gone on to form stalactites. Drops of the solution falling from these on to the floor evaporate similarly and stalagmites build up. Evaporation underground goes on slowly and the rate of 'growth' is very small. If a stalactite is examined it will be found to consist mostly of calcium carbonate with some calcium sulphate and possibly other minerals which were held in solution.

Evaporation and decomposition in the open take place much more rapidly and many visitors to caves in limestone areas have left hats, umbrellas, etc., exposed to the action of this hard water. As a result, such objects have become covered with a layer of calcium carbonate and are said to be 'petrified'.

Exercise 35

Try these mentally:

A. If the solubility of salt at 100°C and 40°C = 39·6 and 36·6 respectively, how much salt will separate on cooling as saturated solution containing 50 cm³ of water from 100°C to 40°C?

B. The solubility of sodium nitrate at 100°C is 247 and at 20°C is 31. If a saturated solution containing 25 cm³ of water is cooled from 100°C to 20°C, how much sodium nitrate will separate?

1 Name two substances in each case which cause (a) temporary, (b) permanent hardness in natural waters. Describe, with essential detail, how you would compare the hardness of two different samples of tap water. Explain how temporary hardness may be removed (i) in the home (ii) commercially. (S)

2 From the figures given draw solubility curves for

Temp.	Potassium sulphate	Potassium chlorate(V)
0	7·4	3·2
10	9·2	4·5
20	11·1	6·5
25	12·0	8·0
30	12·8	9·5
40	14·8	13·2
50	16·5	17·5
60	18·6	23·7
70	20·7	31·5

potassium sulphate and potassium chlorate(V) on the same axes.

3 What do you understand by hardness of water? How does water become temporarily hard? Explain why such hardness is removed by boiling.

Name and write the formula of any *one* substance, which may be found in tap water, causing permanent hardness. Give any two methods by which permanent hardness can be removed. (J)

4 A series of experiments gave the following results:

Temperature	5	15	25	35	40
Solubility	3·5	5	7	10	12

Temperature	50	60	70	80
Solubility	17·3	24	32	42·5

From these values draw a solubility curve for the substance, and from this curve make what deductions you can about the variability of the solubility with temperature and of the ease or otherwise of crystallizing the substance from water.

5 Explain carefully why tap water sometimes contains calcium hydrogencarbonate. State and explain TWO disadvantages of using such water in the home.

Give TWO ways, other than boiling, by which temporary hardness may be removed in the home.

If 10 dm³ of tap water contain 1·62 g of calcium hydrogencarbonate, what is the theoretical weight of the precipitate formed when the solution is boiled until all the bicarbonate is decomposed? (S)

6 The following results (which are the average of a large number of results) were obtained when testing identical quantities of water with soap flakes added one at a time and shaken after the addition of each flake, until a lather lasting at least one minute was obtained.

Type of water	Column 1 Number of flakes required to make a lather	Column 2 Number of flakes required to make a lather after the water has been boiled and then allowed to cool	Column 3 Number of flakes required to make a lather after water has been boiled with sodium carbonate and then allowed to cool
Distilled water	2	2	
Tap water	7	4	
Distilled water with calcium sulphate	20	20	

(a) Fill in the results you would expect in the final column.
(b) Account for the facts that boiled tap water required fewer soap flakes while the calcium sulphate solution remained unchanged after boiling.
(c) Give reasons for the figures you have put in column 3.
(d) (i) Give the equation for the reaction between the calcium sulphate and sodium carbonate.
 (ii) Give the equation for the reaction due to boiling. (NR)

7 You are given a mixture of coarse sand, fine chalk and common salt. Plenty of water is added and the mixture is stirred vigorously and then allowed to stand for a minute.

What happens to (i) the sand, (ii) the chalk, and (iii) the salt?

Describe how you would obtain a sample of pure dry salt from the above mixture. Give a fully labelled diagram of the apparatus you would use. (WM)

8 Give a simple explanation of how water becomes temporarily hard.

Say what chemical changes occur in the formation of this kind of hardness.

Explain how hardness in water can be a serious nuisance in:
(a) laundry work, and
(b) running a domestic hot water system.

Give a simple explanation of what we mean by calling a substance a 'detergent'. (EM)

9 Describe how you would:
(a) determine the solubility of common salt at room temperature;
(b) expel, collect and measure the air that is dissolved in water (give a diagram).

Account for the difference (if any) between ordinary air and the air collected in (b). (L)

10 What is meant by the solubility of a substance in water?

From the following readings draw the solubility curve for potassium chloride, on the graph paper provided.

Temp. °C								
0	10	20	30	40	50	60	70	80
Solubility (g/100 g water)								
28	31	34·5	37·5	40	43	46	49	51

Use the curve to answer the following questions:
(a) The solubility of potassium chloride at 65 °C is ....
(b) The solubility of potassium chloride at 5 °C is ....
(c) The mass of potassium chloride precipitated when a saturated solution containing 100 g water is cooled from 65 °C to 5 °C is ....
(d) The temperature at which 20 g of water will dissolve 9 g of potassium chloride is .... (SR)

11 How is hard water formed in nature?

Starting from calcium carbonate, describe briefly how you would prepare a sample of temporarily hard water.

Describe TWO methods by which each kind of hard water may be softened.

How would you show that a specimen of hard water contained both kinds of hardness? (O)

12 Calcium carbonate is said to be insoluble in water; and the solubility of sodium chloride at room temperature is given as 36 g per 100 g water.

(a) What mass of sodium chloride should there be in 340 g of a saturated solution at room temperature?
(b) 100 g calcium carbonate and 100 g sodium chloride are mixed, and then well shaken with 200 g water.
 (i) The resulting liquid should be a saturated solution of sodium chloride, but should contain no dissolved calcium carbonate. Describe how you could use it to determine the solubility of sodium chloride and to prove that calcium carbonate is insoluble.
 (ii) The solid left undissolved should contain both sodium chloride and calcium carbonate. Describe how

you would obtain from it pure dry samples of each compound. (C)

13 Arrange calcium carbonate, calcium chloride and calcium sulphate in order of increasing solubility in water.

How would you use soap solution to make a rough comparison of the hardness of two waters?

Starting with lime, how would you prepare some water exhibiting temporary hardness? (L)

14 (a) Explain what you understand by *a saturated solution*.

(b) The solubility of potassium nitrate at room temperature is given as 25 g of potassium nitrate per hundred g of water. Each of the following liquids has 40 g of powdered potassium nitrate stirred with it at room temperature for a long time:

(i) 80 g of water;
(ii) a solution of mass 75 g and containing 15 g of dissolved potassium nitrate;
(iii) a solution consisting of 5 g of potassium nitrate dissolved in 40 g of water.

What mass of the powder would you expect to dissolve in each case? All necessary reasoning should be clearly shown.

(c) Describe carefully how you would obtain pure clean crystals of potassium nitrate from a powdered sample that was contaminated with chalk.

[NO diagrams are required in this question.] (C)

36
Electric Currents

If a pipe contains water as in Fig. 36.1(a) one way of making the water flow is to tip the pipe (Fig. 36.1(b)). We should say that water flows from a place where it has high potential energy (the top of the tube) to a place of lower potential energy (the bottom). In some respects the flow of electricity in a wire may be likened to the flow of water in a pipe. Of course we shall not obtain an electric current by tipping a wire on end. What we must do is to arrange that electricity will have lower potential energy at one end of the wire than the other. We must set up a 'potential difference' between the ends of the wire. This may be achieved in a number of ways but the two most important are (a) by connecting to the ends of the wire the terminals of a chemical cell (represented in Fig. 36.2 by its conventional circuit diagram symbol) and (b) by connecting to the two ends of the wire a dynamo or generator (this is effectively what we do when we plug an appliance in to the electricity mains supply).

We shall discuss the structure of chemical cells and dynamos later. But first we must say something about the units in which we measure potential differences and currents. We say that the potential difference between two points in a wire is *one volt* if one joule of potential energy is lost (changed into heat) when one unit of electric charge (1 coulomb) passes between them.

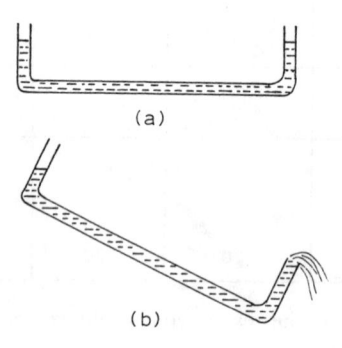

Fig. 36.1. Flow of water in a pipe

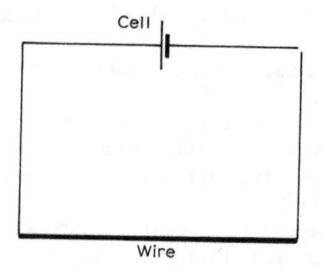

Fig. 36.2. A simple circuit

If we return again to our water analogy of Fig. 36.1, as long as the tube is kept full in some way, water will flow out of it at a steady rate. We might collect the water issuing in a measuring cylinder for a certain period of time. Supposing 100 cm^3 were collected in 10 s, then we should say water was flowing through the tube at a rate of 10 cm^3/s. This rate of flow corresponds in the electrical case to the current which we measure in *amperes*. The total quantity of water collected, 100 cm^3 in the case we have quoted, corresponds to a quantity of electricity or charge which as we have indicated in our definition of the volt, may be measured in coulombs.

The ampere is defined in terms of the strength of the magnetic field which the current produces and a precise definition of it is beyond the scope of this book. However the moving coil galvanometer which we will describe in Chapter 45 measures current by means of its magnetic effect and may be constructed so that it gives a deflection of a pointer which is proportional to the current through it, i.e. if a current of one ampere gives a deflection of 10° then two amperes will deflect through 20° and so on.

The *coulomb* is defined as the charge which passes when a current of one ampere flows for one second.

Experiment 36.1. Experiments to Investigate the Factors Affecting the Current in a Conductor.

Connect a 3 volt torch bulb to one accumulator or dry cell (Fig. 36.3(*a*)) and note how brightly it glows. Now connect a second cell in the circuit (joining the negative terminal of one cell to the positive terminal of the other as in Fig. 36.3(*b*)). You will see that the bulb lights up much more brightly.

Adding a second cell increases, in fact probably doubles, the potential difference being applied to the lamp, and clearly this has the effect of increasing the current through it. By how much the current has been increased we shall have to investigate later.

Now connect another similar bulb in the circuit as shown in Fig. 36.3(*c*). The two bulbs are now connected in *series*. You will see that the brightness of the two bulbs is back to about that observed with one bulb and one cell. Adding a second bulb has reduced the current. We say that we have increased the resistance of the circuit. In effect we have provided a more difficult path for the charge to take through the conductors and consequently we have reduced the rate at which it flows.

Now we must investigate more carefully how the current in a conductor is related to the potential difference we apply to it and what exactly we mean by the resistance of the conductor.

Experiment 36.2. Make a coil by winding about 4 or 5 metres of cotton covered 26 standard wire gauge (s.w.g.) constantan wire on a piece of cardboard tube or any other convenient spool. Connect this coil in a circuit with an ammeter reading up to about one ampere and, in the first instance, one accumulator (Fig. 36.4). Note the reading of the ammeter. Now add more cells in the circuit tabulating ammeter readings and the number of cells. A typical set of results is shown in Table 36.1.

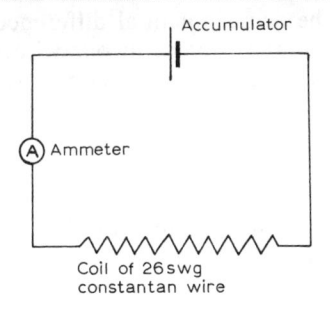

Fig. 36.4. Experiment to verify Ohm's Law

A graph of these figures would be as shown in Fig. 36.5.

The straight line graph we obtain passes through the origin of the graph which we should expect since no current will flow when there are no cells in the circuit. The main conclusion we may come to from our graph is that under the conditions of our experiment the current in the wire is proportional to the number of cells in the circuit, i.e.,

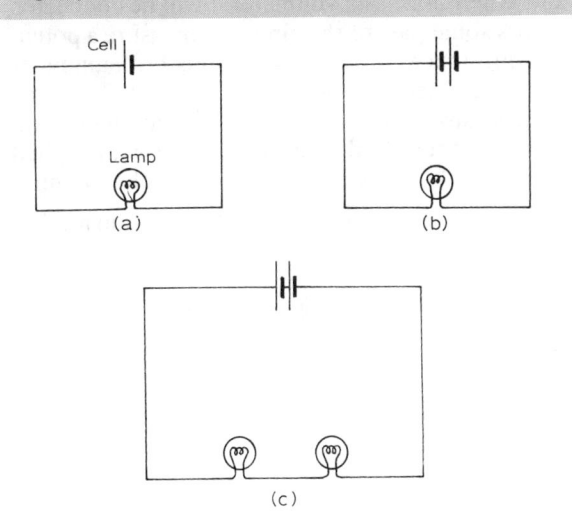

Fig. 36.3 (*a*) 1 cell and 1 lamp
        (*b*) 2 cells and 1 lamp
        (*c*) 2 cells and 2 lamps

Table 36.1

No. of cells	Current (A)
1	0·15
2	0·31
3	0·44
4	0·60
5	0·74

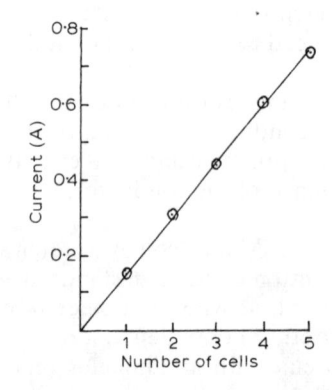

Fig. 36.5. Graph of current and no. of cells

with two cells we obtain twice the current produced by one, with three cells three times the current and so on. If we assume that each cell produces the same potential difference then we may say that the current in our wire (i) is proportional to the potential difference applied to it (V).

$$i \propto V$$

This important relationship was first stated by the German scientist, Ohm, in 1826, and is known as Ohm's law. We must be careful however in generalizing from our results. If the number of cells used was further increased a larger current would flow and the wire would become hot. Then the figures would no longer plot on our straight line. Also we should not be surprised if our law failed to work with some things which allow currents to flow. In fact we have tested it with one type of material which we might call a 'metallic conductor'. We might also have tested semiconductors, liquids and gases, all of which under some circumstances pass currents but may or may not obey Ohm's law.

So a full statement of Ohm's law is:

For a metallic conductor in the form of a wire at constant temperature the current passing is proportional to the potential difference applied across its ends.

As we saw on page 4 we may rewrite $V \propto i$ as $V = $ (a constant) $\times i$

The constant (usually represented by R) is known as the resistance of the wire and depends on the material, its dimensions and the temperature.

The resistance of a wire is usually measured in ohms and is given by:

$$R \text{ ohms} = \frac{\text{Applied potential difference in volts}}{\text{Current in amperes}}$$

$$R = \frac{V}{i} \quad (1)$$

This equation may be rearranged to the forms:

$$V = R \times i \quad (2)$$

$$\text{or } i = \frac{V}{R} \quad (3)$$

The Voltmeter

It follows from what we have said above that the current flowing in the coil of a moving coil galvanometer is proportional to the potential difference applied to it. But we pointed out on page 209 that the deflection of the coil is proportional to the current. Hence we may use our moving coil instrument to measure potential differences as well as currents. It should be noted that when used as an ammeter the instrument will be connected in the circuit to measure the current in it, and when used as a voltmeter it will be connected across some part of the circuit to measure a potential difference. A circuit containing both ammeter and voltmeter is shown in Fig. 36.6. It should be noticed also that a moving coil galvanometer will normally show a full scale deflection for an applied potential difference of a few millivolts (1 millivolt = $\frac{1}{1000}$ volt) and for most uses certain modifi-

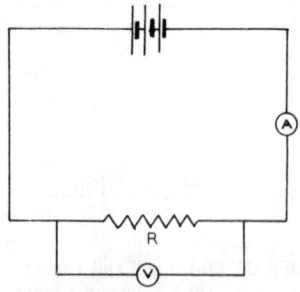

Fig. 36.6. Measurement of resistance (1)

cations, which will be described later, have to be made.

Measurement of Resistance

The resistance of the coil we used in our experiment to verify Ohm's law may be measured by repeating that experiment but this time connecting a voltmeter (e.g. 0–12 V) across the coil (Fig. 36.6). In this case values of the current through R and the potential difference across it for 1, 2, 3, etc. cells in the circuit must be measured. The cells in the circuit must be measured. The value of the resistance may then be found by calculating $\frac{V}{i}$ in each case and averaging the results.

An alternative method is to start with a battery of say 6 cells and include in the circuit a rheostat or variable resistance (Fig. 36.7). By selecting different settings of this, a number of values of V and i may be obtained which may be treated as in Table 36.2.

> **Experiment 36.3. Factors Affecting the Resistance of a Wire.**
> Find by the second method suggested above the resistance of a number of lengths of wire. If possible test 100 cm and 200 cm lengths of about 26 s.w.g. constantan, the same lengths of a thinner constantan wire, i.e. one of higher s.w.g. number, and of either 26 s.w.g. copper wire or the same gauge as the thinner constantan wire.

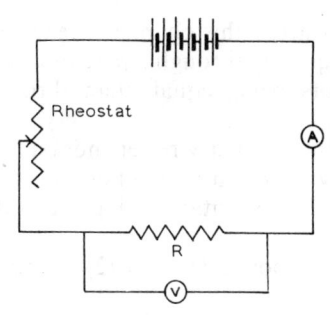

Fig. 36.7. Measurement of resistance (2)

The sort of results you may obtain are shown in Table 36.3.

The results tabulated show that:
1. The resistance of a wire depends on its length. In fact if we double the length we double the resistance.

Resistance $\propto$ length (l).

2. The resistance of a wire depends on its cross sectional area. A thick wire has a lower resistance than the same length of a thinner one of the same material. Although these results do not show it, if we double the cross sectional area of a wire we will halve its resistance.

$$\text{Resistance} \propto \frac{1}{\text{Area of cross section}} \left(\frac{1}{a}\right)$$

The lower resistance of a thick wire is readily appreciated if you recall the water analogy with

Table 36.2

No. of cells	Voltmeter reading (V V)	Ammeter reading (i A)	$\frac{V}{i} = R\,\Omega$
1	2.0	0.15	13.3
2	4.0	0.31	12.9
3	6.0	0.44	13.7
4	8.0	0.60	13.3
5	10.0	0.74	13.5

Table 36.3

Wire	V (V)	i (A)	R (Ω)
100 cm 26 s.w.g. constantan	2.0	0.68	2.9
200 cm 26 s.w.g. constantan	2.0	0.35	5.7
100 cm 30 s.w.g. constantan	2.0	0.33	6.1
200 cm 30 s.w.g. constantan	2.0	0.18	12.1
100 cm 26 s.w.g. copper	0.2	2.0	0.1
200 cm 26 s.w.g. copper	0.2	1.0	0.2

which we started this chapter. Clearly water will flow more quickly through a pipe of large bore, all other factors being equal, than through one of small bore.

3. The resistance of a wire depends on its material. A copper wire has a very much lower resistance than one of constantan having identical dimensions.

Combining points (1) and (2) above we have:

$$R \propto \frac{l}{a}$$

This may be rewritten

$$R = \frac{\rho l}{a}$$

ρ (the Greek letter 'rho') is a constant for the material known as the resistivity. It will be seen that $R = \rho$ when $l = 1$ and $a = 1$.

The resistivity of a material is thus the resistance of a specimen of it 1 m long and 1 m² in cross section. It will be seen from Fig. 36.8 that this is *not* necessarily the same as saying that ρ is the resistance of 1 m³.

Fig. 36.8. Resistivity

In case (a) the resistance of the specimen is by definition ρ. In case (b) we have halved the cross sectional area which doubles the resistance and doubled the length which doubles the resistance again. Thus the resistance in case (b) is 4ρ.

The low resistivity of copper makes it the most suitable material for use in electrical circuits as connecting wire. In most of the circuits we have talked about and will talk about, the resistance of connecting wire may be made so low as to be neglected by comparison with the rest of the circuit.

The resistivity of silver is even lower than that of copper, but its cost makes it unsuitable for all but the most specialized applications as a connecting material.

Resistances in Series and Parallel

It is very common to connect resistances together in electrical circuits. If we restrict ourselves to considering only two resistances there are only two ways in which this can be done and the possible arrangements are shown in Fig. 36.9.

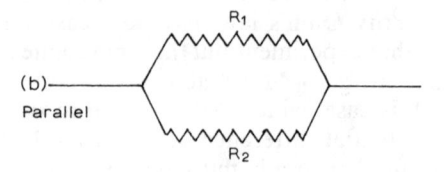

Fig. 36.9. Resistances in series and parallel

The nett effect of these arrangements is most conveniently investigated using standard resistances. These are coils of wire made to have known values of resistance, often a whole number.

Experiment 36.4. Take resistances of 1, 2 and 3 ohms if available and connect them two at a time in each of the circuits shown in Fig. 36.10. Note the voltmeter and ammeter readings in each case, taking care that the rheostat setting is such that the current does not exceed about 0·5 A.

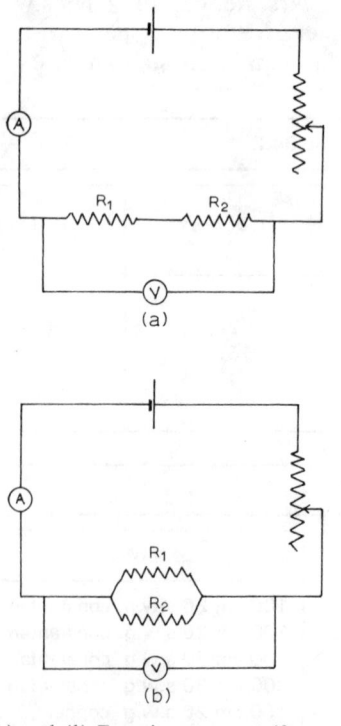

Fig. 36.10. (a) and (b) Experiments to verify expressions for series and parallel resistances

Table 36.4 and Table 36.5

		V (V)	i (A)	Effective Resistance $\frac{V}{i}\Omega$
SERIES	1 and 2 ohms	1·2	0·4	3
	1 and 3 ohms	1·2	0·3	4
	2 and 3 ohms	1·2	0·25	5

		V (V)	i (A)	Effective Resistance $\frac{V}{i}\Omega$
PARALLEL	1 and 2 ohms	0·3	0·45	0·66
	1 and 3 ohms	0·3	0·4	0·75
	2 and 3 ohms	0·6	0·5	1·2

The results should be tabulated as in Tables 36.4 and 36.5.

The effective resistance calculated from $\frac{V}{i}$ is the single resistance that would pass the same current in the circuit as is passed by R_1 and R_2. In case (a) the effective resistance R is clearly given by the simple expression $R_1 + R_2$. This can be extended to the case of more than two resistances; and for the resistances in series in general: $R = R_1 + R_2 + R_3 +$ etc.

In case (b) the relation between the effective resistance and R_1 and R_2 is not obvious. From the table of results you will be able to verify that it is:

$$\frac{1}{R} = \frac{1}{R_1} + \frac{1}{R_2}$$

Again, this works for more than two resistances in parallel and a general expression is:

$$\frac{1}{R} = \frac{1}{R_1} + \frac{1}{R_2} + \frac{1}{R_3} + \text{etc.}$$

We have shown experimentally that these expressions for resistances in series and parallel work, but they are in fact simple theoretical consequences of Ohm's law.

You should note that in the parallel case the effective resistance will always be less than any of the individual values. This you would expect, since you could liken two or more conductors in parallel to one conductor which would of course be larger in cross sectional area than each of them and hence lower in resistance.

Ammeters and Voltmeters

We shall point out later (Chapter 45) that moving coil galvanometers will generally be milliammeters or microammeters. We must consider how we might modify these instruments to measure larger currents and also voltages. It is convenient to consider one particular type of instrument in very common use. This gives a full scale deflection of the pointer when a current of 15 mA (0·015 ampere) passes through it and has a resistance of 5 ohms. We can say immediately that Ohm's law in the form $V = iR$ will tell us the potential difference to be applied in order to cause this current 0·015 A to flow. It is $0·015 \times 5 = 0·075$ V. So the instrument will only measure currents up to 0·015 A and potential differences up to 0·075 V. For currents or voltages much in excess of these values the instrument is likely to be damaged.

Let us suppose first that we wish to modify the instrument to give a full scale deflection for a current of 1·5 A. Clearly most of this current must not be allowed to pass through the galvanometer itself. We must therefore 'by-pass' it in some way. We connect a resistance in parallel with the galvanometer so that the current splits up in the way shown in Fig. 36.11.

It only remains for us to calculate the value this resistance R (which we call a shunt) must have in order for the current to split in the required ratio. The potential difference between A and B must, as we have seen, be 0·075 V to produce a current of 0·015 A in G. This same potential difference causes a current of 1·485 A in R. Thus from Ohm's law:

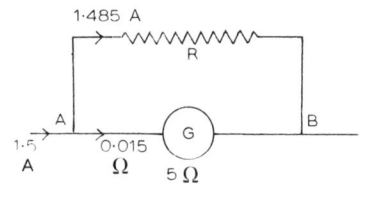

Fig. 36.11. Converting a galvanometer to an ammeter

$$\frac{V}{i} = R$$

$$\frac{0 \cdot 075}{1 \cdot 485} = R$$

$$R = \frac{5}{99}$$

or $\cdot 0505\ \Omega$.

So if a shunt of this value is connected in parallel with the galvanometer it will cause a full scale deflection for a current of $1 \cdot 5$ A. The shunt resistance is much less than that of the galvanometer since the shunt has to carry a much bigger fraction of the current. Also the nett resistance of the ammeter now is less than either of the resistances in parallel, i.e. less than $0 \cdot 0505\ \Omega$. This low value is desirable since, when we introduce an ammeter into a circuit to measure the current, we would prefer not to alter the current appreciably in the process. If an ammeter did have a high resistance, putting it in a circuit would probably reduce the current considerably.

Giving our galvanometer a shunt has changed its current measuring range but it will still show a full scale deflection for an applied potential difference of $0 \cdot 075$ V. Its voltage range is as it was. Suppose we wish to modify our instrument to make it a voltmeter reading up to 15 V. We must arrange that when we apply this potential difference the current which passes is only the $0 \cdot 015$ A required for maximum deflection.

The technique now is to give the galvanometer a series resistance, $Rs\ \Omega$, such that $(Rs + 5)\ \Omega$ (the total value of the resistance) passes a current of $0 \cdot 015$ A for an applied potential difference of 15 V (Fig. 36.12).

Thus from Ohm's law:

$$\frac{V}{i} = R$$

$$\frac{15}{\cdot 015} = Rs + 5$$

$$1000 = Rs + 5$$

$$Rs = 995\ \Omega$$

So with a series resistance of 995 Ω our

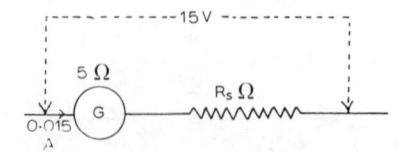

Fig. 36.12. Converting a galvanometer to a voltmeter

galvanometer becomes a voltmeter reading up to 15 V and having a net resistance of 1000 ohms. Again this high value is desirable since when we connect our voltmeter across some component in a circuit we do not want it to take any appreciable current from the circuit. If it does take current it will be altering the thing we are trying to measure. In fact ideally we would like a voltmeter to have infinite resistance so that it would pass no current at all. This is clearly not possible with a moving coil instrument but they will often have values much higher than the 1000 ohms quoted in this example.

For the type of instrument we have been discussing the manufacturers supply a range of shunts and series resistances whose values are calculated to enable the galvanometer to be used in quite a number of current and voltage ranges. They may be clipped or screwed in position as shown in Fig. 36.13.

It is very convenient to have a number of shunts and series resistances mounted in the same box as the moving coil galvanometer in such a way that the required range may be selected simply by turning a switch.

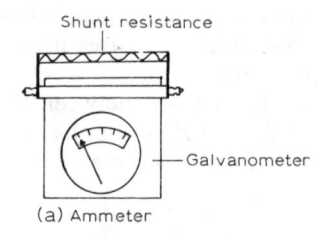

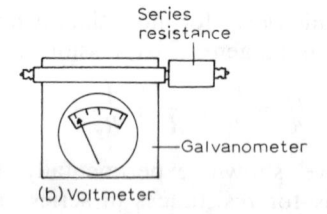

Fig. 36.13. Shunt and series resistances

Internal Resistance and Electromotive Force

Experiment 36.5. Set up the circuit shown in Fig. 36.14 with both of the switches, S_1 and S_2 open. Note the reading on the voltmeter. This is the potential difference across the cell when it is delivering only a very small current to operate the voltmeter. Now close switch S_1. The ammeter will indicate that a current is flowing in the circuit and the voltmeter should show a drop in reading. Now close also switch S_2. The ammeter will show a

Fig. 36.14.

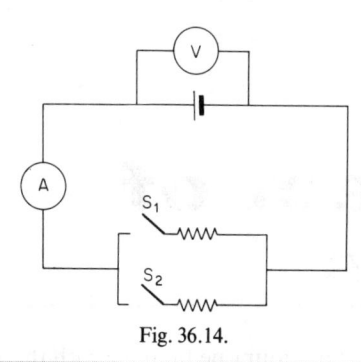

larger current in the circuit and the voltmeter will indicate a further drop.

This experiment suggests that the terminal potential difference of a cell is a variable quantity depending on the current which the cell is giving. The terminal potential difference of a cell when it is giving no current is, on the other hand, a constant, depending only on the chemical composition of the cell. This quantity is called the 'electromotive force' (EMF) of the cell. By simply connecting a high resistance voltmeter across a cell we obtain a reading which is nearly, but not quite, the EMF.

You may understand the drop in terminal potential of a cell when it starts to deliver a current, if you consider the simple circuit in Fig. 36.15. The current in the circuit is, of course, passing through both the resistance, R, and the cell. If we assume that the cell has a resistance also which we call r, the total resistance of the circuit is $(R + r)$ ohms.

Thus since the total potential difference in the circuit is the EMF of the cell, (E volts), we may write Ohm's law for the complete circuit as:

$$E = i(R + r)$$
$$= iR + ir$$

If we connect a voltmeter across the cell in Fig. 36.15 we measure just iR which is clearly less than E.

The quantity, r, in this equation is known as the internal resistance of the cell. For a lead acid ac-

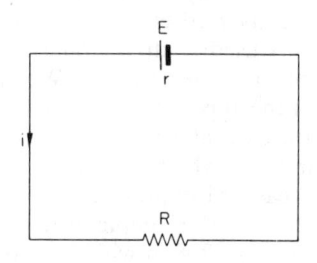

Fig. 36.15.

cumulator this tends to be very small and so, unless the current taken from it is very large, the drop in terminal potential difference is small. This is why a dry battery is suggested for Experiment 36.5. The internal resistance is an important quantity, since it decides what is the maximum current which the cell can deliver. The very low internal resistance of a lead-acid battery makes it possible for the very large current needed to operate the starter motor of a car (perhaps 50 amperes) to be obtained.

Exercise 36

1 State Ohm's Law and describe a simple experiment to illustrate it.

In a wireless set, a voltage of 50 V is applied to a circuit in which a resistance of 8 MΩ is in series with two resistances of 20 and 30 MΩ which are in paralellel. Find the currents in μA flowing through each resistance.
[1 MΩ = 1,000,000 Ω]   (O)

2 Find the resistance of a piece of wire 100 cm long and the area of cross-section 0·25 mm². (Resistivity of the material of the wire is 10^{-6} Ω m.)   (J)

3 Two resistance coils, one of 3 ohms and the other of 5 ohms resistance, are joined (a) in series, (b) in parallel. Find the resultant resistance in each case.   (L)

4 Two resistances of 20 Ω and 30 Ω are connected in parallel. Calculate (a) their combined resistance, (b) the current in the 20 Ω resistance, if the potential difference across the combination is 18 V.   (S)

5 What is the essential difference in construction between (i) a moving coil ammeter and (ii) a moving coil voltmeter. A moving coil ammeter has a resistance of 2 ohms and the pointer is fully deflected by a current of 1 A. How would you convert the meter to read up to 10 A?   (S)

6 An ammeter has a resistance of 700 Ω and is calibrated over the range 0 to 3 A. It is required to use it to measure currents up to 10 A. What resistance should be inserted as a shunt? Explain why it would be difficult to adapt this ammeter for use as a voltmeter.   (O)

7 Describe how you would find the resistance of a 12-volt lamp under its correct working conditions. Draw a diagram of the circuit you would use.

You are provided with two galvanometers each of which has a resistance of 5 Ω and gives a full-scale deflection with a current of 0·01 A. Explain how you would convert one of the galvanometers into an ammeter to read up to 3 A, and the other into a voltmeter to read up to 12 V. Calculate the values of any resistances required.

37
The Heating Effect of Electric Currents

Of the electrical energy produced in power stations, almost all is converted either to kinetic energy in electric motors used to drive domestic and industrial machines or to heat energy. We must discuss now some of the applications of the heating effect of an electric current and at the same time consider the units in which quantities of electrical energy are measured in practice.

If we return to the simple water analogy we introduced on page 208, it will be clear that water flowing down the pipe is losing potential energy. Some of this is at first converted into kinetic energy, but eventually it will all be converted into heat. In the case of electricity flowing in a conductor between the ends of which a potential difference exists heat and kinetic energy may also be produced. If the conductor happens to be the coil of an electric motor much of the energy will be kinetic, but if it is just an ordinary piece of wire all the energy will appear as heat. We have defined the volt (page 208) as the potential difference between two points if one joule of potential energy is lost when one coulomb of charge passes between them. Thus if a potential difference of V volts exists between the ends of a conductor and Q coulombs of charge pass between them, then $V \times Q$ joules of heat and kinetic energy are produced. For a steady current of i amperes flowing for a time t seconds:

Q coulombs $= i\,A \times t\,s$;
∴ energy liberated $= V \times i \times t\,J$.

We have defined *power* (page 78) as the rate at which work is being done. The power in an electrical circuit is the rate at which electrical energy is being dissipated or the rate at which heat or kinetic energy is being produced. If the current and voltage are constant this will be

$$\frac{\text{energy liberated}}{\text{time}}$$

i.e. $\dfrac{V \times i \times t}{t}$ J/s or W

$= V \times i$ W

You will no doubt be familiar with the watt as a unit of electrical power. The average electric light bulb for domestic use is rated at 100 W. If the mains voltage is 240 V.

$100 = 240\,i$
$i = $ current in amperes through the bulb

Hence $i = \dfrac{100}{240}$ A $= 0.42$ A.

A small electric heater may have power of 1 kilowatt (1 kW = 1000 W) and it will therefore pass a current of 4·2 A on 240 V mains.

The latest nuclear power stations have power outputs of hundreds of megawatts (1 MW = 1 million watts) but since the voltage produced by the generator may be 11 000 V and there will be several generators the current will be of the order of 1000 A from each generator.

Payment for Electrical Energy

The consumer of electrical energy will pay the local electricity board for the quantity of energy he consumes. He might pay for the number of joules of electrical energy he used but in fact the unit with which the electricity board works is the kilowatt-hour, known as the Board of Trade Unit or just a 'unit'. This is the energy dissipated in an appliance with a power of 1000 W working for 1 hour.

Such an appliance dissipates 1000 J/s and hence 3600 × 1000 J/hour.

$$1\,\text{kW h} = 3 \cdot 6 \times 10^6 \text{J}$$

It should be noted that if the appliance being used is for example an electric motor, it will be less than 100% efficient and consequently the mechanical energy produced will be less than the electrical energy put in, but it is the latter for which the user pays the electricity board.

The cost of electricity to the domestic consumer is usually about 1p per 'Unit' or kW h.

You will see then that if you accidentally leave a 100 watt lamp on overnight (say for 10 hours) you will consume only 1 unit of electricity.

$\frac{1}{10}$ kW for 10 hours = 1 kW h

This will cost you only about a penny or so, but leaving a 3 kW electric heater on for the same period of time, you will consume 3×10 kW h and at 1p a unit the cost will be 30p.

The Heat Produced in a Conductor

We have seen that the energy dissipated in a conductor is $V \times i \times t$ J. If all the electrical energy is changed into heat and we apply Ohm's law to the conductor ($V = iR$) we have:

$$\text{Heat produced} = iR \times i \times t \text{ J}$$
$$= i^2 Rt \text{ J} \qquad (1)$$

Equation (1) is a mathematical statement of a law which was found experimentally by the English scientist, James Prescott Joule (1818–1889). He showed that the quantity of heat produced in a given conductor is proportional to the square of the current and to the time for which it flows.

$$\text{Heat} \propto i^2$$
$$\text{and Heat} \propto t$$

Experiment 37.1. An Experiment to Verify Joule's Law.

Using the apparatus shown in Fig. 20.1, set up a circuit containing a rheostat and an ammeter as well as the immersion heater and 12 V supply. With 500 g of water in the saucepan, adjust the rheostat to give a convenient value of the current (say, 1 A). Pass a current for five minutes, stirring from time to time, and note the rise in temperature. Starting with 500 g of cold water each time repeat the experiment with the same current flowing for ten and fifteen minutes. Now adjust the rheostat settings so that the current is doubled and note the temperature rise in five minutes. Repeat the experiment with three times the initial current. The following table shows a typical set of results for such a series of experiments.

The heat supplied in the table below is calculated from the fact that the specific heat capacity of water is 4200 J/kg°C. So if 0.50 kg of water are heated through 4°C the heat supplied is

$$0.50 \times 4200 \times 4 \text{ J}.$$

The first table shows that the heat produced is proportional to the time for which the current flows:

$$H \propto t$$

The second table shows that the heat produced is proportional to the current squared:

$$84 : 336 : 756 = (1)^2 : (2)^2 : (3)^2$$
$$H \propto i^2$$

Use of the Heating Effect in the Home

Domestic heating by electricity is becoming very popular, having the big advantages of cleanliness and ease of operation over traditional solid fuel burning heaters. It is however at present considerably more expensive than solid fuel.

Until quite recently the most common type of domestic electric heater was the radiant type (Fig. 37.1). A spiral of wire wound on an earthenware rod is heated to redness by the current and heat is emitted in the form of radiation into the room. A concave polished metal reflector is commonly mounted behind the rod element to produce a beam of radiant heat. This kind of heater is good if heat is required quickly but the benefit is felt only by someone situated directly in front of the fire.

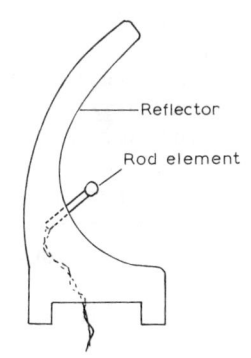

Fig. 37.1. 'Radiant' electric heater

Current (A)	Time (minute)	Rise in temperature °C	Heat supplied (J)
1	5	4	8400
1	10	8	16 800
1	15	12	25 200
1	5	4	8400
2	5	16	33 600
3	5	36	75 600

Much better for producing overall room heating over a long period of time is the electric convector heater, one type of which is illustrated in Fig. 37.2. In this case the air around the heating element is warmed and hence it rises and cool air enters at the bottom to take its place. In this way a flow of warm air is produced.

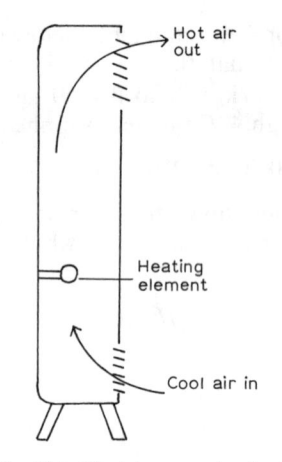

Fig. 37.2. Electric convector heater

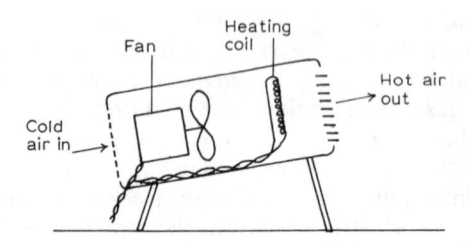

Fig. 37.3. Electric fan heater

Another very convenient form of electric heater is the fan heater (Fig. 37.3). In this case a flow of air over the heating element is produced by an electric fan.

Storage Heaters

One of the many problems faced by the electricity supply companies is the wide variation in demand for electrical energy during a day. Consumption is very low during the night and reaches peaks at about breakfast and tea times. At these times during very cold weather the supply company may have difficulty in meeting the demands of the consumers. It is not an easy matter to shut down or start up generating equipment at short notice and so clearly there is a surplus of electrical energy available at night which the electricity company would like to store in some way. Schemes have been launched in recent years to try to cope with this problem. One is the system of block storage heaters. The conductors of a heating element are embedded in a massive slab of concrete or brick. A current flows in the conductors during the night and the slabs are slowly heated up. Early in the morning the current is automatically switched off by a time switch and the slabs during the day slowly lose their heat to the surroundings. In order to encourage people to use this form of heating the electricity company charges much less for this 'off peak' electrical energy than for electricity consumed during the day.

Electric Lighting

Fig. 37.4 shows an electric filament lamp. This consists essentially of a coil of tungsten wire contained in a glass vessel from which the air has been removed and replaced by a very small quantity of an inert gas, for example argon.

A current passes through the coil via the contacts and the thick lead-in wires, and it is raised to white heat. Thus, although most of the electrical energy dissipated in the coil will be liberated by it in the form of radiant heat, some of it will become light energy. Tungsten wire is chosen for the coil on account of its high melting point. The argon gas is included in the vessel to prevent oxidation and evaporation of the filament.

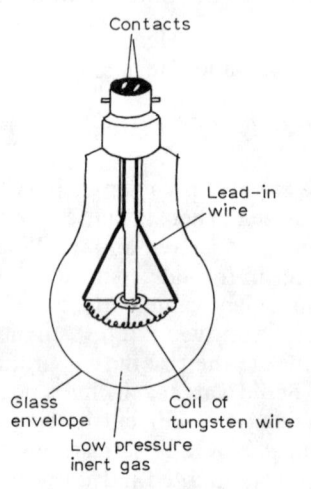

Fig. 37.4. Filament lamp

The luminous efficiency of a lamp $\left(\dfrac{\text{light energy out}}{\text{electrical energy in}}\right)$ is very low and the lamp manufacturers have taken much trouble in trying to increase this. Much of their work has been on the structure of the filament itself and this is now commonly a 'coiled coil'.

Lamps of a much higher luminous efficiency (i.e. much less heat produced) are fluorescent lamps. The disadvantage of these for domestic use is the relatively high initial cost of the tube and the equipment that goes with it. They are widely used commercially particularly when very good illumination is essential, since on account of their length they produce little shadow.

Domestic Electrical Wiring. The Earth and the Fuse

Most houses have two distinct mains electricity wiring systems. They are referred to as the lighting and the power circuits. All the lights in a building will connect in parallel as shown in Fig. 37.5, so that the full mains voltage is applied to each lamp and the failure of one lamp does not affect the others. You may have encountered the effect of connecting lamps in series with decorative 'fairy lights'. In this case all of the lamps go out when one of the string fails.

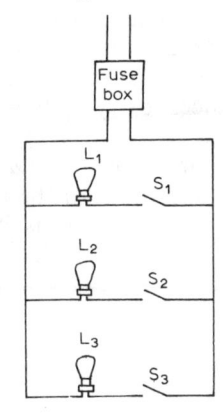

Fig. 37.5. Lighting circuit (diagrammatic)

The cables used in a lighting circuit are generally relatively 'light', i.e. the copper conductors are quite thin, and are not intended to carry currents much in excess of 5 A. Since, as we saw on page 216 a 100 W lamp on 240 V mains takes less than 0·5 A, thin cable is adequate as long as only lamps are used on the circuit. If, however, appliances like electric flat irons which take much larger currents are used on a lighting circuit there is a serious risk that the cables may get hot and a fire result.

The function of the fuse in this lighting circuit is mainly to prevent this overloading of the cables. It consists of a short length of wire included in the circuit, whose cross sectional area and material are chosen so that when the current in it exceeds 5 A it will melt and hence break the circuit. In practice the fuse wire may be of a low melting point alloy of tin and lead and fixed in a porcelain holder.

The power circuit in most modern buildings is based on the system known as the 'ring main'. The cable used may carry a current of up to 15 A although in practice the fuses used are for a maximum current of 13 A. A simplified ring main circuit is shown in Fig. 37.6. The cable used is a three core cable, the three cores being referred to as 'live', 'neutral' and 'earth'. The mains electricity

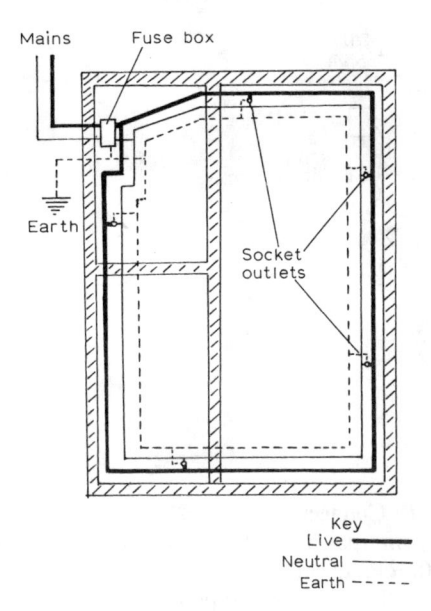

Fig. 37.6. A household 'ring main' circuit (diagrammatic)

supply is generally an alternating current so we cannot refer simply to positive and negative as we have in our consideration of D.C. circuits. In practice the neutral wire is maintained at about 'earth' potential and the 'live' lead is alternately positive and then negative with respect to it. In a ring main power circuit a complete loop of the three core cable runs around, for example, the ground floor of a house and three-pin sockets are connected to the ring in convenient places. It will be seen that if an appliance, an electric heater for example, is connected to a socket on the ring, current can flow from the fuse box in both directions round the ring to it. We have in effect two live and two neutral leads in parallel connecting the appliance to the fuse box. Thus although the cable may pass only 15 A safely the ring has a capacity of 30 A and this will be the limit of the fuse in the fuse box. This 30 A fuse will prevent the cable becoming overheated but it is little use in preventing damage or injury if a faulty appliance is connected to the mains. Such an appliance may normally pass a current of less than one amp. but the main fuse will not fail unless the current exceeds 30 A, so clearly some additional protection is needed. This is provided by including a fuse in the plug attached to the appliance (Fig. 37.7). This fuse is usually a 'cartridge' fuse, i.e. the fuse wire is contained in a small glass tube with metal ends which clip into the plug. The value of this fuse should be chosen to suit the appliance, e.g., 13 A for 3 kW electric heater, 2 A for 400 W flat iron, etc.

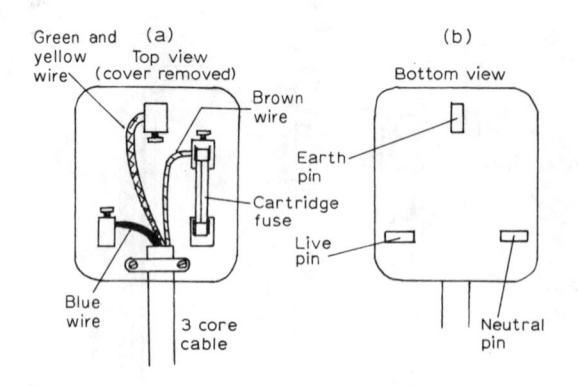

Fig. 37.7. A 13 A plug

The Earth Connection

The principal hazard experienced in the use of domestic electrical appliances is if the metal case becomes 'live'. This happens quite easily if, for example, the mains leads pass through a hole in the case with a sharp edge which may cut the insulation of the leads. In order to minimise this hazard it is usual to connect the case of the device via the third pin of the plug and the earth wire of the cable to a water main somewhere near the main fuse box. Now if the case becomes 'live' there is a low resistance path for the current from the live wire to earth. Thus a large current should pass and 'blow' the fuse.

The following then are useful rules for safety in using electrical appliances:
1. Do not use a device fitted with a three-core cable on a two-pin socket or lighting circuit.
2. If the power circuit is of the ring main type see that each appliance has the correct fuse. (Most electrical appliances carry a label giving their 'wattage').
3. If the fuse fails in ordinary use *do not* replace it with a larger one and hope for the best.
   Do assume that the appliance is faulty and have it checked.

Exercise 37

1 Describe an experiment to find out the relationship that exists between the current flowing through a conductor and the heat produced in the conductor. Draw a diagram of the apparatus you would use.

How much heat is produced by a 1000-watt electric fire which is switched on for one hour? By what processes is this heat transferred from the fire to the room?

2 A lamp marked 6 V 24 W is connected to the terminals of a 6 V accumulator of negligible internal resistance. Find the resistance of the lamp when so used. State what would happen to the brightness of the lamp if a resistor of 4·5 Ω was inserted (*a*) in series, (*b*) in parallel with the lamp. (Assume that the lamp filament resistance is unchanged.) (*O*)

3 An electric iron rated at 750 W is used on a 250 V supply; what current does it take? If the iron has a mass of 2 kg how long will it take to heat it from 20 °C to 270 °C, assuming that no heat is lost?
[Specific heat capacity of iron = 460 J/kg °C]

4 The filament of a lamp consists of a wire which has a resistance of 960 Ω when the lamp is connected across 240 V mains. Calculate (*a*) the wattage of the lamp, (*b*) the cost per hour of using the lamp if the electricity costs 5p per kWh. (*C*)

5 An electric motor operating on 250 V mains passes a current of 1·5 A when it is delivering 200 W of mechanical power. Find (i) the efficiency of the motor, and (ii) the cost of running it for 1 hour if the electricity costs 4p per unit.

6 Explain the use made of the heating effect of an electric current in (i) a fuse wire, (ii) an electric filament lamp, (iii) a bimetal thermostat used with an immersion heater. (*C*)

7 A current of 0·5 A flowing through a wire produces 20 J of heat in $\frac{1}{2}$ minute. Find the resistance of the wire. (*J*)

8 Resistances of 120 Ω and 240 Ω are connected (i) in parallel, (ii) in series, across 240 V mains. What power is generated in each resistance? (*S*)

9 (*a*) Explain the purpose of (i) a fuse wire, (ii) earthing, in an electrical circuit.

What value fuse wire is normally used in the lighting circuit of a house? What might happen if a fuse wire of (iii) too high, (iv) too low, a value were used?

38
Using Electrical Energy in Chemical Changes

One of the chief means by which the chemist provides the necessary energy for reactions to take place is by supplying heat and so increasing the temperature. Heat energy is converted into chemical energy and we know that in many reactions the reverse occurs and chemical energy is changed into heat energy. If a little fairly concentrated sodium hydroxide solution is put into a test-tube containing a thermometer and concentrated hydrochloric acid gradually added while stirring, a considerable change of temperature will be noted as neutralization comes about.

$$HCl + NaOH \rightarrow NaCl + H_2O$$
$$\text{or } H^+ + OH^- \rightarrow H_2O$$

The heat evolved when one mole of sodium hydroxide is neutralized by one mole of hydrochloric or nitric acid is 57 kJ and this is the same if potassium hydroxide is used. This value is called the *Heat of Neutralization* of a strong acid by a strong base. With dibasic acids such as sulphuric, 0.5 mole of acid produces the same quantity of heat energy. The amount of acid involved is that quantity which produces one mole (1 g) of hydrogen *ions*.

Let us now see if electrical energy, also used by scientists, is able to bring about chemical changes and, conversely, whether chemical energy can be converted into electrical energy.

Conduction of Electricity by Solids

Elements. We know that most metals will conduct electricity very well and that most cables used for carrying a current are made of copper or other suitable metal. Most non-metals are non-conductors, but one, carbon (in the form of graphite or gas carbon), is almost as good as most metals.

Compounds. Samples of two very different solids, such as lead bromide and either paraffin wax or sealing wax, will be examined and an attempt made to pass a current (D.C.) through them.

Experiment 38.1. The Action of Electricity on (a) Wax and (b) Lead(II) Bromide.

The apparatus of Fig. 38.1 is used. A crucible is half filled with wax which is melted by warming it. As it cools down, two nickel wires or rods, pushed through corks to act as holders, are inserted and held there as the wax solidifies. When this is cold, the circuit is completed via an ammeter ($1\frac{1}{2}$ A) to a source of direct current, either a power pack or three 2-V accumulators. When the switch is closed *no current passes*. If the crucible is warmed until the wax is again molten the ammeter will still show that no current is passing.

Repeat the experiment using lead(II) bromide, melting this and then inserting the electrodes. When the bromide has solidified close the switch; *no current passes*. Warm the crucible gently and you will see that at the moment when the molten substance touches *both* the electrodes the ammeter needle will suddenly swing over to show that about $1\frac{1}{2}$ A of electricity are passing. If the source of heat is removed the current will cease to flow immediately the lead bromide solidifies around one electrode.

Let the current flow for about 10–15 minutes and notice that a small puff of brown vapour comes off every few seconds from the anode. This is bromine, but the amount is insufficient to cause any discomfort; even so, it is wisest not to breathe

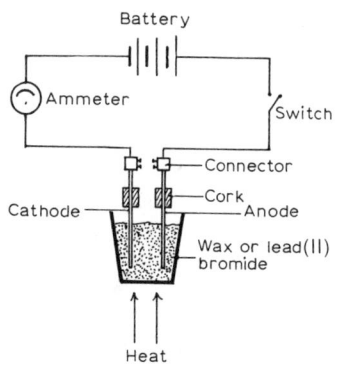

Fig. 38.1. Electrolysis of a fused solid

it. Evidently, decomposition is taking place. Remove the electrodes and examine them. There should be a small deposit of lead on the end of the cathode, further evidence of decomposition. Try to mark a piece of paper with it. The anode will most likely be eaten away to a fine point.

$$PbBr_2(l) \rightarrow Pb(s) + Br_2(g)$$

This experiment can be carried out with common salt, but a very high temperature is needed to fuse it.

Why should wax behave as a non-conductor and lead(II) bromide as a conductor? What differences are there between these substances? If you refer to chapter 6 you will recall that most compounds of carbon are covalent ones (wax is a hydrocarbon), but that salts such as lead bromide and sodium chloride are electrovalent or ionic. Can this have anything to do with the ability of the substance to conduct electricity? Let us now see how liquids in general, other than molten solids, behave towards electricity.

Conduction of Electricity by Liquids

Experiment 38.2. The Action of an Electric Current on Liquids.

Replace the crucible of the last experiment by a 150 cm³ beaker or crystallizing dish and, in place of the nickel wires, use pieces of sheet nickel about 1 cm wide by 5 cm long. These can be connected to short pieces of nickel wire by piercing two holes near one end, threading the wire through and hammering flat (Fig. 38.2). Next, take a slice of cork and push the wires through so that the electrodes are held about 3 cm apart. If deserved, the ammeter can be replaced by a 12 V bulb.

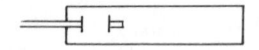

Fig. 38.2. Attaching wire to electrode

Place distilled water in the beaker so that about 4 cm of the electrodes are covered when these stand in the dish (Fig. 38.3). Close the switch; does any current pass? Change the distilled water for tap water; is there any current recorded now? Add two drops of sulphuric acid; does this have any effect? Repeat, using the following liquids and washing the electrodes after each test: alcohol, tetrachloromethane (carbon tetrachloride), liquid paraffin; dilute solutions of sugar, hydrochloric acid, sodium hydroxide, sodium sulphate, acetic acid or vinegar and ammonia. Record the results; these should be similar to those in Table 38.1.

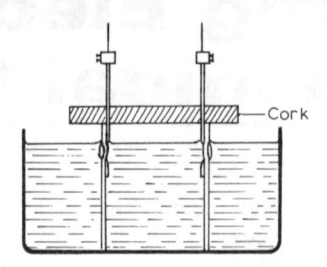

Fig. 38.3. Electrolysis of a solution

Table 38.1

Liquid	Current
distilled water	none
tap water	very slight
acidulated water	large
ethanol	none
tetrachloromethane	none
liquid paraffin	none
sugar solution	none
hydrochloric acid	large
sodium hydroxide	large
sodium sulphate	large
ethanoic acid	fairly small
ammonia	fairly small

Can you see any differences between the liquids which conduct and those which do not?

The latter are all covalent compounds, but those which are good conductors are electrovalent. All mineral acids, alkalis and their salts in solution are good conductors and can be *electrolyzed*; organic and other covalent liquids do not conduct. We know that electrovalent substances all contain ions bearing electric charges and it is these ions which carry the current between the electrodes and complete the circuit.

Pure liquid hydrogen chloride is a covalent liquid and a non-conductor, but in solution it joins with the water molecules to form ions and so conducts. Water is also covalent, only about one molecule in over 500 million being ionized, so pure water is also a non-conductor. The presence of dissolved substances in water increases its conductivity considerably. Ethanoic acid and ammonium hydroxide do not contain many ions and so are poor conductors compared with solutions of the mineral acids. Later in this chapter we shall consider what actually happens when certain solutions are electrolyzed.

We have now seen that an electric current can bring about a chemical change, i.e. electrical energy can be converted into chemical energy. Is the reverse action also true? Chemical change is due to the movement of atoms or molecules or ions and this causes chemical energy. Can chemical energy give rise to electrical energy?

Experiment 38.3. Conversion of Chemical Energy into Electrical Energy.

Prepare two electrodes, about 10 cm × 6 cm, one of zinc and the other of copper and weigh them. Connect them to a milliammeter, the zinc to the negative and the copper to the positive terminal. Place them in a beaker of dilute sulphuric acid. Bubbles of gas soon appear at both the plates, but mostly at the copper one. After fifteen minutes remove the electrodes and weigh them; you will find that the zinc plate has lost in weight. While the bubbles are coming off (these are hydrogen bubbles) notice that a small electric current is recorded on the ammeter. A chemical change has taken place.

$$Zn(s) + H_2SO_4(aq) \rightarrow ZnSO_4(aq) + H_2(g)$$

This experiment shows that chemical energy can be converted into electrical energy. Two practical applications of this are the dry battery or simple cell and the accumulator. (See Chapter 39.)

When electrolysis takes place decomposition occurs and products, either solid or gaseous, are set free. Is there any connection between the amount of a product obtained and the quantity of electricity used?

Experiment 38.4. To find the connection between the volume of a gas liberated and the quantity of electricity used.

The apparatus or cell in which electrolysis is carried out is called a *voltameter* (not to be confused with *voltmeter*) and the three-limbed one in which you have no doubt seen acidulated water being electrolyzed is called a Hoffman voltameter (Fig. 38.4).

Set up this apparatus and connect it to a 6 V battery via an ammeter and a variable rheostat. Pass a current of about 1 A through for about 10 minutes with the taps open to ensure that the water becomes saturated with dissolved hydrogen (only a little) and oxygen (a fair amount). Close the taps and note the time taken for 10 cm³ of hydrogen to collect at the cathode (use a stop-watch) and then

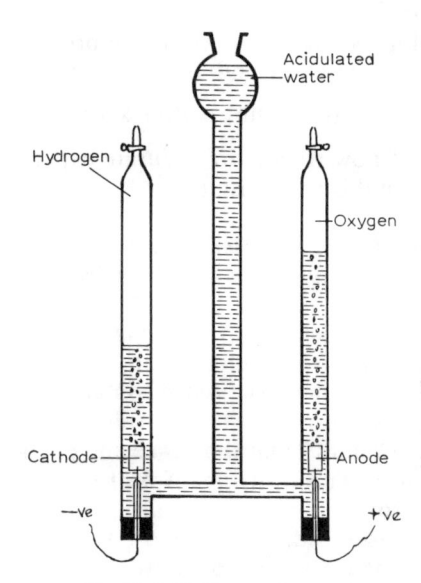

Fig. 38.4. Hoffman voltameter

switch off. During the action it may be necessary to adjust the rheostat in order to maintain the current constant at 1 A. Read also the volume of oxygen collected at the anode. This should be 5 cm³ showing that the volumes are in the ratio suggested by the equation

$$2H_2O \rightarrow 2H_2 + O_2$$

(refer back to Gay Lussac and Avogadro).

If the time taken for the 10 cm³ of hydrogen to be evolved is t s let the current (still 1 A) run for a further t s. Note the additional volume of hydrogen liberated. Adjust the current to 2 A (or 0.5 A) and let it run for t s. Again note the volume of hydrogen. The table below shows figures obtained in such an experiment.

Time (t) (in s)	Current (i) (in A)	$i \times t$	Volume (v) (in cm³)	$\dfrac{v}{it}$
80	1	80	10	0.125
161	1	161	20	0.124
80	2	160	20.1	0.126
80	0.5	40	5	0.125
160	0.5	80	10.1	0.127

These figures show that there is some definite connection between the mass of hydrogen liberated (mass being proportional to volume here as conditions are constant) and the quantity of electricity passing. This latter value is measured in coulombs and is defined as the product of the current in ampères and the time in seconds.

$$q \text{ (coulomb)} = i(A) \times t(s)$$

We shall now try a more accurate experiment to determine this relationship.

Experiment 38.5. To Find the Factors Determining the Mass of a Substance Deposited During Electrolysis.
Thoroughly clean (by washing in water after rubbing with emery cloth) and dry the copper cathode of a voltameter which contains copper(II) sulphate solution as the electrolyte. Weigh the electrode on a chemical balance and connect it in a circuit as shown in Fig. 38.5. *A* is an ammeter reading to 1·5 A and *B* a battery of two or three accumulators. Adjust the rheostat *R* so that the meter reads 1 A and keep it steady at that value for a period of time, measured by means of a stop clock, of 15 minutes or thereabouts. Remove the cathode, wash it gently in water and then in alcohol, dry and reweigh. The experiment should be repeated a number of times with different currents and periods of time as suggested in the table. This is most conveniently done by means of a set of simultaneous experiments, each group in the class taking a different current and period of time. The table at bottom of page shows a typical set of results for a number of experiments.

Column 3 of the table shows values of the quantity of electricity passing in each case. As before

$$q = i \times t$$

You will observe that the values in column 7, of the mass of copper divided by the quantity of electricity, are approximately constant.

$$\frac{M}{q} = \text{constant}$$

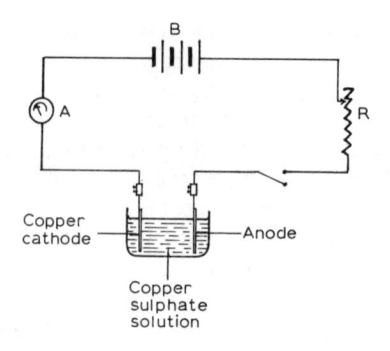

Fig. 38.5. Determining Faraday's First Law

$$\text{or } M = \text{constant} \times q$$
$$\text{or } M = \text{constant} \times i \times t$$

This result is known as Faraday's First Law of Electrolysis which states:

The mass of a substance liberated or deposited during electrolysis is proportional to the current and to the time for which it flows.

The average value of all the readings in the last column gives a gain of 0·000 33 g of copper per coulomb. If this experiment is repeated using aqueous silver nitrate in place of the copper(II) sulphate solution it will be found that the average gain in mass per coulomb is 0·001 118 g; similarly the mass of hydrogen set free by the passage of one coulomb of electricity is 0·000 010 43 g.

It should be obvious that these masses are—electrochemically speaking—equivalent to each other as they are all liberated by one coulomb; they are often referred to as the electrochemical equivalents of the elements concerned.

Measurement of Current by means of a Voltameter
It has been pointed out previously that in most laboratories the most accurate measurements that can be made are those of mass, made with a chemical balance. Time intervals or, perhaps, 15

1	2	3	4	5	6	7
Current (A) i	Time (s) t	Quantity of electricity (C) q	First mass of cathode (g)	Second mass of cathode (g)	Gain in mass of cathode (g) M	Gain in mass / quantity of electricity
1·0	900	900			0·31	0·00034
1·0	1200	1200			0·31	0·00032
1·0	1500	1500			0·50	0·00033
1·5	900	1350			0·44	0·00033
1·5	1200	1800			0·59	0·00033
1·5	1500	2250			0·75	0·00033

minutes—measured with a stop-clock—may be accurate to one part in about one thousand. If the electrochemical equivalent of a metal is known, electric current may be measured by finding the mass deposited in a certain period of time.

$$i(A) = \frac{M(g)}{e.c.e. \times t(s)}$$

In this way an ammeter may be calibrated with the aid of a voltameter.

Although our unit of current, the ampere, is now defined in terms of the magnetic effect, a unit of current has been defined in terms of the chemical effect. The 'international ampère' is the current which will deposit 0·001 118 g of silver in one second in a silver voltameter.

In Experiment 38.5 columns 3 and 6 record the quantities of electricity needed to deposit definite masses of copper, and from these figures the number of coulombs necessary to deposit one gram-atom of copper can be determined. If this is done for a number of elements we get figures as in Table 38.2.

The figure, 96 540 C/mole, is called one *Farady*. The factor by which it is multiplied in the table is, as you may have realized, the valency of the element. The figures illustrate Faraday's second law which states:

The quantities of electricity required to liberate one gram-atom (i.e. one mole) of different elements bear a simple relationship to one another.

As we have just seen, this relationship is expressed by a number, the valency, and so is always a small whole number. A gram-atom of any element always contains 6×10^{23} atoms (called the Avogadro Constant). It should now be obvious that one Faraday of electricity will always deposit or liberate 6×10^{23} atoms of any monovalent element, but that two Faradays will be necessary to deposit this number of atoms of copper and three Faradays will be required to set free 6×10^{23} atoms of aluminium.

We can express this second law of Faraday's in a somewhat different way by writing

The number of atoms of different elements liberated by a fixed quantity of electricity bear a simple relationship to each other.

EXAMPLE (a). A current of 1·5 A running for 4 minutes sets free 0·75 g of a metal. What mass will be deposited by a current of 2 A running for 5·5 minutes?

$1·5 \times 4 \times 60$ C deposit 0·75 g
$2 \times 5·5 \times 60$ C deposit

$$\frac{0·75 \times 2 \times 5·5 \times 60}{1·5 \times 4 \times 60}$$
$$= ·375 \text{ g}$$

EXAMPLES (b). What volume of hydrogen collected at r.t.p. will be set free in the same time as 0·83 g of lead (relative atomic mass = 207) is deposited from a solution of one of its salts?

Lead is divalent
∴ 2 g-atoms of hydrogen will be set free in the same time as 1 g-atom of lead.
i.e. while 207 g of lead are deposited, 2 g of hydrogen will be set free. This has a volume of 24 dm³ at r.t.p. 24 000 cm³.
∴ while 0·83 g lead is deposited, the volume of hydrogen set free will be $\frac{0·83}{207} \times 24\,000$ cm³ = 96·2 cm³.

96 540 C liberate 1 g-atom or mole of (say) chlorine from a solution of a chloride, i.e. one Faraday will liberate 6×10^{23} atoms of chlorine, each one of which has been involved with one electron in the change

Table 38.2

Element	Atomic mass	Coulombs needed to deposit or liberate 1 gram-atom of the element
H	1	96540
Cl	35·5	96540
Ag	108	96540
O	16	193080 = 2 × 96540
Pb	207	193080 = 2 × 96540
Cu	63·5	193080 = 2 × 96540
Cr	52	289620 = 3 × 96540
Al	27	289620 = 3 × 96540

$$Cl^- - e^- \rightarrow Cl$$

So, in all, 6×10^{23} electrons are concerned during the passage of one Faraday of electricity. For this reason, we can consider one Faraday to be a *mole of electrons* since one mole of any substance contains this same number of particles.

EXAMPLE (c). The same current, passed through solutions of nickel(II) sulphate and copper(II) sulphate, deposits 0·445 g of copper. What mass of nickel will be set free? (relative atomic mass of copper = 63·5 and of nickel = 58·7)

Copper and nickel have the same valency, two, and so the number of g-atoms of each deposited in the same time will be the same.

$\dfrac{0·445}{63·5}$ g-atom of Cu is deposited in the same time as $\dfrac{x}{58·7}$ g-atom of Ni.

As these values must be the same

$$\frac{x}{58·7} = \frac{0·445}{63·5}$$

from which $x = \dfrac{0·445 \times 58·7}{63·5} = \underline{0·411 \text{ g}}$

EXAMPLE (d) What mass of lead (relative atomic mass = 207) will be deposited from a solution of lead(II) nitrate by the same current which liberates 1·5 g of chromium (relative atomic mass = 52) from a solution of chromium(III) sulphate?

The valency of lead in lead(II) nitrate is two and that of chromium is three.

1 Faraday will deposit 207/2 g of lead or 52/3 g of chromium so the actual masses deposited must be in this ratio.

i.e. $\dfrac{\text{Mass of lead}}{\text{Mass of chromium}} = \dfrac{207/2}{52/3} = \dfrac{207 \times 3}{52 \times 2}$

the mass of lead $= \dfrac{207 \times 3 \times 1·5}{52 \times 2}$ g

$= \underline{8·969 \text{ g}}$

Electrodes

In most cases the best electrodes, i.e. the most suitable ones, are made of materials unlikely to be affected by the products of an electrolysis or by the electrolyte itself. Platinum is the best in most cases, but it is expensive and nickel is often used in its place. However, when halogens are evolved, most metals are attacked (as in Experiment 38.1) and so, the anode should then be made of carbon (from a dry battery), which is not affected by chlorine or bromine. Carbon can, in fact, be used generally instead of platinum or nickel except as an anode when oxygen is being evolved; it then gets eaten away if the current is passed for a considerable time. Copper is also suitable, but not when used in solutions of copper salts (see page 231). We shall see later that the products obtained may be affected by the materials from which the electrodes are made.

How Electrolysis occurs

When two electrodes connected to a battery are immersed in an electrolyte such as a solution of copper chloride (moderately concentrated) a current flows. Electrons pass from the negative plate of the battery to the cathode which is thus negatively charged and so can attract the positive copper(II) ions in the solution. When these reach the plate they receive electrons from it to form neutral copper atoms which are deposited as the metal on the electrode. At the same time, the negative chloride ions are attracted to the positively-charged anode. Here they give up their electrons which travel back to the battery to complete the circuit (Fig. 38·6). The chloride ions, by loss of an electron, become neutral chlorine atoms. For every copper(II) ion discharged at the cathode there must be two chloride ions discharged at the anode or there will be insufficient electrons returning to the battery to balance those lost by it. For every bivalent ion receiving electrons there must be another bivalent ion (or two monovalent ones) losing electrons. In ionic equations the number of electrical charges on one side must balance those on the other.

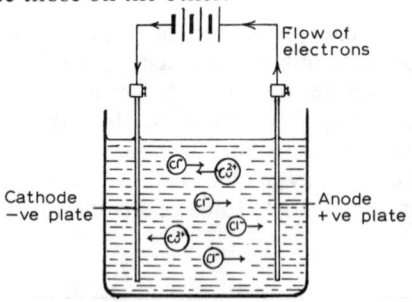

Fig. 38.6. Movement of ions in copper chloride solution

Ionization of Solutions

The following ionic equations illustrate the ways in which *solutions* ionize.

$HCl \rightleftharpoons *H^+ + Cl^-$
$HNO_3 \rightleftharpoons *H^+ + NO_3^-$

*hydrogen ions are usually 'hydrated' in solution, i.e. joined with a molecule of water as the hydroxonium ion, H_3O_M. Correctly, we should write

$$HCl + H_2O \rightleftharpoons H_3O^+ + Cl^-$$
and $$2H_2O \rightleftharpoons H_3O^+ + OH^- \text{ etc.,}$$

but at this stage it will suffice if we still use H^+ to represent the hydrogen ion.

$H_2SO_4 \rightleftharpoons H^+ + HSO_4^- \rightleftharpoons 2H^+ + SO_4^{2-}$
$NaOH \rightleftharpoons Na^+ + OH^-$
$NaCl \rightleftharpoons Na^+ + Cl^-$
$NH_4OH \rightleftharpoons NH_4^+ + OH^-$
$H_2O \rightleftharpoons H^+ + OH^-$

Nitrate and sulphate ions are not discharged from solution (see below); when these are present it is the hydroxide ions from the water which give up their electrons and oxygen is set free.

$$4OH^- \rightarrow 2H_2O + O_2 + 4e^-$$

This explains why the usual product at the anode is oxygen.

Factors affecting the Products of Electrolysis

If the solutions listed in Table 38.3 are electrolyzed it will be found that the products obtained are as shown. Both electrodes are assumed to be of platinum, nickel or carbon unless it is stated otherwise.

These results suggest that there are several factors which influence the nature of the products evolved. The most important of these are considered here.

1. *Position of the Element or Ion in the Electrochemical Series*

As a general rule, the more readily a compound is formed the more difficult it is to decompose it back into its elements, e.g. sodium rusts in air to form the oxide, but no amount of heat on sodium oxide will give back the sodium. Similarly, the more easily an atom forms an ion, the less readily will the ion be discharged from a solution. Look at Table 38.4 which shows the positions of certain ions in the electrochemical series.

Normally, (but see below) only one ion is liberated at a time at an electrode and if a current is passed through a solution containing several of the above ions, the cation and anion set free will be those *lowest* in the table: e.g. hydrogen is usually evolved in preference to sodium being deposited from a solution which also contains hydrogen ions. In a similar way, hydroxide ions give up their electrons at the anode before chloride or sulphate ones.

This is known as *the selective discharge of ions*. As a rule, the change which requires the least energy to bring it about is the one which takes place.

2. *Concentration of the Electrolyte*

If very dilute brine or hydrochloric acid is

Table 38.4

Cations	Anions
Most electropositive	
Na^+	SO_4^{2-}
H^+	NO_3^-
Cu^{2+}	Cl^-
Ag^+	OH^-
Least electropositive	

Table 38.3. Products of Electrolysis under varying Conditions

Electrodes	Solution	Products	
		Cathode	Anode
Pt, Ni or C	dil. NaCl	H_2	O_2
Pt, Ni or C	dil. NaOH	H_2	O_2
Pt, Ni or C	dil. HCl	H_2	O_2
Pt, Ni or C	dil. H_2SO_4	H_2	O_2
Pt, Ni or C	$CuSO_4$	Cu	O_2
Pt, Ni or C	Ag_2SO_4	AG	O_2
Pt, Ni or C	conc. HCl	H_2	Cl
Pt, Ni or C	conc. NaCl	H_2	Cl
Hg cathode	conc. NaCl	Na	Cl
Cu anode	$CuSO_4$	Cu	anode loses Cu
Ag anode	Ag_2SO_4	Ag	anode loses Ag

electrolyzed the anode product is almost pure oxygen. As the concentration of the solution is increased the percentage of chlorine mixed with the oxygen gets larger and a concentrated solution of either gives mainly chlorine with only a little oxygen. Concentration therefore affects the anion set free. (chloride here rather than hydroxide) and may override the natural discharge outline above.

3. Nature of the Electrodes

The material from which the electrodes are made may affect the products liberated (see Table 38.3). Normally, solutions containing sodium ions liberate hydrogen at the cathode, but if a mercury cathode is used—as in one method employed for the production of sodium hydroxide (see p. 259—sodium is discharged instead. Less energy is required for the change

$$Na^+ + Hg + e^- \to Na/Hg \text{ amalgam}$$

than for the one

$$H^+(aq) + e \to H \to H_2(g).$$

Similarly, if a silver anode is used in a solution of silver nitrate (or a copper one in copper(II) sulphate solution), it needs less energy for the metal anode to lose electrons and go into solution than for the hydroxyl ions to be discharged, i.e.

$$Ag(s) \to Ag^+(aq) + e^-$$

or $$Cu(s) \to Cu^{2+}(aq) + 2e^-$$

occurs.

It is now time we looked at some specific cases and considered what actually does take place on electrolysis.

Experiment 38.6. The Electrolysis of Various Solutions and Identification of the Products.

The voltameter to be used is very simple. It consists only of a dish with a central porous partition made of folded blotting paper or (better) of a beaker-mat cut to shape to give a fairly tight fit. This allows the free passage of ions while preventing the mixing of the products of electrolysis. The electrodes are carbon rods (from torch batteries) set in alkathene to support them. The alkathene is melted and then poured into a folded filter paper in a funnel; the carbon rod with wire attached is already in place (see Fig. 38.7; a wide-stemmed funnel needs to be used.)

The solutions to be used in turn are dilute sulphuric acid (acidulated water), dilute brine, dilute sodium hydroxide, concentrated brine, dilute sodium sulphate and copper(II) sulphate

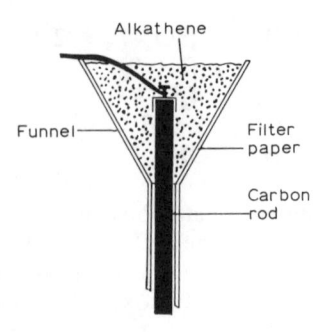

Fig. 38.7. Making an electrode

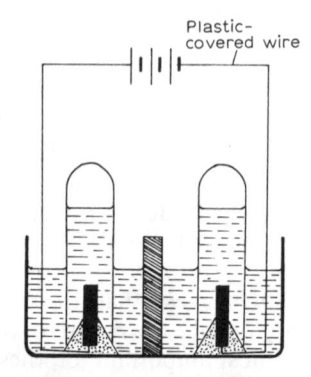

Fig. 38.8. A simple voltameter

and the tubes and dish are filled as in Fig. 38.8. In all cases except the last a little universal indicator or neutral litmus should be mixed with the solution. Although oxygen evolved at the anode will attack the carbon the current will not be running for long enough for much of the rod to be eaten away. (Nickel electrodes can be used instead if desired.) Two 4 V accumulators should be sufficient.

Alternatively, a simple voltameter of the type shown in Fig. 38.9 may be used. The electrodes are carbon rods or nickel foil fastened to nickel wire. The rods may be passed directly through holes in the rubber bungs, but the wire should be passed through narrow glass tubing and the ends of the tube sealed to make the joint water-tight.

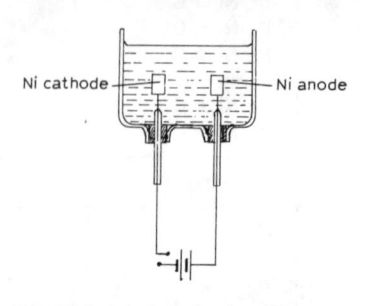

Fig. 38.9. Another simple voltameter

The gases liberated at the anode and cathode can easily be identified if test-tubes of them are collected; hydrogen by the small explosion it makes when burned, oxygen by its ability to relight a glowing splint and chlorine by its colour and bleaching action. In addition the indicator will tell us something about the products in each compartment. At any given time there is a constant number of hydrogen ions and hydroxide ions in a solution due to the ionization (small though it is) of some water molecules

$$H_2O \rightleftharpoons H^+(aq) + OH^-(aq).$$

When hydrogen ions are discharged this equilibrium is disturbed and more water ionizes to try and make good this loss. This leaves an excess of hydroxide ions around the cathode and the solution becomes less acidic or more alkaline here. Similarly, discharge of hydroxide ions at the anode leaves an excess of hydrogen ions in the anode compartment and the indicator there will show an increased acidity. Chlorine given off will bleach the litmus.

Deposition of a metal coating on the cathode can easily be recognised. The electrolysis of copper(II) sulphate solution should be carried out also with copper as the anode.

Dilute Sulphuric Acid (acidulated Water)

H_2SO_4

Electrodes: carbon, nickel or platinum

Ions present:
from the H_2SO_4 H^+ SO_4^{2-}
from the H_2O H^+ OH^-

Action at the Cathode	Action at the Anode
Hydrogen ions move to cathode and are discharged. $$4H^+ + 4e^- \rightarrow 2H_2$$ Discharge of hydrogen ions causes *decrease in acidity* around cathode.	Hydroxide and sulphate ions move to anode where hydroxide ions are discharged. $$4OH^- \rightarrow 2H_2O + O_2 + 4e^-$$ Discharge of hydroxide ions causes *increase in acidity* around anode.

Net result: 2 volumes of hydrogen and 1 volume of oxygen are liberated. The water decomposes but the amount of sulphuric

$$2H_2O \rightarrow 2H_2 + O_2$$

acid remains unchanged. Litmus around the cathode becomes purple or blue; universal indicator becomes less red (possibly orange or yellow) than around the anode.

Sodium Hydroxide Solution NaOH

Electrodes: carbon, nickel or platinum

Ions present:
from the NaOH Na^+ OH^-
from the H_2O H^+ OH^-

Action at the Cathode	Action at the Anode
Sodium and hydrogen ions move to cathode where the hydrogen ions are discharged. $$4H^+ + 4e^- \rightarrow 2H_2$$ Discharge of Hydrogen ions causes *increase in alkalinity* around cathode.	Hydroxide ions are discharged at anode. $$4OH^- \rightarrow 2H_2O + O_2 + 4e^-$$ Discharge of hydroxide ions causes *increase in alkalinity* around anode.

Net result: 2 volumes of hydrogen and 1 volume of oxygen are liberated. The water decomposes

$$2H_2O \rightarrow 2H_2 + O_2$$

Sodium Chloride (i) dilute solution

Electrodes: carbon, nickel or platinum
Ions present:
  from the NaCl Na^+ Cl^-
  from the H_2O H^+ OH^-

Action at the Cathode	Action at the Anode
Sodium and hydrogen ions move to cathode where the hydrogen ions are discharged. $$4H^+ + 4e^- \rightarrow 2H_2$$ Discharge of hydrogen ions causes *increase in alkalinity* around cathode.	Chloride and hydroxide ions move to anode where the hydroxide ions are selectively discharged. $$4OH^- \rightarrow 2H_2O + O_2 + 4e^-$$ Discharge of hydroxide ions causes *increase in acidity* around anode.

Net result: 2 volumes of hydrogen and 1 volume of oxygen are liberated. The water decomposes but the amount of sodium

$$2H_2O \rightarrow 2H_2 + O_2$$

hydroxide remains unchanged. Litmus in the anode compartment becomes purple and universal indicator there becomes less blue, possibly blue green or green. A little chlorine may also be evolved. The solution as a whole remains neutral, but litmus (purple) becomes blue round the cathode (universal indicator green → purple) and both indicators become red around the anode.

(ii) concentrated solution

Electrodes: carbon anode; carbon, nickel or platinum cathode
Ions present:
  from the NaCl Na^+ Cl^-
  from the H_2O H^+ OH^-

Action at the Cathode	Action at the Anode
Sodium and hydrogen ions move to cathode where the hydrogen ions are selectively discharged $$2H^+ + 2e^- \rightarrow H_2$$ Discharge of hydrogen ions causes *increase in alkalinity* around cathode.	Chloride and hydroxide ions move to anode where the increased concentration of chloride ions causes them to be discharged in preference to hydroxide ions $$2Cl^- \rightarrow Cl_2 + 2e^-$$ Presence of hydroxide ions not discharged causes *increase in alkalinity* around anode.

Net result: 1 volume of hydrogen and 1 volume of chlorine discharged (plus a little oxygen). Na^+ and OH^+ ions left in solution which becomes one of sodium hydroxide. Litmus (purple) becomes blue and universal indicator (green) becomes purple throughout the solution, but the indicators are bleached near the anode by the chlorine evolved. Result may be represented as

$$2NaCl + 2H_2O \xrightarrow{electrolysis} NaOH + H_2 + Cl_2$$

Sodium sulphate solution Na_2SO_4

Electrodes: carbon, nickel or platinum
Ions present:
  from the Na_2SO_4 Na^+ SO_4^{2-}
  from the H_2O H^+ OH^-

Action at the Cathode	Action at the Anode
Sodium and hydrogen ions move to cathode where hydrogen ions are selectively discharged $$4H^+ + 4I^- \rightarrow 2H_2$$ Discharge of hydrogen ions causes *increased alkalinity* at cathode.	Sulphate and hydroxide ions move to anode where hydroxide ions are selectively discharged $$4OH^- \rightarrow 2H_2O + O_2 + 4e^-$$ Discharge of hydroxide ions causes *increased acidity* at anode.

Net result: 2 volumes of hydrogen and 1 volume of oxygen are evolved as in the electrolysis of water. Litmus, originally purple, becomes blue at the cathode and red at the anode; universal indicator changes from green to purple at the cathode and red at the anode. In the rest of the solution the colours remain unchanged. The amount of sodium sulphate stays constant.

Copper(II) Sulphate solution $CuSO_4$
Electrodes: (i) carbon, nickel or platinum anode and cathode
(ii) carbon, nickel or platinum cathode, copper anode
Ions present: from the $CuSO_4$ Cu^{2+} SO_4^{2-}
from the H_2O H^+ OH^-

(i) with a 'neutral' anode

Action at the Cathode	Action at the Anode
Hydrogen and copper(II) ions move to cathode where the copper(II) ions are preferentially discharged $$Cu^{2+} + 2e^- \rightarrow Cu$$ Copper is deposited	Hydroxide and sulphate ions move to anode where the hydroxide ions are preferentially discharged $$4OH^- \rightarrow 2H_2O + O_2 + 4e^-$$ Oxygen is liberated

Net result: The cathode receives a coating of copper and oxygen is set free at the anode. As the copper(II) and hydroxide ions are removed the solution of copper(II) sulphate becomes less concentrated and finally is replaced by one of sulphuric acid. Hydrogen is then evolved at the cathode.

(ii) with a copper anode

Action at the Cathode	Action at the Anode
Hydrogen and copper(II) ions move to cathode where the copper(II) ions are preferentially discharged $$Cu^{2+} + 2e^- \rightarrow Cu$$ Copper deposited	Hydroxide and sulphate ions move to anode. There are three possibilities here: (a) Hydroxide ions discharged (b) Sulphate ions discharged (c) $Cu \quad Cu^{2+} + 2e$ The last reaction requires the least energy and so is the one that takes place; the copper anode goes into solution as *copper(II) ions.

Net result: The anode gradually dissolves as the copper changes to copper(II) ions which go into solution. Copper(II) ions in the solution travel to the cathode on which copper is deposited. The copper(II) sulphate solution remains constant in strength. When the anode is all 'eaten away' the current ceases to flow.

*These ions are hydrated and exist as $(Cu.4H_2O)^{2+}$ ions.

Industrial Applications of Electrolysis

(i) Extraction of metals

Metals high in the electrochemical series cannot be obtained from their ores by reduction by chemical means; this has to be brought about by electrolytic methods.

Sodium (see p. 251), calcium, magnesium and aluminium (p. 252) are obtained in this way.

(2) Refining of metals

When metals are obtained from their ores they are rarely pure, but contain varying amounts of impurities derived from the original ore. Refining is often carried out by electrolysis and this is the case with copper.

The impure copper ingot is made the anode in a bath of dilute sulphuric acid (10%) containing about 3-4% of copper(II) sulphate. The cathode is a thin sheet of pure copper. A heavy current is passed and as the anode goes into solution pure copper is deposited on the cathode. Inactive metals such as gold, silver and platinum which may be present in the anode, collect as a muddy deposit on the bottom of the vessels. Active metals such as iron, zinc and tin remain in solution as ions in the electrolyte.

(3) Electroplating

The deposition of a coating of a metal such as nickel, chromium, cadmium, copper and even silver or gold on the surface of iron is frequently adopted in order to give a better appearance, improve wearing properties or prevent corrosion. An alloy such as stainless steel does not corrode, but is expensive and a coating of chromium or nickel may be cheaper. Iron, which is to be chromium plated, first has coats of copper and then nickel deposited on it, each being about 0·002 cm thick. Silver plating has an undercoat of copper.

In carrying out the plating, maximum adhesion of the metallic layer is achieved by:
  (i) cleaning (pickling) the object in acid or alkali,
  (ii) having a weak solution of the electrolyte,
  (iii) using a low current and low voltage.

The object to be plated is made the cathode and the metal to be deposited forms the anode. The electrolyte contains ions of the metal which is to form the coating.

In recent years it has been found possible to deposit both plastics and rubber upon metal surfaces, e.g. kitchen plate racks, thus preventing corrosion and, if necessary, providing insulation. Particles of these substances in colloidal solution carry electric charges and so can be attracted towards an electrode such as a metal rack.

(4) Electrotyping

A newspaper usually has to be printed from several machines at the same time and this can only be done if the pages of type or illustrations are duplicated. It would be costly and time consuming to prepare several sets of each. Instead, wax impressions of each original are made. These are then coated with graphite (a good conductor) and made the cathode in a cell containing copper(II) ions. A layer of copper is then deposited on them. When this is reasonably thick the wax is melted and the copper impression backed by a metal sheet to give strength. This process is known as electrotyping, the copies being called stereos. In the case of a book, the original type can then be broken up for further use. A similar process is employed to make copies of medals or low-relief plaques.

(5) Anodizing

With a low current and a low voltage most metals (except platinum) tend to go into solution if they are used as anodes in electrolysis. At higher voltages, 4–6 V, these metals are rendered passive (see p. 136) by a coating of the oxide being formed while oxygen is being evolved from the solution. This film 'seals' the metal and so prevents its solution. If aluminium is used as the anode in a bath of chromic acid, an oxide film is deposited which is very adherent—the aluminium is 'anodized'. If a dye is added to the bath the film is coloured. Many aluminium articles (e.g. pans and kettles) are coloured in this way.

Summary

Only electrovalent substances conduct an electric (d.c.) current. They must be fused or in solution.

Water, although a poor conductor itself, plays a vital part in electrolysis.

With solutions of metal ions above hydrogen in the electrochemical series hydrogen is usually evolved at the cathode on electrolysis. If the solution contains metal ions below hydrogen then the metal is deposited.

The usual anodic product from a solution is oxygen, but this may be affected by the concentrations of the solution or the nature of the electrodes.

The mass of an element liberated during electrolysis depends on the amount of electricity passing, the gram atomic mass of the element and its valency.

Ions receive electrons from the cathode and give them up at the anode. The change $Cl^- \to Cl$ (atom) is oxidation (loss of electron) and this occurs at the anode. The production of hydrogen (gas) from hydrogen ions is one of reduction (gain of electrons) and occurs at the cathode.

Electro-chemical Series

We have frequently talked about the electrochemical series and here is a simple method by which we can determine the order in the series of some metals. It is limited only by the number of suitable metals available.

Experiment 38.7. Arranging the Metals in order of activity.

Cut a sheet of copper as shown in Fig. 38.10, about 30 cm long, and fasten it to a piece of wood. On the opposite side of the batten are attached strips of magnesium, aluminium, zinc, iron, tin and lead and these are connected together. Magnesium wire or ribbon can be wound round a piece of wood

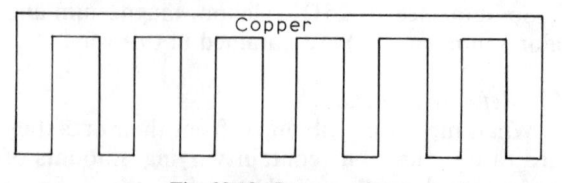

Fig. 38.10. Copper strip

as may tin foil if sheet tin is not available (Fig. 38.11).

The wires from the two sides are connected to a small voltmeter and a 250 cm³ beaker of bench hydrochloric acid is brought up under each pair of metals in turn and the voltage noted. It is found to increase from Cu/Cu (zero) to Mg/Cu and this is the order in which the metals occur in the electrochemical series.

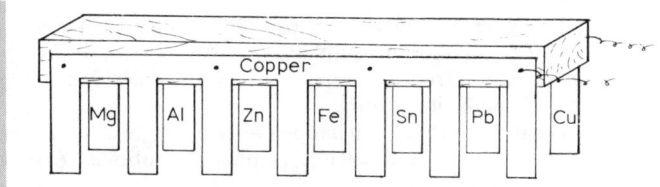

Fig. 38.11. Metals arranged on a wooden block

Exercise 38

1 Draw a simple diagram of a cell in which you could electrolyze concentrated hydrochloric acid.

Label clearly on your diagram:

(a) The two electrodes as anode and cathode, showing which is positive and which negative, and the materials from which they are made.

(b) the products formed at each electrode.

Now attempt to write simple ionic equations to show how the two products are formed.

Why would you not make the electrodes in this cell of the metal zinc?

Why is the metal sodium not obtained when a *solution* of common salt is electrolyzed? (*EM*)

2 Give the formulae of the ions present in a solution of copper(II) sulphate, and state the sign and number of charges carried by each ion. Describe and explain what happens when copper sulphate solution is electrolyzed, using copper electrodes. (*OC*)

3 An electric current is passed for a moderate time between platinum electrodes dipping into (i) a solution of copper sulphate, (ii) water containing a little sulphuric acid. State what you would observe in each case, name the substances appearing at the anode and at the cathode, and say how the composition of each liquid changes. (*C*)

4 State Faraday's laws of electrolysis.

Describe and explain what happens when a direct current is passed through dilute sulphuric acid, using platinum electrodes. Sketch a suitable apparatus with which this can be done and name the gaseous products collected at each electrode. (*J*)

5 (a) Describe what you would observe, and explain what happens, when two platinum plates connected to the terminals of an electric battery are dipped into (i) water containing a little sulphuric acid, (ii) a solution of copper(II) sulphate.

(b) In the second case, what difference would it make if the positive plate were made of copper instead of platinum?

(c) What is this process called, and what names are given to the solution and to each of the metal plates? (*C*)

6 State Faraday's laws of electrolysis.

Describe and explain what happens when a direct current is passed through a solution of copper(II) sulphate, using electrodes of (a) copper, (b) platinum. (*J*)

7 What happens when water containing a little sulphuric acid is electrolyzed? Sketch the apparatus you would use to demonstrate the truth of your statements. How can the observations be explained?

How does water react with (a) calcium oxide, (b) anhydrous copper(II) sulphate? (*OC*)

8 A current of electricity is passed through two cells fitted with platinum electrodes and connected in series. Cell A contains sodium sulphate solution, cell B copper(II) sulphate solution. Draw and label a diagram of the cells and connexions. Name the products formed at each of the electrodes.

State Faraday's laws of electrolysis. (*O*)

9 What is meant by the term *electrochemical equivalent*? Draw a labelled circuit diagram of the apparatus you would use to determine the electrochemical equivalent of copper experimentally. Make a list of all the measurements you would take.

Calculate the time required to deposit 0·5 g of copper in a copper voltameter using a current of 1·5 A.

(Electrochemical equivalent of copper: 0·000 329 g/C.) (*S*)

10 (a) A 6 V d.c. supply is applied across a 6 V bulb joined in series with copper electrodes dipping into a beaker containing copper(II) sulphate (cupric sulphate) solution. When a solution conducts electricity, a reaction can usually be seen taking place at the electrodes but, for the circuit described, the bulb glows despite the fact that no action seems to occur in the beaker. Explain fully how current passes through the solution. What experimental evidence can be used to support your theory?

(b) If the beaker and copper electrodes in the series circuit in (a) are replaced by a crucible filled with powdered lead(II) chloride crystals into which dip carbon electrodes, the bulb fails to light.

If the lead(II) chloride is now melted, the bulb glows and a gas, which bleaches moist litmus paper, is evolved at the anode. There appears to be no deposit on the cathode, but a layer of metal is found in the bottom of the crucible after the experiment.

Account for these observations as fully as you can.

(*C*)

11 Describe and explain (*a*) what happens when a solution of sodium sulphate is electrolyzed, using platinum electrodes; (*b*) ONE test by which the sulphate radical can be identified.

What, if anything, will happen when a sample of water containing dissolved calcium sulphate is (i) boiled, (ii) treated with a solution of sodium carbonate? (*OC*)

12 (*a*) Distinguish between (i) an atom, (ii) a molecule, (iii) an ion.

(*b*) How would you show that a solution of hydrogen chloride in benzene contained no ions?

(*c*) Draw a diagram of an apparatus you would use to electrolyze acidulated water. Show clearly
  (i) the electrodes and their polarity,
  (ii) the products of the electrolysis,
  (iii) the approximate voltage used,
  (iv) the direction of the current.

(*d*) Give the formulae of the ions present in (*c*), together with their charges.

13 When a porcelain basin containing lead(II) chloride crystals has two carbon rods placed in it, these carbon rods being joined in series with a 6 V accumulator and an ammeter, no current flows. But when the basin is heated to 500°C the lead(II) chloride melts, the ammeter gives a reading, a greenish yellow gas, which bleaches moist litmus, bubbles off the anode and a pool of molten metal collects under the cathode.

What is the greenish yellow gas?
What metal collects under the cathode?
This experiment suggests that the liquid contains particles that possess electric charges. What are their formulae?
Why do the crystals of lead(II) chloride not conduct electricity? (*C*)

14 When platinum electrodes that are connected to a 6V accumulator are dipped into water containing a little sulphuric acid, effervescence occurs at both electrodes.

What gas is evolved at the anode? What gas is evolved at the cathode?
Write the formulae of the ions that are present in the solution
Which ions migrate towards the anode?
Give the equations for the reactions that occur at the electrodes.
In a certain time, it is found that 10 cm³ of gas are evolved at the anode. What volume of gas is evolved at the cathode in the same time and under the same conditions?
What differences would be seen if lead electrodes are used in this experiment in place of platinum?
What would be observed if the plates were now connected to a voltmeter?
Which plate should be joined to the positive terminal of the voltmeter? (*L*)

39
Chemical Cells

Although almost all the electrical energy consumed in industry and in the home is produced by dynamos at power stations, it is sometimes convenient to operate electrical appliances without connecting them to a mains electricity supply point. Some examples of this are: portable radio receivers, pocket torches, all the electrical equipment in motor vehicles, electrical apparatus in rockets and satellites. In all these instances, and many others which you can probably think of, chemical batteries are used to provide the electrical energy. Batteries are devices which cause a direct conversion of chemical energy into electrical energy. In a power station the starting point is also chemical energy, stored in the coal or oil fuel. This is changed into heat by combustion which subsequently produces mechanical energy in the turbine and electricity in the dynamo (Fig. 39.1). Many types of battery have been used since the first one was demonstrated in 1799 by Volta. Most of these are of no practical importance today and we shall concern ourselves mainly with those which are

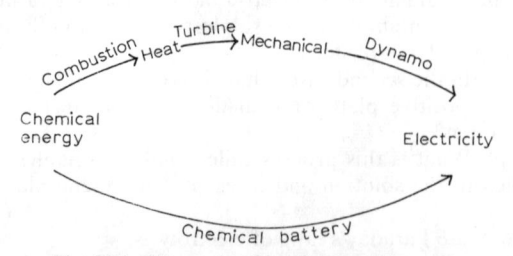

Fig. 39.1. The production of electrical energy

widely used. However, we shall first carry out experiments with a cell which, although of no practical significance, does show us some of the difficulties involved in the production of an efficient battery.

Experiment 39.1. Set up a *Simple Cell* as shown in Fig. 39.2 consisting of zinc and copper plates and dilute sulphuric acid. Connect the plates to a 1·5 V torch bulb. You will probably observe that the bulb glows dimly for a short while before going out. Now add some potassium dichromate(VI) (dichromate) to the acid in the cell. You will see that the bulb glows quite brightly and continues to do so for some time. While the bulb is still glowing disconnect it and replace it by a voltmeter. You should get a reading a little over 1 volt. If you reconnect the bulb in circuit you will find that in due course it gets dimmer before finally going out.

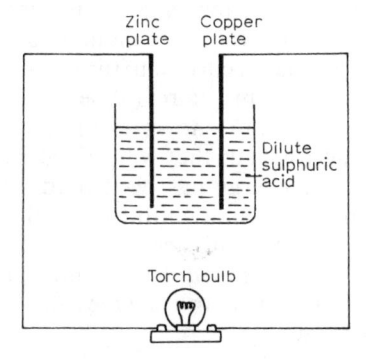

Fig. 39.2. A simple cell

This experiment shows that when the cell is set up, a potential difference is established between the two plates. This potential difference causes a current fo flow when the cell is connected in a complete circuit. In fact we call the copper plate the positive electrode and the zinc plate, the negative. We say that the current flows from positive to negative. Both of these last statements are matters of convention. We might just as easily have labelled things the other way round.

The experiment also shows one of the principal defects of this and most other cells. In order that the copper plate shall remain at a positive potential, positive charges must be continuously arriving at it in the cell, similarly negative charges must be arriving at the zinc electrode (Fig. 39.3). These charges are attached to hydrogen and sulphate ions of the dilute sulphuric acid which ionizes thus:

$$H_2SO_4 \rightarrow 2H^+ + SO_4^{2-}$$

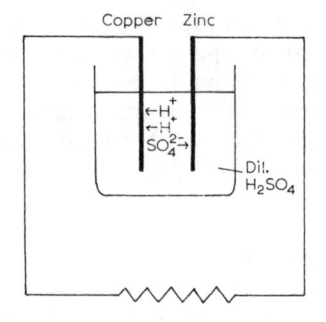

Fig. 39.3. Movement of ions in a simple cell

Do not be confused by the fact that the copper plate at which the hydrogen ions are liberated is referred to here as the *positive* plate, whereas positive hydrogen ions are, of course, always attracted to a *negative* plate when a battery is in the circuit. In the case considered here the plates are originally neutral, only becoming positive or negative when the ions are liberated because of the potential difference between the plates.

Thus hydrogen is liberated at the copper plate and a layer of bubbles of the gas forms on it. This effect is known as *polarization*. You should note that this is the opposite of electrolysis in which positive hydrogen ions are liberated at the cathode or negative electrode. This layer of gas causes the cell to cease operating in two ways. It is an insulator and it therefore very much increases the resistance of the cell. Also it acts as an electrode of a cell with the zinc plate and the dilute sulphuric acid, the voltage of which is opposed to that of the copper-zinc cell. Thus the net voltage of the cell is greatly reduced. We are now in a position to see why adding potassium dichromate(VI) ($K_2Cr_2O_7$) restored the cell. It acts as an oxidizing agent, some of its oxygen combining with the hydrogen gas to give water. In this way the layer of gas is prevented from forming. The potassium dichromate(VI) is known as a depolarizer. Of course the effect of the depolarizer will only last quite a short while and the cell will again cease to function.

When you set up your simple cell you will probably observe a great deal of gas being liberated at the zinc plate and this will occur whether a current is being taken from the cell or not. This is hydrogen gas being evolved and zinc at the same time is going into solution as zinc sulphate. So if the cell is left set up for any length of time the zinc electrode will dissolve completely. It is clear that in any practical cell it is essential that the chemical reactions in the cell should only take place while it is in use. If the zinc in our simple cell were completely pure, this would be the case. In practice, the zinc itself, impurities in it and the acid, form tiny cells in the electrode which operate continuously. This is known as local action. It may be prevented either by using pure zinc or by amalgamating the plate with mercury. If it is rubbed with mercury it is given a coating which prevents the acid reaching the impurities.

The Leclanché Cell

The dry battery used in torches, cycle lamps and many other applications is a version of the Leclanché cell. It represents a very big improvement on the simple cell we have been describing. The wet Leclanché cell (very little used today) is shown in Fig. 39.4. It consists of zinc and carbon electrodes with a solution of ammonium chloride as the electrolyte. Once again the problem of polarization arises. Hydrogen collects at the carbon (positive) electrode. This is therefore surrounded with manganese dioxide powder which oxidizes the hydrogen to form water. The manganese(IV) oxide depolarizer is contained in a porous pot which surrounds the carbon electrode. The current through the cell would not readily pass through manganese(IV) oxide alone; it is therefore mixed with powdered carbon and the porous pot is sealed with a layer of pitch.

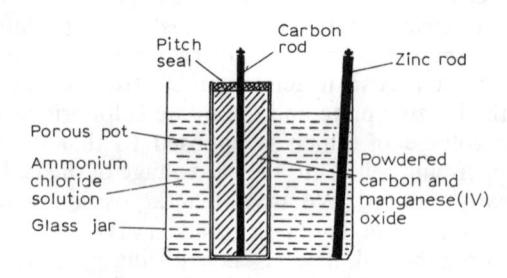

Fig. 39.4. A 'wet' Leclanché cell

The dry cell (Fig. 39.5) is chemically identical to the wet Leclanché. In it the negative zinc electrode is the case itself. The electrolyte is a paste of ammonium chloride and the carbon rod is surrounded by a muslin bag containing the powdered carbon and manganese(IV) oxide. The whole cell is sealed with a layer of pitch. Both the wet and dry Leclanché cells produce a maximum potential difference of 1·5 V. In many applications a larger potential difference than this is required. In these cases a number of cells are joined together in series. For many uses in electronics, 'high tension' batteries are used. They may consist of 80 cells (120 V) or more.

Dry batteries have two important limitations. The maximum current that may be taken from

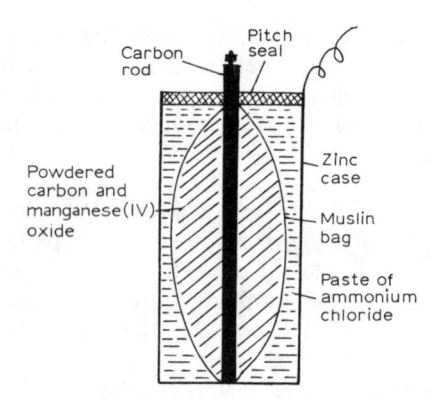

Fig. 39.5. A dry Leclanché cell

them is limited to, at most, an ampere or two. Also when the chemical energy stored in them has been used, they can only be thrown away.

Accumulators

On account of the limitations mentioned above, dry batteries are quite unsuitable for use in motor vehicles. The starter motor alone in a motor car requires a very large current to turn it. A car battery is thus of an entirely different type. The Leclanché and simple cells are known as primary cells. As soon as they are set up they will give current. When all the stored energy has been used up there is nothing that can be done to restore them. A car battery is a lead-acid accumulator which is a secondary or storage cell. The essential differences between the two types will emerge from the following experiment.

Experiment 39.2. Set up a cell consisting of two lead plates in sulphuric acid diluted to a relative density of about 1·2. Connect it in circuit with 1·5 V torch bulb (Fig. 39.6). You will find that the bulb does *not* light up. Now disconnect the bulb and connect the cell instead to the terminals of a low voltage D.C. supply (about 5 V) (Fig. 39.7).

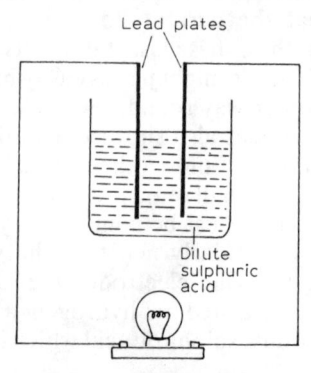

Fig. 39.6. A lead-acid secondary cell—discharging

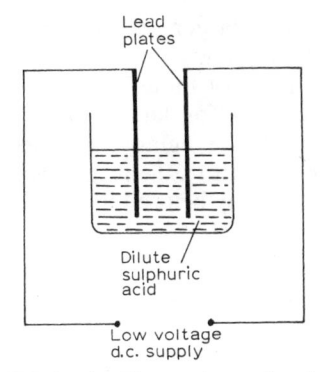

Fig. 39.7. Lead-acid secondary cell—charging

A current will flow through the cell causing gas to be given off at the plates. Pass the current for five or ten minutes after which time test the cell again by connecting it to the torch bulb. It should glow quite brightly. Test the cell also with a voltmeter. It should show a reading approaching 2 V.

You can now see why the lead-acid cell is known as a secondary or storage cell. It is only when it has been charged by passing a current through it that it will give a current, but the charging process may be repeated an indefinite number of times. When the cell is charged it is effectively storing electrical energy.

If you look at the plates of your accumulator after charging you should see that one of them (the one that was connected to the positive terminal of the low voltage supply) has been changed from grey to dark brown. It is the positive electrode of the accumulator and it has been coated by electrolysis during the charging process with brown lead(IV) oxide (PbO_2).

Commercial Lead-Acid Accumulators and their Maintenance

The plates of a car battery do not start life as solid lead as did those in our experimental arrangement. The positive plate consists of a grid packed with a paste of lead(IV) oxide. A number of positive and negative plates are used in each cell as shown in Fig. 39.8. Positives and negatives are prevented from touching by wooden separators. The whole thing is contained in a glass (or with car batteries, rubber vessel filled with sulphuric acid.

Acid Density

During discharging, i.e. when a current is being taken from the battery, both electrodes become coated with lead(II) sulphate. This means that sulphate ions are taken from the electrolyte with the result that its density drops. This change in density of battery acid provides a means of telling how well charged the battery is. When fully charged the density should be 1.25 g/cm^3; when discharged it may drop to 1.15 g/cm^3.

Care of Lead-Acid Accumulators

(a) The liquid in an accumulator should be kept topped up. Liquid lost due to evaporation will be mostly water since sulphuric acid does not readily evaporate. Thus distilled water needs to be added.
(b) A lead-acid cell should not be short circuited. If it is, a very large current is taken from it causing it to become hot and the plates to buckle.
(c) An accumulator should not be left in a discharge condition for a long period of time. Permanent damage will result if it is.

Fuel Cells

We mentioned in discussing the simple cell that a layer of hydrogen on the copper plate acted as an electrode to form a cell with the zinc plate and sulphuric acid. Research has been carried on in recent years to investigate the possibility of making useful cells with gaseous electrodes. The advantage here would be that gas could be continuously replaced as it is used up. In fact the gas supplied is acting as a fuel but the conversion is directly into electricity and not via heat and mechanical energy as in Fig. 39.1. It is therefore possible that much greater efficiency may be obtained by generating electricity in this way than by other methods. One of the big problems about fuel cells is in making good electrical contact with gases. Fig. 39.9 shows

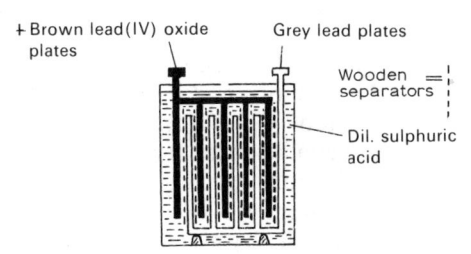

Fig. 39.8. The structure of a lead-acid accumulator

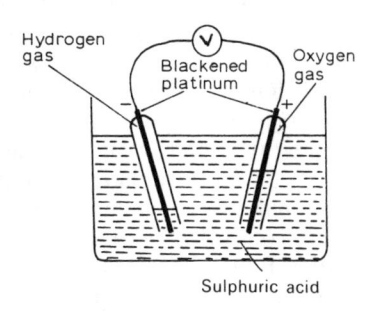

Fig. 39.9. The principle of the Grove fuel cell

the principle of a fuel cell described as long ago as 1839 by Sir William Grove. The gases used were hydrogen and oxygen, the electrolyte sulphuric acid and strips of blackened platinum were used to make contact with the gas. Grove found that such a cell produced about 1 V but that only a minute current could be obtained. Much modern research has been devoted to replacing the blackened platinum strips with porous materials which make possible a much larger area of contact with the electrolyte and gases.

Although it is not thought likely that fuel cells will ever be used for large scale generation of electricity, they are being developed for use in manned spacecraft. Scientists also hope that eventually they may be used in some vehicles such as milk floats which are at present propelled by electricity obtained from accumulators.

Exercise 39

1 Draw a labelled diagram of a Leclanché cell (dry type). State ONE advantage and ONE disadvantage of this type of cell. Give reasons for the use of accumulators rather than dry cells for a motor-car battery. *(C)*

2 Describe a simple voltaic cell and by reference to the cell explain the terms *local action, polarization, back e.m.f. due to polarization*. How is the effect of polarization reduced in a Leclanché cell. *(C)*

3 Describe and explain a method of finding out how well-charged a car battery is. What care and maintenance does such a battery require?

4 What are the advantages of fuel cells? What are the difficulties involved in these cells and what practical applications may they have?

40
Standard Solutions; Volumetric Analysis and pH

Saturated solutions are rarely used except when growing crystals. For most chemical reactions and particularly in volumetric analysis more dilute ones are most suitable and it is convenient to have some standard of comparison.

Any solution in which a known mass of solute is dissolved in a definite volume of liquid is called a *standard solution*, e.g. 6 g/dm³ (litre).

Solutions are often made by dissolving the relative molar mass (gram-formula mass) of the solute in water and making the volume up with more distilled water to one litre. Such a solution is termed a *molar solution*. Equal volumes of molar solutions contain an equal number of 'molecules' or the correct number of ions to give this number of 'molecules'. In considering reactions between acids, bases, carbonates, metals you must be familiar with the equation for such reactions as these tell us the number of moles of the different reagents involved. We shall need to make use of the molar masses shown in the table on p. 239.

A molar solution, then, contains one mole of solute per dm³ of solution and a 2 molar solution would contain 2 moles per dm³; however, these solutions are too concentrated for most reactions involving neutralization and ones containing only one-tenth or one-twentieth of a mole per dm³ are more often used, these would be 0·1 molar and 0·05 molar respectively.

Hydrochloric acid	HCl	36·5
Nitric acid	HNO_3	63
Sulphuric acid	H_2SO_4	98
Lithium hydroxide	LiOH	24
Sodium hydroxide	NaOH	40
Potassium hydroxide	KOH	56
Ammonium hydroxide	NH_4OH	35
Sodium hydrogencarbonate	$NaHCO_3$	84
Sodium carbonate	Na_2CO_3	106

Learn the following relationships which we shall use in our calculations:

(a) Molarity $= \dfrac{\text{Grams per dm}^3}{\text{Molar Mass}}$

(b) Molar mass $= \dfrac{\text{Grams per dm}^3}{\text{Molarity}}$

(c) Grams per dm^3 = Molarity × Molar mass

Using the molar masses given in the table and the above relationships try to solve the following simple problems; you may find that you can do them in your head, in any case only a minimum of writing should be put down.

A. What mass of sulphuric acid per dm^3 would make a solution of molarity: (a) 1; (b) 0·1; (c) 0·25; (d) 0·4?

B. What mass of solute will be present in the following volumes of sodium hydroxide solution of the given concentrations: (a) 600 cm^3 M; (b) 500 cm^3 0·1M; (c) 700 cm^3 0·8M; (d) 250 cm^3 0·12M?

C. What volumes of solution will be represented by the following: (a) 5·6 g M potassium hydroxide; (b) 6·125 g 0·5M sulphuric acid; (c) 5·3 g 0·1 M sodium carbonate; (d) 6·3 g 0·1M nitric acid; (e) 7·3 g 0·5M hydrochloric acid?

D. What mass of solute per dm^3 is represented by the following: (a) 2M lithiumhydroxide; (b) 0·4M hydrochloric acid; (c) 0·8M sodium hydroxide; (d) 0·08M sodium carbonate; (e) 0·05M sodium hydrogen carbonate?

Experiment 40.1. Preparation of a decimolar solution of sodium carbonate.

Sodium carbonate has a relative molar mass of 106 and a decimolar solution of it will therefore contain 10·6 g per dm^3 or 2·65 g in 250 cm^3.

Take several grams of anhydrous sodium carbonate and heat it for a few minutes in a dish to drive off any moisture and then allow it to cool in a desiccator. Transfer it to a dry weighing bottle and find the exact mass of the bottle and its contents. Sodium carbonate must now be transferred by a spatula from the bottle to a clean dry beaker until the new mass is 2·65 g less than the original. Add some warm distilled water to the powder and stir until all goes into solution. Now pour it via a funnel into a 250 cm^3 volumetric flask, washing out the beaker and funnel and transferring the washings to the flask. Add distilled water to the flask until it nearly reaches the mark on the neck. When the solution is quite cold add a little more water from a dropping pipette or wash-bottle so that the volume of solution is exactly 250 cm^3.

The acid to be neutralized by the sodium carbonate may be hydrochloric or sulphuric. Neither can be conveniently weighed out as the former is volatile and the latter is hygroscopic and absorbs water even while it is being weighed. It is better to calculate the volume of acid containing the required mass and to run this out from a burette (one burette can serve all the class).

Experiment 40.2. Preparation of a Standard Solution of Sulphuric Acid.

The relative density of pure sulphuric acid is 1·84 (check this figure on the bottle from which the acid is obtained) and a decimolar solution of it will contain 9·8 g per dm^3. This mass is contained in a volume of about 5·4 cm^3. Run about 5·5 cm^3 of the concentrated acid from a burette into a 1 dm^3 flask containing a little distilled water (why?) and then make the solution up to the dm^3 mark. Shake to ensure thorough mixing. This solution will be approximately decimolar.

Keep these solutions and let us have a look at the method used for finding the molarity of a solution of unknown concentration. In all cases a definite volume (usually 25 cm^3 or 10 cm^3) of one reagent is placed in a conical flask and the other reagent added gradually from a burette until neutrality is achieved. This method is known as *titration*.

Let us consider the reaction between sodium hydroxide and hydrochloric acid.

$$NaOH + HCl \longrightarrow NaCl + H_2O$$
$$\text{1 mole} \quad \text{1 mole}$$
$$\text{40 g} \quad \text{36·5 g}$$

From the equation we see that one mole of sodium hydroxide neutralizes one mole of hydrochloric acid. A molar solution of the alkali contains 40 g of it per dm^3 of solution. A molar solution of the acid contains 36·5 g of it per dm^3 of solution. Therefore, one dm^3 of molar sodium hydroxide will neutralize one dm^3 of molar hydrochloric acid, i.e. equal volumes of the two solutions will neutralize each other.

This means that in titrating one of the solution

against the other the volumes used will be in the inverse ratio to the molarities, i.e.

Volume NaOH × molarity NaOH =
                    volume HCl × molarity HCl

If three of the four quantities here are known, then the fourth can be calculated.

Let us now look at the reaction between sodium carbonate and sulphuric acid.

$$Na_2CO_3 + H_2SO_4 \longrightarrow Na_2SO_4 + H_2O + CO_2$$
$$\text{1 mole} \quad \text{1 mole}$$
$$\text{106 g} \quad \text{98 g}$$

Again, one mole of the alkali neutralizes (or is equivalent to) one mole of the acid, i.e.

$$106 \text{ g } Na_2CO_3 \equiv 98 \text{ g } H_2SO_4$$

so 1 dm³ M $Na_2CO_3 \equiv$ 1 dm³ M H_2SO_4 and any volume M Na_2CO_3. $\equiv$ the same volume M H_2SO_4 or:

Volume Na_2CO_3 × molarity Na_2CO_3 $\equiv$
      Volume H_2SO_4 × molarity H_2SO_4.

This follows (as above) since if the alkali is twice as concentrated (2M) only half the volume will be required.

Study the following examples before completing the experiment.

(a) *If 20 cm³ of 0·1 M hydrochloric acid neutralize 25 cm³ of aqueous potassium hydroxide, what is the molarity of the latter and how many grams of potassium hydroxide are there in one dm³ of the solution?*

As hydrochloric acid and potassium hydroxide react

$$KOH + HCl \longrightarrow KCl + H_2O$$

one mole of the alkali neutralizes one mole of the acid; we can therefore write

$$20 \times 0\cdot1M \text{ (acid)} = 25 \times m \text{ (alkali)}$$
or $\qquad m = \underline{0\cdot08 \text{ M}}$

i.e. one dm³ of solution contains 0·08 mole of potassium hydroxide or

$$0\cdot08 \times 56 \text{ g} = \underline{4\cdot48 \text{ g}}$$

(b) *25 cm³ of 0·1 M sodium hydroxide neutralize 21 cm³ hydrochloric acid. Find the concentration of the latter and calculate how much water must be added to a dm³ of it in order to make it exactly decimolar.*

$$NaOH + HCl \rightarrow NaCl + H_2O$$
$$25 \times 0\cdot1M \text{ (alkali)} = 21 \times m \text{ (acid)}$$
$$m = \underline{0\cdot119M}$$

or

Grams per dm³ = molarity × molar mass

$$= 0\cdot119 \times 36\cdot5 = \underline{4\cdot34 \text{ g}}$$

If 4 cm³ of water are added to each 21 cm³ of the acid the solution will become exactly decimolar as then the neutralizing volumes will be equal. Therefore, the amount of water to make one dm³ of the solution decimolar will be

$$4 \times \frac{1000}{21} = 190 \text{ cm}^3$$

It can also be calculated in the following way: 1000 cm³ 0·119M HCl $\equiv$ V cm³ 0·1M HCl from which $\qquad$ V = 1190 cm³
so, the water to be added = $\underline{190 \text{ cm}^3}$

(c) *If 25 cm³ of 0·15 M sodium carbonate are neutralized by 30 cm³ of sulphuric acid find the molarity of the acid.*

$$Na_2CO_3 + H_2SO_4 = Na_2SO_4 + H_2O + CO_2$$

As one mole of sodium carbonate reacts with one mole of sulphuric acid we can write

$$25 \times 0\cdot15M = 30 \times m$$

$$m = \left(\frac{25 \times 0\cdot15}{30}\right) M = \underline{0.125M}$$

Experiment 40.3. Finding the Molarity of the Sulphuric Acid.

Place 25 cm³ of the decimolar sodium carbonate solution into a clean conical flask and add 3–4 drops of a suitable indicator such as methyl orange which will go yellow. Fill the burette with the acid, making certain that the tube below the tap is also full. Note the level of the meniscus. Run the acid into the flask fairly quickly shaking the flask with the other hand, until the indicator *just* turns a reddish orange. Read off the volume used and take this is a guide for the next more accurate titration.

Repeat these titrations using fresh 25 cm³ portions of the alkali and running in almost the full amount of the acid needed for neutralization and then add the last small amount a drop or two at a time. Do this until two successive readings do not differ by more than 0·2 cm³. Enter your results as below (using your own figures) and work out the molarity of the acid.

Final reading Initial reading (cm³)	24·8 0·0	49·1 24·8	24·0 0·0	48·2 24·0
Acid used (cm³)	24·8	24·3	24·0	24·2
	rough		average	

$$= 24\cdot1 \text{ cm}^3$$

We have 25 (carbonate) × 0·1 M
$$= 24\cdot 1 \text{ (acid)} \times m$$
From this
$$m = \frac{25 \times 0\cdot 1}{24\cdot 1} M$$
$$= 0\cdot 104 M$$

The reactions we have so far looked at are ones in which one mole of A reacts with one mole of B, but this is not always the case. Consider the reaction between sodium hydroxide and sulphuric acid.

$$\underset{\text{2 moles}}{2NaOH} + \underset{\text{1 mole}}{H_2SO_4} \to Na_2SO_4 + 2H_2O$$

Here 80 g of sodium hydroxide are equivalent to or react with 98 g sulphuric acid
i.e. 2 dm³ of M NaOH ≡ 1 dm³ M H_2SO_4 so any volume of M NaOH ≡ half the same volume of M H_2SO_4.

Had we used M calcium hydroxide and M hydrochloric acid we should find from the equation

$$Ca(OH)_2 + 2HCl \to CaCl_2 + 2H_2O$$

that any volume of M $Ca(OH)_2$ ≡ twice the same volume of M HCl.

Let us look at further examples.

(d) *If 25 cm³ of 0·1 M sodium carbonate are neutralized by 20 cm³ of hydrochloric acid, what is the molarity of the acid?*

$$\underset{\text{1 mole}}{Na_2CO_3} + \underset{\text{2 moles}}{2HCl} \longrightarrow 2NaCl + H_2O + CO_2$$

Here 1 mole of the carbonate reacts with 2 moles of the acid.

25 cm³ of 0·1M Na_2CO_3 contain $0\cdot 1 \times \frac{25}{1000}$
$$= 0\cdot 0025 \text{ mole } Na_2CO_3$$

∴ 20 cm³ of the acid must contain $2 \times 0\cdot 0025$
$$= 0\cdot 005 \text{ mole hydrochloric acid}$$

1 dm³ of the acid will contain $\frac{1000}{20} \times 0\cdot 005$
$$= 0\cdot 25 \text{ mole}$$

The acid is therefore 0·25 molar.

(e) *If 10 cm³ of 0·12 M sodium hydrogencarbonate require 6 cm³ of sulphuric acid for complete neutralization, what is the molarity of the acid and its concentration in grams per dm³?*

$$\underset{\text{2 moles}}{2NaHCO_3} + \underset{\text{1 mole}}{H_2SO_4} \to Na_2SO_4 + 2H_2O + 2CO_2$$

10 cm³ of 0·12M $NaHCO_3$ contain $\frac{10}{1000} \times 0\cdot 12$ mole of $NaHCO_3$ = 0·0012 mole.

As the number of moles of the acid reacting is only half that of the alkali we can say

6 cm³ of H_2SO_4 contain $0\cdot 5 \times 0\cdot 0012$ mole
$$= 0\cdot 0006 \text{ mole } H_2SO_4$$

1 dm³ of the acid will contain

$$\frac{1000 \times 0\cdot 0006}{6} \text{ mole } H_2SO_4$$

$$= 0\cdot 1 \text{ mole acid} = \underline{9\cdot 8 \text{ per dm}^3}$$

The method of working out the results may seem rather long and complicated, but it can be quite simply expressed in the formula:

$$\frac{\text{Volume} \times \text{molarity}}{\text{No. of moles reacting}} \text{ (base)}$$

$$= \frac{\text{Volume} \times \text{molarity}}{\text{No. of moles reacting}} \text{ (acid)}$$

this can be transposed

$$\frac{\text{Volume} \times \text{molarity (acid)}}{\text{Volume} \times \text{molarity (base)}}$$

$$= \frac{\text{No. of moles reacting (acid)}}{\text{No. of moles reacting (base)}}$$

In examples (a), (b) and (c) the right-hand side of this expression equals one as the moles of acid and base reacting are the same. So, in (a) molarity of base

$$= \frac{\text{volume} \times \text{molarity (acid)}}{\text{volume of base}}$$

$$= \frac{20 \times 0\cdot 1}{25} M = \underline{0\cdot 08 M}$$

in (b) molarity of acid

$$= \frac{\text{volume} \times \text{molarity (base)}}{\text{volume of acid}}$$

$$= \frac{25 \times 0\cdot 1}{21} M = \underline{0\cdot 119 M}$$

in (c) molarity of acid

$$= \frac{25 \times 0\cdot 15}{30} M = \underline{0\cdot 125 M}$$

in (d) the ratio = 2:1
so molarity of acid

$$= 2 \times \frac{\text{volume} \times \text{molarity (base)}}{\text{volume of acid}}$$

$$= \frac{2 \times 25 \times 0\cdot 1}{20} M = \underline{0\cdot 25 M}$$

in (e) the ratio = 1:2
so molarity of acid
$$= \frac{1}{2} \times \frac{10 \times 0 \cdot 12}{6} M = \underline{0 \cdot 1 M}$$

Make use of this formula by all means, but do try to understand how it is derived. Always write down the equation for the reaction you are concerned with.

Calculations concerning molarity are really quite simple; get some practice with the following ones. As before, try to do them mentally if you can.

E. How much water must be added to 20 cm³ of M lithium hydroxide in order to make it exactly (a) 0·1 M; (b) 0·01 M; (c) 0·001 M?
F. How much water must be added to 50 cm³ of M sulphuric acid in order to make it exactly (a) 0·5M; (b) 0·2M; (c) 0·1M?
G. What volume in cm³ of the following solutions will neutralize one dm² of 0·1M sulphuric acid? (a) M sodium hydroxide; (b) 0·5M potassium hydroxide; (c) 0·025M lithium hydroxide; (d) 0·125M sodium carbonate.
H. What volumes of 0·05M sulphuric acid will neutralize: (a) 20 cm³ 0·1M sodium hydroxide; (b) 5 cm³ M potassium hydroxide, (c) 25 cm³ 0·25M calcium hydroxide; (d) 10 cm³ 0·4M ammonium hydroxide?
I. What will be the molarity of a solution of hydrochloric acid if 10 cm³ of it are neutralized by (a) 10 cm³ M lithium hydroxide; (b) 10 cm³ 0·1M sodium hydrogencarbonate; (c) 25 cm³ 0·5M calcium hydroxide; (d) 15 cm³ 0·2M sodium carbonate?

Reactions of Acids with Metals

Look at this reaction:
$$\underset{\text{1 mole}}{Mg(s)} + \underset{\text{1 mole}}{H_2SO_4(aq)} \rightarrow MgSO_4(aq) + H_2(g)$$

From the equation we see that one mole of sulphuric acid dissolves one mole of magnesium, i.e. 24 g of magnesium will react completely with one dm³ of molar sulphuric acid and in so doing one mole of hydrogen molecules is liberated (this has a volume of 22·4 dm³ at s.t.p. or approximately 24 dm³ at room temperature and atmospheric pressure). We can use this fact to find the atomic mass of a metal.

(f) *0·706 g of a divalent metal dissolves completely in 30 cm³ of 0·98 M sulphuric acid. What is the relative atomic mass of the metal?*

30 cm³ of the acid contain $\frac{30}{1000} \times 0.98$ mole of acid = 0·0294 mole

Since the metal is divalent one g-atom or mole of it will dissolve in one mole of sulphuric acid; therefore

0·0295 mole of the acid dissolves 0·706 g of the metal and 1 mole of acid dissolves

$$\frac{0 \cdot 706}{0 \cdot 0294} \text{ g of the metal} = 24 \text{ g.}$$

∴ The relative atomic mass of the metal is 24

(g) *What mass of zinc (relative atomic mass = 65·4) will dissolve in 60 cm³ of 2·2 M hydrochloric acid and what approximate volume of hydrogen will be liberated? (Take the molar volume of hydrogen to be 24 dm³ at room temperature.)*

$$\underset{\text{1 mole}}{Zn(s)} + \underset{\text{2 moles}}{2HCl(aq)} \rightarrow ZnCl_2(aq) + H_2(g)$$

1 dm³ of 2M HCl reacts with 1 mole of zinc.
1 dm³ of 2·2M HCl will react with 2·2/2
    = 1·1 mole of zinc.
60 cm³ of 2·2M HCl will react with

$$1 \cdot 1 \times \frac{60}{1000} \text{mole} = 0 \cdot 066 \text{ mole.}$$

The mass of zinc = 0·066 × 65·4 g
    = 4·32 g.

As 1 mole of zinc causes the liberation of 1 mole of hydrogen molecules, 0·066 mole will liberate 0·066 mole of hydrogen.

The volume of hydrogen = 0·066 × 24 dm³
    = 1·584 dm³

(h) *1·004 g copper(II) oxide dissolve in 24 cm³ of 1·05M hydrochloric acid. What is the relative atomic mass of copper if that of oxygen is 16?*

$$\underset{\text{1 mole}}{CuO} + \underset{\text{2 moles}}{2HCl} \rightarrow CuCl_2 + H_2O$$

2 moles HCl dissolve 1 mole CuO
∴ 24 cm³ of 1·05M HCl will dissolve $\frac{24}{1000} \times \frac{1 \cdot 05}{2}$
mole CuO = 0·0126 mole
0·0126 mole of CuO has a mass of 1·004 g
1 mole of CuO will have a mass of $\frac{1 \cdot 004}{0 \cdot 0126}$ g = 79·5 g

As the relative atomic mass of oxygen is 16 and the molar mass of copper(II) oxide is 79·5, the relative atomic mass of copper must be

$$79 \cdot 5 - 16 = \underline{63 \cdot 5}$$

Using the following short table of relative atomic masses and valencies try to work out mentally the answers to the following problems.

	Aluminium	Iron	Magnesium	Zinc	Tin
Relative Atomic Mass		56	24	65·5	118·5
Valency	3	2	2	2	2

J. What mass of iron will dissolve in 100 cm³ of M hydrochloric acid?

K. What mass of zinc will dissolve in 250 cm³ of 0·1M sulphuric acid?

L. If 3·2g of a divalent metal dissolve in 800 cm³ of 0·2M hydrochloric acid what is its atomic mass?

M. 30 cm³ of 2M potassium hydroxide dissolve 0·54g of aluminium. What is the atomic mass of the metal?

N. What volume of 0·1 M hydrochloric acid will just dissolve 1·8 g of magnesium?

O. What volume of 0·5M sulphuric acid will just dissolve 23·7 g of tin?

Hydrogen Ion Concentration

At the neutral or equivalence point there are no free hydrogen or hydroxide ions in a solution.

H^+ (from acid) + OH^- (from alkali) → H_2O (neutral)

or $2H^+$ (from acid) + CO_3^{2-} (from carbonate) → $H_2O + CO_2\uparrow$

or H^+ (from acid) + HCO_3^- (from hydrogencarbonate) → $H_2O + CO_2\uparrow$

Small differences in the acidity of alkalinity of a solution, especially near the neutral point, are of great importance in biological and electrolytic processes, in farming, gardening, brewing and baking and in the manufacture of many substances. Such differences can be detected in solution by the use of a complex indicator such as Universal Indicator which gives a complete spectrum of colour changes from red (strong acid) to purple (strong alkali), the neutral point being indicated by a yellow-green colour (as in distilled water).

The concentration of hydrogen ions is expressed by what is known as the pH system. A 0·1M solution of hydrochloric acid has 0·1 (10^{-1}) g of hydrogen ions per dm³, and is said to have a pH of 1. If this solution is diluted ten times we get a 0·01 M solution which contains 0·01 (or 10^{-2}) g of hydrogen ions per dm³

0·1 M hydrochloric acid	1·0	urine	7·4
lemons	2·3	sea water	8·2
apples	3·0	sodium hydrogencarbonate	8·4
tomatoes	4·0	borax	9·2
saliva	6·4	lime water	12·3
milk	6·6	0·1 M sodium hydroxide	13·0

The pH is now 2. Pure water contains only about 10^{-7} g of hydrogen ions per dm³ and so has a pH of 7. Solutions with a pH greater than 7 (up to 14) are alkaline.

The average values of the pH for some familiar substances are shown in the table.

Experiment 40.4. Determining the pH of some common Substances.

An indicator called Universal Indicator or Wide-Range Indicator has been developed which changes colour from red (very acidic) through yellow, green (neutral), and blue to purple (very alkaline) as the pH of a solution increases. This can be used either in solution or as strips of paper.

Dip a strip of indicator paper in distilled water, tap water and dilute solutions of vinegar, household ammonia, soap, soda, sodium hydrogencarbonate detergent, bleach, borax and in other common substances such as milk and notice the colour produced. The covers of the booklets of papers carry colour charts so that you can see immediately what the pH of any solution is. Draw up a list of the substances in order of decreasing acidity.

Indicator paper is very useful in determining the degree of acidity of soils. Some plants will only grow if the pH is within certain limits, e.g. azaleas require an acidic soil such as clay. Gardeners' 'Lime' is often used to correct too much acidity in a soil. This may be powdered limestone or chalk or it may be slaked or hydrated lime. Let us see the effect of adding these to a slightly acidic solution.

Experiment 40.5. Correcting Acidity.

Take about 10 cm³ of water and add to it 2 drops of bench hydrochloric acid. Test with indicator paper and note the pH. With a small spatula add as much chalk as will lie on the bent end; stir and when effervescence has ceased again dip in the paper. Repeat with successive small portions until there is no further change. You will find that the pH will increase to pH 7 (neutral) but no further. Chalk itself is a neutral substance (being insoluble in water) so it cannot do more than neutralize the acid.

Repeat the whole experiment using slaked lime in place of chalk. This is an alkaline substance and eventually the solution becomes distinctly alkaline with a pH of about 11. Evidently addition of slaked lime to an acidic soil must be carefully controlled.

In each of the above cases draw a graph showing how the pH varies with successive additions of each substance.

Exercise 40

1 25 cm³ of a solution of sodium carbonate are neutralized by 20 cm³ of a decimolar solution of hydrochloric acid. Calculate (i) the molarity of the sodium carbonate solution, (ii) the number of grams of sodium carbonate in one dm³ of it.   (*OC*)

2 25 cm³ of 0·06 M sodium carbonate were neutralized with hydrochloric acid and the readings obtained by the experimenter were 24·6, 24·2, 24·1 and 24·2 cm³. If the maximum permitted error is 0·2 cm³ calculate the molarity of the acid.

3 If 24·6 cm³ of 0·12M hydrochloric acid are neutralized by 22·8 cm³ of sodium hydroxide, what mass of the latter is present in each dm³ of the solution?

4 What mass of magnesium (relative atomic mass = 24) will dissolve in 75 cm³ of 1·05M hydrochloric acid?

5 What mass of divalent zinc (relative atomic mass = 65·4) will dissolve in 400 cm³ of 2·3M hydrochloric acid?

6 (*a*) How many cm³ of decimolar hydrochloric acid would be required to neutralize 20 cm³ of a solution of sodium hydroxide containing 6 g of sodium hydroxide per dm³?

(*b*) How many cm³ of decimolar potassium hydroxide will neutralize 25 cm³ of hydrochloric acid containing 1·46 g hydrochloric acid per dm³?

(*c*) What is (i) the formula mass of nitric acid, (ii) the molarity of a solution which contains, per dm³, 10 g of a monovalent acid of formula mass 80?   (*OC*)

7 (*a*) What do you understand by a decimolar solution of (i) sulphuric acid, (ii) potassium hydroxide, and what mass per dm³ of the reagent does each solution contain?

(*b*) What is the molarity of a solution of (i) 7 g of sulphuric acid per dm³, (ii) 7 g of potassium hydroxide per dm³?

(*c*) 25 cm³ of a solution of sulphuric acid neutralize 30 cm³ of a solution of sodium hydroxide containing 5 g per dm³ of the alkali. Calculate its molarity.   (*O*)

8 A piece of metal X with a mass of 0·9 g is dissolved completely in 100 cm³ of molar hydrochloric acid, leaving excess acid which is found to require 25 cm³ of molar sodium hydroxide for its neutralization. If the valency of X is two, what is its atomic mass?   (*OC*)

41
The Periodic Table; Order among the Elements

Let us summarize some of the facts we have so far learned about the atom.

> Atoms consist of protons, neutrons and electrons.
> Protons carry a positive charge of electricity and electrons a negative one; neutrons are not charged.
> The nucleus contains the protons and neutrons (these particles are also referred to as nucleons) which are equal in mass; each is as heavy as about 1840 electrons.
> The nucleus is extremely small compared with the size of the atom, but most of the mass of the atom is concentrated there.
> The electrons move in shells or orbits around the nucleus, but at different distances from it. (These orbits are often called 'energy levels' because when an electron moves from one level to another energy is either given out (perhaps as light) or taken in (usually in the form of heat).)
> These outer (valency) electrons are the ones concerned in chemical changes, the nucleus never being effected.
> There are never more than eight electrons in the outermost shell.
> The number of protons increases regularly from one in the hydrogen atom to 104 in the atom of kurchetovium and it equals the number of electrons in the atom considered.
> Every atom of any particular element must contain the same number of protons, but the atoms of different elements always have a different number of protons.

Scientists have long known that many elements resemble each other and that certain 'families' can be picked out, e.g. the *alkali metals*, lithium, sodium, potassium, rubidium, caesium and francium or the *halogens* fluorine, chlorine, bromine, iodine and astatine. There are smaller groups such as magnesium, calcium, strontium, barium (the *alkaline earths*) and arsenic, antimony and bismuth which also have properties that are very much alike.

From the middle years of the nineteenth century scientists had tried to arrange the elements in an order which would bring out this resemblance and the order adopted was usually that of their atomic masses. This was not very successful for two reasons; many elements had not been discovered and many atomic masses had not been calculated with any great accuracy. It was only in 1869 that a Russian scientist, Dmitri Mendeleev and a German, Lothar Meyer, independently succeeded in this task. Mendeleev arranged the elements in rows according to their atomic masses and found a family resemblance in the vertical columns. Where an element was obviously out of place (according to its properties) he did not hesitate to question the accuracy of the atomic mass determination and moved it to a more appropriate position, so leaving a gap for some as yet undiscovered element. Let us look at some of the rows as set out by Mendeleev.

Group:	I	II	III	IV	V	VI	VII	VIII		
	Li	Be	B	C	N	O				
	Na	Mg	Al	Si	P	S	Cl			
	K	Ca		Ti	V	Cr	Mn	Fe	Co	Ni
	Cu	Zn			As	Se	Br			
	Rb	Sr	Y	Zr		Mo		Ru	Rh	Pd
	Ag	Cd	In	Sn	Sb	Te	I			
	Cs	Ba								

What do you notice?

The families mentioned and underlined do fit in the same vertical columns.

The elements in any group or column all have the same number of electrons in the outer shell. (When Mendeleev first drew up his table he knew nothing at all about electrons, they had not been discovered.)

The number of outer electrons is the same as the group number except for group VIII.

All the elements can have a valency equal to the group number. This can be seen by writing down the formula of suitable oxides in, say, row 2. (Remember that oxygen has a valency of two.) You must, however, bear in mind that the valency of an element is not always the same and that many elements have more than one valency, e.g. phosphorus has valencies of three and five, sulphur of two, four and six, chlorine of one and seven. The valencies indicated above are not necessarily the principal one.

Formula	Na_2O	MgO	Al_2O_3	SiO_2	P_2O_5	SO_3	Cl_2O_7
Valency of element	1	2	3	4	5	6	7

Mendeleev predicted that all the gaps he left in the table would eventually be filled by elements whose properties he was able to forecast with scientific accuracy. These elements were later discovered and his predictions were true in every case.

Some 20–30 years later a whole new family of gases was discovered to be present in the air, the 'noble' gases, neon, argon, krypton, xenon often referred to as 'argonons'. There were no gaps in the table where all these might be fitted, but they did very conveniently form a complete column which could be placed at the beginning without disturbing any of the others; it was called group O and two other elements, helium and radon, also belong to it.

With the discovery of the structure of the atom and the knowledge that the number of protons or electrons differed in each element it was seen that all the elements in the periodic table were in the order of increasing atomic number, $H = 1, He = 2, Li = 3$, etc. This itself was a splendid vindication of Mendeleev's boldness in leaving gaps where existing elements did not fit and moving others (e.g. iodine and tellurium) where the atomic mass order seemed to put them in the wrong column.

This 'short form' of the Periodic Table is not much used nowadays because it is obvious that there are many 'strangers' in the groups which do not really resemble the families we have mentioned; e.g. copper is not like sodium, chromium bears little relationship to sulphur and there is not much resemblance between chlorine and manganese in their chemical behaviour. Bohr, who did so much work on atomic structure, planned a new type of table called the 'long' or 'extended' type and a modern version of this is shown below. All the families are still there in the same columns but the strangers have been removed to more suitable places. Notice that hydrogen is given an indefinite place at the top of the table; in some of its properties it resembles the metals in group I (can you think of any of these properties?) and in others it behaves like the halogens of group VII. After the first two rows in which the increasing number of electrons fills up the outer shell to eight before starting a new one we get two rows with a large number of elements (scandium to zinc and yttrium to cadmium) separating the two halves of the table. Most of these have two electrons in their outermost layer and the additional ones enter the shell below this and build this up to 18. These metals are called transitional ones and there is only a very gradual change in the properties with increasing atomic number.

In the next two rows we get further groups of elements which differ even less from each other. Once again the number of electrons in the outer shell remains stationary while further building up goes on inside the atom. These two series of elements are known as the 'lanthonons' and 'actinons', and their place in the periodic table is indicated.

For many years uranium was thought to be the element with the highest atomic number and by 1940 only three gaps still remained in the table, elements number 43, 85 and 87. These three are now known: 43 is technetium, so called because it was produced by technical means artificially and has not been found naturally; 85 is astatine, a halogen, and 87 is francium, an alkali metal. Many new elements have since been produced artificially by nuclear physicists. All are heavier than uranium, but only one or possibly two have since been detected in the earth's crust. They all closely resemble each other and are collectively known as 'actinons' or 'transuranic' metals. One hundred and four elements are now known, although the last one has not yet been given an agreed name; it is quite likely that other heavier ones will eventually be prepared but the extraction of each new one gets much more difficult and involves a far lengthier process. No matter how many additional elements are prepared, it is quite certain that they will all fit into the periodic table devised by Mendeleev so long ago and extended and modified by Bohr and other workers in the sub-atomic field.

Avogadro Constant

The number of atoms in one gram of hydrogen,

PERIODIC TABLE OF THE ELEMENTS
(LONG FORM)

I	II											III	IV	V	VI	VII	O
																	2 He
3 Li	4 Be					1 H						5 B	6 C	7 N	8 O	9 F	10 Ne
11 Na	12 Mg											13 Al	14 Si	15 P	16 S	17 Cl	18 Ar
19 K	20 Ca	21 Sc	22 Ti	23 V	24 Cr	25 Mn	26 Fe	27 Co	28 Ni	29 Cu	30 Zn	31 Ga	32 Ge	33 As	34 Se	35 Br	36 Kr
37 Rb	38 Sr	39 Y	40 Zr	41 Nb	42 Mo	43 Tc	44 Ru	45 Rh	46 Pd	47 Ag	48 Cd	49 In	50 Sn	51 Sb	52 Te	53 I	54 Xe
55 Cs	56 Ba	57* La	72 Hf	73 Ta	74 W	75 Re	76 Os	77 Ir	78 Pt	79 Au	80 Hg	81 Tl	82 Pb	83 Bi	84 Po	85 At	86 Rn
87 Fr	88 Ra	89† Ac	104 Ku														

Lanthanons*	58 Ce	59 Pr	60 Nd	61 Pm	62 Sm	63 Eu	64 Gd	65 Tb	66 Dy	67 Ho	68 Er	69 Tm	70 Yb	71 Lu
Actinons†	90 Th	91 Pa	92 U	93 Np	94 Pu	95 Am	96 Cm	97 Bk	98 Cf	99 Es	100 Fm	101 Md	102 No	103 Lw

These two series are included after elements 57 and 89.

Group I. The alkali metals
Group II. The alkaline earths
Group VII. The halogens
Group O. The argonons or noble gases

in 12 grams of carbon, in 16 grams of oxygen or in 200 grams of mercury is exactly the same. This number is extremely large and equals approximately six hundred thousand million million million, i.e. 6 followed by 23 noughts! It is written as 6×10^{23} and is called the Avogadro Constant or Avogadro Number. It is the number of atoms in one gram-atom of any element; it is the number of molecules in one gram-molecule of any covalent compound such as carbon dioxide, CO_2, or water, H_2O; it is the number of ions in one gram-ion of chloride ions, Cl^-, or copper(II) ions, Cu^{2+}, or ammonium ions, NH_4^+; it is the number of ion-pairs in one formula mass of an ionic compound such as sodium chloride, Na^+Cl^-, or copper(II) sulphate, $Cu^{2+}SO_4^{2-}$. Because all these differently named units contain the same number of particles one common name has been adopted for all of them, the MOLE. Remember then, one mole conains the Avogadro Constant of particles, 6×10^{23}.

Periodicity

Let us look at the elements of the second period in the periodic table (sodium—argon) and at their melting and boiling points. These are the figures—but do not try to learn them!

Fig. 41.1 shows these values on a graph plotted against the atomic number or position of the element. You will notice that the two 'curves' are similar. In each case there is a general rise in the melting or boiling points as far as Group IV (silicon) and then a fall to the end of the period. The elements to the left of silicon are metallic and all are solids at room temperature, whilst those that follow have low melting and boiling points. These values are influenced by the attraction between the atoms or molecules in an element; if this is strong then a considerable amount of heat energy will be needed to make the atoms move apart and so form a liquid or gas. It would seem that silicon atoms are bonded together very strongly indeed.

Another physical characteristic that depends on this bonding is the *atomic volume* of the element. This is the volume in cubic centimetres of the element in the solid state and is calculated by dividing the atomic mass of the element by the solid density. Fig. 41.2 is a graph of the atomic volume plotted against the atomic number of the element, i.e. against its position in the Periodic Table. Two curves are shown; one for the elements lithium–fluorine and the other for the corresponding elements in the next row, sodium–

Element	Na	Mg	Al	Si	P	S	Cl	Ar
M. pt. °C	98	650	660	1410	45	119	−102	−189
B. pt. °C	883	1100	2400	2360	280	444	−34	−186

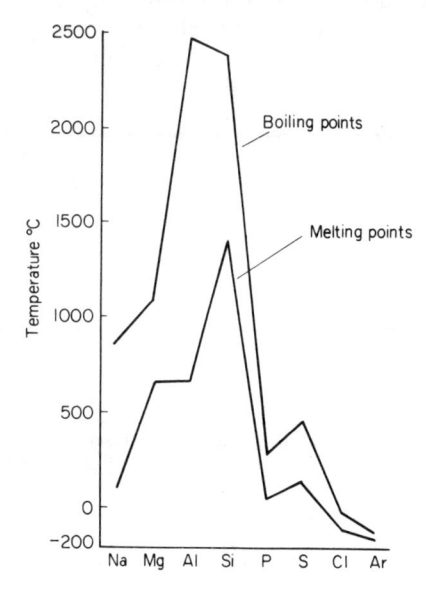

Fig. 41.1.

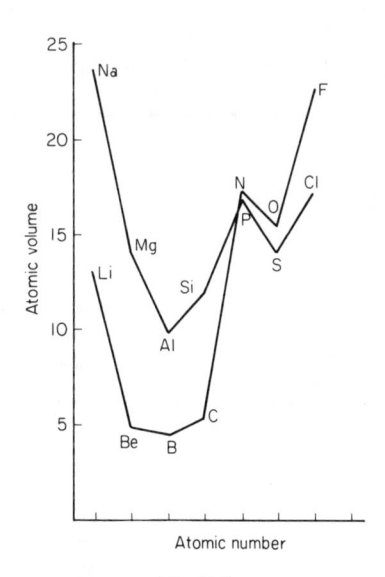

Fig. 41.2.

chlorine. Notice the similarity in their shape. The alkali metals are light and have larger atomic volumes than other metals. The atomic volume decreases to the middle of the table and then rises again. These three properties so far considered are said to be 'periodic', i.e. the curves for each successive row across the table are similar.

We ought to look now at how the chemical properties vary along a similar period.

Na_2O	MgO	Al_2O_3	SiO_2	P_2O_5	SO_3	Cl_2O_7
strongly basic	basic	amphoteric	weakly acidic	acidic	strongly acidic	

Here again is the list of oxides given earlier in the chapter along with the nature of the oxide. You should recall that basic oxides are soluble in acids to give salts and, in fact, some of them dissolve in water to form alkaline solutions. Acidic oxides dissolve in alkalis to form salts and generally in water also to form acids. Amphoteric oxides are soluble in both acids and alkalis to form salts.

Examine the Periodic Table closely and imagine that a diagonal line is drawn from the top left corner of the space marked boron (B) to the bottom right corner of the space for polonium (Po). Put a ruler or straight edge across the table to represent this line. Can you recognize any difference between the elements to the left of the line and those to the right? Those to the left are predominantly metals and those to the right non-metals.

What is the difference between a metal and a non-metal? One cannot always go by physical properties. Chemical properties must be considered.

Metals form positive ions and do not form compounds with each other. By transferring the outermost electrons to an atom of a non-metal, ionic (electrovalent) compounds are formed:

$$Na. + \cdot \overset{..}{\underset{..}{Cl}} \cdot \longrightarrow Na^+Cl^-$$

Non-metals form negative ions by losing their outer electrons, but they are also able to join up with other non-metals by sharing pairs of electrons to form covalent compounds or molecules:

$$H. + \cdot \overset{..}{\underset{..}{Cl}} \cdot \longrightarrow H \cdot \overset{..}{\underset{..}{Cl}} \cdot$$

Ionic compounds are usually crystalline solids with a fairly high melting points whereas covalent ones are generally liquids or gases or solids which melt easily. Other examples of ionic compounds are copper(II) sulphate, $CuSO_4$ ($Cu^{2+}SO_4^{2-}$), sodium hydroxide, NaOH (Na^+OH^-) and potassium nitrate, KNO_3($K^+NO_3^-$); covalent examples are ammonia, NH_3(a gas), ethanol, C_2H_5OH (a liquid) and phosphorus pentachloride, PCl_5 (a solid).

Electrovalent compounds are built up into giant structures or lattices of ions and such compounds will conduct electricity (and decompose) if they are molten or in aqueous solution. Heating or dissolving in water loosens the forces holding the ions together and the ions are then able to move freely about. Covalent compounds form either simple molecules such as those of water or complex ones such as those of many of the plastics. Some atoms, such as those of carbon in the form of diamond, also build up into giant structures by covalent bonding (see p. 141). Covalent compounds do not conduct electricity at all.

Exercise 41

1 The following table is a list of a few elements with some of their properties; the elements are labelled A, B, etc., to H.

Element	Relative atomic mass	Atomic number	Density in kg/m^3 at s.t.p.	Boiling point in °C	Elect. conductivity	Action with air	Action with water
A	12	6	2250	4200	Good	Smoulders giving a colourless odourless gas	Reacts with steam at high temperatures
B	14	7	1·26	−196	Poor	None, except at high temperatures	None
C	16	8	1·44	−183	Poor	None except at high temperatures	None
D	19	9	1·51	−188	Poor	None	It reacts

Element	Relative atomic mass	Atomic number	Density in kg/m³ at s.t.p.	Boiling point in °C	Elect. conductivity	Action with air	Action with water
E	21	10	0·945	−246	Poor	None	None
F	23	11	970	890	Good	Burns leaving a white ash	Violent effervescence
G	24	12	1740	1110	Good	Burns leaving a white ash	Very slow reaction
H	27	13	2700	2470	Good	Burns if finely divided	Unreactive

(a) Which of the above could be an alkali metal (i.e. like sodium or potassium)? Give a reason for your choice.

(b) Which of the remaining ones could be metals? Give TWO reasons for choosing these.

(c) Which one could be carbon? Give TWO reasons for choosing this one.

(d) Using the symbols (letters) given plus the appropriate electric charge, write down the formulae of the ions that are formed by the following elements. The element D; the element F.

(e) Which element could be a noble (i.e. inert) gas? Why?

(f) Which could be a noble (i.e. inert) gas? Why?

(g) What is the valency of element H? Suggest a use for this element and give a reason for your suggestion.

(h) Draw a diagram of an atom of element A, showing its nucleus and the arrangement of electrons outside. Represent.

(i) electrons by ·, (ii) protons by +, (iii) neutrons by N.

Which element, if any, is a noble metal? Give a reason for your answer.

(i) Calculate the volume in cubic metres at s.t.p. of 1 kg-atom of (a) element D, (b) element G. (L)

2 The following is a list of eight elements and some of their properties:

Element (use these letters in your answers)	Atomic number	Atomic mass	Melting point in °C	Boiling point in °C	Density at stp in kg/m³	Electrical conductivity	Action when heated in air
A	3	7	180	1330	530	Good	Burns leaving a white ash
B	26	56	1535	3000	7900	Good	Becomes superficially oxidized
C	16	32	113	445	2000	Very poor	Burns giving a pungent gas
D	10	20	−249	−246	0·945	Very poor	Unaffected
E	78	195	1769	3830	21 000	Good	Unaffected
F	6	12	3730	4200	2250	Good	Smoulders giving a colourless odourless gas
G	7	14	−210	−196	1·26	Poor	Reacts only at high temperatures
H	80	201	−39	357	13 600	Good	Slowly forms a red solid of formula type 'HO'

(a) Which of the elements are metals? Give a reason for choosing these.

(b) Which of these could be an alkali metal (i.e. like sodium)?

(c) Which element could be carbon? Give **two** reasons for choosing this one.

(d) What is the valency of H? Give a common scientific use of this element.

(e) At s.t.p. 1 dm³ of D has a mass of 0·893 g and 1 dm³ of G a mass of 1·25 g.

What would these values be in dm³ per g-atom? Suggest from these answers the number of atoms of D in one molecule and the number of atoms of G in one molecule.

(1 mole of gas at s.t.p. occupies 22·4 dm³). (C)

42
The Winning of Metals

Most people, if asked the question 'What is a metal?' would mention some of the following properties as being associated with metals: shiny surface, high density, hardness, high melting point, malleability, ductibility and good condition of heat and electricity. For most metals this is true, but there are many exceptions. Sodium and lead are only shiny if freshly cut; sodium and potassium are light enough to float on water and are very soft; lithium is also light, but it is much harder than sodium. Bismuth is a poor conductor and is too brittle to be drawn into wires; tin melts at 232°C. On the other hand, the non-metal carbon (*as diamond*) is extremely hard, has a high melting point and (*as gas carbon*) is a good conductor of electricity; one form of silicon has a very metallic appearance.

It is easier to distinguish metals from non-metals by consideration of their chemical properties. Metals, on burning, form basic oxides which neutralize acids to give salts. If the oxides are soluble in water they form alkaline hydroxides. Non-metallic oxides, if soluble in water, always form acids and will neutralize alkalis to produce salts, i.e.

$$Ca \xrightarrow{+O_2} CaO \xrightarrow{+H_2O} Ca(OH)_2$$
$$S \xrightarrow{+O_2} SO_2 \xrightarrow{+H_2O} H_2SO_3$$

When metallic elements form more than one oxide there is a tendency for the basic nature to decrease as the proportion of oxygen increases, e.g. MnO is basic and forms manganese(II) (Mn^{2+}) salts, but Mn_2O_7 is acidic and forms manganate(VII) (MnO_4^-).

Metals form metallic ions by *loss* of electrons

$$\underset{\text{11 electrons}}{Na} - e^- \longrightarrow \underset{\text{10 electrons}}{Na^+}$$

and so can be considered to be oxidized (see p. 120). Non-metals, however, form simple ions by addition of electrons, i.e. they are reduced.

$$\underset{\text{19 electrons}}{Cl} + e^- \longrightarrow \underset{\text{18 electrons}}{Cl^-}$$

If a piece of metal is placed in a solution of one of its salts (e.g. zinc in zinc sulphate solution) a very small potential difference can be detected between the metal and the metallic ions in the solution. This potential difference varies with different metals; when it is largest the metals (Na, Ca, K) are very *electropositive* and vice versa. If a list is drawn up with the most electropositive ones at the top and the least electropositive at the bottom we get a table, called the electrochemical series, which is identical with the one, based on the 'activity' of the metals, which we have already considered and so often made use of (see p. 1). One cannot, of course, place a piece of sodium in a solution of one of its salts, but other methods are available for finding the potential difference in such a case.

Abundance of the Elements in Nature

Over 75% of the earth's crust is made up of the non-metals silicon and oxygen (mainly in the combined state) and no one metal is present in any great amount. The most abundant is aluminium (7·3%), mostly in clays and other complex aluminosilicates. Fig. 42.1 shows the relative abundance of the commoner elements and you will see that thirteen elements make up over 99%.

Extraction of the Metals

Strongly electropositive elements such as calcium and sodium are more active than ones like copper and mercury and more readily form compounds.

The change Ca → CaO requires less energy and therefore occurs more readily than Hg → HgO and the reverse change HgO → Hg is much more easily brought about than CaO → Ca. In general, the more easily a compound is formed from its elements, the less easily is it decomposed.

Metals, other than feebly electropositive ones such as gold and silver, do not often occur in the free state in nature, but generally as oxides, chlorides, carbonates or sulphides. The two latter compounds are usually roasted to form the oxides before the metal is extracted. The following table

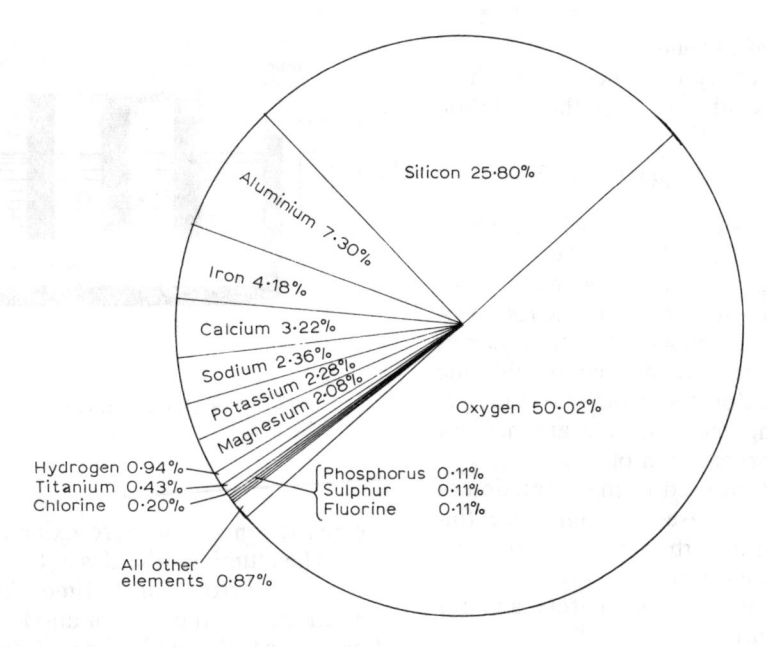

Fig. 42.1. Relative abundance of elements by mass in the earth's crust (*U.S. Geological Survey*)

shows the usual method of decomposing a metallic ore to give the metal.

The amount of energy required to bring about these reductions is greatest for the metals at the top of the table and this is one reason why they were not 'discovered' until many many centuries after iron and copper; they had to await the time when electrical energy became available. Sodium was first isolated by Davy in 1807 by the electrolysis of fused sodium hydroxide.

K
Li
Ca
Na
Mg } chloride fused and then electrolyzed

Al } oxide fused and electrolyzed

Zn
Fe } ore roasted to give oxide and then this is reduced with coke

Cu
Hg } sulphide roasted to give oxide and then metal.

Ag
Au } Found in the native state as the metals.

Extraction of Sodium

Sodium is obtained by one of two methods; (*a*) from the fused chloride in a single process, (*b*) from brine by a two-stage process.

Let us look at the former method.

The apparatus used is called a Down's Cell and is shown in Fig. 42.2. As salt melts at a comparatively high temperature, 800°C, a little impurity

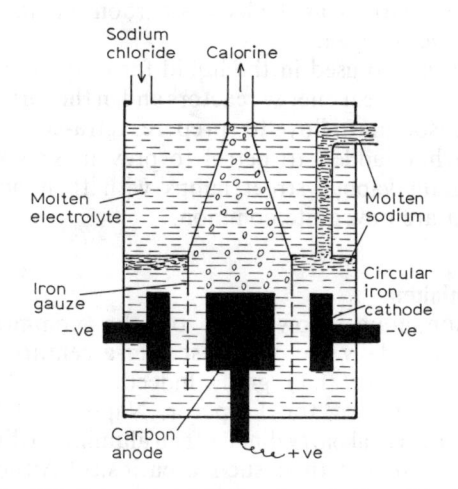

Fig. 42.2. Extraction of Sodium from Sodium Chloride

such as calcium chloride or strontium chloride is added and this lowers the melting point to about 650°C. The vessel used is circular and is made of iron lined with firebrick. The iron cathode is also circular and surrounds the graphite anode.

On electrolysis, chlorine is liberated at the anode (why is graphite used?) and led away, while sodium, set free at the cathode, rises to the top of the fused salt inside a tube and is collected.

$$2Na^+Cl^- \rightarrow 2Na^+ + 2Cl^-$$
$$2Na^+ + 2e^- \rightarrow 2Na$$
$$2Cl^- - 2e^- \rightarrow Cl_2$$

or $2NaCl \xrightarrow[\text{fused}]{\text{electrolyzed}} 2Na + Cl_2$

Properties and Uses of Sodium

Sodium is a soft, easily tarnished metal which rapidly decomposes cold water with the evolution of hydrogen.

$$2Na + 2H_2O \rightarrow 2NaOH + H_2$$

The heat evolved is sufficient to melt the metal. If the hydrogen is to be collected, the sodium is placed in a sodium spoon or wrapped in a piece of lead foil and small pin-holes made in the foil. As it is so high in the electrochemical (or activity) series sodium will readily remove oxygen or chlorine from the oxides or chlorides of other metals; i.e. it is a strong reducing agent. Its uses are many; it is concerned in the preparation of

sodium cyanide, NaCN, used in the extraction of gold and silver and for 'case-hardening' steel (increasing the amount of carbon in the outer layers so as to make them harder)

sodamide, $NaNH_2$, used in detonators and for making some dyestuffs

sodium peroxide, Na_2O_2, used for purifying the air in submarines as it absorbs carbon dioxide and releases oxygen.

It is also used in the liquid form to 'cool' the fuel in nuclear energy reactors and in the form of a lead/sodium alloy to prepare tetra-ethyl lead, which is added to petrol to prevent knocking. Sodium vapour street lamps with their amber light are now quite common.

Aluminium

Aluminium, now one of the commonest metals in daily use, was, just over a century ago, much scarcer than gold. Indeed, when Queen Victoria visited the Emperor, Napoleon III, in Paris the royal party dined off an aluminium dinner service; it was then such a curiosity! Although compounds containing aluminium are abundant (see p. 250), the amount of energy needed to extract the metal is extremely large and this extraction has only been carried out on the large scale since cheap electricity became available. The first small amounts of it were obtained by reducing the chloride with sodium

$$AlCl_3 + 3Na \rightarrow Al + 3NaCl$$

The oxide is too stable to be reduced by coke, but it will dissolve in a substance called cryolite (Na_3AlF_6) at about 1000°C and can then be electrolyzed. This was not discovered until 1886.

The ore, bauxite, also contains iron oxide, silica and titanium(IV) oxide and these impurities have first to be removed. The apparatus used for

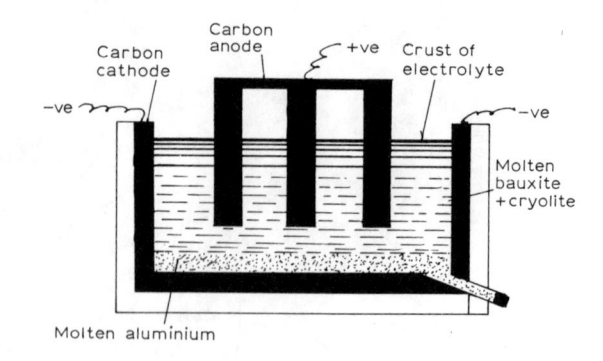

Fig. 42.3. Extraction of aluminium

the reduction of the pure oxide is shown in Fig. 42.3. The aluminium settles at the bottom and can be drawn off from time to time while the liberated oxygen attacks the carbon anode to form carbon dioxide. At the end, the cryolite remains unchanged chemically so the reaction is essentially

$$2Al_2O_3 \xrightarrow{electrolysis} 4Al + 3O_2$$

Properties and Uses of Aluminium

Aluminium is a very good conductor of heat and electricity and does not corrode as a protective film of oxide forms on the surface. This can be thickened by anodizing and then coloured. The metal is soluble in hydrochloric acid,

$$2Al(s) + 6HCl(aq) \rightleftharpoons 2AlCl_3(aq) + 3H_2(g)$$

but is hardly attacked by nitric or dilute sulphuric acids. It is also soluble in fairly concentrated sodium hydroxide solution to form sodium aluminate.

It is used:

in the light alloys magnalium and duralumin, as a constituent of 'Thermite' which is used for the reduction of iron(III) oxide, the molten iron formed being utilized for welding, for kitchen saucepans, frying pans, etc.,

for coating the mirror of the giant telescope at Mt. Palomar in the U.S.A.,

for the bodies of London Underground trains, for 'silver' paint and paper, for milk bottle tops, for tubes for dentifrice,

for television masts.

Iron

Iron is the most widely used of all metals for constructional purposes and has been known for over 4000 years.

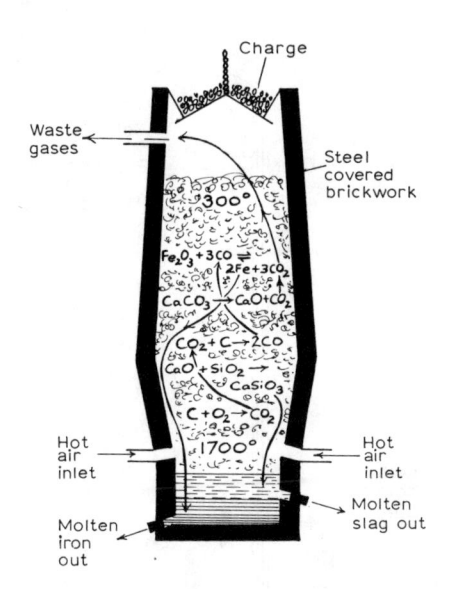

Fig. 42.4. Extraction of Iron in the blast furnace

Extraction of Iron

The chief ore is the oxide, Fe_2O_3, which occurs as the mineral haematite. The production of iron (pig-iron) from the ore is carried out in blast furnaces which are about 25 m high and 6 m in diameter at the widest part.

The ore is roasted, crushed and mixed with coke and limestone (5:2:1) and fed into the top of the furnace shown in Fig. 42.4. Heat is supplied by blowing preheated air or oxygen in at the bottom through tuyères causing the coke to burn.

$$C + O_2 \rightarrow CO_2$$

The carbon dioxide is reduced to the monoxide as it passes up through the bed of coke

$$C + CO_2 \rightarrow 2CO$$

and this reduces the iron(III) oxide to the metal at a temperature of about 800 °C.

$$Fe_2O_3 + 3CO \rightleftharpoons 2Fe + 3CO_2$$

The temperature at the widest part of the furnace is well below the melting point of pure iron (1530 °C), but as the iron passes down the furnace it absorbs some carbon and its melting point is lowered.

The limestone combines with impurities in the ore, such as silica, to form a readily fusible slag

$$CaCO_3 + SiO_2 \rightarrow CaSiO_3 + CO_2$$

which floats on top of the molten iron and prevents it from being oxidized by the incoming air at the bottom. (This slag finds use in the manufacture of cement and in road making.) Some of the carbon dioxide is changed to the monoxide and helps the reduction of the ore. Both the slag and the iron are tapped off at intervals.

The waste gases issuing from the top of the furnace contain both carbon monoxide and dioxide and the considerable amount of heat left in them is not lost, but is transferred to the incoming air which is thus preheated to about 600 °C.

The furnaces are not allowed to cool down, but are run continuously for years. For each Mg of iron made the raw materials are approximately 2 Mg of ore, 800 kg of coke, 400 kg of limestone and about 4 Mg of air or the equivalent amount of pure oxygen. The iron run off is called 'pig iron' and contains about 4–5% of carbon and small amounts of other impurities such as manganese, silicon, sulphur and phosphorus. It is used for the production of cast iron, wrought iron and steel.

Cast Iron

Pig-iron is mixed with scrap iron, coke and limestone and heated strongly by hot air or oxygen. The metal is poured into sand moulds in which it expands on cooling to give an exact shape of the mould. Gratings for drains, man-hole covers, kitchen stoves, radiators and much of the heavy framework of machinery are made in this way.

Steel

Cast iron contains a fairly high percentage of carbon (3–4%) and about 1% of phosphorus, manganese and sulphur. The actual amounts of the last three elements depend on the kind of ore used, the quantity of limestone mixed with it and the temperature of the blast.

In order to change iron into steel the amounts of all impurities, particularly of carbon, must be reduced. The figures below represent the *average* composition of impurities in pig-iron from the furnace and in a mild steel. In other kinds of steel the percentage of carbon varies from about 0·07–1·4%.

	Pig-Iron	Mild Steel
C	4·0	0·15
Si	0·85	0·03
S	0·08	0·05
P	1·6	0·05
Mn	1·0	0·5

The essence of the process for the removal of these elements is one of oxidation and this is carried out mainly by the Open Hearth process (Fig. 42.5).

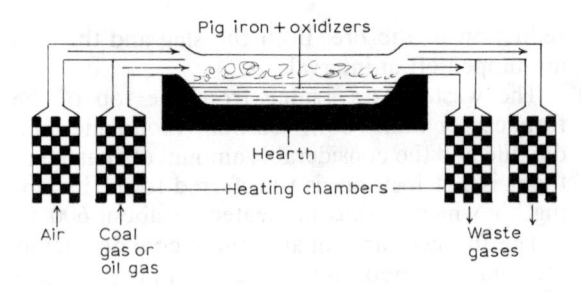

Fig. 42.5. Open-hearth process

The furnace consists of a shallow bath capable of holding from 60–300 Mg of metal at a time. On each side of it is a set of two heating chambers; each set consists of one to preheat the coal gas or atomized oil used and the other to preheat the air or oxygen. These chambers are filled with open chequered brickwork. The gas and air are led through one set of chambers to the furnace bed where they burn with a very fierce flame. The hot spent gases then pass through the other set of chequers in order to heat them. When the first set begins to cool the direction of the flow is reversed and the gas and air again heat up before combustion occurs.

In order to cope with the tremendous heat of the flame the furnace is built of special bricks in a steel casing. The conversion of many furnaces to use atomized oil in place of coal gas has resulted in even greater temperatures and higher output.

The pig-iron plus large quantities of steel scrap and some lime are melted on the hearth. Most of the silicon and much of the carbon are oxidized by the oxygen in the flame and the remaining silicon, phosphorus, sulphur and a controlled amount of carbon must be removed by the addition of oxidizers such as iron oxide and more lime. If the steel is required for a particular purpose other metals such as tungsten (for hardness), nickel or chromium (for stainless steel) are added to the molten alloy as it is run out.

Smaller amounts of steel are made by the Bessemer process (Fig. 42.6). In this, a steel vessel (a converter) lined with dolomite (a carbonate of magnesium and calcium) is heated. Lime is then added and molten pig iron poured in. Oxygen under pressure is then blown through holes in the bottom of the converter. As the blast is forced through the molten metal flames appear at the mouth and these grow and increase in brilliance until one vast sheet of flame is produced. The silicon, manganese and carbon rapidly oxidize and after about twenty minutes all the carbon has gone, but the blast is kept on a little longer to oxidize all the phosphorus.

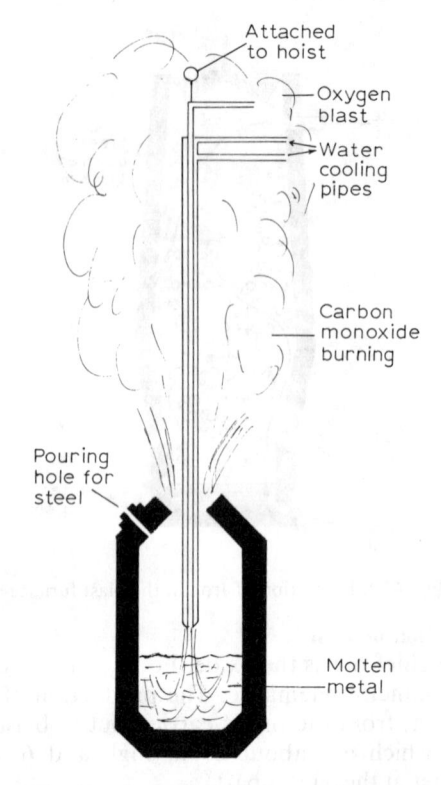

Fig. 42.6. Bessemer converter: L-D process

As all the carbon is removed a calculated amount must now be returned to the metal; this is done by adding some ferro-manganese, an alloy of iron, manganese and carbon. This also helps to bring the steel into good condition by converting any surplus iron oxide into manganese oxide which floats off in the slag. A recent modification of this method, developed in Austria, is the L-D process. In this, the oxygen is blown through a long water-cooled lance on to the surface of the molten metal.

A still more recent method now in use in the United Kingdom transforms molten iron from the blast furnace to steel in a fraction of a second; although this method is being used more and more it has not yet completely replaced the older methods and it may be some years before this happens. By then, even better methods may be available. In it, the impure metal at $1300\,°C$ pours down through a circle of oxygen jets which shatter it into a ring of very fine droplets and raise the temperature to more than $2000\,°C$. The impurities are rapidly burnt away, the amount of carbon being strictly controlled.

Steels of special composition are sometimes made in an electric furnace which produces an intense heat.

Properties of Iron

Iron is soluble in dilute acids to form iron(II) salts. When warmed with concentrated sulphuric acid sulphur dioxide is evolved, but concentrated nitric acid oxidizes the metal surface and renders the iron *passive*; it no longer dissolves in dilute acids.

It is a silvery metal when pure and melts at 1530°C. It is magnetic and if finely divided or in the form of wire it burns in oxygen to form the tri-iron tetroxide, Fe_3O_4

$$3Fe + 2O_2 \rightarrow Fe_3O_4$$

The same compound is obtained when steam is passed over the heated metal. This process was once used commercially for the production of hydrogen (Lane's process)

$$3Fe(s) + 4H_2O(g) \rightleftharpoons Fe_3O_4(s) + 4H_2(g)$$

Experiment 42.1. The Action of Steam on Iron.
Take a hard-glass test-tube and put in it about one cm depth of water. Add wisps of asbestos or mineral wool until all the water is soaked up. Hold the tube horizontally and with a spatula place a heap of iron filings in the tube close to the asbestos. Clamp the tube in this position and insert a cork and delivery tube leading to a test-tube standing in water (Fig. 42.7). Heat the iron and the air will be driven out of the tube. The water on the wool will be changed to steam which then reacts with the iron to give hydrogen. If a light is applied to the collecting tube after it has been removed from the water there will be the 'pop' characteristic of a mixture of hydrogen and air. Remove the reaction tube before water rises into it as it cools down.

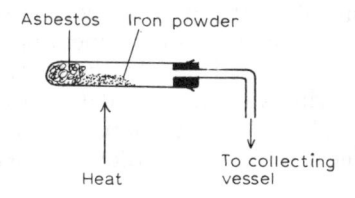

Fig. 42.7. Oxidation of Iron by steam

In the presence of both air and water iron rusts to form hydrated iron(III) oxide. The actual mechanism by which rusting takes place is complicated and at this stage a short explanation must suffice. Try the following experiments.

Experiment 42.2. The Rusting of Iron.
(a) Take a long eudiometer tube and make the inside wet before dropping in a spoonful of iron filings. Rotate the tube and shake out the surplus metal. Stand the tube, mouth down, in a jar of water and notice the volume of air enclosed (raise the tube before measuring so that the water levels inside and outside are the same; why is this done?) If left for some days the water will gradually rise up the tube. What does this suggest? How much air, expressed as a percentage by volume, is used up as the iron rusts?
(b) Set up three flasks as in Fig. 42.8 and have available some clean iron nails. In A the nail is in contact with *dry air only*. Boil the water in flask B and when steam is issuing briskly insert the stopper and remove the bunsen. Here the nails are in contact with *water only*. In flask C the alkali absorbs any carbon dioxide present in the air and so the nails are not in contact with this gas.

Leave for some time before examining.
(c) Put some iron filings in an evaporating basin and warm in a steam oven to ensure dryness. Weigh the basin and contents along with a wooden spill. Add water occasionally during the next week and stir to expose fresh iron to the moisture. Remove excess water by drying once more in the oven and again weigh.

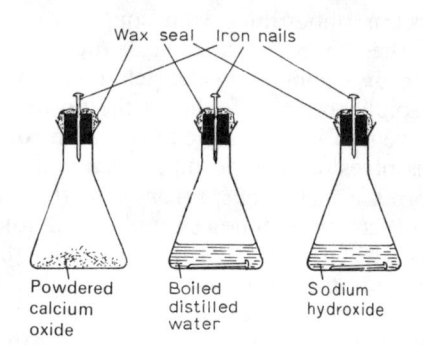

Fig. 42.8. The action of air and water on Iron

These experiments show that:
(i) iron gains in mass when it rusts (why?)
(ii) iron removes about one-fifth (20%) of the air when it rusts in an enclosed volume of damp air—this is the oxygen content.
(iii) iron exposed to dry air only does not rust.
(iv) iron exposed to air-free water does not rust (even after several years) although it may become dulled.
(v) iron does not rust in the presence of air and moisture if carbon dioxide is first removed.

The factors which cause or help rusting to take place are:
(vi) the presence of both air and moisture.
(vii) an acidic atmosphere such as is caused by

carbon dioxide forming carbonic acid with the water present.

The prevention of rusting costs millions of pounds a year. All iron or steelwork exposed to air or water has to be painted, enamelled, greased, alloyed with non-corrosive metals such as nickel or chromium or covered with coatings of zinc, tin or shellac.

When iron is in contact with another metal and the two are immersed in water containing some dissolved salts (sea-water) or acidic gases (rain water), the more electropositive metal, i.e. the one higher in the electrochemical series, begins to corrode. This can be demonstrated in the following way.

Experiment 42.3. How other Metals affect the Rusting or Iron.

Stir some powdered gelatine into about 100 cm^3 of hot water so as to make a fairly viscous solution. Add about four drops of potassium hexacyanoferrate(III) solution and one drop of phenolphthalein. If the latter goes pink add *just sufficient* dilute hydrochloric acid to make the solution colourless. Pour this into three shallow (petrie) dishes (or test-tubes). Take three clean nails and attach one to some magnesium ribbon and wind copper wire round one of the others. Place each of the nails into a separate dish containing the gelatine, cover and cool rapidly to set. The dishes should be left for about twenty-four hours and then examined. When iron dissolves or corrodes in contact with water it first forms iron(II) (Fe^{2+}) irons and these give a deep blue colour with potassium hexacyanoferrate (III). This will be seen in the dish containing the iron by itself, but in the one where it is attached to the copper the colour will be deeper and more extensive, showing that corrosion has gone farther. In the third tube there should be little if any blue colour, but the indicator will have turned pink where the magnesium has dissolved in preference to the iron.

The sacrificial action of other metals connected to iron by which rusting is prevented is made use of in industry. Galvanizing is common; in this method sheets of iron are coated with metallic zinc and if a hole or break occurs it is the zinc which goes into solution. Small objects such as bolts are often heated with powdered zinc which penetrates the metal to produce the same effect; this is known as sheradizing. Iron coated with a thin layer of tin is used for food cans and further protection may be given by a layer of shellac. Tin is more resistant to the action of acids in canned foods than zinc would be, but if a scratch occurs it is the iron which quickly rusts. Steel pipes of framework and the hulls of ships continually in contact with water may be connected to blocks of magnesium; the magnesium corrodes away first and it is cheaper to replace this than effect the repairs to the ironwork which would otherwise be necessary. Copper or brass bolts cannot be used with iron joists or plates exposed to water or the iron will rust. Two dissimilar metals in contact in the presence of water usually produce the conditions necessary for the rusting of one or the other to take place.

Zinc

The main ores of zinc are the sulphide, zinc blende and the carbonate, calamine, and most of these processed in Britain come from Australia. They are first roasted to form the oxide and then reduced to the metal.

$$2ZnS(s) + 3O_2(g) \rightarrow 2ZnO(s) + 2SO_2(g)$$
$$ZnCO_3(s) \rightarrow ZnO(s) + CO_2(g)$$

Two modern methods of bringing about this reduction are in use at the Avonmouth works of the Imperial Smelting Corporation.

The oxide is powdered, mixed with ground coal and anthracite and pressed into briquettes. These are charged into vertical retorts at a temperature of about 600–700 °C. Zinc vapour distils over and is cooled, giving 8 t of zinc per retort every day.

$$ZnO(s) + C(s) \rightarrow Zn(s) + CO(g)$$

A more recent method involves the use of a blast furnace with a capacity of more than 100t of zinc a day. The oxide, in lumps, is mixed with coke and fed into the furnace. Whereas with iron the metal is tapped at the bottom of the furnace, zinc is volatile at the temperature at which the oxide is reduced and leaves with the waste gases. The zinc vapour is condensed by spraying with molten lead. When the mixture cools the zinc floats to the surface and overflows; the lead is then re-used.

Many thousands of tonnes of zinc are used in the United Kingdom each week for making into battery cups, foil, weather stripping, lithographic plates, organ pipes, roofing tiles, stencils and for lining wooden packing cases and tea chests. Zinc alloy anodes are used for cathodic protection against rusting, particularly of ships' hulls and ballast tanks and pier installations.

Lead

The chief ore of lead is the sulphide, galena,

PbS. One method by which the metal is extracted from it is as follows:

The sulphide is mixed with quartz and roasted in a current of air. Oxidation occurs.

$$2PbS(s) + 3O_2(g) \rightarrow 2PbO(s) + 2SO_2(g)$$

Some lead(II) sulphate is formed, but this reacts with the quartz to give lead(II) silicate. The products of roasting are then mixed with coke and limestone and heated in a blast furnace. Here the oxide is reduced to the metal by the coke and the carbon monoxide formed.

$$PbO(s) + C(s) \rightarrow Pb(s) + CO(g)$$
$$PbO(s) + CO(g) \rightarrow Pb(s) + CO_2(g)$$

The silicate forms a slag by reacting with the limestone.

$$CaCO_3 + PbSiO_3 + CO \rightarrow Pb + CaSiO_3 + 2CO_2$$

Refining of the metal has to be carried out and the final stage involves electrolysis, with the impure lead as the anode and a pure lead cathode.

As zinc and lead ores are usually found together much lead is obtained during the extraction of zinc.

The pliability of lead has caused it to be used since the time of the Romans for plumbing, but it is not entirely satisfactory with very soft water. Lead(II) hydroxide may be formed and act as a cumulative poison. Hard water forms a protective layer of lead(II) carbonate or sulphate. Because of its resistance to corrosion lead is used for covering cables and some roofs. Lead, alloyed with antimony, is used in accumulators, while other alloys include solder and type metal. Much lead (about 25% of the amount produced) goes to make the carbonate and red lead which are ingredients of paints. Lead tetra-ethyl is added to petrol to prevent knocking.

Copper

Copper is most often found as the impure sulphide and this is roasted to form copper(I) oxide

Table 42.1 Alloys

Main Metals and other Elements	Particular Properties	Uses
IRON—carbon 0·5–1·4%	malleable and can be welded	bridges, pistons, projectiles, springs, drills
—Cr < 3%	very hard	files and gears
—Cr 13%	stainless	cutlery
—Cr 13–20% Ni 8%	stainless	dairy and food processing equipment
—C 9·1% Cr 2–10% W 4–25%	very hard	high speed tools
—Si 16%	acid-proof	plumbing and chemical ware
COPPER—Sn up to 12% (Bronze)	hard wearing	medals
—Zn 20% Ni 1%	hard wearing	coins
—Zn up to 30% (Brass)	hard wearing	screws, cartridges
—P 0·3% Sn 4–6% (Phosphor bronze)	hard wearing	watch springs
ALUMINIUM—Cu 10% (Aluminium bronze)	hard wearing	aircraft parts
—Mg 10% (Magnalium)	light but strong	aircraft parts
—Cu 4% Mg 1% Mn 1% (Duralumin)	weight saving, strong as steel	aircraft plates, parts for cars, prefabricated houses
LEAD—Sn up to 50% (Solder)	low m.pt. and good adhesion to metals	soldering metals
—Sn and Sb	expands slightly on solidifying	type metal
BISMUTH—Sn 24% Pb 27%	m.pt. below 100°C.	electric fuses, sprinkler plugs

$$2Cu_2S + 3O_2 \rightarrow 2Cu_2O + 2SO_2 \rightarrow$$

The oxide then reacts with more of the sulphide to give impure copper

$$Cu_2S + 2CuO \rightarrow 6Cu + SO_2$$

If you add both equations together and then simplify you will see that the overall reaction is just

$$Cu_2S(s) + O_2(g) \rightarrow 2Cu(s) + SO_2(g)$$

The metal then has to be refined electrolytically as described on page 232.

The uses of copper are manifold, particularly in the electrical industry and for making cooking utensils and boilers where good conductivity—either electrical or thermal—is important.

Alloys

We do not very often come across pure metals in our daily life outside the laboratory with the exception of lead, copper and aluminium. In general, metals are mixed with each other and sometimes with non-metals to form alloys which are more suitable for particular purposes. Let us look at some alloys and their uses (but do not attempt to learn all the facts given in the table on p. 257!).

(Questions on this chapter will be found at the end of Chapter 43.)

43
Compounds of Metals

Oxides and Hydroxides

The common oxides of the metals sodium, calcium, zinc, iron and copper have the formulae Na_2O, CaO, ZnO, Fe_2O_3, Fe_3O_4 and CuO. *Sodium* 'rusts' readily on exposure to air and forms the oxide (or burns, with a yellow flame if heated to give both the monoxide and the peroxide).

$$4Na + O_2 \xrightarrow{\Delta} 2Na_2O$$

$$2Na + O_2 \xrightarrow{\Delta} Na_2O_2$$

The monoxide is readily soluble in water to form caustic soda, or sodium hydroxide.

$$Na_2O + H_2O \rightarrow 2NaOH$$

The oxide cannot be reduced to the metal by any chemical reducing agent and the hydroxide is not decomposed by strong heat. As these compounds are so readily formed we would expect them to be stable.

Calcium, a brittle metal, slowly turns to white calcium oxide when left in the air. Burning the metal in oxygen or strongly heating the carbonate is a better method.

$$2Ca + O_2 \xrightarrow{\Delta} 2CaO$$

$$CaCO_3 \xrightarrow{\Delta} CaO + CO_2$$

The oxide formed becomes incandescent at a high temperature (limelight) but is not decomposed. When a little water is added to calcium oxide it is converted into steam by the heat of the slaking (a considerable amount of energy is released) and the solid swells and crumbles to give the white hydroxide.

$$CaO + H_2O \rightarrow Ca(OH)_2 + Heat$$

This process is reversible, heating driving off the water taken in.

Sodium and calcium both react vigorously with water to liberate hydrogen and form hydroxides. Calcium, being denser than water, sinks to the bottom and if an inverted, short-stemmed funnel is placed over the metal the hydrogen may be collected in a test-tube (use distilled water). The water under the funnel becomes cloudy showing that the hydroxide is not very soluble in water.

$$Ca(s) + 2H_2O(l) \rightarrow Ca(OH)_2(aq) + H_2(g)$$

The solution is usually called limewater and the solid slaked lime.

Sodium hydroxide is a white deliquescent solid, usually obtained in sticks, pellets or flakes, but *calcium hydroxide*, like the oxide, is a dry white powder.

Both aqueous sodium hydroxide and limewater are alkaline to indicators, the former the more strongly so. A solution of sodium hydroxide is used in the laboratory for absorbing acidic gases such as carbon dioxide and sulphur dioxide, and for precipitating insoluble hydroxides and so detecting the pressence of certain metallic ions in a solution. Both alkalis will liberate ammonia from ammonium compounds with which they are heated and limewater is used for detecting the presence of carbon dioxide. In industry sodium hydroxide plays a part in the preparation of sodium itself, of rayon and of soap, in the refining of petroleum and in the removal of paint and grease.

Slaked lime is much used in the building industry for the preparation of lime mortar in which it is mixed with sand and water. Calcium carbonate and silicate are formed as the mixture dries out. It is now being replaced by a cement mortar which is harder and more durable.

Calcium oxide is produced industrially by allowing chalk or limestone to fall down a vertical kiln, or through a rotating inclined kiln, which is being heated to about 1500°C. by a flame from oil, coal or natural gas or producer gas.

$$CaCO_3(s) \rightarrow CaO(s) + CO_2(g)$$

When calcium oxide and coke are heated strongly in an electric furnace calcium carbide is formed. From this can be obtained the gas ethyne, C_2H_2, which is used in making many plastics and other organic compounds.

Experiment 43.1. Production of Calcium Oxide from Chalk.

Place some chalk in a crucible and arrange it in a small muffle furnace (i.e. one 'wrapped up' with fireclay or asbestos to conserve heat—Fig. 43.1). Heat the crucible gently at first and then very strongly for about twenty minutes. Allow it to cool down and then add a little water to the powder. Much heat will be evolved as the quicklime formed is slaked.

Industrially sodium hydroxide is produced almost entirely by the electrolysis of brine, chlorine and hydrogen being the other products. Both are collected for use and this is the main source of

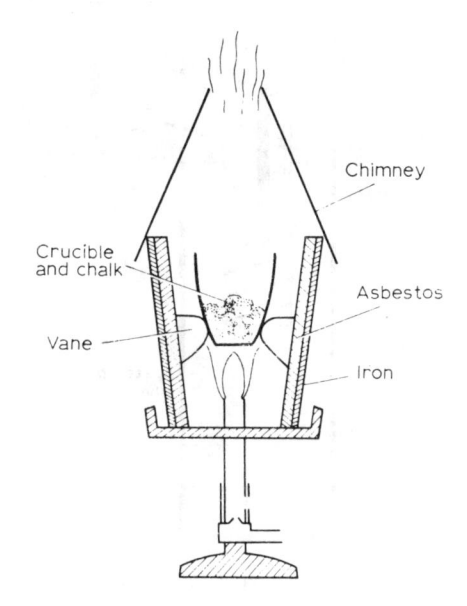

Fig. 43.1. Muffle Furnace

chlorine (see p. 26). The demand for chlorine is very great and continually growing whereas few new fields in which sodium hydroxide can be used have been found in recent years. In fact, some sodium hydroxide is now converted into sodium carbonate instead of the other way round as was the case when the lime-soda process was used.

The electrolysis may be carried out in one cell (Nelson or Hooker cell) or in two cells (Kellner-Solvay method). In the former the brine is contained in a porous asbestos cell covered with steel mesh serving as the cathode. A carbon rod in the brine is the anode. The hydrogen and sodium ions are attracted to the cathode where the hydrogen is set free. At the anode the chloride ions lose electrons to form chlorine gas which is evolved. The solution contains both sodium and hydroxide ions and so is one of sodium hydroxide.

In the 2-stage process the cathode is a mercury one and in this case sodium is liberated to form an amalgam with the mercury which then flows into a second cell containing water. Here the amalgam reacts to give sodium hydroxide and hydrogen.

$$2\,Na/Hg + 2H_2O \rightarrow 2NaOH + 2Hg + H_2$$

No hydrogen is evolved in the first cell as the reaction

$$Na^+ + Hg + e^- \rightarrow Na/Hg$$

requires less energy than the one for the discharge of hydrogen

$$H^+ + e^- \rightarrow H \rightarrow H_2(g)$$

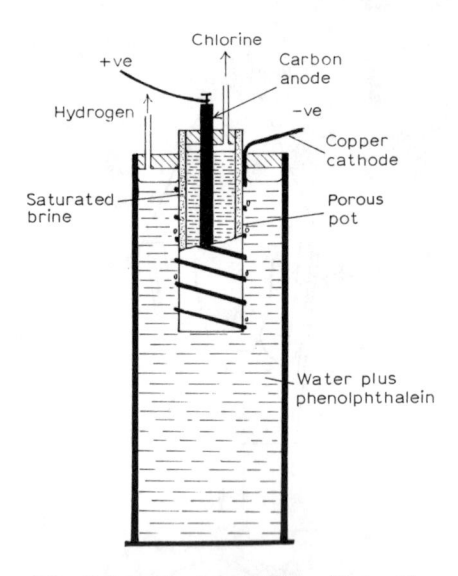

Fig. 43.2. Production of Sodium Hydroxide

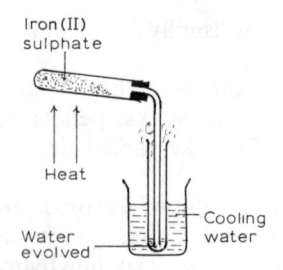

Fig. 43.3. Action of heat on Iron(II) Sulphate

Experiment 43.2. Preparation of Sodium Hydroxide (electrolytic process).

The apparatus in Fig. 43.2 represents a Nelson Cell. Copper wire wrapped round the 2 cm diameter porous pot forms the cathode. When the current is switched on bubbles of gas are seen to form on the wire and hydrogen is evolved. Chlorine can soon be detected by its smell and bleaching action on damp litmus paper. The phenolphthalein in the water turns pink as sodium hydroxide is produced.

Zinc and copper oxides are best made by heating the carbonates or nitrates strongly. Copper(II) oxide is black and zinc oxide yellow when hot and white when cold and this is one way of identifying it.

$$ZnCO_3(s) \rightarrow ZnO(s) + CO_2(g)$$
$$2Cu(NO_3)_2(s) \rightarrow 2CuO(s) + 4NO_2(g) + O_2(g)$$

Iron(III) oxide is best obtained by heating crystals of iron(II) sulphate very strongly.

Experiment 43.3. The Action of Heat on Iron(II) Sulphate.

These pale green crystals contain water of crystallization and on heating (Fig. 43.3) this is lost.

$$\underset{\text{green}}{FeSO_4 \cdot 7H_2O} \rightarrow \underset{\text{white}}{FeSO_4} + 7H_2O$$

Further strong heating decomposes the white anhydrous solid to evolve sulphur dioxide and sulphur trioxide and leaves red-brown iron(III) oxide.

$$2FeSO_4(s) \rightarrow Fe_2O_3(s) + SO_2(g) + SO_3(g)$$

The fumes dissolve in the water of crystallization already given off to form sulphurous and sulphuric acids.

$$SO_2 + SO_3 + 2H_2O \rightarrow H_2SO_3 + H_2SO_4$$

Divide the solution into two parts; to one add potassium permanganate solution and to the other some dilute hydrochloric acid followed by barium chloride solution. What do you expect to happen? If you have forgotten refer to pp. 68 and 73.

Zinc oxide is used in making white paints, celluloid, rayon, metal polish, tyres and antiseptic ointments (e.g. on adhesive plaster); copper(II) oxide plays a part in the production of blue glass and iron(III) oxide, when finely ground, is known as jeweller's rouge and is used for polishing metallurgical specimens and silverware. It is also a pigment, 'red oxide paint', used on metalwork.

As these oxides are insoluble in water the corresponding hydroxides must be made by double decomposition. A solution of zinc sulphate, copper(II) sulphate or iron(III) chloride is added to one of sodium hydroxide.

$$Zn^{2+}(aq) + 2OH^-(aq) \rightarrow Zn(OH)_2(s) \text{ white}$$
$$Cu^{2+}(aq) + 2OH^-(aq) \rightarrow Cu(OH)_2(s) \text{ pale blue}$$
$$Fe^{3+}(aq) + 3OH^-(aq) \rightarrow Fe(OH)_3(s) \text{ red-brown}$$

If iron(II) sulphate solution is used instead of iron(III) chloride, iron(II) hydroxide is precipitated.

$$Fe^{2+}(aq) + 2OH^-(aq) \rightarrow Fe(OH)_2(s) \text{ dirty green}$$

When pure, e.g. prepared in an atmosphere of hydrogen, iron(II) hydroxide is white, but in the air it rapidly becomes oxidized and eventually changes to iron(III) hydroxide. The difference in colour of the hydroxides gives us a means of distinguishing between iron(II) and iron(III) compounds in solution. The oxide corresponding to $Fe(OH)_2$ is iron(II) oxide, FeO. This is a very unstable compound and you are not likely to encounter it.

When these hydroxides are heated strongly the products are zinc, copper(II) and iron(III) oxides.

Tri-iron tetroxide (Fe_3O_4) is formed when iron is burnt in oxygen or heated strongly in steam. The presence of *one* of the substances gives ghe black oxide; if *both* are present iron(III) oxide is formed (in the hydrated form, as rust).

All the metal oxides and hydroxides mentioned dissolve in dilute acids to form solutions from which, by careful evaporation, salts may be obtained.

$$ZnO + H_2SO_4 \rightarrow ZnSO_4 + H_2O$$
$$CuO + 2HCl \rightarrow CuCl_2 + H_2O$$
$$2Fe(OH)_3 + 3H_2SO_4 \rightarrow Fe_2(SO_4)_3 + 6H_2O$$

Tri-iron tetroxide is only slightly soluble in dilute acids and behaves as though it is a mixture of iron(II) and iron(III) oxides, i.e. $FeO.Fe_2O_3$. With hydrochloric acid it forms a mixture of iron(II) and iron(III) chlorides.

Zinc oxide and hydroxide behave differently from the others mentioned in that they are also soluble in aqueous sodium hydroxide or potassium hydroxide solution to form a zincate.

$$ZnO + 2NaOH \rightarrow \underset{\text{sodium zincate}}{Na_2ZnO_2} + H_2O$$
$$Zn(OH)_2 + 2NaOH \rightarrow Na_2ZnO_2 + 2H_2O$$

They are said to be amphoteric (see p. 118) and have the property of being soluble in dilute acids and in alkalis. Consequently, in preparing zinc hydroxide by double decomposition it is best to add the zinc sulphate to the caustic alkali to ensure that the latter is not in excess or the precipitate may not be formed. Aluminium oxide and hydroxide behave similarly.

Lead(II) oxide (lead monoxide) exists in two forms, a yellow one called massicot and an orange one known as litharge. It is prepared by heating lead, lead(II) nitrate, lead(II) carbonate or dilead (II) lead(IV) oxide (tri-lead tetroxide, red lead) fairly strongly.

$$2Pb(s) + O_2(g) \rightarrow 2PbO(s)$$
$$2Pb(NO_3)_2(s) \rightarrow 2PbO(s) + 4NO_2(g) + O_2(g)$$
$$PbCO_3(s) \rightarrow PbO(s) + CO_2(g)$$
$$2Pb_3O_4(s) \rightarrow 6PbO(s) + O_2(g)$$

If the preparation is carried out in a test-tube the oxide fuses into the glass, forming some lead silicate. Lead(II) oxide is used in making heavy lead glass (flint glass).

The oxide is insoluble in water and only slightly soluble in cold hydrochloric acid to form lead(II) chloride (solubility only 0·7 g per 100 g of water); it is more soluble when the acid is heated as the solubility of the chloride increases to 3. It will readily dissolve in nitric acid but not in sulphuric (why do you think this is so?). If the chloride or sulphate is to be made from the oxide the nitrate should first be prepared in solution and then the required salt formed by double decomposition.

$$Pb(NO_3)_2(aq) + H_2SO_4(aq) \rightarrow PbSO_4(s) + 2HNO_3(aq)$$

Lead(II) hydroxide can be precipitated from a solution of lead(II) nitrate or lead(II) ethanoate (acetate) by adding aqueous sodium hydroxide. It is white and dense, but like zinc hydroxide it dissolves if excess alkali is added. Another amphoteric hydroxide is aluminium hydroxide, $Al(OH)_3$ (see table on p. 321).

All these hydroxides form oxides when heated strongly. Both iron(II) and iron(III) oxides, $Fe(OH)_2$ and $Fe(OH)_3$, form iron(III) oxide, Fe_2O_3; any iron(II) oxide, FeO, formed will immediately oxidize to the other one.

$$4Fe(OH)_2 + O_2 \rightarrow 2Fe_2O_3 + 4H_2O$$

Carbonates and Hydrogencarbonates

The commonest natural carbonate is undoubtedly *calcium carbonate* which exists as limestone, marble, chalk and calcite in many parts of the world and is the main constituent of the shells of sea creatures such as oysters and crabs, of bones (with calcium phosphate) and of egg-shells.

It is insoluble in pure water, but dissolves in rain water (which is acidic due to the presence of carbon dioxide in solution) to form calcium hydrogencarbonate.

$$CaCO_3 + CO_2 + H_2O \rightarrow Ca(HCO_3)_2$$

This dissolving of the rock has caused many caves in the limestone areas of Britain, of the Pyrenees and of central Europe. The hydrogencarbonate is the main cause of temporary hardness in water (see p. 204).

The uses of calcium carbonate are numerous. From it is obtained calcium oxide (see p.204) and if it is mixed with clay before the roasting takes place we get cement; a mixture of this substance with water sets hard. Crushed chalk is the basis of many polishing powders and toothpastes and of putty and whiting or whitewash. Limestone, although subject to weathering by wind and rain, is employed for building purposes and it is one of the raw materials in the manufacture of glass and of iron. It also makes up most of the ballast in which railway sleepers are bedded and small chippings mixed with tar make good road surfaces. Marble has been produced from chalk by the action of great heat and pressure over millions of years and is much harder. It is used for ornamental building work such as floors and pillars in churches and town halls and is the chief stone used by sculptors. It is often coloured by iron oxide and other minerals present in it.

Sodium carbonate is the most important industrial carbonate and is exported from Britain to countries in all parts of the world. It is made by the ammonia-soda process which is outlined in a flow diagram in Fig. 43.4. The total world production is of the order of five million tons per annum.

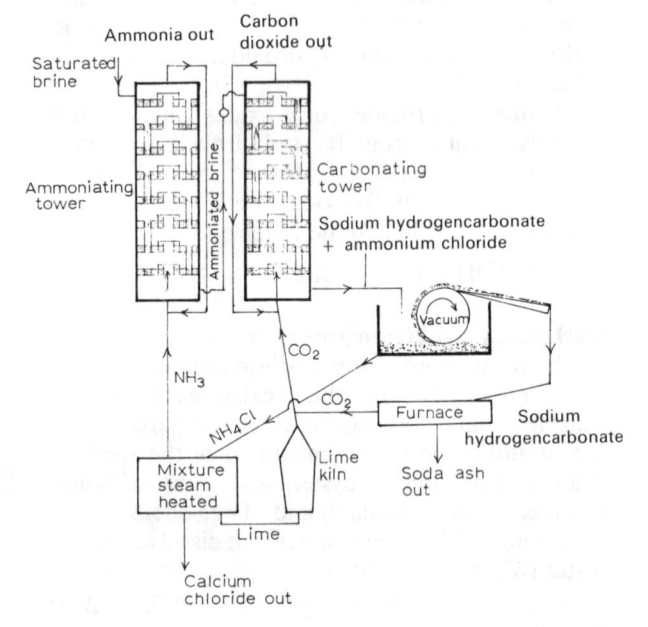

Fig. 43.4. Solvay process for Sodium Carbonate

Ammonia-Soda or Solvay Process

Brine is saturated with ammonia, any hydroxides precipitated from impurities present being allowed to settle. This solution is pumped to the top of the Solvay tower up which carbon dioxide is passing. This reacts with the ammoniated brine in this way:

$$NH_3(g) + NaCl(aq) + H_2O(l) + CO_2(g) \rightarrow NH_4Cl(aq) + NaHCO_3(s)$$

Conditions are so arranged that two-thirds of the salt is converted into sodium hydrogencarbonate (bicarbonate); this, being only sparingly soluble in water is precipitated whilst the ammonium chloride remains in solution. The hydrogencarbonate is separated on a rotating vacuum filter and roasted to give the anhydrous carbonate or soda ash.

$$2NaHCO_3(s) \xrightarrow{\Delta} Na_2CO_3(s) + CO_2(g) + H_2O(g)$$

The carbon dioxide liberated provides part of the gas needed for the first reaction, the rest being obtained by heating limestone in a kiln.

$$CaCO_3(s) \xrightarrow{\Delta} CaO(s) + CO_2(g)$$

Ammonia is recovered from the ammonium chloride solution by heating it with lime from the kiln.

$$2NH_4Cl(aq) + CaO(s) + H_2O(l) \xrightarrow{\Delta} 2NH_3(g) + CaCl_2(aq) + 2H_2O(g)$$

The only waste material is calcium chloride for which, as yet, there are no large-scale uses.

To produce washing soda crystals the soda ash is dissolved in hot water to give a saturated solution and allowed to cool and crystallize.

Sodium carbonate crystallizes as the decahydrate, $Na_2CO_3.10H_2O$, washing soda. When exposed to the atmosphere it effloresces, i.e. much of the water of crystallization is removed and the crystals become opaque, being covered with a white powder, the monohydrate.

$$Na_2CO_3.10H_2O \rightarrow Na_2CO_3.H_2O + 9H_2O$$

This change does not occur if the air around already contains a fair percentage of moisture as in many tropical (or even temperate) countries. In extreme cases this percentage may rise as high as 6% or even more. In parts of Kenya and Malaysia sodium carbonate is definitely deliquescent and this is also the case in other equatorial countries.

On heating washing soda crystals strongly they decrepitate, i.e. break up noisily, and the final product is anhydrous sodium carbonate. *No further decomposition occurs and no carbon dioxide is given off.* The only common carbonates not decomposed by heat are those of sodium and potassium.

$$Na_2CO_3.10H_2O \xrightarrow{\Delta} Na_2CO_3 + 10H_2O$$

Addition of dilute acid to the crystals liberates carbon dioxide and the volume of gas produced is the same as that obtained by adding the acid to the residue left after the same weight of crystals has been heated. Sodium carbonate is very soluble in water. It gives an alkaline solution with a pH of about 11–12. Alkaline solutions of sodium hydroxide (caustic alkali) or sodium carbonate (mild alkali) have a soapy feeling; they remove the protective oils from the surface of the skin.

You will remember that carbonates are formulated by substituting two atoms of sodium or one of calcium for the two atoms of hydrogen in carbonic acid, H_2CO_3, i.e. Na_2CO_3 or $CaCO_3$. When, however, only one atom of hydrogen is substituted hydrogencarbonates are formed, i.e. $NaHCO_3$ or $Ca(HCO_3)_2$. (We cannot write $Ca\frac{1}{2}HCO_3$ so the whole formula is doubled.) Bicarbonates are now called hydrogen-carbonates, eg. sodium hydrogencarbonate.

When carbon dioxide is passed into lime water calcium carbonate is first formed as a precipitate and this then dissolves to give the hydrogencarbonate. Similarly, when this gas is passed into sodium hydroxide solution the carbonate is first formed, but as this is soluble we cannot be sure when this stage of the reaction is complete. So, in preparing the carbonate in the laboratory we first of all make the hydrogencarbonate.

> *Experiment 43.4. Preparation of Sodium Carbonate and Sodium Hydrogencarbonate.*
> (a) Carbon dioxide from a Kipp's is passed through a U-tube or Drechsel bottle containing cotton wool to hold back any acid spray carried over by the gas. It is then passed into 25 cm³ of a fairly concentrated solution of sodium hydroxide (about 15 g per 100 cm³) for an hour or more to make sure that excess gas is used. Sodium hydrogencarbonate is formed.
>
> $$NaOH + CO_2 \rightarrow NaHCO_3$$
>
> and will eventually precipitate out owing to its low solubility. (Industrially, it is obtained by the Solvay process.)
> (b) Take some of the precipitated hydrogencarbonate and heat. Water and carbon dioxide come off and anhydrous sodium carbonate is left.
>
> $$2NaHCO_3(s) \rightarrow Na_2CO_3(s) + CO_2(g) + H_2O(g)$$
>
> Dissolve some of this soda ash in a little hot water and evaporate until crystallization is about to commence. Leave and crystals of washing soda will form.

If carbon dioxide, produced and cleaned as in (a) is passed into a saturated solution of sodium carbonate the hydrogencarbonate will eventually precipitate out.

$$Na_2CO_3(aq) + H_2O(l) + CO_2(g) \rightarrow 2NaHCO_3(s)$$

If calcium carbonate (precipitated chalk) is to be prepared in a similar way it is best to pass excess carbon dioxide into limewater to give the soluble hydrogencarbonate and then heat this solution to decompose it.

$$Ca(HCO_3)_2 \xrightarrow{\Delta} CaCO_3(s) + H_2O(l) + CO_2(g)$$

This may seem a roundabout way as the carbonate is first precipitated, but filtration is much easier if it is obtained from the hydrogencarbonate.

Calcium hydrogencarbonate is so easily decomposed that it has only been obtained in the solid state at low temperatures.

The main uses of sodium carbonate include the manufacture of soap and soap powders, rayon, textiles and paper. Most of it, however, is fused with limestone and sand to make ordinary soft glass. It is also employed in water softening.

$$CO_3^{2-}(aq) + Ca^{2+}(aq) \rightarrow CaCO_3(s)$$

Sodium hydrogencarbonate (sodium bicarbonate) is called an acid salt (see p. 71) because only half the hydrogen in carbonic acid has been replaced. Nevertheless it is slightly alkaline in solution (pH = 8 – 9). As it is such a mild alkali it can be used to neutralize acids spilled on the skin or present in excess in the stomach and causing indigestion. It is a constituent of baking powder and is itself known as baking soda. Addition of it to a cake mixture causes the latter to rise when heated as steam and carbon dioxide are given off. This action must be completed before the cake is removed from the oven or the steam bubbles will condense back to water and make the cake 'heavy' or 'sad'. Sodium carbonate would be left in the cake, but this is prevented by the addition of tartaric acid or cream of tartar to the hydrogencarbonate.

Addition of dilute acids (except carbonic) to carbonates or hydrogencarbonates causes the evolution of carbon dioxide.

$$Na_2CO_3(s) + H_2SO_4(aq) \rightarrow Na_2SO_4(aq) + H_2O(l) + CO_2(g)$$

or $CO_3^{2-}(aq) + 2H^+(aq) \rightarrow H_2O(l) + CO_2(g)$
and $HCO_3^-(aq) + H^+(aq) \rightarrow H_2O(l) + CO_2(g)$

If a solid acid such as citric or tartaric is mixed with sodium hydrogencarbonate nothing happens as long as the mixture is kept dry. Addition of water dissolves the acid and then the hydrogen and hydrogencarbonate ions react together as above to release carbon dioxide with considerable effervescence. Such is the action of water on 'health salts' (which may also contain sugar and Epsom salts.)

The acid-soda type of fire extinguisher (Fig. 43.5) contains a small bottle of sulphuric acid and a large amount of sodium hydrogencarbonate solution. Breaking the bottle or causing the stopper to fall out mixes the solutions and the carbon dioxide evolved forces out a jet of the solution up to nine

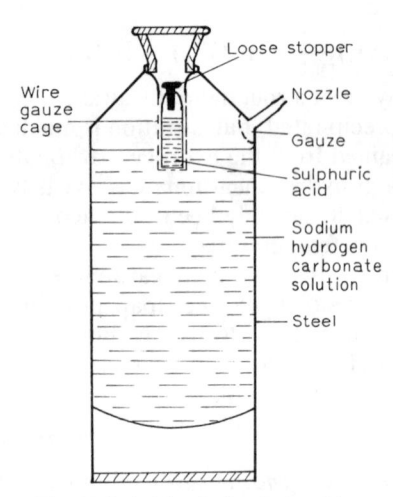

Fig. 43.5. Acid-soda fire extinguisher

metres long. In foam fire extinguishers the acid is replaced by a solution of aluminium sulphate (acidic in action) and a foaming agent. When the two liquids are mixed a foam of aluminium hydroxide, water and bubbles of carbon dioxide effectively 'blankets' any fire.

*Experiment 43.5. The action of a Foam Fire Extinguisher.

Make a fairly concentrated solution of aluminium sulphate and place it in a tap funnel as shown in Fig. 43.6. The gas jar contains a saturated solution of sodium hydrogencarbonate plus a little detergent or (better) saponin—obtained by grinding horse-chestnuts under water and filtering. The solution should not occupy more than three-quarters of the jar. Put some cotton wool on an empty sand-bath standing in a metal tray and moisten it with ethanol or benzene and then set it alight. Run the aluminium sulphate solution into the jar and close the tap. A jet of liquid issues from the nozzle and can be played on the fire to extinguish it.

Extinguishers which have been more recently developed and which are now in use may contain:
(a) a highly-expanded foam made from a detergent and air under very high pressure. The water content is low so little damage is done.
(b) 'light water'. Here a carbon-fluorine compound is used as the foaming agent and this foam can actually wet the burning oil and spread out as an aqueous film on *top* of the oil and so exclude air; hence its name, *light water*. It is forced out by air or nitrogen at about 13 atmospheres pressure.

Chlorides

The most important sodium compound is undoubtedly *sodium chloride*. No other, except the nitrate, exists in very large amounts anywhere in the world, although sufficient sodium carbonate crystallizes out on the north-east shores of Lake Chad, south of the Sahara in Africa, to make its utilisation commercially feasible. From the beginning of the world common salt has played an important part in our lives. People who work in hot climates or sweat profusely at their jobs (e.g. miners, stokers, foundry workers) lose much salt and need to replace this by extra sodium chloride in their food. The hydrochloric acid in the stomach, which is used for digesting food, is produced from this salt.

Sea water contains about 3% of sodium chloride by weight and in hot countries, such as India, the water is trapped at high tide in large 'pans'; solar heat is sufficient to evaporate the water away leaving the salt. It is not economical to use other forms of energy for this purpose. In other countries the chloride is obtained from underground deposits (formed in past time by the evaporation of underground seas) either by mining—one salt mine still exists in Britain, in Cheshire, and there are many in Europe—or by forcing water down to dissolve the salt and pumping up the brine so formed. The water is then evaporated, usually by steam coils. The steam formed from the brine is passed to a further evaporator also containing brine so that its heat may not be lost and so on. This is called 'multiple effect evaporation'.

Salt, as mined, varies in colour from white to

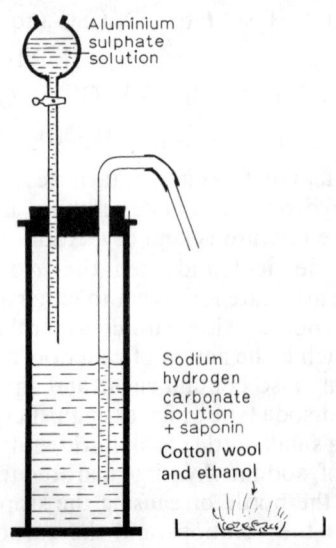

Fig. 43.6. Foam fire extinguisher

brown due to the presence of iron(III) oxide. It is known as rock salt. It is not decomposed by heat (m.pt. 800°C); this is as expected when we consider the readiness with which sodium and chlorine combine to form it. It is not very soluble in water (36–40 g per 100 g water) and crystallizes normally in cubes or prisms. When quite pure it is not deliquescent, but it may contain traces of magnesium chloride which soon make it become damp. Addition of calcium phosphate avoids this and makes it free-running. Table salt which has been 'iodized' is used by people suffering from iodine deficiency, but is available for all.

> *Experiment 43.6. To obtain pure Salt from Rock Salt.*
> Grind some rock salt in a mortar and transfer it to a test-tube. Add water and shake for a while to obtain a saturated solution. Filter off the undissolved solid and add a little concentrated hydrochloric acid to the filtrate. White sodium chloride is precipitated as salt is almost insoluble in this acid.* Filter again and wash the residue on the paper with a little water to remove any traces of acid. Dry and taste.

Sodium chloride is the raw material for the preparation of most sodium compounds and many other substances. Amongst these are:
sodium (by electrolysis of the fused chloride)
sodium hydroxide (by electrolysis of cold brine)
chlorine and *hydrogen* (by electrolysis of cold brine)
sodium chlorate(I) (by electrolysis of cold brine with the electrodes close together); this is used as a bleach in Milton, Parazone, Domestos, etc.
hydrogen chloride (by heating with concentrated sulphuric acid or from hydrogen and chlorine) and *hydrochloric acid*
sodium carbonate (Solvay process)
bleaching powder (from chlorine)
soap (from sodium hydroxide)
sodium hydrogencarbonate, *rayon*, etc.

More than eighty million megagrams of salt are produced annually in the world.

Calcium chloride is the 'waste' material of the Solvay process and as yet there are no large industrial uses for it. Some is used for the preparation of calcium metal (electrolysis of the fused chloride). The chief physical property of the anhydrous or fused salt is that before it forms a solution it readily absorbs water and for this reason it is made use

---

*The concentrated acid has a strong attraction for water and dissolves in it leaving insufficient water to keep the salt in solution, hence the precipitation.

$$CaCl_2 \rightarrow CaCl_2.6H_2O$$

of in the laboratory for the drying of gases (not ammonia) or solids (in a desiccator). It is used also for removing traces of water from organic liquids. Sprinkled on the surface of dirt roads in dry weather it helps keep down the dust and its deliquescent nature similarly helps to delay the setting of concrete in hot dry weather.

Iron forms both *iron(II)* and *iron(III) chlorides*.

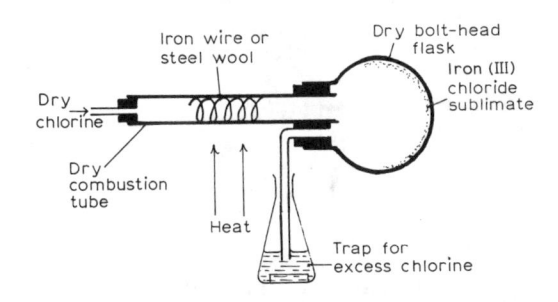

Fig. 43.7. Preparation of Iron(III) Chloride

> *Experiment 43.7. Preparation of Anhydrous Iron(III) Chloride.*
> Set up the apparatus of Fig. 43.7, making sure that all the glassware has been previously heated and allowed to cool in a desiccator to remove all moisture. Heat the iron gently and pass a current of dry chlorine through the tube, absorbing excess in a tower containing calcium hydroxide or in a flask containing stick sodium hydroxide in ammonium hydroxide. The iron soon glows brightly and the reaction continues even if the flame is removed.
>
> $$2Fe + 3Cl_2 \rightarrow 2FeCl_3$$
>
> The iron(III) chloride formed sublimes and is deposited as black crystals on the cold surface of the flask. If any moisture is present the colour may be yellow-brown.

Iron(II) chloride may be made by:
(a) dissolving iron wire or filings in dilute hydrochloric acid (see iron(II) sulphate for similar details). This method gives pale green crystals of the hydrated salt,

$$Fe + 2HCl + 6H_2O \rightarrow FeCl_2.6H_2O + H_2$$

(b) heating iron wire in a current of dry hydrogen chloride (as for iron(III) chloride). White anhydrous iron(II) chloride is formed. An alternative method is described below.

Experiment 43.8. Preparation of Anhydrous Iron(II) Chloride.

A mixture of dry potassium chloride and dry potassium hydrogensulphate in the bottom of a hard test-tube (Fig. 43.8) is heated strongly. Above the mixture is a plug of asbestos carrying a layer of iron filings. This is heated and the iron glows to form white anhydrous iron(II) chloride.

$$KCl + KHSO_4 \rightarrow K_2SO_4 + HCl$$
$$Fe + 2HCl \rightarrow FeCl_2 + H_2$$

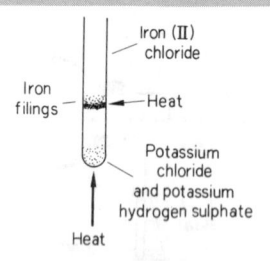

Fig. 43.8. Preparation of Iron(II) Chloride

Both chlorides are very soluble in water to give acidic solutions; iron(II) chloride forms a pale green and iron(III) chloride a yellow solution.

Sulphates and Hydrogensulphates

Sodium sulphate crystallizes with ten molecules of water and is known as Glauber's salt. It can be made by neutralizing sodium hydroxide with sulphuric acid and crystallizing from the resulting solution.

$$2NaOH + H_2SO_4 \rightarrow Na_2SO_4 + H_2O$$

If only half as much sodium hydroxide (or twice as much acid) is taken *sodium hydrogensulphate* is formed instead.

$$NaOH + H_2SO_4 \rightarrow NaHSO_4 + H_2O$$

For details of the preparation of these two salts see p. 70. The differences between acid and normal salts were explained there. Sodium hydrogensulphate behaves very much like sulphuric acid and if it is heated *very* strongly with common salt it is converted into the normal sulphate.

$$NaHSO_4(s) + NaCl(s) \rightarrow Na_2SO_4(s) + HCl(g)$$

Sodium sulphate crystals are sold as a purgative and the anhydrous substance is used in the manufacture of glass, paper and textiles.

Calcium sulphate is found in nature as the minerals anhydrite, $CaSO_4$, and gypsum, $CaSO_4.2H_2O$. It is only slightly soluble in water and so can be precipitated from solutions of calcium salts by the addition of sulphuric acid.

$$Ca^{2+}(aq) + SO_4^{2-}(aq) \rightarrow CaSO_4(s)$$

This slight solubility causes water to be permanently hard (see p. 204).

Anhydrite is an important source of sulphur dioxide which is used in the manufacture of sulphuric acid. When gypsum is heated to 120°C it loses three-quarters of its water of crystallization and forms plaster of Paris, $(CaSO_4)_2.H_2O$.

$$2CaSO_4.2H_2O \rightarrow (CaSO_4)_2.H_2O + 3H_2O$$

When this is mixed with water the reverse change gradually takes place and an interlocking mass of gypsum crystals is formed. Expansion occurs and so plaster of Paris is useful for making casts of footprints and tracks and moulds of various kinds. It is used to provide firm supports for broken limbs, is the main constituent of blackboard chalk and is present in some cements and wall plasters.

Copper(II) sulphate crystallizes as blue vitriol, $CuSO_4.5H_2O$ and is easily prepared in the laboratory.

Experiment 43.9. Preparation of Copper(II) Sulphate Crystals from Copper(II) Oxide

50 cm³ of bench sulphuric acid is warmed to about 60°C while copper(II) oxide is stirred in until no more will dissolve. The mixture is then filtered and the filtrate evaporated until crystallization sets in. Blue crystals form and should be scraped on to a filter paper in a Buchner funnel fitted to a filter flask (Fig. 43.9) which is connected to a filter pump. Wash with a *little* water to remove any acid and dry on a porous plate or between filter papers.

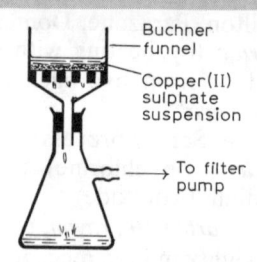

Fig. 43.9. Filtering under suction

When copper(II) sulphate crystals are subjected to moderate heat the water of crystallization is driven off to leave anhydrous copper(II) sulphate.

$$\underset{\text{blue}}{CuSO_4.5H_2O} \rightarrow \underset{\text{dirty white}}{CuSO_4} + 5H_2O$$

The reaction is reversible and is used as test for water. When water, or any liquid containing water, is added to the anhydrous powder a bluish (possibly bluish-green) colour is produced.

Strong heat on the sulphate decomposes it to black copper(II) oxide.

Copper does not dissolve in nor liberate hydrogen from dilute sulphuric acid. Its position in the electrochemical series is too low for this to happen. In industry, however, the sulphate is prepared by exposing scrap copper to the action of dilute sulphuric acid and blowing air through.

$$Cu + H_2SO_4 + [O] \rightarrow CuSO_4 + H_2O$$

This method can be imitated in the laboratory by having hydrogen peroxide as the oxidizing agent present. (Copper, even in minute amounts, catalytically decomposes hydrogen peroxide to oxygen and water.)

Experiment 43.10. Preparation of Copper(II) Sulphate from Copper.

Place a few copper turnings in a crucible or evaporating dish and cover with 2M sulphuric oxide and notice the evolution of oxygen and the blue colour developed. When all the copper has dissolved evaporate until crystallization starts.

Copper(II) sulphate is used in agriculture for preventing blight on potatoes and (with lime) as Bordeaux mixture for spraying fruit trees. The growth of algae on the sides of swimming baths is slowed down by the addition of copper(II) sulphate crystals. Copper(II) sulphate solution forms the electrolyte in copper plating and it is also used in calico printing and as a wood preservative. If a piece of wood is soaked in aqueous copper(II) sulphate and then placed in a solution of potassium chromate(VI), insoluble copper(II) chromate(VI) is precipitated in the pores of the wood which is made much more resistant to attack by bacteria in the soil.

Iron(II) sulphate, $FeSO_4.7H_2O$, is known as green vitriol (the third vitriol being zinc sulphate, $ZnSO_4.7H_2O$.) The word 'vitriol' comes from the Latin, 'vitriolum', meaning a piece of glass, and the different crystals do look like small pieces of coloured glass.

Experiments 43.11. Preparation of Iron(II) Sulphate.

One method of preparing iron(II) sulphate was described on p. 71. This method differs in the precautions taken to prevent oxidation which readily takes place.

Place about 5 g of iron filings or reduced iron in a small flask fitted with a Bunsen value (Fig. 43.10) which maintains an atmosphere of hydrogen inside the flask. Cover the metal with dilute sulphuric acid and warm slightly. When effervescence ceases (there should still be undissolved iron) filter the suspension into a test-tube containing about 10–20 cm³ of ethanol. Iron(II) sulphate crystallizes out more quickly than it would from the aqueous solution alone.

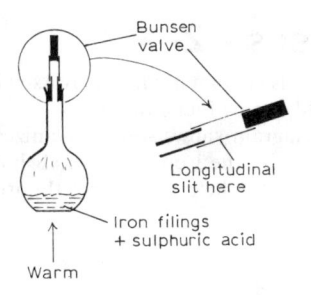

Fig. 43.10. Preparation of Iron(II) Sulphate

Earlier in the chapter details were given of the action of heat on iron(II) sulphate crystals. Sulphuric acid was originally produced by this method, hence its old name, *oil of vitriol*.

Iron(II) sulphate absorbs nitrogen monoxide to form nitroso iron(II) sulphate, a brown compound, and this action is the cause of the brown colour produced in the test for nitrates. Iron tonics, often recommended to those suffering from anaemia, may contain iron(II) sulphate. Mixed with gallic or tannic acid and a blue dye iron(II) sulphate is present in writing ink. On exposure to air oxidation occurs to give a black iron(III) compound; hence the change which occurs when writing ink is left for some time.

Nitrates

All metal nitrates are soluble in water and can be prepared in solution by reacting dilute nitric acid with the oxide, hydroxide or carbonate of a metal, e.g.

$$CuO + 2HNO_3 \rightarrow Cu(NO_3)_2 + H_2O$$
$$NaOH + HNO_3 \rightarrow NaNO_3 + H_2O$$
$$PbCO_3 + 2HNO_3 \rightarrow Pb(NO_3)_2 + H_2O + CO_2$$

Reference was made on p. 138 to the action of heat on nitrates and to the method of detecting them in solution.

Sodium nitrate, $NaNO_3$, (Chile saltpetre) is found in large quantities in the desert regions of northern Chile and is a valuable nitrogenous fertilizer. From it is made *potassium nitrate*, KNO_3, (saltpetre) used in gunpowder along with charcoal and sulphur (sodium nitrate is not suitable as it is deliquescent). *Lead(II) nitrate*, $Pb(NO_3)_2$, forms anhydrous crystals and it is used in the laboratory for the preparation of nitrogen dioxide.

Exercises 42–43

1 (a) Why is sodium not found by itself in the ground whereas gold can be found alone?

(b) The diagram shows some galvanized iron, where the zinc covering the iron has a small hole in it. How does the zinc still help to protect the iron at X from rusting?

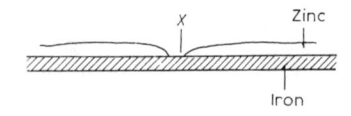

(c) What are the essential steps in the making of steel from pig iron (no diagram)? What is the main difference in the composition of iron from the blast furnace and steel? (*MR*)

2 Name: (a) a metal which occurs in the earth as its sulphide;

(b) a metal which is used extensively for making domestic water piping;

(c) a metal which is not attacked by dilute hydrochloric acid;

(d) The gas given off when sodium reacts with water;

(e) the chief ore of lead.

Explain why: (f) Lead is not normally used for making nails.

(g) aluminium is preferred to iron for making saucepans;

(h) magnesium is used in the making of photographic flash bulbs;

(i) ball-bearings and gear-wheels are not normally made of zinc;

(j) copper is preferred to iron for making coinage. (*EM*)

3 (a) Write an account with diagram of the way in which iron is obtained from iron-ore.

(b) Give three physical properties of tin that illustrate its metallic character.

(c) What is copper used for electric wiring?

(d) Why is aluminium and its alloys used for aircraft construction?

(e) What metal is used to make containers for radioactive materials? (*WM*)

4 (a) Give the names of TWO alloys which contain lead.

(b) Explain very briefly why lead is used (i) in making pipes, (ii) in some hydrometers as lead shot. (*S*)

5 Clean iron nails are placed in each of four jars, A, B, C, and D. Jar A contains dry air, jar B ordinary tap water, jar C boiled water. Jars A, B and C are then carefully sealed, but jar D is left open to the atmosphere. Describe the changes you would expect to see after a few days. What conclusions would you draw from these results? (*WR*)

7 State THREE characteristic *chemical* properties of metals which distinguish them from non-metals.

What is an *alloy*? Name any TWO common alloys and the substances present in each of them.

Give, in each case, TWO important uses of (a) lead, (b) copper, stating the property or properties on which the uses depend. (*J*)

7 (i) Name the ore from which aluminium is obtained.

(ii) Name the compound of aluminium obtained by purifying this ore.

(iii) Name the substance used to lower the melting point of the compound given in (ii).

(iv) State in one word the process used to extract aluminium from the molten mixture. (*S*)

8 Given some lead(II) oxide (PbO), describe in detail how you would prepare crystals of lead nitrate.

Describe and explain what is observed when (a) crystals of lead(II) nitrate are heated, (b) dilute hydrochloric acid is added to their solution in water. (*J*)

9 Name the FOUR materials used in a blast furnace producing pig iron. Describe, mainly by means of equations, the principal reactions taking place in the furnace, and name ONE by-product of the process.

Describe and explain, stating any necessary conditions, the reaction between iron and (i) steam, (ii) hydrochloric acid.

Calculate the theoretical mass of crude iron ore containing 30 per cent of iron(III) oxide needed to produce a quantity of pig iron containing 1000 kg of the pure metal. (*S*)

10 Describe briefly the steps you would take in preparing samples of the following substances:

(a) Lead(II) oxide, starting with lead(II) carbonate.

(b) Lead(II) nitrate crystals, starting with lead(II) oxide.

(c) hydrated copper(II) sulphate ($CuSO_4 \cdot 5H_2O$) given copper(II) oxide.

11 What is an alloy? In each case name and give ONE use of an alloy which contains (a) copper, (b) lead, (c) aluminium. State clearly the main advantage gained by alloying each of the above metals.

Describe in essential detail how you would proceed in the laboratory to convert copper foil into pure black copper(II) oxide.

A metal *M*, with relative atomic mass 70, forms an oxide containing 25·5% of oxygen. Find the formula of the oxide. (*S*)

12 Aluminium is obtained from its ore by electrolysis. What do you mean by electrolysis?

What is the chief ore of aluminium?

Why is aluminium made by this method and not by the blast furnace as for iron?

Name two other metals also obtained by electrolysis.

Now draw a clear, simple diagram to show how aluminium is obtained by electrolysis. Label the anode, the cathode, the electrolyte. Give the name of the electrolyte. Say on the diagrams what the electrodes are made of, and indicate where the aluminium is formed, and any other products which are obtained. (*EM*)

13 What three substances are put into a blast furnace for making iron?

Explain briefly why:

(a) This metal has to be manufactured in enormous quantities annually, and

(b) how the crude iron is converted into steel, and
(c) why the metal has to be converted into steel if it is to be used for making tools like saws, chisels, etc.

Explain briefly how, from a solution of zinc sulphate, you would obtain some pure, dry crystals of the solid. *(EM)*

14 Give the chemical name and the common name of a compound of sodium in each case which is:
(a) efflorescent,
(b) deliquescent, and absorbs carbon dioxide from the air,
(c) anhydrous, yet gives water vapour on heating,
(d) present in sea water. *(S)*

15 When fused sodium chloride is electrolyzed, how do the products differ from those obtained by the electrolysis of sodium chloride solution?

Describe how each of these processes is used in industry and for what purposes. *(O)*

16 Given a solution of sodium hydroxide, and any other necessary materials, describe how you would prepare (a) pure, dry crystals of sodium nitrate, (b) a sample of clean, dry sodium hydrogencarbonate.

Describe and explain the action of heat on each of these substances.

State ONE use each for sodium nitrate and sodium hydrogencarbonate. *(S)*

17 Outline the Solvay Process for the production of sodium hydrogencarbonate.

Write the equation for the action of heat on sodium hydrogencarbonate. Calculate the mass of residue left after heating 21 g of the salt.

How would you distinguish qualitatively between sodium hydrogencarbonate and washing soda? Explain the use of sodium hydrogencarbonate as baking soda. *(L)*

18 Describe and explain how sodium hydroxide and chlorine can be manufactured by the electrolysis of brine. Draw a sketch of a suitable cell for this process.

What do you know of the reactions which can take place between a solution of sodium hydroxide and (a) carbon dioxide, (b) a solution of zinc sulphate? (For each reaction, mention the conditions under which it takes place, write an equation for it, and name the products.) *(OC)*

19 From what compound of sodium is metallic sodium isolated, and how is the extraction carried out?

How is sodium carbonate made industrially from common salt?

What is soap? Indicate in a few words how it is made. *(C)*

20 A small piece of magnesium is added to a test tube containing dilute sulphuric acid.
(a) Describe what happens and give an equation to show the changes which occur.
(b) More magnesium is added and the reaction finally stops. Name ALL the substances remaining in the test tube.
(c) How would you obtain crystals of the dissolved salt?

(d) How much of the salt would be obtained if 5 g of the metal combined with the acid?
(e) If a solution of the salt is added to sodium carbonate solution an insoluble substance forms. Give an equation to show what happens.
(f) When hydrochloric acid is poured on to the substance it disappears. Explain what happens. *(SR)*

21 If you are given a solution of copper(II) sulphate and a solution of sodium carbonate how would you obtain:
(a) a dry sample of copper(II) carbonate,
(b) a sample of copper(II) oxide.

Using the copper(II) oxide explain what you would do to obtain:
(i) copper(II) chloride,
(ii) copper(II) sulphate. *(SR)*

22 What would you OBSERVE when each of the following substances is heated: (a) tri-lead tetroxide, (b) sodium hydrogencarbonate, (c) lead(II) nitrate, (d) ammonium nitrate, (e) mercury(II) oxide, (f) iron(II) sulphate crystals? Write equations representing the reactions. *(J)*

23 Give FOUR uses for metallic copper and in each case name the particular property upon which the use depends.

Describe how you would prepare well-crystallized copper sulphate(II) from copper, and copper(II) oxide from the sulphate. *(O)*

24 How does steel differ from pig-iron (a) physically, (b) chemically? Describe briefly how pig-iron is converted into steel.

What happens when (a) carbon monoxide is passed over heated iron(III) oxide, (b) steam is passed over red-hot iron, (c) iron is treated with dilute hydrochloric acid, (d) iron(II) sulphide is treated with dilute hydrochloric acid? *(OC)*

25 (a) Describe, with essential detail, how you would obtain clean, dry samples of the following:
(i) precipitated chalk using lime water,
(ii) crystalline zinc sulphate using zinc oxide,
(iii) lead(II) chloride using crystalline lead(II) nitrate.
Any other necessary materials may be used.

(b) What mass of sodium hydroxide is needed to neutralize exactly dilute sulphuric acid containing 19·6 g of the pure acid? *(S)*

26 What is *observed* (a) when zinc is heated strongly in air, (b) when excess sodium hydroxide solution is slowly added to a solution of zinc sulphate? Give equations for the reactions involved.

Describe with essential experimental detail how you would prepare a sample of dry, crystalline zinc sulphate, starting from zinc oxide. What is the maximum amount of this salt (of formula $ZnSO_4 \cdot 7H_2O$) that could be obtained from 9 g of the oxide? *(OC)*

27 (a) Describe what would be observed and explain your observations, giving equations, when sodium hydroxide solution is
(i) heated with solid ammonium chloride,
(ii) added to a solution of copper(II) sulphate,
and when dilute hydrochloric acid is added to

(iii) solid sodium sulphite,
(iv) sodium carbonate crystals.

(b) A solution of a compound P gave a white precipitate when added to barium chloride solution, and a green precipitate with sodium hydroxide solution.
 (i) Name the compound P, and
 (ii) give equations for the two reactions. (L)

28 Rewrite the following elements in their correct order in the activity series, putting the most reactive at the top of the list and the least reactive at the bottom: Aluminium (Al), Calcium (Ca), Copper (Cu), Hydrogen (H), Magnesium (Mg), Sodium (Na), Zinc (Zn).

A student half-filled four beakers with tap water and, at the same time, placed a clean iron nail in each. Before placing the nails in the water, he twisted strips of two different common metals A and B tightly round two of the nails (beakers (ii) and (iii) respectively) and he pushed a third nail through a piece of granulated tin (beaker (iv)). The other nail was placed by itself in beaker (i).

The student left the beakers undisturbed for several days and then examined them. He found that the nail in (i) was beginning to corrode, that in (ii) was very corroded, that in (iii) showed no sign of corrosion and the nail in (iv) was more corroded than (i), but less corroded than (ii). When (iii) was examined more carefully, a white gelatinous deposit was seen on B.

When fresh strips of A and B were connected, one to the positive and the other to the negative terminal of a voltmeter, and then placed, so that they did not touch each other, in a beaker of dilute sulphuric acid the voltmeter showed a steady reading of about 1 volt.

Suggest the identities of the elements A and B.

What was the purpose of (i) in the first experiment?

Which of the metals, A or B, was joined to the positive terminal of the voltmeter?

After a while, the reading of the voltmeter in the second experiment, gradually dropped towards zero. Why did this occur?

Tin-plate (i.e. sheet iron coated with tin) is often used for making containers in which food is preserved. It appears from the first experiment that iron coated with B might be more effective in withstanding corrosion, if used for this purpose. Suggest why B is not employed in practice.

Name another common metal (i.e. not iron) that is used for making containers for preserving food and drink. (C)

29 The equation that represents the reaction between steam and red-hot iron is as follows:

$$3Fe + 4H_2O \rightleftharpoons Fe_3O_4 + 4H_2.$$

What volume of hydrogen (measured at s.t.p.) is obtained from 0·84 g of pure iron when all the iron is oxidized according to the above equation?

A different oxide of iron from the one in the above equation contains 30% of oxygen by mass. Calculate its simplest formula.

If this second oxide of iron is allowed to react with hydrochloric acid and the mixture is then filtered, what observation would you expect to make upon adding sodium hydroxide solution to the filtrate? Illustrate your answer with an equation. (C)

30 A compound X is heated in a tube, a colourless dry gas Y being collected in a glass syringe. The laboratory temperature was 25°C. The test-tube and compound X were weighed before and after heating. The results are shown below.

Mass of test-tube and X before heating = 19·147 g
Mass of test-tube and contents after heating = 19·047 g
Volume of gas collected in the syringe = 50 cm³.

(a) What is the mass of (i) 50 cm³, (ii) 1 dm³ (1000 cm³) of the gas Y?

(b) Below are the masses of 1 dm³ of a number of gases at 25°C.

Gas	Mass of 1 dm³ at 25°C
Hydrogen	0·08 g
Ammonia	0·71 g
Oxygen	1·38 g
Carbon dioxide	1·83 g
Sulphur dioxide	2·66 g

(i) On the evidence obtained which gas is likely to be Y?

(ii) Give a chemical test to confirm the nature of Y.

(iii) Name one substance which would release gas Y upon heating.

(c) (i) Name a substance which when heated would release one of the gases, other than Y, shown above. Indicate the gas formed.

(ii) Give a chemical test to confirm the nature of the named in (c)(i). (L)

44
Ferromagnetism

The ability of certain iron ores to attract other pieces of iron was known in ancient times, but the first really systematic study of this was carried out by William Gilbert, physician to Queen Elizabeth I and a native of Colchester. He was aware that mariners had for a long time been using pieces of this magnetic ore, known as lodestone or magnetite, in navigation. They had found that if freely suspended or placed on a cork floating on water the lodestone would always set in a certain direction. Gilbert observed that the magnetic attractions of a bar of lodestone were not evenly spread over the material, but were generally concentrated in two regions which he called the poles of the magnet.

He showed that the line joining these poles in a freely suspended piece of lodestone pointed north and south: the pole pointing to the north he called the *north seeking pole* of the magnet (usually abbreviated to the north pole of the magnet). Gilbert observed also that pieces of iron placed near lodestone became magnets. For our investigations however it will be more convenient to use the powerful permanent magnets which may now be produced by magnetising special alloys; these may or may not contain iron.

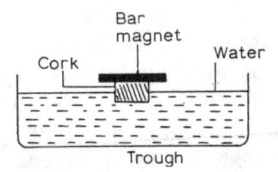

Fig. 44.2. The action of a magnetic compass

Experiment 44.1. To Illustrate the Simple Properties of Magnets.

(a) Place a steel knitting needle on the bench and stroke it several times with a bar magnet as shown in Fig. 44.1. Now dip the needle into iron filings and notice that many of them will be picked up and attached to the ends of the needle. The experiment shows that we may magnetize a specimen by stroking and that the poles so produced are at its ends.

(b) Place the bar magnet on a cork floating on water as shown in Fig. 44.2. The magnet will turn until it points north and south. This experiment, besides illustrating the action of a compass, tells us something about the strength of magnetic poles. If one of the poles was stronger than the other the magnet would not only rotate but would also move towards the north or the south. Since this does not happen, we are led to believe that the two poles of our bar magnet are of the same strength.

(c) Take the bar magnet used in the above experiment off its cork and mark its north seeking pole. Now place another bar magnet floating on the cork and when it has come to rest pointing north and south bring the north pole of the first magnet carefully towards the north pole of the floating one. You will see that the pole of the floating magnet is repelled. Repeat the experiment bringing the north and south poles of the first magnet in turn towards the two poles of the floating one as in Fig. 44.3.

You will notice that the force which occurs in each case is given by the following rule:

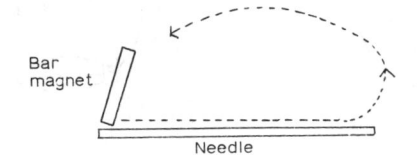

Fig. 44.1. Magnetizing a needle by stroking

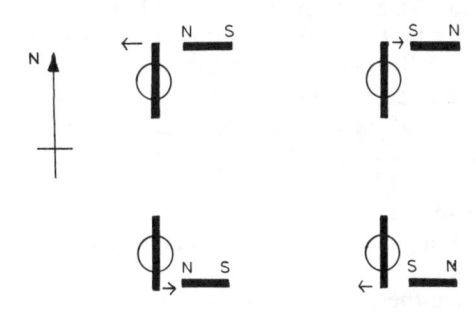

Fig. 44.3. Law of force between magnetic poles

Similar poles (i.e. two norths or two souths) repel each other: dissimilar poles (i.e. a north and a south) attract each other. (See Fig. 44.3).

If we wish to test a bar of steel in order to determine whether or not it is magnetized we must see if we can obtain *repulsion* between it and a magnet as in the above experiment. We may obtain attraction whether our bar of steel is magnetized or not.

Experiment 44.2. Making a Bar Magnet.
We have already shown that we may magnetize a bar of steel by stroking it with a magnet. A more satisfactory method, however—and that used to produce permanent magnets—is an electrical one.

Place the specimen to be magnetized inside a solenoid or coil of wire with many turns which is connected to a battery to produce a fairly large direct current flowing in it (Fig. 44.4).

If when looking towards the end of the solenoid the current is flowing clockwise then a south pole will be obtained at that end. If the current is going anti-clockwise then a north pole will be obtained.

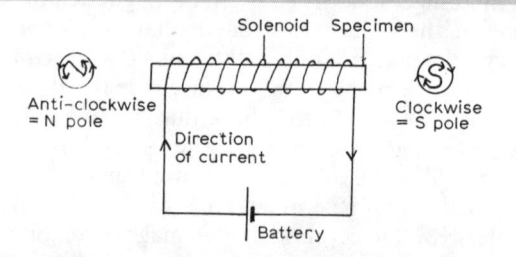

Fig. 44.4. Magnetizing an Iron bar in a solenoid

Experiment 44.3. Demagnetising a Bar
Place the bar magnet in a solenoid which is connected to an alternating current supply. A rapidly-changing current flowing first one way and then the other is passing through coil and at the same time the direction of magnetization of the bar is rapidly changing. Now slowly withdraw the bar from the solenoid moving it along the axis of the coil in a straight line until it is a long way from it (Fig. 44.5). If you now test the bar you should find that it will show an attraction at both ends of the compass needle, indicating that it has been demagnetized. The same result could also be brought about by heating the magnet to red heat.

Lines of Force
Although in theory the magnetic effects of a magnet extend outwards to an infinite distance, in practice these effects may be detected over only a limited region around it. This region we call the

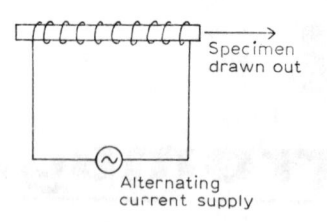

Fig. 44.5. Demagnetizing

field of the magnet. A graphic way of describing a magnetic field is in terms of what Michael Faraday (1791–1867) called 'lines of force'. These may be seen most easily in the following experiment.

Experiment 44.4. To Determine the Lines of Force of a Magnet.
(*a*) Place a bar magnet on the bench and cover it with a piece of drawing paper. Now sprinkle iron filings evenly over the paper. This may conveniently be done by shaking them from a tin with a perforated lid. If the paper is now tapped lightly the iron filings will show the pattern of the lines of force. The experiment may be repeated with a horseshoe magnet and with two bar magnets placed (i) with similar poles close together and (ii) with dissimilar poles together. The sorts of patterns we can expect to obtain are shown in Fig. 44.6.

We may get some idea of the significance of these lines of force from the following experiment.

(*b*) Fill a trough with water and mount a bar magnet so that it rests in the surface of the water.

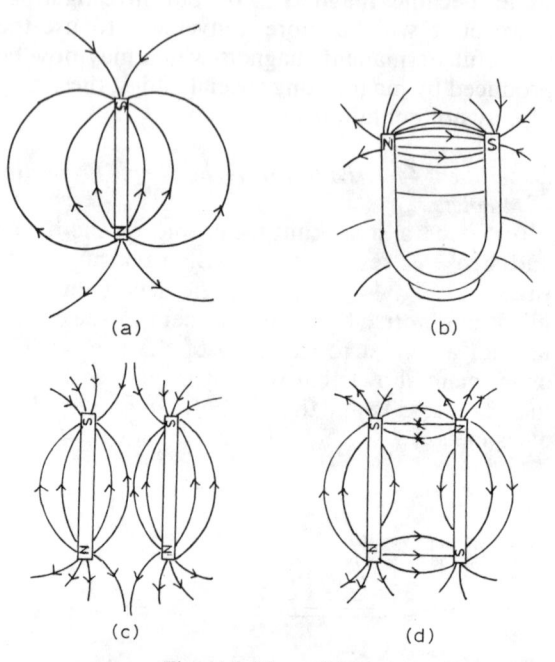

Fig. 44.6. Lines of force

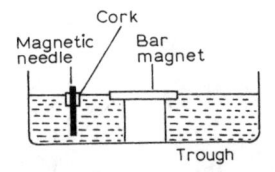

Fig. 44.7. The meaning of a magnetic line of force

Magnetize a steel needle and push it through a cork (Fig. 44.7) so that it can be made to float with one end just above the surface. If the cork is placed close to one pole of the bar magnet it will be observed to move along a path which will probably end at the other pole. In fact, the path should coincide approximately with one of the lines of force shown in Fig. 44.6(a).

By arranging the needle as we have done with one pole much farther from the bar magnet than the other, we have to some extent achieved the effect of an unattached pole. Hence we may define magnetic lines of force as the paths traced out by an unattached north pole. With this definition, we can see that the direction of our lines of force in the field of a bar magnet, for example, will be away from the north pole (like poles repel) and towards the south pole.

A more satisfactory method of plotting lines of force—and one which gives a permanent record of the pattern obtained—uses a plotting compass. This consists of a small magnetic needle mounted on a pivot in a case with glass faces above and below the needle.

(c) Place a bar magnet on a piece of drawing paper and pencil in its outline. Now put a plotting compass close to one end of the magnet and make two dots to mark the positions of the two ends of the compass needle. Move the plotting compass away from the magnet so that one end of the pointer is next to the dot which previously marked the other end. Now make another dot to mark the new position (Fig. 44.8). Repeat this process until the other end of the magnet is reached. Now join up the dots you have plotted. In this way by starting at different points nearer the end of the magnet a complete pattern of lines of force may be obtained.

Magnetic Induction

If the end of a bar magnet is dipped into a dish

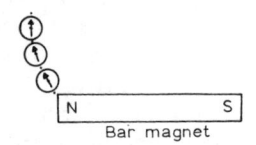

Fig. 44.8. Plotting lines of force with a plotting compass

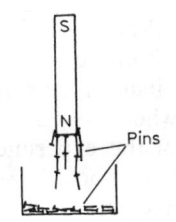

Fig. 44.9. Magnetic induction

containing steel pins a large number of them will be picked up (Fig. 44.9). Many of these will not be touching the magnet or even very near to it. The pins will in fact form chains as shown. If one of these chains is removed from the magnet it will probably fall apart.

This is explained by assuming that each pin, when in contact with the magnet or with another pin which is touching the magnet, itself becomes a magnet. It must also be assumed that the pins lose their magnetism when removed from the vicinity of the magnet. We say that the pins have been magnetized by induction.

Care of Permanent Magnets

Permanent magnets always have a tendency to lose their magnetism. This process is speeded up if the magnet is treated roughly or heated. In any case the magnet's own field tends to demagnetize it. This effect may be minimised by storing magnets correctly. Bar magnets should be stored in pairs as shown in Fig. 44.10 with two pieces of soft iron known as keepers placed across the ends. A horseshoe magnet should also be provided with a keeper as in Fig. 44.11. In each case the effect of the keepers is very much to reduce the field around the magnets and hence to lessen this demagnetizing effect.

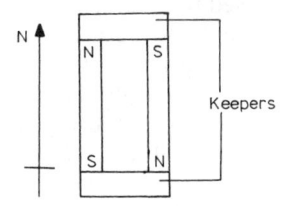

Fig. 44.10. Keepers

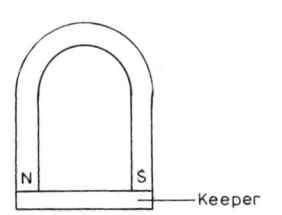

Fig. 44.11. A keeper with a horseshoe magnet

Theories of Magnetism

Scientists have been very successful in the last thirty years in producing a satisfactory explanation of what happens when a bar of iron is magnetized. Although much of the experimental work which led to this theory is beyond the scope of this book we can get some idea of it and how it explains the elementary properties we have discussed.

Towards the end of the last century some scientists, notably Weber and an English scientist, Sir James Ewing, believed that a bar of iron in the demagnetized state consisted of a large number of tiny permanent magnets arranged in a random way so that there was no overall magnetic effect. They suggested that when the bar was magnetized these so-called 'molecular magnets' were lined up so as to produce a north pole at one end and a south pole at the other. It is clear from Fig. 44.12, for example, how this would explain magnetization by stroking. It is also easy to see how heat or rough treatment might cause the elementary magnets to revert to their former random arrangement.

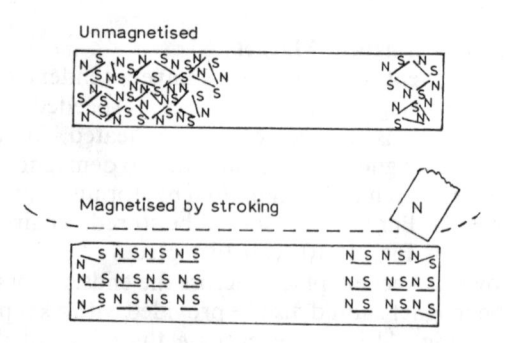

Fig. 44.12. Molecular theory of magnetism

The current theory bears resemblances to the molecular theory, but the elementary magnets which are believed to exist in an unmagnetized bar of iron are thought to be much larger than the molecules of Ewing's theory. In fact they are referred to as magnetic domains and consist of between 10^9 and 10^{15} atoms. Iron, like most metals, is a crystalline material and the domains are subdivisions of the crystals. The simplest possible arrangement of domains in a crystal is shown in Fig. 44.13(a). The arrows indicate the direction in which the domain is normally magnetized. Fig. 44.13(b) shows what happens when a small magnetizing field is applied in the direction of the arrow above. The boundary wall between the domains magnetized in the 'right' direction and those in the 'wrong' direction moves downwards. The top two grow at the expense of the others. In Fig. 44.13(c) the magnetizing field has been increased and the four domains become only two and finally in Fig. 44.13(d) with a still larger magnetizing field the whole crystal is magnetized in the direction of the applied field. This final stage fits in with the observed fact that there is a limit to the extent to which we can magnetize a specimen. We have in fact reached 'saturation'.

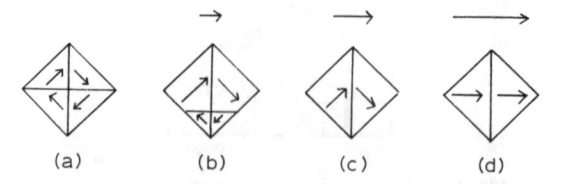

Fig. 44.13. Magnetic domains during the process of magnetisation

The unmagnetized arrangement of domains shown in Fig. 44.13(a) corresponds to the two bar magnets with keepers shown in Fig. 44.10 in that it is an arrangement with a minimum external field. Scientists have developed techniques to enable them actually to see under a high-powdered microscope the boundary walls between the magnetic domains and their movement when a magnetizing field is applied. The domain arrangement shown in Fig. 44.13(a) is the simplest possible one and in practice the arrangements are almost always more complicated than this.

Magnetic Materials

The development of the domain theory in the last thirty years has been largely responsible for the production of a wide range of magnetic materials for all sorts of applications. However, magnetic materials can be divided roughly into two groups which we shall call 'hard' and 'soft'. We can distinguish between these by means of the following experiment.

Experiment 44.5. A Comparison of Steel and Soft Iron as Magnetic Materials.

Connect in series with a battery and a switch two solenoids as nearly identical as possible. Place similar bars of steel and soft iron in the solenoids (Fig. 44.14) and clamp them in a vertical position with dishes of pins beneath them. Switch on the current and bring the dishes up to the ends of the metal bars so that some pins become attached to them. You will probably observe that the soft iron has collected considerably more pins than the steel bar. Now switch off the current. Most of the pins collected by the soft iron will drop off (possibly all of them) while a considerable number will remain on the steel bar.

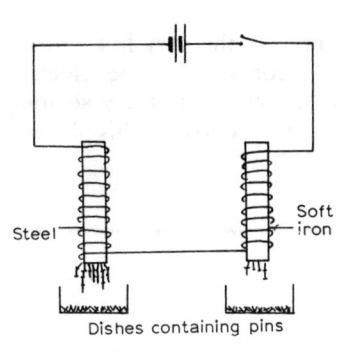

Fig. 44.14. Magnetic materials. When current is switched off iron retains few pins

The iron bar has properties which are characteristic of 'soft' magnetic materials, i.e. it is readily magnetized quite strongly, but also it loses almost all its magnetism when the magnetizing field is removed. Steel, on the other hand, is a typical 'hard' magnetic material. It requires a stronger magnetizing field to magnetize it strongly, but it retains its magnetism to a much greater extent than does iron.

Soft magnetic materials are used for the cores of electromagnets and transformers about which we shall have more to say later. Hard magnetic materials are used for making permanent magnets. Although we have mentioned steel as typical of this class, there are many alloys which make better permanent magnets than steel, notably some containing aluminium and nickel.

Terrestrial Magnetism

The earth's magnetic field was mentioned earlier in this chapter and we must now look a little more carefully at what kind of a field it is and try to explain its origin. William Gilbert (see p. 271) noted that the mariners' compass needle often did not remain in a horizontal plane, but in the northern hemisphere dipped towards the north and in the southern towards the south. He noticed also that this tendency of the compass needle to dip increased as the observer moved farther away from the equator. In fact the general characteristics of the earth's magnetic field are those which would be obtained if there were a bar magnet (short in length relative to the earth's diameter) at the centre of the earth with its axis close to the axis of rotation. The lines of force that would be produced by such a magnet are shown in Fig. 44.15 and these, as magnetic surveys have shown, closely resemble the lines of force which actually exist. From the diagram it may be seen that the lines of force cut the earth's surface at different angles. There are two

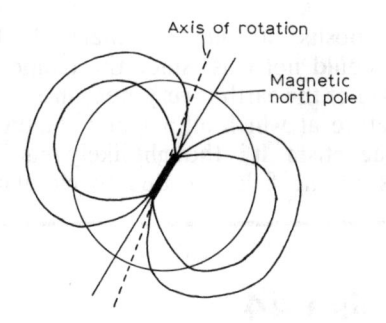

Fig. 44.15. The magnetic field of the earth

places, called the magnetic poles, where the lines of force make a right angle with the surface of the earth. These points are on the axis of our imaginary bar magnet. At places lying on what might be called the magnetic equator the lines of force are parallel to the surface of the earth and there the compass needle has no tendency to dip. The angle between the line of force and the horizontal is known as the *angle of dip* and, of course, it varies from place to place, being 90° at the magnetic poles and zero on the magnetic equator.

Since the geographical north pole and the magnetic north pole do not coincide, the magnetic compass needle will not point to 'true' north. In order to specify the direction of the line of force of the earth's field at a particular place it is necessary to know not only the angle of dip but also the angle between the direction indicated by the compass needle and the geographical north. This angle is referred to as the *angle of declination* (Fig. 44.16). It varies from place to place and also slowly from time to time, since the magnetic pole is not a fixed point on the earth's surface. The value of declination in London at present is about 8° west (i.e. the compass needle points 8° west of geographical north). It is believed that in the seventeenth century the declination in London was about zero and that it reached a maximum of 24°W. in the early nineteenth century.

The origin of the earth's magnetic field has not yet been satisfactorily explained, but it can be said with certainty that it is NOT due to the imaginary bar magnet we have been talking about nor to any

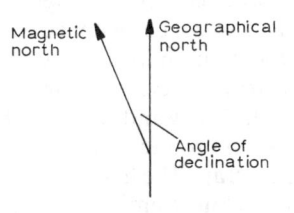

Fig. 44.16. Declination

other deposits of magnetic material. The bar magnet could not exist since the temperature at the centre of the earth is very much in excess of the temperature at which all magnetic materials lose their magnetism. It is thought likely that the main features of the field are due to circulations of molten material in the very hot core of the earth which really constitute large electric currents. More than this we cannot say so long as man's knowledge of the centre of this planet remains so scanty.

Exercise 44

1 You are provided with three metal bars. One is a magnet, another is capable of being magnetised and the third is non-magnetic. How could you determine the nature of each bar? You may use any other apparatus you choose except another magnet or compass needle.

Draw diagrams to explain what is meant by (a) the angle of variation (declination), (b) the angle of dip (inclination), at a place on the surface of the earth. (J)

2 Describe how you would show by experiment which of the two materials, iron and steel, is better for (a) a permanent magnet, (b) an electromagnet. State, with reasons, the conclusions you would reach. (C)

3 If you have a permanent magnet, describe:

(a) How you would make a piece of steel (e.g., a sewing needle or a razor blade) into a satisfactory compass needle? How would the poles on your compass needle compare with the pole of the magnet used?

(b) How you would turn a large nail into an electromagnet and how you would you find between the direction (a) to test its polarity. What connection would you find between the direction of current flow and the polarity of the electromagnet?

(c) How you would remove the magnetism from a carpenter's chisel (magnetised accidentally) by an electrical method? (NR)

45
Electromagnetism and the Motor Effect

In the previous chapter we saw how a bar of steel could be made into a permanent magnet by placing it in a solenoid carrying a current. We also saw how a bar of soft iron became strongly magnetized in a solenoid as long as the current was flowing, but it lost most of its magnetism as soon as the current was switched off. A magnet of this kind which can, as it were, be switched on and off, has advantages for many purposes over a permanent magnet. Electromagnets of various kinds are widely used in industry and research. For example, much of the steel produced in the steel industry is obtained from scrap. The handling of scrap iron on a large scale is almost entirely by means of large electromagnets on cranes.

An electromagnet consisting of a solenoid wound on a straight iron bar may be shown, by

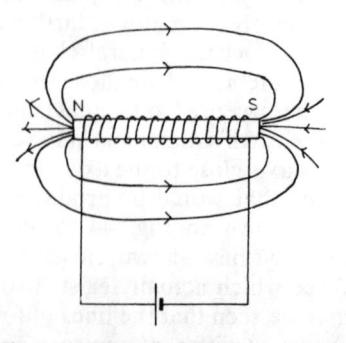

Fig. 45.1. Magnetic field of a solenoid

means of plotting compasses, to have a magnetic field similar to that of a permanent bar magnet (Fig. 45.1). 'Horseshoe' electromagnets are often used in research where a very intense magnetic field is required. (Fig. 45.2).

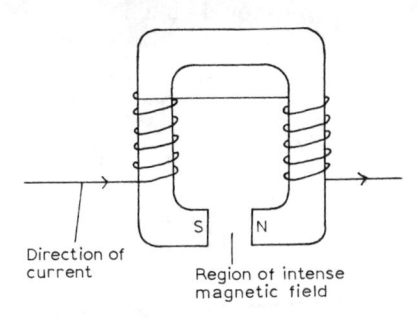

Fig. 45.2. A horseshoe electromagnet

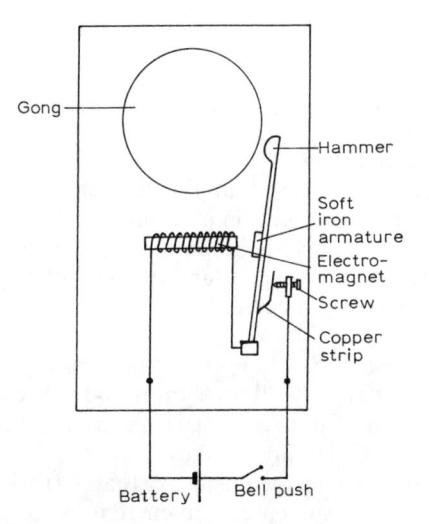

Fig. 45.3. An electric bell

The Electric Bell and the Relay

Many everyday devices also use electromagnets, for example the electric bell shown in Fig. 45.3. When the circuit is completed by closing the bell-push switch, a current flows through the electromagnet, causing it to attract the soft iron armature. Thus the hammer strikes the gong and at the same time the copper strip is pulled away from the screw contact. The circuit is broken and the armature is released from the electromagnet. This process is repeated very rapidly and a continuous ringing of the bell is produced.

Another similar and very widely used electromagnetic device is the relay. It is, in fact, an electrically operated switch. In this case when a current passes through the solenoid one part of a soft iron armature is attracted to it, causing another part of the armature to push together two contacts to complete another circuit. Relays are very flexible devices. Sometimes they may contain many pairs of contacts and they may just as easily be used to break circuits as to complete them.

Moving Iron Ammeters

Most instruments for measuring electric currents use their magnetic effects to do this. The repulsion type moving iron ammeter illustrated in Fig. 45.5 is one such instrument. The current to be measured is passed through the solenoid and the two soft iron bars A and B become magnetised with similar poles adjacent to one another. Bar A is fixed but bar B is attached to one end of a pointer which is pivoted on an axle at the centre of the solenoid. Thus the repulsion between A and B causes the pointer to move over the scale which may be calibrated in amps. As in all current measuring instruments, the moving iron ammeter must include something to provide a couple opposing the turning effect of the repulsion between the pieces of iron. This may be a pair of spiral springs but is often, as illustrated in Fig. 45.5, just a weight rigidly fixed to the pointer below the axle. If there were no such 'control' the pointer would show a full scale deflection for almost any current. The control also causes the pointer to return to zero when the current is switched off. This kind of 'gravity' control means that the instrument may only be used in a vertical position.

It should be noted that whichever way the current passes through the solenoid, there will always

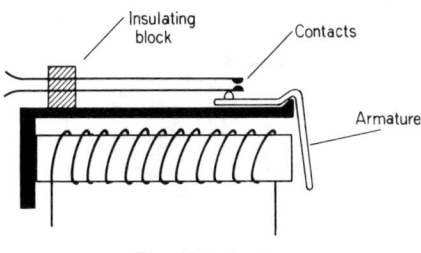

Fig. 45.4. A relay

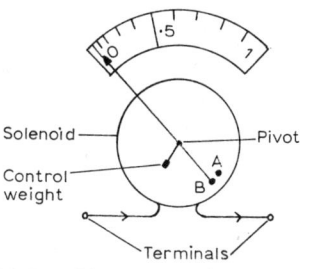

Fig. 45.5. Repulsion type moving iron ammeter

be a repulsion between the iron bars and hence the instrument may be used to measure alternating currents. It has the disadvantage, however, that the divisions on the scale are not evenly spaced and the scale will generally have to be marked off by comparing it with an instrument of another type.

The Magnetic Field around a Conductor carrying a Current

Experiment 45.1. Stretch a long piece of wire over a pivoted magnetic needle with the wire lying in the same direction as the needle. Connect the wire to a low voltage D.C. supply and switch on the current. You will observe the needle to be deflected as shown in Fig. 45.6.

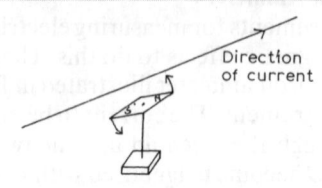

Fig. 45.6. Magnetic effect of a current in a straight wire

This simple experiment, first carried out by the Danish scientist Oersted in 1820, shows that there is a magnetic field around a straight wire carrying a current.

Experiment 45.2. Lines of Force.
Arrange a straight wire so that it is vertical with a horizontal cardboard or wooden table surrounding it about half way up (Fig. 45.7). Pass a current through the coil by connecting it to a battery or other low voltage D.C. supply. Sprinkle iron filings over the table and give it a light tap. A pattern of lines of force similar to that shown in Fig. 45.7 should be obtained, consisting of concentric circles centred on the wire. The direction of the lines is given by Maxwell's corkscrew rule. This states that if an ordinary right handed screw were being moved in the direction of the current then the way in which it would be turned would indicate the direction of the lines of force.

Repeat the experiment with a flat circular coil instead of the straight wire. Fig. 45.8 shows the sort of pattern you should obtain.

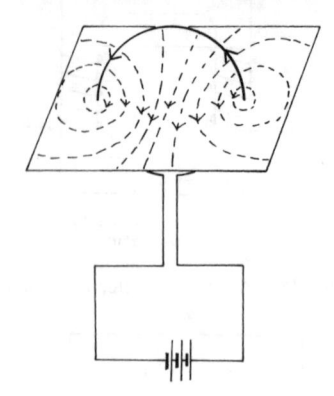

Fig. 45.8. Lines of force around a plane circular coil

The Motor Effect

Since we observed in the last chapter that a magnet exerts forces on other magnets close to it, it is to be expected that some kind of force will exist between a magnet and a neighbouring conductor carrying a current.

Experiment 45.3. Clamp two brass rods in a horizontal position so that a third rod *AB* can rest across them. Connect the rods in circuit with a low voltage D.C. supply as shown in Fig. 45.9. Now bring up a horseshoe magnet so that *AB* rests in the pole gap and switch on the current. You will see *AB* roll along the other two rods. Note the direction of the current (from positive to negative of low voltage supply), the direction of the field (from north to south of the magnet) and the direction of motion of the rod *AB*. Reverse, in turn, the direc-

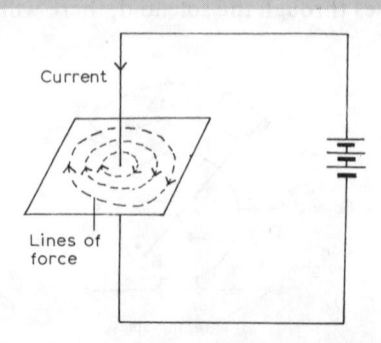

Fig. 45.7. Lines of force due to a straight wire

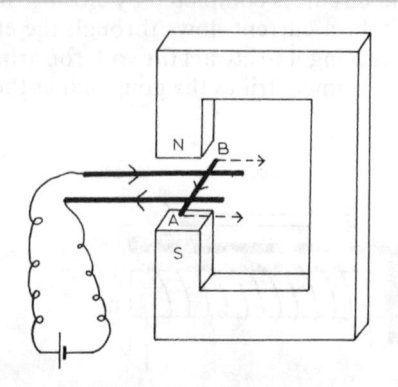

Fig. 45.9. The Motor Effect

tions of the current and the magnetic field and note the direction of motion in each case. You will observe that the force which causes the rod to roll is not a simple attraction to or repulsion from either of the magnet poles.

In fact in the experiment above the field, the current and the direction of motion are mutually at right angles to one another. A rule which enables one to forecast the direction of motion in a case of this sort is known as Fleming's Left Hand rule. It states: 'If the thumb, first and second fingers of the left hand are placed mutually at right angles to each other (Fig. 45.10), the first finger indicating the direction of the field, the second finger the direction of the current, then the thumb will indicate the direction of the motion of the conductor'.

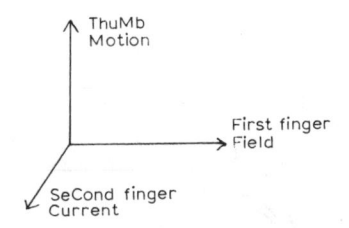

Fig. 45.10. Fleming's Left Hand rule

The Moving Coil Galvanometer

Probably the most common type of current measuring instrument is the moving coil galvanometer which depends for its action on the motor effect. In its simplest possible form it consists, as shown in Fig. 45.11 of a rectangular coil of a few turns suspended or mounted on an axle between the poles of a horseshoe permanent magnet. The current to be measured is passed round this coil and it will flow down one side and up the other. Applying Fleming's Left Hand rule to the vertical sides of the coil it will be seen that they will experience equal and opposite forces in the directions indicated in Fig. 45.11. These will cause the coil to turn and the pointer to move over the scale. The necessary opposition to the turning effect of the current which we referred to in connection with the moving iron ammeter (page 277) is provided in the case of the suspension type instrument shown in Fig. 45.11 by the 'torsion' of the suspension wire. This is the tendency of a wire, fixed at one end and twisted at the other, to untwist itself. The most suitable material for the suspension is a thin wire or flat strip of phosphor-bronze. Instruments of this kind may be made very sensitive, measuring currents less than a micro-ampere (a millionth of an ampere). They have the disadvantage, though, that they are fragile, the suspension wire being easily broken, and also that in many cases they must be carefully levelled before use so that the coil hangs symmetrically between the pole pieces. If a less sensitive but more robust instrument is required the coil will be fixed to an axle mounted on jewelled bearings. In this case the control will be provided by a pair of spiral springs one at each end of the axle. If the instrument is made as in Fig. 45.11 it can easily be seen that the divisions on the scale will not be equal in size. Fig. 45.12 shows in top view the coil in three different positions. Although the forces on the coil may be the same in cases (a) and (b) their turning effect or moment will clearly be less in case (b) since moment is the force multiplied by the perpendicular distance from the axle to the line of action of the force and this distance in case (b) (d_2) is less than in case (a) (d_1).

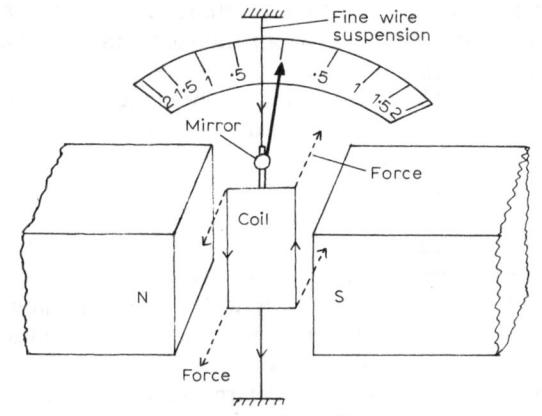

Fig. 45.11. Simple moving coil galvanometer

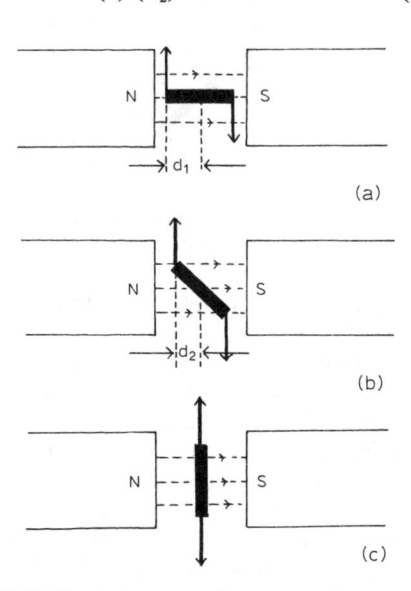

Fig. 45.12. Turning moment decreases as the coil rotates

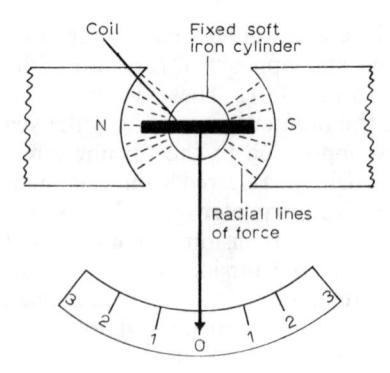

Fig. 45.13. Galvanometer with a radial field

In fact the turning effect decreases as the coil turns, from a maximum in case (a) to zero in case (c).

The moving coil galvanometer may be made to have an evenly divided scale by modifying it as shown in Fig. 45.13. The poles of the magnet are made concave and the coil is made to rotate about a fixed cylinder of soft iron. This has the effect of changing the lines of force into a radial patterns and hence as may be seen from Fig. 45.14 a constant turning effect is maintained as the coil turns up to the limit shown in Fig. 45.14(b).

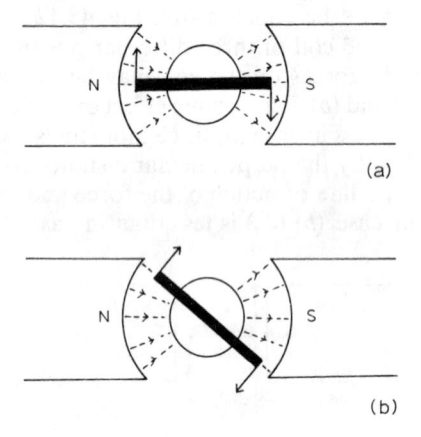

Fig. 45.14. The turning moment remains constant as the coil rotates

Moving coil galvanometers as described here will probably be micro-ammeters or milliammeters (1mA = 1 thousandth of an ampere). However as we have seen in Chapter 36 they may be modified to measure much larger currents as well as voltages. The principal limitation on their use is the fact that they will not measure alternating currents. As the direction of the current in the coil reverses the forces on it reverse also. Hence the pointer will tend to move first one way and then the other. Most alternating currents change direction so frequently that the pointer will be quite unable to cope and will show zero deflection.

Electric Motors

We have seen that a rectangular coil mounted on an axle in a magnetic field, as in the case of a moving coil galvanometer, will experience forces that will cause it to turn. But the maximum rotation of a galvanometer coil we could obtain is 180°. In making an electric motor we must achieve a continuous rotation of the coil. This is done in a simple motor by supplying the current to the coil by way of a pair of brushes and a split ring commutator. As will be seen from Fig. 45.15 this will reverse the direction of the current in the coil when it reaches the position shown in Fig. 45.12(c) causing it to turn through a further 180° when the current will reverse again.

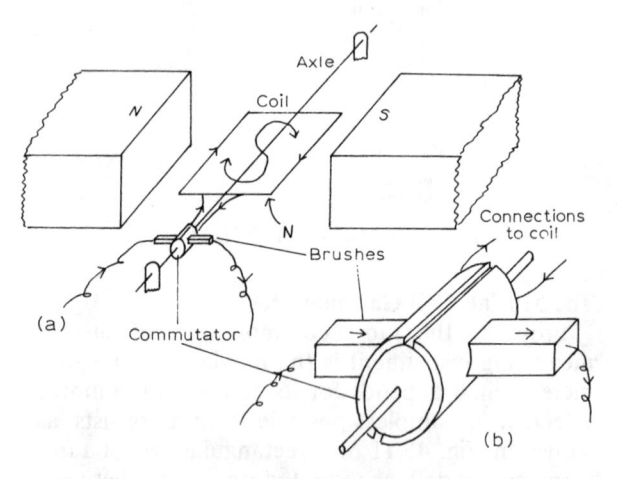

Fig. 45.15. A simple electric motor

Another way of looking at this is to consider that the coil has a North and South face determined by the direction of the current as indicated in Fig. 45.16. The coil will then rotate so that the North face is moving away from the North pole of the magnet, being repelled by it.

The split ring will in practice be made of brass or copper mounted on an insulating cylinder fixed to the axle of the motor. The brushes may be of graphite which, besides being a good conductor, will produce little friction against the commutator. The brushes are usually held in position against the split ring by means of spiral springs.

Electric motors are very widely used and range in size from the relatively small ones used in domestic appliances like sewing machines, washing machines, etc., to the very large ones used for driving industrial machinery and electric locomotives.

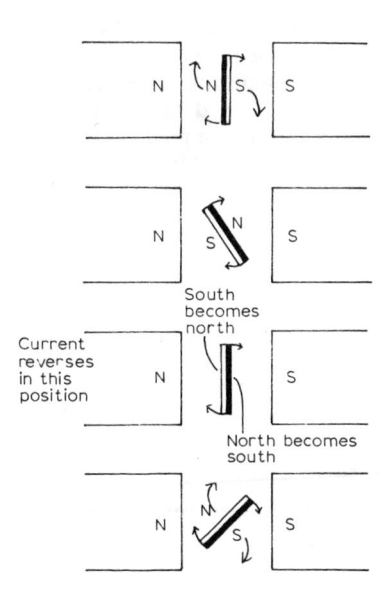

Fig. 45.16. Reversing the current in the coil every 180° produces continuous rotation

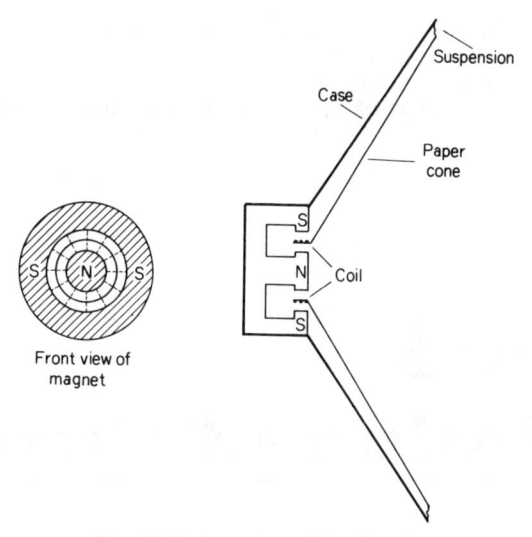

Fig. 45.17. A moving coil loudspeaker

In all these cases the magnetic field will be provided by an electromagnet rather than a permanent one and they will be a lot more complicated in detail than the one we have discussed, but in basic principle they will be similar to it.

The Moving Coil Loudspeaker

Another device of considerable practical importance which depends for its action on the motor effect, is the moving coil loudspeaker. This is the part of almost all radio and television receivers and record players which actually produces the sound. The permanent magnet of a loudspeaker (Fig. 45.17) consists of a cylindrical pole S in the form of a ring. The lines of force between these poles, indicated by dotted lines in the diagram, form a radial magnetic field. The moving coil, which is cylindrical in shape, is supported in a position to the coaxial with the magnet pole pieces. The coil is fixed to a cone of paper or some other material which is light but rigid. This cone is flexibly fixed around its circumference to the metal frame of the loudspeaker.

Small alternating currents with frequencies and amplitudes representing the pitches and loudness of the sounds to be produced, are passed through the coil. This will move at right angles to both the current direction and the lines of force according to Fleming's left hand rule. The coil and the cone will thus oscillate as shown in the diagram. It is, of course, the cone which displaces the air to produce sound.

Exercise 45

1 Describe an experiment to show that there is a magnetic field in the neighbourhood of a wire carrying an electric current.
Draw a large, labelled diagram of an electric bell and describe briefly its mode of working. (O)

2 Explain fully how you would investigate experimentally the nature of the magnetic field set by a current flowing in a coil of several turns of wire.
Draw a diagram of either a moving-coil or a moving-iron ammeter or galvanometer and explain its mode of working. (O)

3 Draw a labelled diagram of a repulsion type moving-iron ammeter and explain its action. Explain the fact that this instrument can be used for measuring an alternating current. (C)

4 Draw a labelled diagram of a simple direct current electric motor and explain how it works. (L)

5 Draw a diagram to show the magnetic field set up by a current flowing down a vertical wire.
Describe an experiment to show that a straight conductor carrying a current and placed in a magnetic field experiences a mechanical force. State a rule to determine the direction of the force.
Name TWO practical applications of this principle. (J)

6 Describe the structure of an electromagnet. How

do the properties of such a magnet differ from those of an ordinary magnet?

Make a drawing of an instrument that employs a magnet and which can be used for measuring electric currents. Explain how the instrument works.

In what units should this instrument be graduated? How could it be adapted to measure larger electric currents than its maximum graduation?

46
Induced Currents

Almost the whole of our industrial civilization depends on the production of electrical energy on a very large scale. At the beginning of the nineteenth century, currents could be produced only in circuits containing chemical cells, and even these were in a very early stage of development. The discovery of the principle upon which the dynamo is based is due to Michael Faraday. In the 1830s he was looking for an effect in current electricity comparable to induction in electrostatics. He was also aware that the magnetic field due to an electric current could produce motion and he wondered if some converse effect might be found in which motion produced an electric current. The effect which he discovered is known as electromagnetic induction.

Experiment 46.1. Experiments to Demonstrate Electromagnetic Induction.

Set up the circuit shown in Fig. 46.1, the coil consisting of several hundred turns of wire, and the centre zero moving coil galvanometer having a minimum sensitivity of a few milliamperes for a full scale deflection. Move the bar magnet into the coil with its north pole entering the coil first. As long as the magnet is moving a deflection of the pointer of the galvanometer will be observed. Note the direction of the deflection. Note also that if the magnet is at rest inside the coil no deflection occurs. Now withdraw the magnet from the coil. You will see that the pointer is deflected in the opposite direction. Repeat the experiment with the south pole of the magnet entering the coil first and again note the direction in which the pointer moves when the magnet moves in and out of the coil.

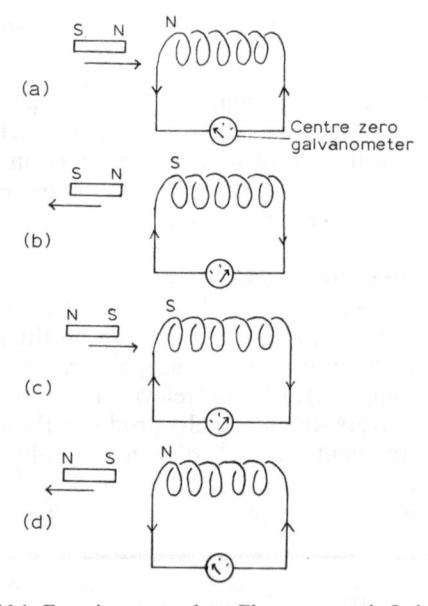

Fig. 46.1. Experiments to show Electromagnetic Induction

In these experiments we have obtained what we call 'induced' currents in the circuit as long as the magnet is moving. The bar magnet has, as we have seen previously, a large number of lines of force in the space around it. Faraday suggested that the condition necessary for the production of an induced current in a circuit is that the magnetic lines of force linking the circuit shall be changing. We shall see later other ways in which this condition may be achieved.

If the way in which the coil is wound is known and the direction of the current in it may be deduced from the movement of the galvanometer pointer, the directions of the current in each case will be seen to be as in Fig. 46.1. In case (*a*) the current flows so as to establish a north pole at the left hand end of the coil (see p. 276) while in case (*b*) a south pole appears at this end. In all cases the polarity of the coil due to the induced current in it is such as to oppose the movement of the magnet. These results are generalized in **Lenz's law** which states that **"the direction of the induced current is such as to oppose the change causing it"**.

If we consider these experiments in terms of energy, we can see why the direction of the induced current is determined in this way. Suppose we repeat these experiments with an unmagnetised bar of steel which is identical in all other respects to our bar magnet. We are apparently doing the same amount of mechanical work in both cases but when we move the bar magnet we obtain some electrical energy which will later appear as heat, while using the unmagnetised bar we obtain nothing. We would, in fact, appear in the first case to be obtaining something for nothing. However Lenz's law tells us that the induced current in the solenoid causes a magnetic field around it which opposes the movement. So in moving our bar magnet we must do extra mechanical work against the magnetic attraction or repulsion of the solenoid and this extra work is the origin of the electrical energy we obtain in our experiments.

You will find that it is possible to increase the size of the induced currents in these experiments in a number of ways, e.g.

(*a*) by moving the magnet more quickly.
(*b*) by using a stronger magnet.
(*c*) by using a coil with more turns.

By more careful experiments along these lines it is possible to verify a second law of electromagnetic induction. This, known as **Faraday's law**, states that **the magnitude of the induced voltage in a circuit is proportional to the rate of change of the number of lines of force linking it.**

Dynamos

The experiments described above show how electrical energy may be produced from mechanical energy. However the in and out movement of the magnet is less convenient in practical devices than a rotation. Most practical dynamos consist therefore of either a magnet rotating in a coil or a coil rotating in a magnetic field and it is the latter arrangement that we shall consider now.

The apparatus shown in Fig. 46.2 is similar to the simple electric motor (chapter 45) but the commutator has been replaced by two 'slip rings' fixed to the axle on which the coil is mounted and each connected to one end of the coil.

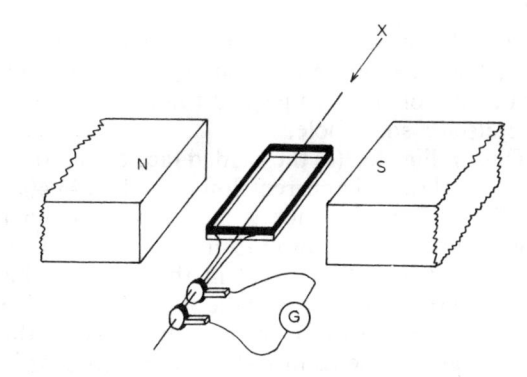

Fig. 46.2. An A.C. dynamo

When the coil is rotated the number of lines of force linking it will be changing and hence a voltage will be induced in it. The coil is part of a complete circuit containing also the slip rings, brushes and a centre zero galvanometer and so a current will flow causing a deflection of the pointer of the galvanometer. If the coil is rotated steadily and slowly by hand it will be observed that the current flows first one way round the circuit and then the other, reversing its direction every half revolution of the coil. In fact our simple dynamo is producing an alternating current. Why this is so may be seen by considering how the laws of electromagnetic induction relate the the arrangement. Fig. 46.3 shows three positions of the coil rotating clockwise in end-on view (viewed in direction X in Fig. 46.2). Lenz's law tells us that the direction of the induced current in the coil will be such as to

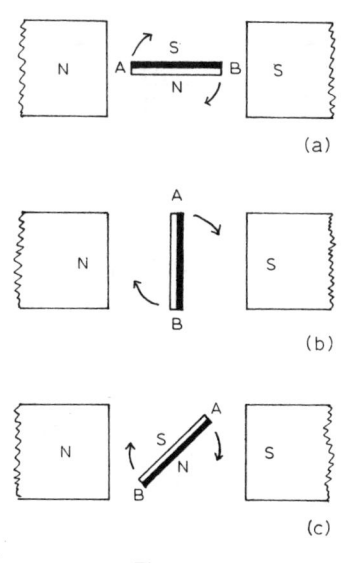

Fig. 46.3.

oppose the rotation. The current will flow so as to make that face of the coil, moving away from the north pole of the magnet and towards the south pole, itself a south pole.

Thus in Fig. 46.3(a) the shaded face of the coil is a south pole and the current flows as in Fig. 46.4(a). In Fig. 46.3(b) this face is momentarily neither moving towards nor away from the south pole of the magnet and the current in the coil is at this instant zero. In Fig. 46.3(c) the shaded face of the coil is moving away from the south pole of the magnet and consequently the current flows as in Fig. 46.4(b) making this face a north pole.

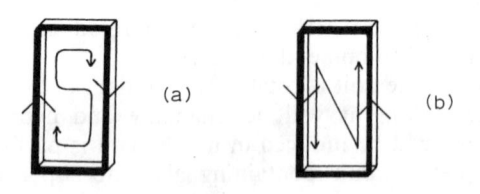

Fig. 46.4.

The size of the induced current depends, according to Faraday's law, on the rate at which lines of force are being cut. You will see that with the coil in a position indicated in Fig. 46.5(a) the two *sides* of the coil A and B are moving at right angles to the lines of force and consequently cutting them at a maximum rate. In case (b) the sides A and B are moving obliquely to the field and thus the rate of cutting lines is reduced. In case (c) A and B are momentarily moving parallel to the lines of force and not cutting them at all. So the induced current will have a maximum value at position (a) of coil

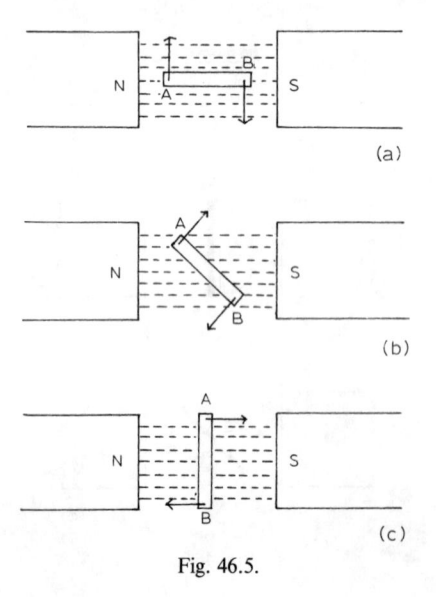

Fig. 46.5.

falling to zero at (c) before reversing its direction as the coil turns further. Fig. 46.6 shows a graph of the variation of the current in the coil with time.

Points (a), (b) and (c) on the curve refer to the positions of the coil in Fig. 46.5. The time T is the time taken for the coil to make one complete revolution. Positive and negative values of the current refer only to its direction.

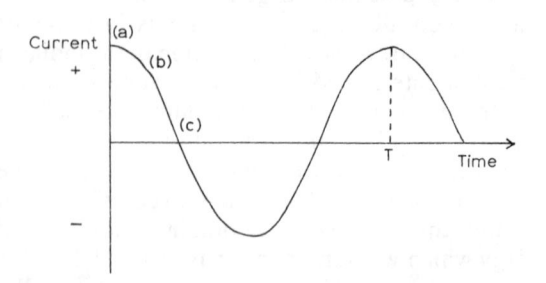

Fig. 46.6. Current-time graph for A.C. dynamo

A dynamo of this kind in which the current is taken from the coil by way of a pair of slip rings is an alternating current dynamo. If we take the electric motor described in Chapter 45 and instead of putting a current into the coil we turn it by hand, clearly it will act as a sort of dynamo. We shall have to take the current from the coil by way of a split ring commutator instead of slip rings. What sort of current should we obtain through a centre zero galvanometer connected to such a dynamo? Fig. 46.7 shows the position of the coil when the current

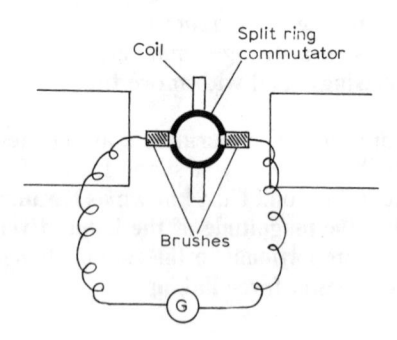

Fig. 46.7

in it is zero and just reversing its direction. It can be seen that at the same time the connections between the commutator and brushes are also changing over. The nett effect of this must be that the current in the galvanometer always flows in the same direction. The current from this D.C. dynamo will vary with time as shown in Fig. 46.8. It is not the steady direct current we should obtain if there was a battery in the circuit instead of the dynamo.

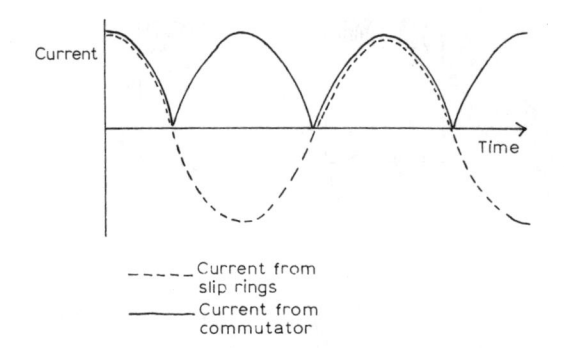

Fig. 46.8.

For electric heating and lighting there is no reason for preferring either alternating or direct current. However for reasons which we shall explain later, for electrical power production on a large scale, A.C. dynamos are almost always used.

The Generation and Distribution of Electrical Energy

Most of the electrical energy generated in the power stations of this country is produced from heat energy given to steam. The heat is produced first in coal or oil burning boilers or nuclear reactors. In the latter case the heat produced by the splitting or fission of the nuclei of the uranium fuel atoms is first transferred to a coolant liquid or gas which is pumped through the reactor (Fig. 46.9). In all British nuclear power stations to date this coolant has been carbon dioxide but in the future it may be, for example, liquid sodium. The heated coolant passes through a heat exchanger in which it gives up as much of its heat as possible to steam which flows in pipes in the exchanger. The high temperature, high pressure steam produced in the heat exchangers of a nuclear power station or in the boilers of a 'conventional' power station passes through steam turbines (see chapter 19). The

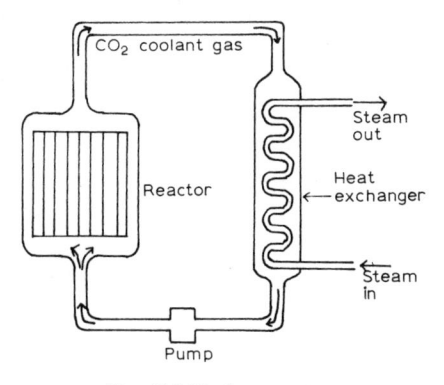

Fig. 46.9 Nuclear reactor

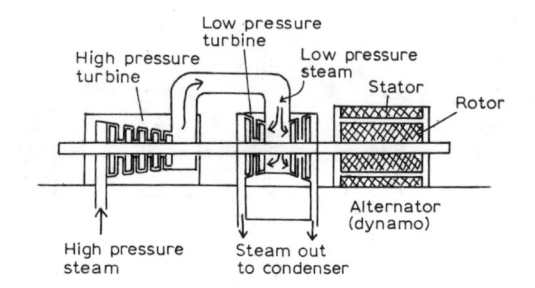

Fig. 46.10. A power station

rotation of the turbine is transmitted by means of a shaft to the roto or the dynamo. (Fig. 46.10).

The magnetic field in power station generators is not of course provided by permanent magnets. Some of the current produced is fed back into coils in the dynamo to produce the field. In practice the field coils are rotated and the induced currents are produced in stationary coils (stators).

The voltage produced in the generator may typically be about 10 000 V and the power 10 MW (10 million watts). This implies a current (i A)—given by

$$10\,000\, i = 10\,000\,000 \text{ (see page 217)}$$
$$\text{or} \quad i = 1000 \text{ A}$$

If the energy is transmitted to the consumer, who may be hundreds of kilometres from the power station, by means of cables connected directly to the generator, a very large amount of energy will be wasted as heat produced in the cable. If the resistance of the cable (R) is 1 Ω the heat produced is $i^2 R$ J/s.

With the value of the current quoted above this will be $(1000)^2 \times 1$ J/s = 1 million J/s or 1 MW.

This represents one tenth of the power output of the generator wasted as heat produced in the cable. This wastage may be reduced (*a*) by making the resistance of the cable lower and (*b*) by reducing the current. The first possibility means making the cable much thicker which involves a number of difficulties including very heavy cost. Reducing the current means, for a given power, increasing the voltage. For example if our 10 MW of power is transmitted at 100 000 V instead of 10 000, the current (i) is given by

$$100\,000\, i = 10\,000\,000 \quad \text{or} \quad i = 100 \text{ A}$$

The power wastage in our 1 ohm of transmission line is now

$$(100)^2 \times 1 \text{ W}$$
$$= 10\,000 \text{ W.}$$

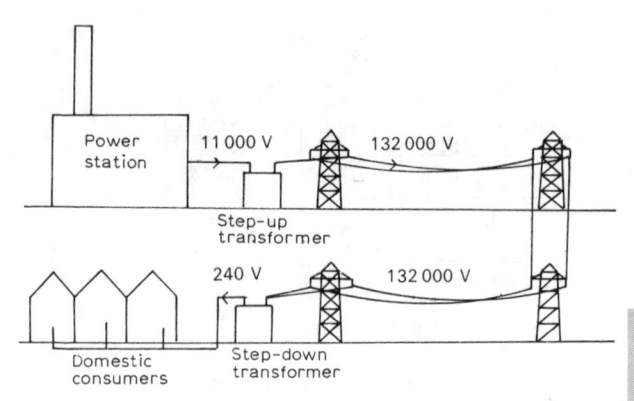

Fig. 46.11. Distribution of electrical energy

So by stepping up the voltage to 100 000 V we reduce power waste by 100 times.

In practice the voltage used in the National Grid system of cables, whereby electrical energy is transmitted over large distances in this country, is 132 000 V (Fig. 46.11).

The tendency in the electricity industry is away from a large number of small power stations and towards relatively few, very large ones. If these are nuclear power stations they must be situated in quite remote places, near the sea, and consequently the distances involved become even greater. The electricity authorities are thus setting up a 'Super Grid' working at over 400 000 V.

Clearly the electrical energy cannot be supplied to the consumer at these very high voltages. The voltage must therefore be stepped down at the 'receiving' end of the cables to an acceptable voltage which, for domestic purposes in this country, will be in the range of 200–250 V.

The Transformer

The devices used for stepping up and stepping down voltages in the Grid system are obviously very important. They are known as transformers and like dynamos are based on the effect of Electromagnetic induction.

Experiment 46.2. Experiments to show the Principle of the Transformer.

The right hand circuit in Fig. 46.12 is the same as the one we used in earlier experiments on electromagnetic induction (page 282) and we previously observed that induced currents were obtained in this circuit if the lines of force linking the coil were changing. The other circuit in Fig. 46.12 consists of another coil C_1 similar to C_2, a battery and a switch. Place the two coils end to end as indicated in the diagram and close the switch S. A momentary deflection of the pointer of G will

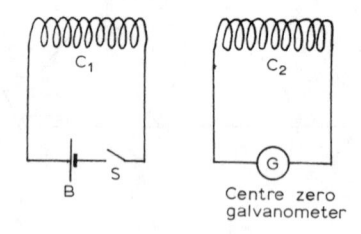

Fig. 46.12

be observed. As long as the switch S remains closed and a steady current flows in C_1 no further movement of the pointer will be observed. Now open the switch S. Once again the pointer shows a 'kick' but in the opposite direction.

When a current flows in C_1 it has a magnetic field associated with it and the lines of force of this field will be linking C_2. If the current is steady the lines of force linking C_2 will not be changing and there will be no induced currents. For a short while only, when the switch is being closed or opened and the field about C_1 is establishing itself or disappearing, do we have the necessary conditions for the production of induced currents.

Now place a bar of soft iron into the two coils as shown in Fig. 46.13 and make and break the circuit containing the battery. The kicks of the galvanometer pointer will be much larger. The presence of the iron bar much increases the number of lines of force and consequently increases the induced currents. Finally replace the battery by a low voltage alternating current supply and the galvanometer by a lamp bulb (Fig. 46.14).

When the switch S is closed the lamp bulb will light up and will stay on until either circuit is broken. In this case the alternating current in C_1 means that the lines of force linking C_2 are con-

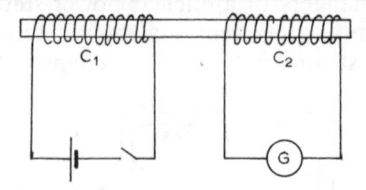

Fig. 46.13

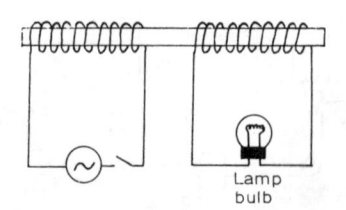

Fig. 46.14

stantly changing and the induced current in C_2 will flow continuously. These induced currents are alternating also and our centre zero moving coil instrument would not have shown a deflection to indicate them.

The arrangement we have in this experiment of two coils linked by a common iron 'core' is a transformer. The coil C_1 to which we apply the alternating potential difference is known as the primary and C_2 in which induced currents are produced, the secondary. In practice transformers are more commonly arranged as in Fig. 46.15. The coils are wound on a rectangular soft iron former.

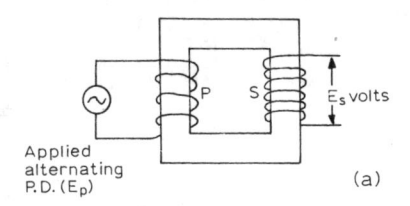

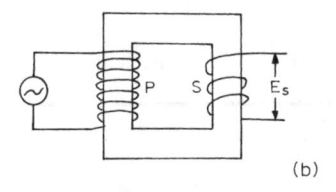

Fig. 46.15. Transformers

Fig. 46.15(a) shows a transformer with more turns in the secondary coil than in the primary. This is a 'step up' transformer which means the induced potential difference in the secondary will be larger than the applied potential difference in the primary. In fact if E_p and E_s are the primary and secondary voltages and n_p and n_s the number of turns in the primary and secondary coils then:

$$\frac{E_p}{E_s} = \frac{n_p}{n_s}$$

Fig. 46.15(b) therefore shows a step down transformer.

The current in the primary and secondary coils will depend on what is connected to complete the secondary circuit, but it should be noted that a transformer will be less than 100% efficient. This implies that the power in the secondary circuit (the power out) will be less than the power in the primary (the power in).

Power out < Power in

The power (in watts) in a circuit is the product of voltage and current in amperes.

Thus $E_s \times i_s < E_p \times i_p$

where i_s and i_p are secondary and primary currents.

Since in practice transformers may be over 95% efficient $E_s \times i_s$ may be very nearly equal to $E_p \times i_p$. One consequence of this is that in a 'step-up' transformer, the current is in fact stepped down.

When transformers are working at very high power and even when they are not, it is very important to make the efficiency as high as possible. The energy wastage arises mainly in two ways. The transformer core is magnetized by the alternating current in the primary coil, first one way and then the other. Work is done in very rapidly magnetizing and demagnetizing the iron former a larger number of times each second. All that can be done to minimize the energy used in doing this is to make the former of as 'soft' a magnetic material as possible. Soft iron has been quoted but in practice alloys more satisfactory for the purpose have been developed. The other source of energy loss is in the formation of 'eddy currents' in the core material. It should be noticed that the changing lines of force arising at the primary are linking not only the secondary coil but also the former. Induced currents circulate in the soft iron in the way indicated in Fig. 46.16. Eddy currents are reduced by

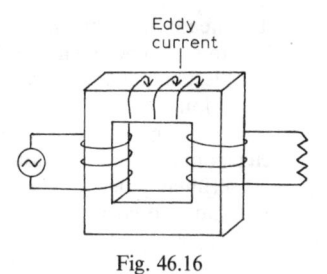

Fig. 46.16

making the former so that it offers as high a resistance as possible to their circulation. This is achieved by making the former not of one solid block of metal but of a large number of plates or laminations, separated from each other by some insulating material like pitch. Fig. 46.17 shows a

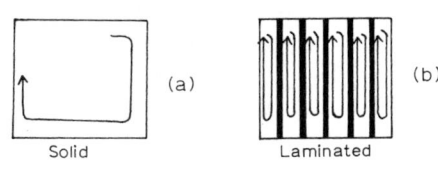

Fig. 46.17

cross section through a block of solid metal and a stack of laminations indicating the paths of eddy currents. The thickness of the laminations is of course much exaggerated.

The Induction Coil

You will have observed that transformers only work with an alternating current in the primary, which is of course the main reason why alternating is preferred to direct current in the electricity supply industry. It is sometimes desirable to obtain high voltage from a low voltage D.C. supply. A device which makes this possible is the 'induction coil' (Fig. 46.18). This is a type of step up transformer, but since the primary circuit contains a D.C. supply the current in the primary must be switched on and off in order that the lines of force linking the secondary shall be changing. This is achieved by including in the primary circuit a 'make and break' device similar to that used in an electrical bell. The primary coil, consisting of relatively few turns of thick copper wire, is wound on a soft iron core, often consisting of a bundle of wires. A bundle of wires rather than a solid core is used for the same reason as a laminated former is used in a transformer, namely to minimise eddy currents. When the switch in the primary circuit is closed the soft iron core becomes magnetized and attracts the armature A. This causes the circuit to be broken at X. The armature is released and the steel spring returns to complete the circuit again. In this way the primary current is rapidly switched on and off. The changing lines of force associated with the primary coil link the secondary which consists of many thousands of turns. The induced voltage in the secondary, determined by the turns ratio, will be large and it is often applied to a spark gap fixed on top of the coils.

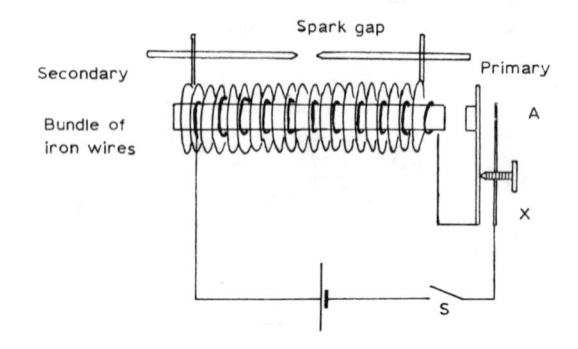

Fig. 46.18. An induction coil

Exercise 46

1 Explain what is meant by *electromagnetic induction*. Describe carefully simple experiments by which you could demonstrate this phenomenon. State the laws of electromagnetic induction.

Describe briefly ONE piece of apparatus that depends for its action on electromagnetic induction. (*O*)

2 Describe an experiment you could perform to show the production of an induced current.

An induction coil is used with a 6 V accumulator to give a potential difference of about 2000 V. Draw a labelled diagram of the coil showing the complete circuit, and explain how the high potential difference is produced.

Describe briefly ONE practical application of such a coil.

3 What are the conditions necessary for the production of an induced current? Draw a labelled diagram of a simple alternating current generator and explain how it works.

Discuss the advantages of using alternating current for long-distance transmission of electrical energy. (*C*)

4 (*a*) Draw a labelled diagram illustrating the essential features of a step-up transformer. What happens to the current in the transformer? State TWO advantages gained by the use of transformers in the transmission of electrical power. Why does the national grid system supply a.c. rather than d.c. electricity? (*S*)

5 Give a labelled diagram showing the principal parts of a transformer that can be used to reduce an alternating current from a high voltage to a lower one.

(*a*) Why has alternating current to be used?
(*b*) What factors determine the output voltage?
(*c*) What is the purpose of having a magnetically 'soft' core?
(*d*) Why is the core laminated? (*EM*)

47
Static Electricity

The fact that some materials like glass, when rubbed with silk, will attract light objects has been known for a long time and such effects were discussed in detail by William Gilbert whom we mentioned in Chapter 44. However, with the development in recent years of plastics like polythene and nylon, these effects of static electricity have become almost everyday experience. You have probably noticed, for example, how quickly gramophone records collect dust or you may have experienced the crackling sound from nylon garments. But the study of static electricity has acquired great scientific importance since it has been established that some of the forces acting between atoms and within atoms are electrostatic in nature.

Experiment 47.1. Electrostatic Forces.

Suspend two inflated rubber balloons by lengths of cotton from stands. Rub them both with your hand and bring them towards each other. You will observe a repulsion between the balloons which holds them apart (Fig. 47.1). We say that the balloons have acquired electric charges by friction.

Bring up towards one of these charged balloons first a rod of polythene which has been rubbed with a piece of woollen cloth and then a strip of perspex which has been charged in the same way. The polythene rod will repel the balloon but the perspex strip will attract it.

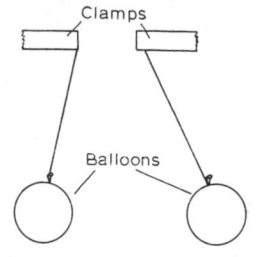

Fig. 47.1. Repulsion between charged balloons

Suspend from a cotton thread a small perspex strip which has been charged by rubbing (Fig. 47.2) and bring up to it in turn the charged polythene and perspex rods. The polythene attracts the suspended

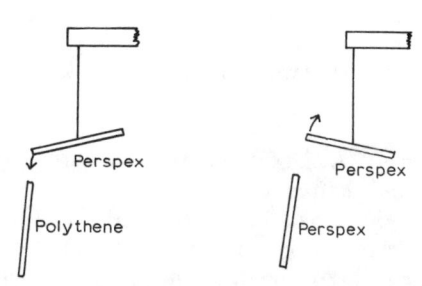

Fig. 47.2. The forces between charged rods

perspex strip but there is a repulsion between the two pieces of perspex. You should note that the polythene attracts both ends of the perspex and there is no question of there being poles as in the case of magnets. The experiments show, however, that there are two types of charges. We call those which the perspex and balloons acquire positive charges and those acquired by the polythene negative.

Similar experiments will show that a glass rod will acquire a positive charge by friction and ebonite a negative one. We have established a law of force between electrostatic charges, similar to the one for magnetic poles.

Like charges repel. Unlike charges attract.

The Gold Leaf Electroscope

The instrument illustrated in Fig. 47.3 is a sensitive device for detecting, and sometimes measuring

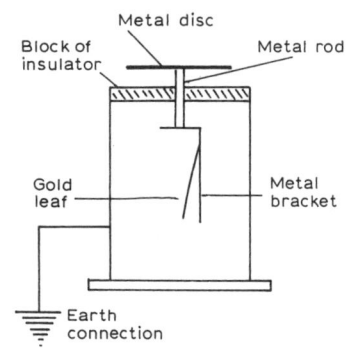

Fig. 47.3. A gold leaf electroscope

289

electrostatic charges. It consists of a metal disc and rod to the lower end of which is attached a small strip of gold leaf. This conducting part of the instrument is mounted in a block of insulating material (e.g. perspex) which forms the lid of a metal case fitted with glass windows so that the gold leaf may be seen. The case usually has a terminal on the outside by which it may be connected to earth. However, as we shall see, wood may be regarded as a conductor in electrostatic experiments and it is usually sufficient simply to place the instrument on the wooden bench to make this earth connection.

Experiment 47.2. Charging the Electroscope by Contact and Testing Charges.

Touch the cap of an electroscope with a charged polythene rod. The leaf will be deflected upwards. It will probably drop a little when the rod is removed but should retain some deflection. We say that the electroscope has been charged by contact. Some of the charge on the polythene rod has effectively been transferred to the insulated metal part of the electroscope. Thus the lower part of the metal rod and the gold leaf have similar charges and will repel each other. The electroscope charged in this way acquires a charge of the same sign as the polythene rod, i.e. negative.

Bring the polythene rod again up to the electroscope, this time without actually touching it (Fig. 47.4). An increase in the deflection of the leaf will be observed as long as the rod is held in position. Now bring in turn a charged perspex rod and an uncharged rod of any kind close to the negatively charged electroscope. A decrease in the deflection will be seen in both cases. Discharge the electroscope by touching the cap with your finger and recharge it positively by contact with a perspex rod. Again bring positive, negative and uncharged rods near to the cap and observe the change in deflection. This time the positive rod causes an increase and the other two a decrease.

These experiments show that if we wish to be certain which charge an object has we must make it bring about an increase in the deflection of the leaf of an electroscope. If it does so, the body must have the same sort of charge as the electroscope.

Experiment 47.3 Conductors and Insulators.

We have already seen that if you touch the cap of a charged electroscope with your hand, the leaf will collapse. We say that the body is a conductor and the charge passes through the body to earth.

Charge the electroscope and touch its cap in turn with some of the following substances held in the hand: copper, iron, carbon, wood, wet and dry cotton, rubber and nylon. You will find that with the metals, carbon and wet cotton the leaf collapses almost instantly, with wood and dry cotton more slowly and with rubber and nylon not at all. With the low voltage normally encountered in current electricity wood and dry cotton are classed with the plastics as insulators. In electrostatics where higher voltages occur they are conductors.

Experiment 47.4. Charging by Induction.

Place two large copper calorimeters or similar metal cans inverted on separate plates of uncharged insulating material, e.g. polythene, and move them until the cans are touching each other as Fig. 47.5(*a*). Hold a charged polythene rod in a position close to one of the cans and slide the other on its plate away from it. Having removed the polythene rod, test the calorimeters by bringing them in turn close to a partially charged electroscope. They must be moved, of course, on the insulating bases without touching the metal. You will see that the can which was nearer the negatively charged rod has acquired a positive charge, and the other a negative charge. Bring the cans back into contact again and test them separately for charge. You should find them both uncharged.

In the experiment we began by giving the cans opposite charges which, since they just cancelled out when the cans touched again, must have been equal in size. This may be explained by considering that the cans to start with contain equal quantities of positive and negative electricity, some or all of which is free to move. The presence of the negatively charged rod will produce a separation of this electricty as indicated in Fig. 47.5. A net positive

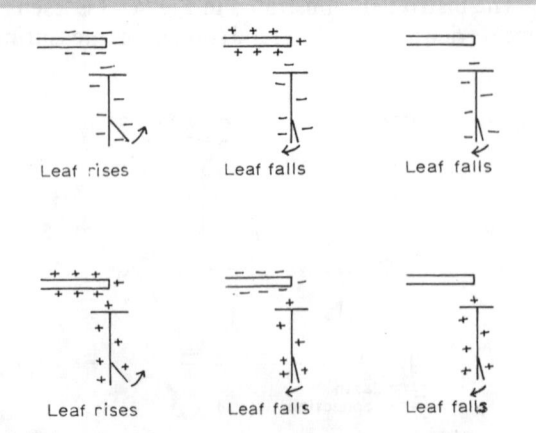

Fig. 47.4. Testing a charged rod with an electroscope

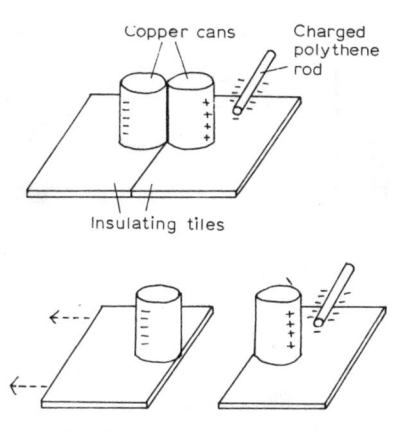

Fig. 47.5. Charging by induction

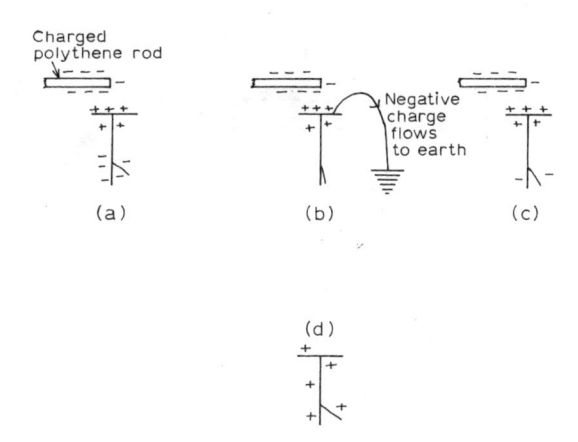

Fig. 47.6. Charging an electroscope by induction

charge in one part of a body could be due to a gain of positive charge or a loss of negative charge. When the cans are separated the cans will retain their charge.

The process we have demonstrated is known as charging by induction. The charges on the cans represent electrical energy gained since if they are earthed and consequently discharged, a momentary small current will flow and some heat will be produced. Where has this energy come from? Certainly work was done in charging the polythene rod but none of this charge will have been lost in the induction process. The origin of the electrical energy is the work done in separating the cans against the attractive force between them caused by their opposite charges.

Experiment 47.5. Charging an Electroscope by Induction.

Bring a charged polythene rod close to the cap of an uncharged gold leaf electroscope. Now touch the cap of the electroscope to earth it, remove the earth connection and finally remove the polythene rod. The leaf of the electroscope will indicate that it has been left with a charge. Test the sign of this charge by bringing up in turn charged polythene and perspex rods. The polythene will cause a decrease and the perspex an increase in the divergence of the leaf. This indicates that the electroscope has a positive charge, opposite in sign to the charging rod.

We can again explain this process by suggesting first that the polythene rod produces a separation of charge in the conducting part of the electroscope (Fig. 47.6(a)). Now when the cap of the electroscope is touched some negative charge, initially repelled to the gold leaf, is able to move further away from the negatively charged polythene rod by flowing to earth. The leaf therefore collapses (Fig. 47.6(b)). When the earth is removed there is a further flow of negative charge to the leaf and a small deflection is observed (Fig. 47.6(c)). But when the rod is removed, the electroscope is left with a net shortage of negative charge or an excess of positive which becomes distributed over the whole of the metal part of it. This is a much more effective way of charging an electroscope than by contact with a charged insulator.

Electrostatic Generators

In recent years, certain machines capable of building up large electrostatic charges have become quite important, particularly in Nuclear Physics research. The most widely used of these is the Van de Graff generator which we shall describe later. We shall now investigate a simple generator which, like the others, depends for its action on the principle of electrostatic induction.

Experiment 47.6. The Electrophorus.

Charge a plate of insulating material, e.g. polythene, by rubbing and place on it a metal disc on an insulating handle (Fig. 47.7 (a)). Remove the disc and test it by bringing it close to a partly charged electroscope. It will probably show a small charge of the same sign as the insulating plate. Now repeat the process but this time momentarily earth the metal disc by touching its top face. This time on testing you should find that the disc has acquired a large charge of opposite sign to the insulating plate. In fact if you bring the charged disc close to a water or gas tap you may see quite a big spark jump across. The charging here is clearly an induction process as explained in Fig. 47.7.

Once again the electrical energy we obtain here

291

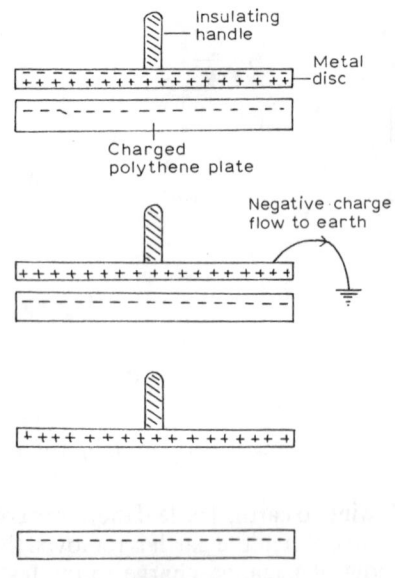

Fig. 47.7. An electrophorus

arises from the work we do in separating the positive charge on the disc and the negative charge on the plate. Since we are not actually taking charge from the polythene plate, the process may be, in principle at least, continued indefinitely without the charge on the insulator decreasing.

Distribution of Charge on a Conductor and the Discharging Action of Points

Experiment 47.7. Place an inverted copper calorimeter on an uncharged insulating plate, and charge it by touching it three or four times with a metal disc charged by an electrophorus as in the last section. Now take a small metal disc on an insulating handle (which we shall call a 'proof plane') and place it in contact with the flat bottom of the calorimeter (Fig. 47.8(a)). Transfer it to the cap of an electroscope and note carefully the deflection of the leaf. Discharge the electroscope and repeat the experiment transferring charge in turn from the curved sides (Fig. 47.8(b)) and the edge of the base (Fig. 47.8(c)), noting the deflections of the leaf in each case. You should find that the largest deflection occurs in the third case and the smallest in the first. Finally turn the calorimeter up, recharge it, and transfer charge to the electroscope from the bottom as in Fig. 47.8(d). Very little or no deflection at all should be observed.

We can conclude from this experiment that charge distributes itself over a conductor so as to produce a maximum concentration at parts of the surface which are sharply curved leaving flat

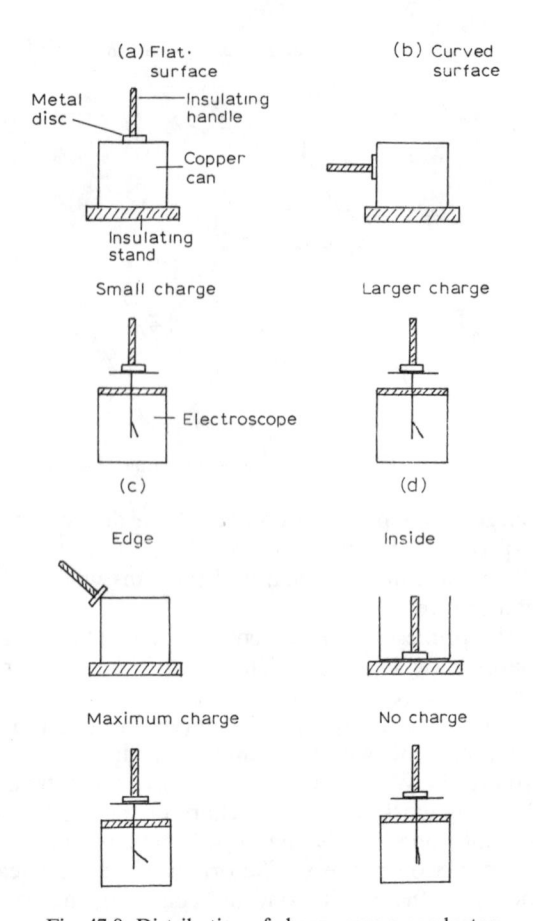

Fig. 47.8. Distribution of charge over a conductor

surfaces with the charge more sparsely distributed. Also there is little or no charge inside a hollow conductor.

The second of these two conclusions has a number of important practical applications. For example, components in electronic apparatus such as valves may be effectively screened from electrostatic fields arising from neighbouring charged objects by completely enclosing them in metal cans.

The first conclusion leads one to believe that a charged pointed conductor would show a large concentration of charge near the point. This does in fact occur and may be demonstrated by the following experiments.

Experiment 47.8. Connect a pin to the conductor of an electrostatic generator, e.g. a Van de Graaff machine and point it at an uncharged gold leaf electroscope placed a few feet from it. (Fig. 47.9). You will see that when the machine is operated, the electroscope shows a charge which it retains even when the generator is switched off and discharged.

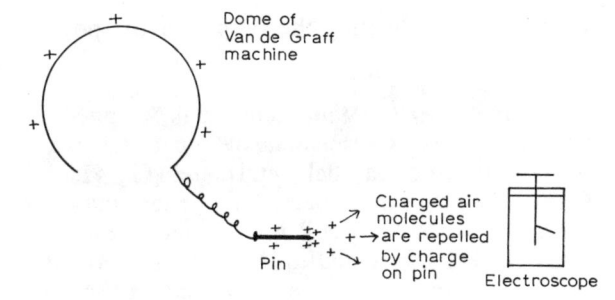

Fig. 47.9. Discharging action of points

If you put your hand a few inches from the pin while the machine is working you will feel a draught.

This discharging action of points may be explained by suggesting that the air molecules close to a point where there is a concentration of charge themselves acquire some of this charge. They are then repelled by the charge remaining on the point and stream away from it.

The effect may also be demonstrated by means of Hamilton's mill. This consists of a number of pins arranged as shown in Fig. 47.10 and balanced on a pivot. It is connected to the conductor of a generator and the repulsion between the charged air molecules and the points causes the mill to rotate.

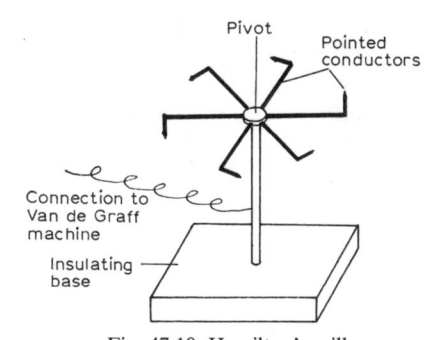

Fig. 47.10. Hamilton's mill

The discharging action of points besides being used, for example, in the operation of the electrostatic generators we have mentioned and in lightning conductors, sometimes causes trouble. In soldering connections in high voltage electrical apparatus it is important to obtain a smooth blob of solder without the spikes that the inexperienced solderer so readily produces.

The Van de Graaff Generator

Our present knowledge of the structure of the nucleus of the atom has been obtained largely by allowing very high speed atomic particles to collide with nuclei causing them to split up. The high speeds are often given to the particles in machines known as accelerators. Large Van de Graaff generators often form the main part of these devices. The particles travel along a tube at one end of which there is a plate which is earthed and at the other a plate connected to the dome of the machine which is given a large charge. The atomic particles being themselves charged will be attracted towards one or other of these plates and will be accelerated to high speeds in the process.

The Van de Graaff generator itself consists of a large metal dome mounted on an insulating stand. A continuous belt of some insulating material such as silk or rubber passes over two rollers, the top one situated inside the dome and the bottom one driven by an electric motor in the direction indicated in Fig. 47.11. Two 'combs' of pointed conductors are situated close to the belt. The one at the bottom is connected to one terminal of a high voltage (10 000 volt) D.C. supply, the other terminal of which is connected to earth. This comb 'sprays' charge on to the belt and the charge is carried up towards the dome. The other comb is inside the dome and connected to it. The charge coming up (assumed positive in this case) produces a separation of charge making the points negative and leaving the dome positive. The points discharge on to the belt, neutralizing the charge on it, and leave the dome with a net positive charge. This process will continue and with large machines sufficient charge may be accumulated on the dome to give it a potential of several million volts.

The Van de Graaff machine is a device for con-

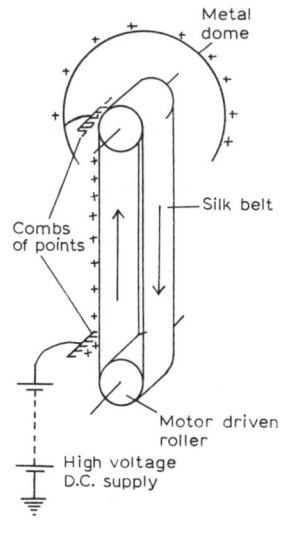

Fig. 47.11. A Van de Graaff machine

verting mechanical energy into electrical energy and it is desirable to consider where exactly the electrical energy 'comes from'. In this case work must be done carrying positive charge upwards against the repulsive force due to positive charge already present on the dome.

Static and Current Electricity
In this chapter we have been assuming that static electricity is related in some way to the kind of electricity we discussed in previous chapters, although we have little evidence for it so far. The next two experiments, however, will show some connection between the two.

Experiment 47.9 The high voltage power supply shown in Fig. 47.12 is a device for converting the mains supply (240V A.C.) to a D.C. supply of 0 to 500 V. It therefore consists of a transformer and a rectifier (this converts A.C. to D.C.). There will also be some sort of rheostat so that the output voltage may be varied. Set the output of the power supply to 0 and connect the terminals to the cap and case of a gold leaf electroscope. Now slowly turn up the output voltage. You will observe that the leaves of the electroscope gradually diverge.

The power supply unit in this experiment is a current electricity device operating from the mains. The gold leaf electroscope is, of course, an electrostatic instrument, but here it is acting like a voltmeter in indicating the output of the power supply.

Experiment 47.10. A table tennis ball wrapped in aluminium foil is suspended by a cotton thread between a pair of parallel metal plates (Fig. 47.13). One of the plates is connected to one terminal of a very sensitive moving coil galvanometer while the other plate is connected to the dome of a Van de Graaff generator. The other terminal of the galvanometer and the base of the generator are connected to a common earth.

Switch on the Van de Graaf machine. You will observe that the ball starts to bounce backwards and forwards between the plates and at the same time the galvanometer shows a small current. The ball is, in fact, picking up from one plate the charge which has been generated by the machine, and ferrying it to the other from where it flows to earth via the galvanometer.

In this experiment we have observed the galvanometer, which is a current electricity instrument, showing a movement of electrostatic charge. These two experiments together indicate that electrostatics is generally about quite high voltages and very small currents.

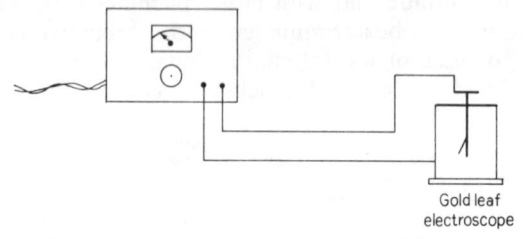

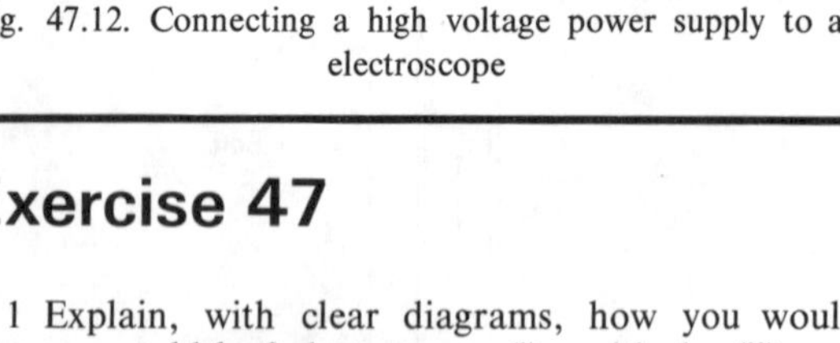

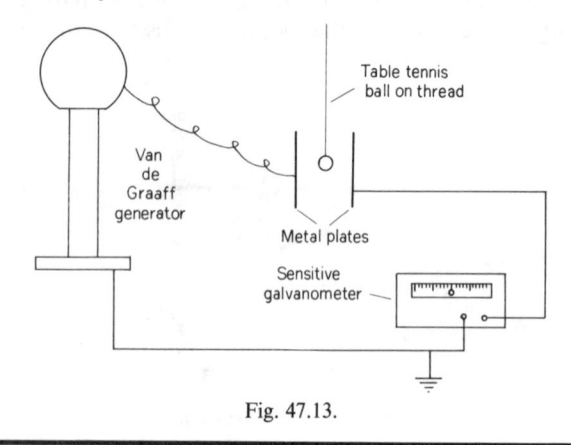

Fig. 47.12. Connecting a high voltage power supply to an electroscope

Fig. 47.13.

Exercise 47

1 Explain, with clear diagrams, how you would charge a gold leaf electroscope (i) positively, (ii) negatively, if you were provided with an insulated brass sphere negatively charged.
If you now had another charged insulated conductor how could you determine the sign of its charge? (*L*)

2 Describe and explain how, with a negatively charged ebonite rod and two insulated uncharged metal spheres, you could obtain a positive charge on one sphere and a negative charge on the other (*J*)

3 Why has one end of a lightning conductor a sharp point? Describe a laboratory experiment to support your explanation. (*J*)

4 A small metal can is insulated by standing it on a slab of insulating material. How would you give it a charge of positive electricity using an ebonite rod and fur? How would you expect the charge to be distributed over its surfaces? What would be the result of holding a sharp-pointed needle near the can?
Explain how a knowledge of this result is used in the

construction of a lightning conductor. (L)

5 Describe an electrophorus, and show with the aid of a series of diagrams how it can be charged.

Explain how a charge of electricity can be given to an insulated body, by means of the electrophorus. (L)

6 Describe *two* experiments to demonstrate that the whole of an electric charge on a charged hollow conductor resides on the outside.

Explain how a charge is transferred completely from a charged body to a hollow conductor.

Describe *two* experiments designed to show the discharging action of points and mention *two* practical uses of the phenomenon. (L)

48
States of Matter; Structure; Reaction Rates; Equilibrium

State of Matter

We have often referred to solids, liquids and gases, the three states in which a substance may occur. What is the difference between them?

Solids keep a fixed shape unless deformed by some physical force such as cutting, breaking or melting. The atoms or molecules of which they are made are packed closely together in an orderly fashion and the solid is not easily compressed.

Liquids take the shape of the container in which they are placed. This need not be closed at the top. One liquid will diffuse (spread out) slowly throughout another as is shown in the next experiment. Liquids can hardly be compressed at all.

Gases take the shape of the container, which must be closed or they will diffuse quickly into the extra space available outside. The molecules are not held tightly together and can move about freely (and rapidly); there is no order in their arrangement at all (there is no real order about the molecules in a liquid, but, at least, they do not move out of the container!). Because there is space between the molecules a gas can be compressed considerably.

Experiment 48.1. Diffusion in Liquids*
  (i) Place a coloured crystal such as copper(II) sulphate or ammonium dichromate(VI) in a 250 cm^3 beaker full of water. Leave for a day.
 (ii) Fill a small tube with a solution of potassium manganate(VII) (potassium permanganate) and stand it upright (or sloping against the side) in a 400 cm^3 beaker. Carefully fill the beaker with water so that the tube is covered. Leave for a day.

After a day or so there will be a uniform colour in each beaker showing that the coloured particles have moved freely throughout the water.

(iii) Get your teacher to place two drops of bromine in a tall gas jar and cover this with a lid. Put a white paper at the back and notice how quickly the red-brown vapour creeps up (against gravity) and fills the jar.

The movement of molecules shown in these experiments is important and we must look more closely at it as the state in which a substance exists depends on the motion of the molecules of which the substance is composed. Why should this be so? Look at fig. 48.1 before you read further.

Solids. The orderly arrangement of the atoms or molecules means that they can only vibrate slightly about a fixed position. If energy is applied to the solid in the form of heat, the molecules will vibrate faster and faster until they break away from their fixed positions; the solid will then melt and become

*Some of the experiments in this chapter repeat ones in Chapter 7 which were probably carried out previously.

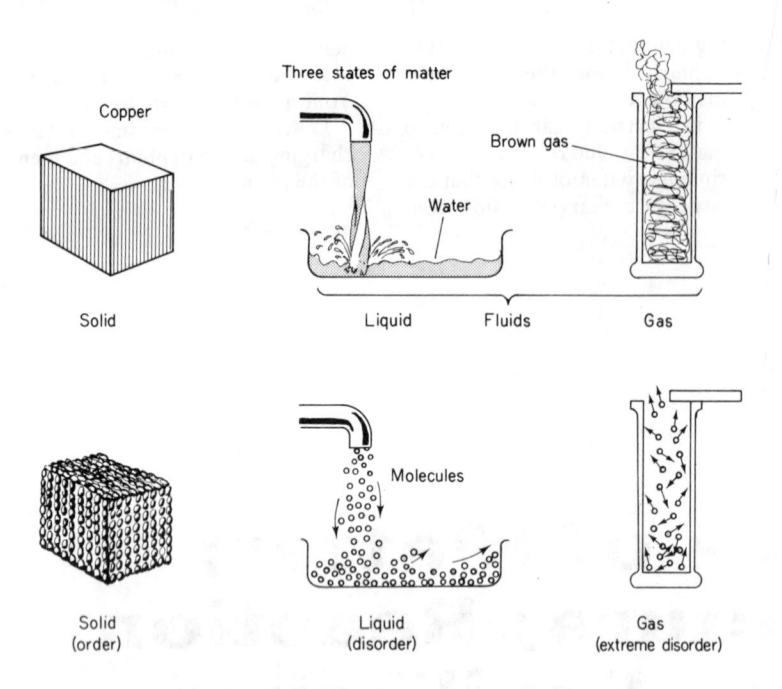

Fig. 48.1. The three states of matter.

a liquid. If we are considering a crystalline solid such as sodium chloride, you will remember that this does not contain any molecules, but is an arrangement of sodium ions and chloride ions, i.e. Na^+ and Cl^-. Because these ions carry opposing charges there is a force of attraction between them which is only overcome when the vibrations get sufficiently vigorous, i.e. at a particular temperature, the melting point of the solid. Ionic crystals melts at a definite temperature (usually fairly high), but covalent solids such as wax soften before they finally melt.

Liquids. In solutions which are ionic the ions no longer hold any fixed positions, but are in a state of disorder, moving around the container. They are, however, still attracted to each other because of their opposing charges. The molecules in a covalent liquid such as ether or benzene do not attract each other in the same way: as a general rule such liquids evaporate fairly readily when left exposed to the air. The molecules then exist in a free-moving gaseous state.

Gases. The molecules here are able to move much more rapidly and do so in straight lines, being deflecting whenever they collide with each other or strike the sides of the container in which they may be confined. There is no order at all in their position and if the container is open the molecules are able to move freely through the air outside. This idea of the movement of particles in the states of matter explains why the disordered gases can be compressed into much smaller volumes and if we compress them sufficiently at a low temperature we can push the molecules close enough together to form a liquid (such as liquid air). In the same way we can reason that the molecules of a liquid and a solid are already so close together that they cannot easily be pushed any closer by pressure. This movement also explains why gases can diffuse through other gases quickly as there is plenty of room between the molecules to allow a fairly free movement. Diffusion in liquids is slower because the molecules are so much nearer to one another.

Ice, water and steam are all the same substance, but 1 g of ice or water will only occupy about 1 cm^3; if this is changed into 1 gram of steam at 100 °C the volume will become about 1700 cm^3. There are still the same number of molecules, but the distances between the molecules must be much greater.

If we cool a gas or vapour we slow down the movement of the molecules and eventually their speed only matches that of molecules in a liquid and the gas liquefies or the vapour condenses. Cool it still more and the further drop in speed changes the substance into a solid.

A very simple experiment can be carried out to illustrate what happens when a solid changes into a liquid and then into a vapour as energy is put in.

Experiment 48.2. Increasing the Energy Applied to a Solid.

Fasten a long piece of rubber tubing to the stem of a large tap-funnel and connect the other end to a foot or hand-pump. Fill about one third of the funnel with pith balls (Rice Krispies or ants' eggs will do equally well), and start pumping *gently*. The balls represent the molecules of a solid such as ice and they move only a little as do the molecules in ice. Now increase the rate of pumping; the "molecules" move much more rapidly up and down, some faster than others. Here the liquid state is represented. Now pump very much more vigorously and the particles become really agitated and widely separated. They may fill the whole of the funnel and the faster-moving ones may even escape into the air. This is just what happens when a liquid changes into a vapour.

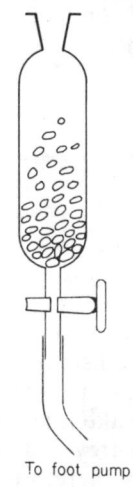

Fig. 48.2. Illustration of the effect of increase of energy applied to a solid.

Almost certainly you will have become much warmer while you were pumping, evidence of the fact that you were using quite a lot of energy in the process. Energy in the form of heat is needed to change a solid into a liquid and then into a vapour.

Structure

Let us now revise what we have learnt about the different types of structure present in elements and compounds and look at some new facts also.

Ionic (electrovalent) bonding

Crystalline salts consist of ions which are held in place by the attraction between those with a positive charge and those negatively charged, e.g. sodium chloride forms a giant structure or lattice of ions as shown in fig. 48.3.

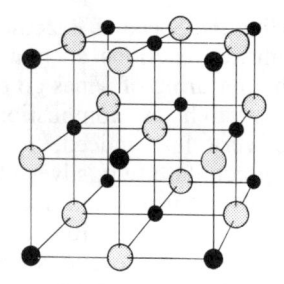

Fig. 48.3. Ionic lattice

This is rigid and only melts at a high temperature (800 °C) because much heat energy is necessary to break down the bonding (attraction) between the ions. Such a solid does not conduct electricity unless this happens and the ions are then free to move about. Dissolving in water has the same effect.

Covalent bonding

(*a*) Carbon in the form of diamond consists of carbon atoms joined together covalently (do you remember what this means?) to form a hard, rigid, giant structure of atoms which does not conduct electricity and has a very high melting point. Graphite is similar although the arrangement of atoms is different from that in diamond. (See Figs. 26.3 and 26.4).

(*b*) Simple molecules such as those of carbon dioxide, methane, hydrogen chloride and alcohol all contain a small number of atoms bonded covalently. They are usually liquids or gases and do not conduct electricity.

Some non-metallic elements also consist of simply molecules, e.g. oxygen, nitrogen, hydrogen and the halogens, which all contain two atoms to the molecule, i.e. they are diatomic. In ozone (O_3) there are three atoms in the molecule, in sulphur eight and in phosphorus four. Where these are solids they all have low melting points.

Free atoms

The argonons (helium, neon etc., the rare gases in the atmosphere) do not form any molecules, but consist simply of free single atoms.

Metallic bonding

Metals form a different kind of structure. Consider sodium: in any piece of sodium there are millions of atoms, each with one electron in its outer shell. These valency electrons are free to move throughout the whole piece of sodium (they are said to be delocalized) and can carry an electric current.

Factors affecting the Speed of Reactions

Some chemical reactions such as the rusting of iron or the maturing of wines go on extremely slowly: others such as combustion and ionic reactions go very fast indeed. Ionic reactions producing a precipitate, e.g. silver chloride from solutions of silver nitrate and sodium chloride, or those involving neutralization of an acid by an alkali go as fast as the solutions can be mixed. The silver chloride precipitation is complete in 10^{-8} s, i.e. in one hundred-millionth of a second after mixing. In this section we shall consider only reactions that go on at a measurable rate under ordinary conditions. There are several factors on which the speed of a reaction depends and by varying them it is possible to control the rate at which it takes place. You can probably think of ways in which this can be done.

Not all the particles in a liquid or vapour (gas) are moving at the same speed, some move faster and possess more energy. Chemical changes always involve the making and breaking of chemical bonds. For this to happen, particles must be in contact and those with the highest speeds are most likely to cause a change to occur on collision. Most atoms or molecules just rebound on colliding, but some will possess enough energy to break some of the bonds between other atoms and so bring about a change.

When a bond is formed between atoms, energy is *released* and the stronger the bond the more energy is given out: to break a bond and bring about decomposition, energy has to be *put in*. When hydrogen and chlorine join to form hydrogen chloride (by burning hydrogen in chlorine or exposing them to strong sunlight), the molecules of the two gases must first split into separate atoms (energy put in—endothermic) and then atoms of hydrogen join up with atoms of chlorine (energy given out—exothermic). For any reaction a balance sheet of energy gains and losses can be drawn up to decide if the overall reaction is endo- or exothermic. Nitrogen does not burn (is inert) because the N_2 bond is very strong and $N-O$ bonds are weak, but fuels burn readily given out much heat because strong $O-H$ and $C-O$ bonds in water and carbon dioxide are formed.

Here is a simple illustration of the speed of moving molecules.

Experiment 48.3. Comparing the Speed of Molecules
Set up horizontally a wide glass tube (say, 2·5 cm bore) about 60 cm or more long and have ready two bungs prepared in this way. With a sharp cork-borer cut a ring in the narrow end of each to take the end of a 3 cm piece of glass tubing about 2 cm bore. Fill the free space in the tubes with cotton wool. Dip one of the tubes in concentrated hydrochloric acid and let it drain. Do the same with the other using concentrated ammonia solution instead. Insert them both simultaneously in the long tube.

In a few minutes white fumes of ammonium chloride will appear somewhere in the region shown. (Fig. 48.4). Why doesn't this happen right in the middle of the tube? Do you think that the fact that the molecular mass of hydrogen chloride is 36·5 and that of ammonia is 17 has anything to do with it? We know that the molecules of different gases move at different speeds and that the ratio of these velocities for ammonia and hydrogen chloride is 1·5 to 1 approximately. Measure the lengths AB and BC and find the ratio $AB:BC$. Does this agree with the ratio of the velocities? Why is it unlikely that there will be a very close agreement.

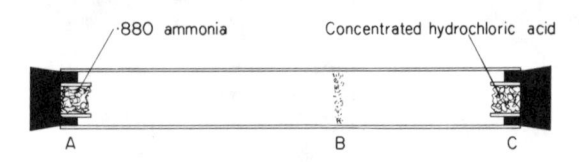

Fig. 48.4. Movement of molecules

The rate at which a reaction goes depends on (*a*) the number of collisions and (*b*) the force of these collisions. What factors do you think would affect the rate and make it faster?

Heating puts in energy and increases the speed of the particles and therefore the force of any collision.

If the reacting substances are gases and are put under pressure there will be more particles in any given volume and therefore more likelihood of collisions.

The same effect can be brought about in solutions by increasing the concentration.

Particle Size

One other factor to be considered in a reaction between a solid and a liquid is the size of the solid particles as this affects the area of the two in contact. Consider this simple example.

A cube with sides of 1 cm has a volume of 1 cm³ and a surface area of 6 cm². If this is cut into 1000 smaller cubes each with a side of 1 mm the total surface area will be 1000×6 mm² = 6000 mm² or 60 cm². The total volume is still 1 cm³, but the area has been increased tenfold. Imagine the great difference in surface area between a lump and a *powder* of equal volume or equal mass.

When dilute hydrochloric acid is added to powdered chalk there is a brisk effervescence as carbon dioxide is given off and the action is soon complete. If the same weight of calcium carbonate in the form of a lump of marble is used the rate is much slower.

If we pour 50/50 sulphuric acid on to powdered salt or rock salt the evolution of hydrogen chloride is much faster in the first case. The reason here is fairly obvious; the powder presents a larger surface area to the acid and so more molecules of the reacting substances are actually in contact at any one time. Let us look at one more example. If a piece of lead is scratched it will gradually become dull as it is converted into lead(II) oxide; the reaction then ceases. However lead can be obtained in the form of a powder (called pyrophoric lead) which takes fire immediately it is exposed to the air: it has, of course, to be kept in an inert atmosphere.

Concentration

When two solutions are mixed any chemical changes which occur do so as a result of the collisions between the molecules or ions of the different liquids. If therefore the concentration of one or both solutions is varied the rate at which these collisions occur will also vary. In a suitable case. doubling the concentration of one solution will approximately double the rate of the reaction, but this is not necessarily so with all reactions.

An experiment will illustrate this point.

Experiment 48.4. To See the Effect of Varying Concentration on the Rate of a Reaction.

For this you will need 6 strips of magnesium ribbon of exactly the same length, say 15 cm. You will also need six different concentrations of hydrochloric acid, prepared as below. (Make sure the acid and water are thoroughly mixed.)

A 50 cm³ of concentrated acid.
B 50 cm³ of concentrated acid + 50 cm³ water.
C 50 cm³ of B + 50 cm³ of water
D 50 cm³ of C + 50 cm³ of water
E 50 cm³ of D + 50 cm³ of water
F 50 cm³ of E + 50 cm³ of water.

If you now pour away 50 cm³ of F you will have 50 cm³ of each solution left.

If we let the concentration of F be 1, then the six concentrations from A–F will be 32, 16, 8, 4, 2 and 1.

To each solution in turn in a flask add one of the strips of magnesium shake and record the time to the nearest second (a stop watch is useful) which elapses before all the metal has disappeared and the bubbles have stopped coming off. Naturally, the time will increase as the acid gets less concentrated (or more dilute). Look at these figures for an actual experiment:

Concentration	32	16	8	4	2	1
Dilution	1	2	4	8	16	32
Time in s	5	11	20	38	79	162

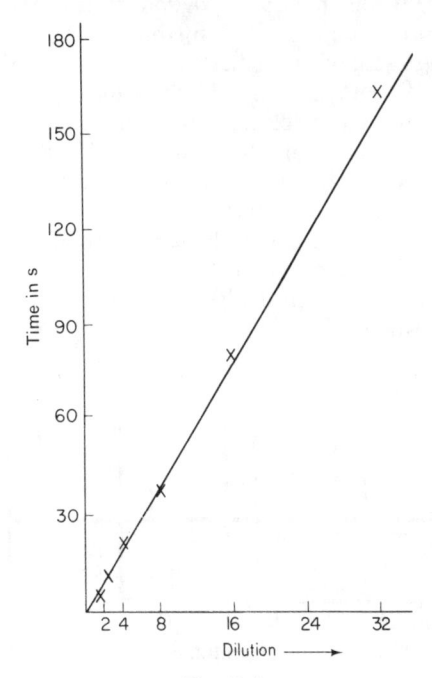

Fig. 48.5.

These figures for time and dilution have been plotted on the graph (Fig. 48.5) and you will see that they all lie close to a straight line through the origin and this points to a direct relationship between the time for the reaction to be complete and the dilution; this is the same as saying that the speed or rate of the reaction is directly proportional to the concentration. (You will learn later that there are reactions where this is not always the case, but we are not concerned with those now.)

Plot the results of your experiment in the same way and see if you get a similar result. It may be more accurate to pool the results of the different class groups and mark them all on one graph. Try to draw a straight line through them so that there are about as many points on one side of the line as on the other (a clear plastic ruler is helpful in doing this).

The equation for the reaction is, of course,

$$Mg(S) + 2HCl(aq) \rightarrow MgCl_2(aq) + H_2(g)$$

A similar experiment can be carried out using six equal amounts of powdered chalk (between 1–2 g) instead of the magnesium.

Here is another experiment to show the effect of varying the concentration.

Experiment 48.5. To See What Effect Varying Concentration Has on the Speed of a Reaction.

Set up four measuring cylinders as in Fig. 48.6. Make up a solution of sodium thiosulphate(VI) containing 2·5 g of the crystals in 100 cm³ of solution. Pour 10 cm³, 20 cm³, 30 cm³ and 40 cm³ respectively of the solution into the four cylinders. Arrange a sheet of white paper behind the vessels and draw a thick black line across it at about the level of the 100 cm³ mark. Have ready four beakers each holding 10 cm³ of bench hydrochloric acid. To the first add 130 cm³ of water, to the second 120 cm³, to the third 110 cm³ and to the last 100 cm³ of water.

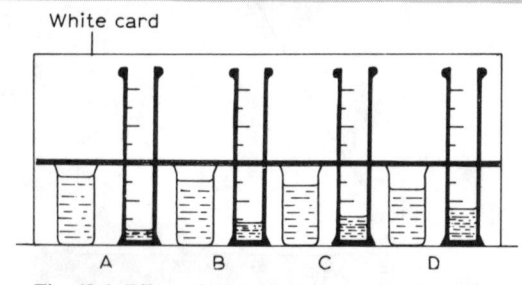

Fig. 48.6. Effect of concentration on rate of reaction

Pour the thiosulphate(VI) in cylinder A into the first beaker and then pour the mixture back into the cylinder as quickly as possible. Have another person start a stop-watch immediately the solution from the cylinder touches the acid in the beaker. Sulphur begins to be precipitated according to the equation.

$$Na_2S_2O_3(aq) + 2HCl(aq) \rightarrow 2NaCl(aq) + SO_2(g) + H_2O(l) + S(s)$$

and the time that elapses before the black line at the back can no longer be distinguished is recorded.

Repeat with the 130 cm³ of acid and the thiosulphate in the second cylinder and so on. The times will be found to be fairly closely in the ratio 1:2:3:4. As the total volume in each case is 150 cm³ the concentrations of sodium thiosulphate(VI) are also 1 : 2 : 3 : 4 while the concentration of the acid remains constant.

This experiment shows there is a connection between concentration and rate of reaction.

This experiment can also be carried out individually using conical flasks as the reaction vessels and standing them over a black cross marked on a white paper. The time for the cross to become obscured by the sulphur formed is noted.

Temperature

50/50 sulphuric acid acts slowly on rock salt and concentrated hydrochloric acid is only slowly oxidised by manganese dioxide in the cold. In these reactions, for making hydrogen chloride and chlorine in the laboratory, the mixture is usually heated so as to increase the rate. A sheet of iron or copper held in a Bunsen flame does not burn and only gradually does oxidation take place. Yet, if iron filings or powdered copper is dropped into the flame each tiny particle glows and forms the oxide. This increase in the speed is not due solely to the difference in size, but to the fact that each filing reaches a much higher temperature than the sheet. Most chemical changes occur more rapidly at higher temperatures and you can no doubt think of others.

An increase of 10°C in the temperature at which a *homogeneous* reaction is being carried out roughly doubles the rate at which it takes place. A homogeneous reaction is one in which the substances concerned are all liquids or all gases. One example would be the combination of hydrogen and chlorine to form hydrogen chloride, another the decomposition of ammonia into hydrogen and nitrogen. A rise in temperature results in more rapid movement of the molecules and an increase in their energy and consequently there is more likelihood that a change will occur. Oil dries by oxidation. If oily rags are piled together the heat produced by this exothermic reaction cannot escape and so the rate of oxidation rapidly grows with the production of even more heat. Eventually the temperature of the rags rises above the ignition point and they take fire. This is a case of spontaneous combustion. Similar fires may occur in hay stacked while damp or in huge piles of coal. Thermometers, inserted in iron pipes, are usually placed at intervals in such a coal stack to give warning of any undue rise of temperature.

**Experiment 48.6. A Demonstration of Spontaneous Combustion.*

Mix 2 g of powdered sulphur very thoroughly with 1 g of powdered zinc. Next place 1 g of fine crystals of potassium manganate (VII) in a crucible on an asbestos slab and then, using a teat pipette, add three drops of glycerol (glycerine) to the crucible. As quickly as possible put the zinc and sulphur mixture into the crucible, completely covering the other ingredients. STAND AWAY FROM THE CRUCIBLE and in about one minute or less the whole mixture will burst into flame.

Cooling below the normal reaction temperature decreases the rate in a similar way. It has been found that if the human body is cooled well below its normal temperature the various chemical processes going on are very much slowed down. Under such conditions it is possible to carry out operations on the body which would be impossible otherwise.

Experiment 48.7. To See What Effect Varying the Temperature Has on the Speed of a Reaction.
Look at the figures you obtained in your experiment with the magnesium and different concentrations of acid and choose a concentration which gave a time of about 100–150 s. Put 50 cm^3 of this mixture into each of six separate flasks.

Place a thermometer in the first flask and note the temperature. Add a 15 cm strip of magnesium ribbon, stir and record the time it takes for all the metal to disappear and the reaction to be completed.

Now warm the second flask to, say, 30°C, remove from the heat and add a fresh piece of magnesium of the same length. Again, record the time. Warm the remaining solutions in turn to, say, 35°, 40°, 45° and 50°. Repeat the experiment with these. (The temperatures need not be exactly these figures as long as you take note of what they are.) When you have all the results draw a graph with the temperature marked horizontally and the time vertically. Remember to mark the *highest* temperature close to the origin and work down to the lowest as the time will be shortest where the acid is warmest. Again the points should lie close to a straight line (particularly if you pool the class results) showing that the time taken is inversely proportional to the temperature, or that the rate is directly proportional to the temperature. It is doubtful whether you will get such good results as in the previous experiment as this relationship may not hold at the highest temperatures.

Catalysts

Many reactions that go on slowly can be hastened by the addition of a suitable catalyst. **A catalyst is a substance which alters the rate of a chemical reaction, but remains unchanged at the end.**

Let us consider a few examples.
(i) Hydrogen peroxide gradually decomposes into water and oxygen

$$2H_2O_2(aq) \rightarrow 2H_2O(l) + O_2(g)$$

But the addition of a little caustic soda appreciably speeds up the evolution. If now a pinch of manganese(IV) oxide is added the rate of the reaction is very considerably increased.

(ii) Nitrogen and hydrogen only combine slowly even at high pressures (this has the effect of pushing the molecules closer together), and without a catalyst the Haber process for the production of ammonia (see p. 129) would be quite uneconomic. The use of finely divided iron not only enables a lower pressure to be used, but also increases the speed of the reaction.

(iii) The Contact process for the production of sulphuric acid is carried out at atmospheric pressure and a temperature of about 450°C, but the rate of formation of sulphur(VI) oxide is very slow unless a catalyst such as platinum or vanadium(V) oxide is present.

(iv) Sodium sulphite slowly oxidizes to the sulphate, but the rate is markedly increased by adding copper(II) sulphate solution even in such minute quantities as one gram in four million dm^3 of the solution!

(v) Potassium chlorate(V) melts when heated, but needs to be heated to a much higher temperature before it will decompose and give off oxygen.

$$2KClO_3(s) \xrightarrow{\Delta} 2KCl(s) + 3O_2(g)$$

If, however, a pinch of manganese(IV) oxide is added to the just molten chlorate(V), oxygen is *immediately* evolved.

If in (i) we weigh the manganese(IV) oxide before adding it to the peroxide and, at the end, recover it by filtration and reweigh, we shall find that its mass (after drying) has not altered. In (v) the potassium chloride formed can be dissolved out in water and the insoluble manganese(IV) oxide recovered, dried and weighed. Again there will be no change.

Equilibrium

Most of the reactions you have studied have been ones which go to completion: this means that all the reactants are used up in forming new substances. However, this is not always the case.

If iron is heated in a current of steam it changes to tri-iron tetroxide (Fe_3O_4) and hydrogen is formed from the steam.

(i) $3Fe(s) + 4H_2O(g) \rightarrow Fe_3O_4(s) + 4H_2(g)$

If this oxide is heated now in a current of hydrogen it forms iron again along with steam.

(ii) $Fe_3O_4(s) + 4H_2(g) \rightarrow 3Fe(s) + 4H_2O(g)$

This, then, is a *reversible reaction* which can be made to go in either direction by choosing the starting materials. The equation may be written

(iii) $3Fe(s) + 4H_2O(g) \rightleftharpoons Fe_3O_4(s) + 4H_2(g)$

The reaction will only go to completion if one of the products is removed. In (i) the hydrogen escapes or is burnt; in (ii) the steam formed condenses to water and is taken away and so takes no further part in the reaction.

If some iron and water are placed in a sealed flask and heated, reaction (i) begins, but as soon as any iron oxide and hydrogen are formed these react together as in (ii). Eventually a stage is reached when the two reactions are both proceeding at the same rate—we have a balanced reaction. This does not mean that the two reactions have ceased; they both continue and we get a state of dynamic equilibrium with all four substances present, but the individual ones are continually reacting together so that the atoms move from one to another and back. We can demonstrate this quite easily with bismuth trichloride dissolved in concentrated hydrochloric acid, and water.

Experiment 48.8. A Reversible Reaction. Add a few drops of the bismuth(III) chloride solution to about 50 cm³ of water in a 250 cm³ beaker. The water becomes opalescent (milky) as insoluble bismuth oxychloride (BiOCl) is formed.

$BiCl_3(l) + H_2O(l) \rightarrow BiOCl(s) + 2HCl(aq)$

If you now add more of the bismuth chloride the milkiness increases, showing that the forward reaction (left → right) is going on. Next add some dilute hydrochloric acid to the beaker; the milkiness disappears as the oxychloride changes back to the chloride. The backward reaction (left ← right) is now proceeding. Evidently the reaction is reversible:

$BiCl_3(l) + H_2O(l) \rightleftharpoons BiOCl(s) + 2HCl(aq)$

The forward reaction is favoured by increasing the amounts of bismuth(III) chloride or water present. i.e. by adding more of either of these substances. The backward reaction is helped by adding more hydrochloric acid. The reaction can be made to go in one direction and then in the other several times in this manner. (The hydrochloric acid is, of course present as hydrogen ions and chloride ions.)

A reaction mentioned in Chapter 14 was the one in the Contact Process for the manufacture of sulphuric acid. This is really a reversible reaction

$2SO_2(g) + O_2(g) \rightleftharpoons 2SO_3(g)$

The yield of sulphur(VI) oxide can be improved by using an excess of either sulphur dioxide or oxygen instead of the quantities suggested by the equation (2:1). Using air is more economical as sulphur dioxide has to be made whereas air can readily be obtained and is used in place of oxygen (it works well although a bigger volume is needed).

Exercise 48

1. Pure chemical substances are made of
A free atoms,
B simple molecules,
C a giant structure of atoms,
D free ions, or
E a giant structure of ions.
Write down the letter which best describes the structure of each of the substances below. (A letter may be used once, more than once, or not at all.)
 (a) Neon.
 (b) Graphite carbon.
 (c) Sulphur.
 (d) Copper(II) sulphate crystals.
 (e) Methane.
 (f) Ethanol. (L)

2. A crystal of iodine was placed in a tube fitted with a cork. To the inside wall was stuck a strip of filter paper marked in centimetres and moistened with freshly-made starch solution. The experiment was designed to investigate the movement of iodine vapour at 20°C. The results obtained are shown on the graph in which the distance moved is plotted against the time taken to move that distance.
 (a) How does the starch indicate the position of the iodine vapour?
 (b) What is the process called by which the iodine

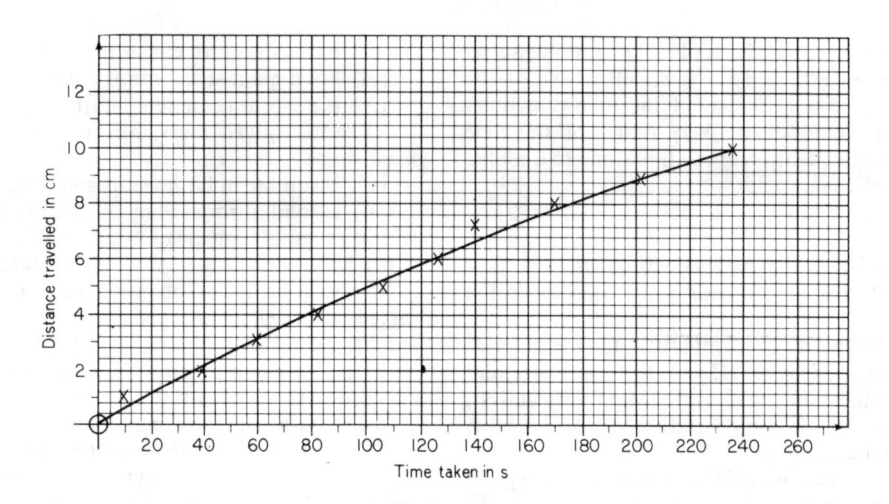

vapour moves?

(c) (i) How far had the iodine vapour travelled in 3 minutes?

(ii) What was the time taken for the iodine vapour to travel 9 cm?

(d) The iodine solid changed to a vapour without melting. What is this physical process called?

(e) The apparatus was set up a number of times using fresh starch and iodine on each occasion and three different experiments were carried out:

Experiment A. The test tube was placed in water at 30°C.

Experiment B. The air was extracted from the test tube.

Experiment C. The tube was placed horizontal.

How would the results of these experiments compare with the original experiment? Briefly explain each in turn, giving the letter of the experiment.

(f) Explain briefly how you would devise an experiment in which the movement of the vapour would be slower. (L)

3 (a) Give an account of an experiment by means of which you could find out whether or not the rate of reaction between marble chips and dilute hydrochloric acid was affected by the area of the surface of the marble chips. Describe the result you would expect in the investigation.

(b) What is *a catalyst*? What effect does a catalyst have on a chemical reaction? Illustrate your answer by brief reference to **two** industrial processes that employ catalysts. (C)

4 A certain type of adhesive used in woodwork is made to 'set' by the addition of a 'hardener' such as ethanoic (acetic) acid. How is the speed of the reaction taking place affected by the temperature of the room? Why should it have any effect at all?

5 Zinc is said to react more quickly than iron with dilute sulphuric acid.

A piece of zinc 1 cm square and 0·05 mm thick of mass 1·25 g and the same mass of iron filings were each placed in 100 cm³ of dilute sulphuric acid.

After 18 hours it was found that 0·6 g of zinc remained but that all the iron filings had reacted with the acid.

Does this experiment show that iron is more reactive in sulphuric acid than zinc? If not design an experiment to demonstrate that your answer is correct. (EM)

6 Describe an investigation which you have performed in the laboratory to show how the concentration of the reactants affects the rate of a chemical reaction.

Explain what you did and the observations which you made. Give an equation for the reaction and explain your results in terms of the number of particles present. (L)

7 (a) What is meant by *dynamic equilibrium*?

(b) Bismuth(III) chloride dissolves in concentrated hydrochloric acid to give a clear colourless solution. When this solution is placed in excess water a white precipitate of bismuth oxychloride is formed and an equilibrium is set up as shown by the equation

$$BiCl_3(aq) + H_2O(l) \rightleftharpoons \underset{white}{BiOCl(s)} + 2H^+(aq) + 2Cl^-(aq)$$

(i) What would be the *approximate* pH of the equilibrium mixture?

(ii) How could you demonstrate that chloride ions were present in the equilibrium mixture?

(c) What would be observed if the following substances in turn were added to separate samples of the equilibrium mixture as shown by the equation in (b)? Explain your observations.

(i) Water.
(ii) Concentrated hydrochloric acid.
(iii) Sodium chloride solution.
(iv) Zinc dust.

(d) Give two examples of chemical reactions in which the substances involved are not in equilibrium during the reaction. (L)

8 Explain with full experimental details how you would show that powdered chalk is more reactive than marble, although both are calcium carbonate.

10 g of powdered chalk took 300 s to 'dissolve' in 100 cm³ of an acid while a lump of marble of the same mass took 700 s. Which of the following times do you think would most likely represent the time taken for 10 g of roughly crushed limestone to dissolve in 100 cm³ of the same acid.
A 200 s   B 350 s   C 550 s   D 750 s
E 900 s?

9 (a) A catalyst is a substance which *alters the rate* of a chemical reaction *without being used up* in the process.
By reference to the reaction involving the decomposition of hydrogen peroxide (H_2O_2), to form water and oxygen, using manganese(IV) oxide (MnO_2) as catalyst, explain how you would demonstrate the truth of the two italicised statements above.

(b) Describe what would be observed, giving an equation, when smouldering charcoal is lowered into a gas jar of oxygen.
What further observations would be made if the product of the reaction is added to
(i) water containing universal pH indicator solution,
(ii) a clear solution of calcium hydroxide.   (L)

(a) From the substances in the table below name *one*, in each case, which
(i) has a structure consisting of ions,
(ii) has a structure consisting of molecules,
(iii) has a giant structure of atoms,
(iv) is a gas at room temperature,
(v) for a given mass, has a fixed volume but no fixed shape,
(vi) is soluble in both water and toluene,
(vii) is solid at room temperature and liquid over the widest range of temperature.

(b) Name one of the salts which would
(i) give a brown colour when dissolved in water and mixed with chlorine water, and another salt which would
(ii) give a white precipitate when dissolved in water and added to silver nitrate solution.   (L)

11 Equal amounts of magnesium wire (0·06 g) were placed in two identical conical flasks A and B. Into flask A were placed 25 cm³ of M hydrochloric acid (HCl) whilst into flask B were placed 25 cm³ of M phosphoric acid. (H_3PO_4). Both acids were at a temperature of 20°C. Immediately the acid was added each flask was connected to a syringe and the volume of gas collected was recorded at regular intervals, and graphs A and B were plotted as shown in the graph at the bottom of the page.

(a) Give the name and formula of the gas formed when magnesium dissolves in both acids.

10 The information set out in the table below gives some physical properties of a number of elements and compounds.

Substance	Melting Point °C	Boiling Point °C	Electrical Conductivity		Solubility in Water
			Solid	Liquid	
Cobalt	1495	2877	good	good	insoluble
Water	0	100	poor	poor	—
Hydrogen bromide	−87	−67	poor	poor	good
Calcium iodide	740	1100	poor	good	good
Potassium chloride	772	1407	poor	good	good
Tin	232	2687	good	good	insoluble
Lithium bromide	547	1265	poor	good	good
Toluene	−95	111	poor	poor	insoluble
Gold	1063	2707	good	good	insoluble

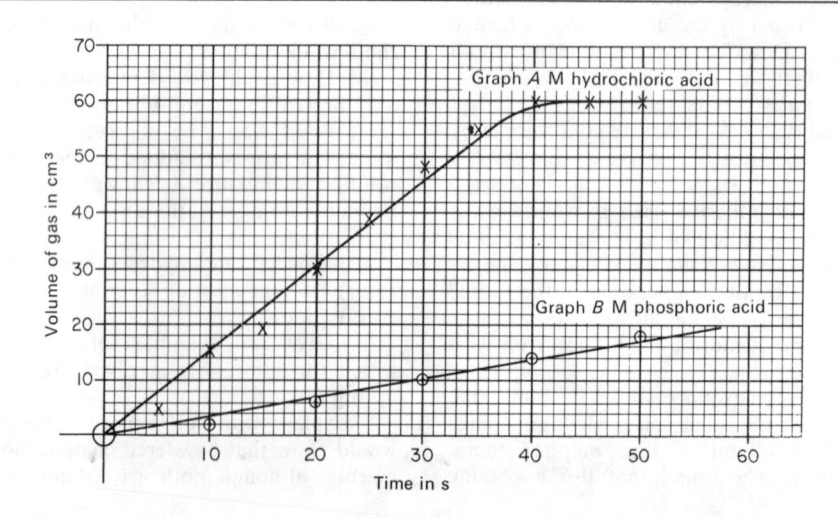

(b) What fraction of a g-atom of magnesium is 0·06 g? (1 g-atom of magnesium has a mass of 24 g.)
(c) What was the maximum volume of gas obtained in reaction A?
(d) (i) What volume of gas had been formed in reaction B after 35 s?
　　(ii) What would be the final volume of gas formed in reaction B?
(e) After how many seconds would half of the magnesium have dissolved in M hydrochloric acid?
(f) Give one visible difference between the contents of the two flasks after 50 s.
(g) Explain briefly why reaction B is slower than reaction A.
(h) Write an ionic equation for reaction A. (L)

49
The Electron; Radioactivity

In Chapters 1, 6 and 41 of this book we mentioned the particles which go to make up an atom of matter. Now we shall consider these particles in a little more detail. First we shall examine some of the properties of electrons and in the second part of the chapter we will study the phenomenon of radioactivity, which has provided scientists with a great deal of their knowledge of the structure of the atomic nucleus.

For our investigations into the properties of electrons a convenient range of apparatus is the set of vacuum tubes manufactured by Teltron Ltd. and the discussion which follows is based on this apparatus.

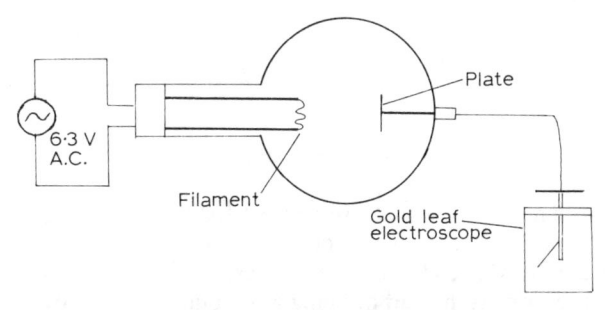

Fig. 49.1. Thermionic diode

Experiment 49.1. The tube needed for this experiment (Fig. 49.1) is known as a diode. It consists of an evacuated glass bulb sealed into which are two electrodes. One of them, the filament, consists of a short length of tungsten wire which may be heated to white heat by connecting it to a low voltage (6·3 V) A.C. supply. The other electrode, the plate or anode, is simply a small circular plate of metal. A metal stand is provided to mount all the tubes in this set. Arrange the apparatus as shown in Fig. 49.1 with the filament connected to its low voltage supply and the plate connected to the cap of a gold leaf electroscope. First give the electroscope a negative charge (by induction with a perspex rod which has been positively charged by rubbing with silk) and switch on the filament current. Watch the gold leaf carefully. You will probably find that its deflection does not change. Now switch off the filament current, discharge the electroscope and charge it this time positively. You can do this by induction with a polythene rod which has been negatively charged by rubbing with fur. Switch on the filament current and again watch the gold leaf carefully. This time the leaf will probably fall fairly rapidly when the filament becomes hot.

The fall in the leaf of the electroscope when positively charged in this experiment could be explained in two ways: 1. The electroscope and the plate of the tube between them give up their excess positive charge in some way, or 2. The electroscope and the plate gain some negative charge. In fact it is suggested that the second of these two explanations is the more likely one. We would now say that particles of negative electricity or electrons are being effectively boiled off

305

the filament and if the plate is positively charged, they will be attracted to it. The experiment described above is not by any means conclusive but we will suggest further experiments later which will confirm or refute the explanation we have given. Since the plate of our diode must be positively charged if there is to be any flow of electricity through the vacuum of the tube, the plate is also known as the anode. Consequently the filament is sometimes referred to as the cathode and the flow of electricity as cathode rays. The liberation of electrons from a heated wire is known as the thermionic effect.

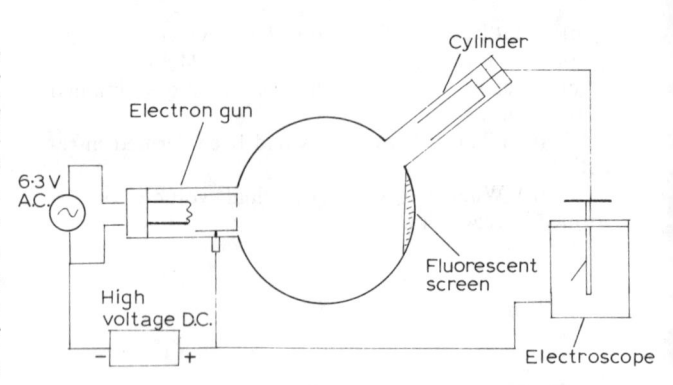

Fig. 49.3. A Perrin tube for collecting electrons

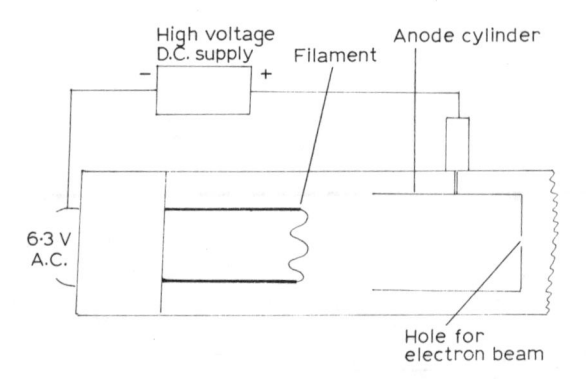

Fig. 49.2. An electron gun

The tubes which are needed for the remaining experiments in this section on the properties of electrons are all basically diodes, but in all of them the anode, instead of being a flat plate, is a hollow cylinder with a small hole as shown in Fig. 49.2. If the anode of such a tube is kept at a high electrical potential with respect to the filament, the electrons boiled off will be strongly attracted to the anode and will consequently have considerable speed when they reach it. Those of the fast electrons which arrive at the hole in the anode will pass through it and produce a fairly narrow beam in the space beyond. Such an arrangement is sometimes known as an electron gun and it forms the basis of many important electronic devices such as television picture tubes and electron microscopes. We shall now describe a number of experiments using electron beams produced in this way, which will help to establish some of the basic properties of electrons.

Experiment 49.2. The tube shown in Fig. 49.3 contains an electron gun of the type described above and when the filament is raised to an appropriate temperature it gives off thermionic electrons. If a high potential difference (a few thousand volts) is applied between the anode and the filament, then a beam of fast electrons is obtained emerging through the hole in the anode. The glass wall of the tube opposite the anode is coated on the inside with a fluorescent material which gives off green light where the electrons strike it. A side tube above the fluorescent screen contains an open hollow metal cylinder connected to a terminal on the outside of the tube. Connect this terminal to the cap of a gold leaf electroscope and apply the low and high voltages to the tube as shown in Fig. 49.3. Now bring a magnet near to the wall of the tube. You will find that by moving the magnet about you can deflect the spot on the screen, almost at will. This deflection of an electron beam by a magnet is obviously a most important effect but we shall study it more carefully later. For the moment, see if you can deflect the electrons to the top of the screen and into the cylinder in the side tube. If you succeed in doing this you should see a charge registered by a deflection of the gold leaf. When you have 'collected some electrons' in this way, disconnect the electroscope from the tube without discharging it. Now determine the sign of the charge you have collected, as described in chapter 45, by bringing up in turn positively and negatively charged rods. In this way you should be able to confirm that we are dealing with negatively charged particles.

Finally in this series of experiments on the properties of electrons, we will consider a little more carefully the deflection of an electron beam in a magnetic field.

Experiment 49.3. The tube needed for these experiments is shown in fig. 49.4. In it the fluorescent screen is mounted almost parallel to the path of the electrons coming from the electron gun, so that their path may be observed along its whole length.

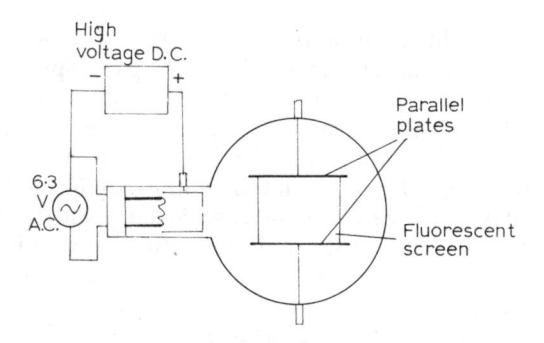

Fig. 49.4. Tube for demonstrating magnetic and electrostatic deflection of electrons

The tube also contains two horizontal metal plates, one above and one below the electron beam, and each connected to a terminal on the outside of the tube. We shall use these plates to investigate the effect of an electric field on the electron beam. But first let us consider again the deflection of the beam by a magnet which we have already observed. You will appreciate that bringing up a bar or horseshoe magnet to the outside of the tube applies to the electron beam a magnetic field which is far from uniform. It will be easier for us to understand exactly how the electrons are being deflected if we can apply a magnetic field in which the lines of force are more or less parellel. This is not easily achieved over such a large volume as that of the tube, but it can be done with a large pair of plane coils mounted as in Fig. 49.5 and connected in series with a battery which will give a current through both coils of a few amperes. A pair of coils used in this way is known as Helmholtz coils, and they produce a magnntic field between them in which the lines of force are straight, parallel and at right angles to the planes of the coils.

Connect the low and high voltages to the electron gun. You should see a straight trace of light along the fluorescent screen in the tube, showing the path of the electron stream. Now pass a current through the pair of Helmholtz coils. The trace on the screen will be bent into a curve which you may be able to recognise as the arc of a circle. Try reversing the direction of the current in the coils (and so reversing the direction of the magnetic field). You will see the trace deflected in the opposite direction. If you have a way of varying the current in the coils do so. You will find that a stronger field makes the curvature of the electron beam sharper.

In order to understand what is happening in this experiment it is helpful to refer back to Experiment 45.3. In that experiment we arranged for a straight wire carrying a current to be situated in a uniform magnetic field with the direction of the current at right angles to the lines of force. We discovered that the wire experienced a force which was at right angles to both the lines of force and to the wire. The situation in our vacuum tube is really very similar to this. The electron flow is effectively an electric current and it is at right angles to the lines of force produced by the Helmholtz coils. We thus expect the electrons to experience a force which is at right angles to their path, and to the field direction. Such a force will cause them to move in the arc of a circle. You may like to apply Fleming's left hand rule to the circumstances of your experiment. You will need to determine the field direction from the direction of the current in the Helmholtz coils. You will also need to bear in mind that when we discussed Fleming's rule we were assuming that the current flowed from the positive to the negative terminal of a battery, i.e. we were thinking of a flow of positive electricity. This means that to apply the rule to our vacuum tube we must regard the 'current' as flowing into the electron gun.

Experiment 49.4. Using the same tube as in the last experiment we may study the behaviour of the electron beam in an electrostatic field. Such a field will be established between the parallel plates of the tube if a high voltage (of the order of a thousand volts) obtained from a power supply, is applied to the plates. When this is done the path of the electron beam as observed on the fluorescent screen, is again deflected, (attracted to the positive plate and repelled by the negative one). However we should notice the essential differences between the deflecting force in this case and that obtained in a magnetic field. Firstly the curved trace is not an arc of a circle. It is in fact a parabola although this will, of course, not be obvious. Furthermore, the lines of force of the electrostatic field may be regarded as straight lines across between the plates. Hence the electrons are being deflected in the plane of the field and not at right angles to it as in the magnetic case.

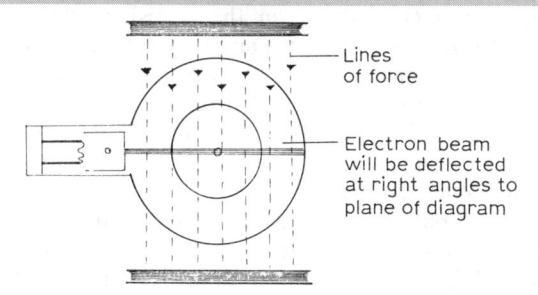

Fig. 49.5. Electron tube fitted with Helmholtz coils (plan view)

To summarize then, we have established the following properties of electrons:

1. They have negative electric charges.
2. They are liberated by a piece of metal if it is raised to a high enough temperature.
3. Electron beams are deflected by electric and magnetic fields.

We shall now describe just two of the many ways in which electron beams are used.

The Cathode Ray Oscilloscope

This is one of the most powerful and versatile of scientific instruments. It is used in almost all branches of Science. In principle it is very much the same as the electron tubes described earlier in this chapter but it has some special features and a rather different shape (Fig. 49.6). It consists of an evacuated tube with a fluorescent screen at one end and an electron gun at the other. This electron gun will be rather more complicated than the one shown in Fig. 49.2, as it will include additional electrodes to focus the beam onto the screen and to control the brightness of the spot. This is done by varying the potential on these electrodes. In addition the electron beam, in going from the electron gun to the screen, will pass between two pairs of parallel metal plates. One pair is vertical, the other, horizontal.

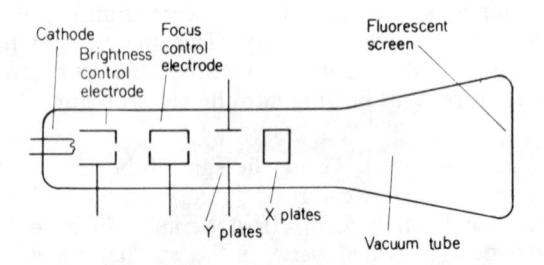

Fig. 49.6. Oscilloscope

If the electron gun potentials are adjusted so that a small bright spot is obtained at the centre of the screen and a potential difference is applied across the horizontal plates, the spot will move upwards or downwards, depending on the polarity of the p.d. These plates are known as the Y plates. A p.d. applied to the vertical plates, (the X plates) will cause a movement of the spot to left or right. The size of the deflection of the spot will depend on the voltage applied. This then is the first application of the CRO. It may be used as a voltmeter, and as such it has many advantages over other types. For example, since the inertia of an electron beam is so very small, the response to a change in voltage will be almost instantaneous. Furthermore, when a potential difference is applied to a pair of plates, no current is taken and so the e.m.f. is measured.

Perhaps the most common use of the CRO is as a graph plotter, with 'time' on one axis, usually the X axis. This is achieved by applying to the X plates a voltage which increases steadily with time up to a certain value and then returns almost instantaneously to its initial value to start again. This causes the spot to travel from left to right across the screen at a constant speed and then to go back to its starting point in a very short time. This arrangement is known as a time base. The time base frequency is usually such that the spot moves so quickly that it seems a line on the screen. However, with most CRO's it is possible to reduce the time base frequency sufficiently for the movement of the spot to be observed. With a suitable time base frequency any potential difference which varies regularly with time may be applied to the plates and the wave form studied. One of the many possible applications of this kind was mentioned in Chapter 34. This involved connecting the output from a microphone to the Y plates. The output from a microphone will be small alternating voltages whose waveforms will be the same as those of the sounds falling on the microphone.

The Television Picture Tube

A black and white television receiver is, in many ways, similar to the cathode ray oscilloscope. The picture tube consists of an electron gun, X and Y deflection plates and a fluorescent screen. However, in a television set time base voltages are applied to both X and Y plates. The Y time base frequency is much lower than the X. The result of this is that the spot moves from left to right of the screen at a steady rate and then back to the left again in a short time, under the action of the X or 'line' time base. At the same time the spot moves down the screen slightly under the action of the much slower Y or 'frame' time base. In this way, a series of horizontal lines or 'raster' is produced on the screen (Fig. 49.7). In the British system at present 625 lines are used and they are traced out in 1/25 th. of a second. If no signal is being received, the raster will be of uniform brightness. A signal is applied to an electrode in the tube which has the effect of varying the brightness of the spot as it moves across the screen. In this way bright and dark patches are created on the screen to make up the picture. Since a completely new picture is formed 25 times in one second, small changes from picture to picture give the effect of movement.

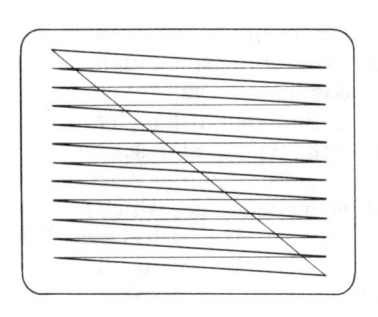

Fig. 49.7. A 'raster'

X-rays

Quite soon after the discovery of cathode rays the German scientist, Röntgen, found in 1895 that with them he could produce a new and very different kind of radiation. By keeping the anode of a diode at a very high positive voltage with respect to the cathode, he was able to make the electrons strike the anode at very high speeds. Some of the energy lost by the electrons on impact was found by Röntgen to be radiated as what he called X-rays. This radiation has the effect of blackening photographic plates and causing fluorescent screens to glow. But the most important property of all possessed by X-rays is probably familiar to you. It is highly penetrating radiation. It passes through human flesh and sometimes even through considerable thicknesses of metal. It has therefore many uses in medicine and industry. The penetrating power, (the hardness of the rays), is determined by the voltage applied to the anode of the diode. A higher voltage results in more penetrating rays. In medical uses for example, the hardness may be adjusted so that flesh is transparent to the radiation while the bones are opaque. In this way X-ray photographs may be obtained to help in the diagnosis of broken bones.

Fig. 49.8 shows a modern form of X-ray tube known as a Coolidge tube. The cathode is a heated filament and the anode, which is bombarded with electrons, is often a block of copper (to conduct away the heat generated) with a tungsten target set in it.

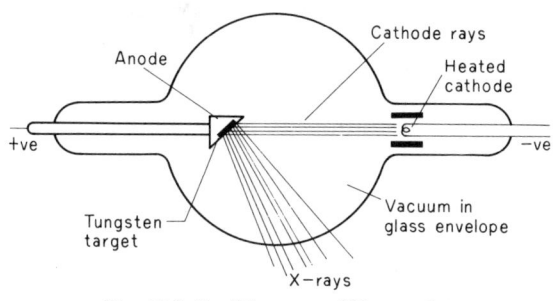

Fig. 49.8. Coolidge type of X-ray tube

Although Röntgen was not immediately aware of the nature of the radiation he discovered, it has subsequently been established that it is, as we mentioned in chapter 32, electromagnetic radiation of short wavelength.

Radioactivity

In 1896 the French scientist Henri Becquerel was experimenting with compounds of the naturally occurring element uranium. During his experiments he accidentally left a crystal of a uranium compound in a drawer on top of a packet containing unexposed photographic film. The film was developed after a few days and was found to be intensely blackened. Becquerel concluded that the blackening was due to a previously unknown type of radiation given off by the uranium which penetrated the packet to reach the film. Becquerel is therefore regarded as the discoverer of what we now call radioactivity.

Scientists after Becquerel, notably Pierre and Marie Curie, found that a number of other elements occurring in nature (radium and thorium, for example) were also radioactive. However the technology which has grown up in recent years based on the use of radioactive materials has employed mostly substances which have been made radioactive by keeping them for a while in a nuclear reactor.

Quite a lot of simple experimental work on radioactivity can now be carried out in school laboratories using naturally occurring substances like uranium and thorium and small quantities of artificially produced radioactive material. However it should be noted that regulations are laid down by the Department of Education and Science concerning the use of radioactive substances in schools and authorization has to be obtained before schools may work with radioactive materials other than naturally occurring ones.

We shall now describe some of the experiments which are possible in schools, but it will be necessary to quote some properties of the radiation without experimental evidence in support.

Experiment 49.5. A gold leaf electroscope is required for this experiment but it will be most effective if the gold leaf-metal rod system of the electroscope is as small as possible and if the metal cap has been removed. Place a small uranium oxide source inside the electroscope case (Fig. 49.9). (Details of the preparation of such sources may be obtained in more specialised literature.) Charge the electroscope and watch the

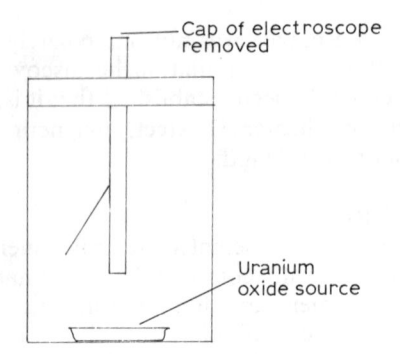

Fig. 49.9. Gold leaf electroscope used for the detection of ionizing radiation

gold leaf over a period of time. You should see that the electroscope is gradually discharged. You may find it necessary to check that the electroscope loses its charge much more slowly in the absence of the uranium oxide.

The effect of the uranium oxide in this experiment is to produce radiation which ionizes the air inside the electroscope case. This means that the radiation knocks electrons out of some of the gas atoms, leaving them with a net positive charge. Hence in the air there is a supply of free positively and negatively charged particles. If the electroscope leaf system is positively charged, it will attract the negative ions (electrons) and its charge will be neutralised. If the electroscope is negatively charged it will be the positive gas atoms which are attracted to the gold leaf.

The Geiger-Muller Counter

Ionization produced by the radiation from radioactive materials is the basis of almost all of the many different devices for detecting and measuring the radiation. Perhaps the most common of these devices is the Geiger-Muller counter. It would not be appropriate to discuss this detector here in detail, but we shall say a little about its structure and operation and we shall go on to describe some experiments in which it may be used.

A Geiger counter (Fig. 49.10) consists of a metal cylinder with a wire along its axis and isolated electrically from it. This is mounted in a glass or metal case containing a gas at low pressure. In use the wire is maintained at a high positive potential (400 volts in the type used in schools) with respect to the cylinder. The counter often has a very thin 'end window' made of mica which permits even the least penetrating radiation to enter. When radiation passes through the volume of gas contained in the cylinder, a very large number of electrons and positive gas ions are formed and when they are attracted to the positive and negative electrodes of the tube they constitute quite a large current. Thus a pulse of current is produced in the circuit for every 'particle' of radiation entering the tube. The most common technique is to use an electronic counter (known as a scaler) to count up these pulses of current over a convenient period of time.

Types of Radiation

Experiment 49.6. The radioactive sources needed for this experiment make up a set, several alternative versions of which are available from the manufacturers. Here we will consider a typical one which consists of 3 sources, all artificially produced radioactive substances, americium 241, strontium 90 and cobalt 60. A long pair of tweezers or a special source handling tool is needed so that the amount of radiation falling on the body of the experimenter is small when the sources are in use.

Connect an end-window Geiger tube to a scaler and observe first that when the correct voltage is applied to the tube, the scaler counts slowly even when there is no radioactive material near the detector. This indicates the background radiation which will be present in all our experiments unless we take elaborate steps to shield the Geiger tube from it. The background radiation is mostly due to cosmic rays and it should not give rise to a count much in excess of 30 counts per minute.

Place the americium 241 source close to the end-window of the G-M tube. If the tube has a sufficiently thin end-window you will observe the scaler showing a count rate substantially above the background. Place a thin piece of cardboard (Fig. 49.11) between the tube and the source and

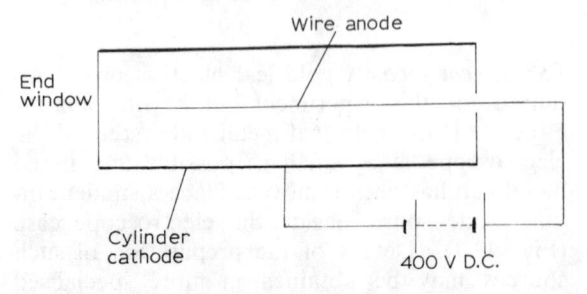

Fig. 49.10. A Geiger-Muller tube

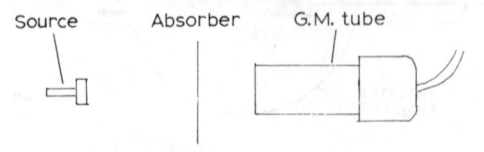

Fig. 49.11. Experiments on absorption of radiation

again note the count rate. It will probably have returned to background level. The americium emits radiation which is readily stopped by even a thin piece of card. A few centimetres of air will also stop the radiation reaching the tube. If the end-window is too thick you may not be able to detect this radiation at all.

Repeat the above process with the strontium 90 source. You will get a much higher count rate which is only a little reduced by introducing a piece of card over the end of the tube. Try instead various thicknesses of aluminium. A few millimetres thickness will be needed to reduce the count to background. Alternatively, perhaps a metre of air will be needed to stop the radiation from this source. Finally try the cobalt 60. Here you will find that not even a centimetre of lead will prevent all of the radiation reaching the tube.

These experiments suggest that we may be dealing with three different types of radiation. These have been named alpha (α), beta (β) and gamma (γ) radiation. That from the americium, the least penetrating, is known as α, the strontium emits β radiation which is of intermediate penetrating power, and highly penetrating γ radiation is given off by the cobalt source. It should be noted that our studies of the absorption of radiation have not established that these three types of radiation are different in nature, nor that there are no other types. Much more detailed studies have shown scientists what the three radiations are. Furthermore it has been shown that they are by far the most common types of radioactivity, although there are others.

Alpha Radiation

The nature of the α radiation was firmly established by the great New Zealand scientist Lord Rutherford in 1909. He showed that it consisted of particles which were the nuclei of helium atoms. In fact the nuclei of the atoms in the alpha source spontaneously shoot off parts of themselves, consisting of 2 protons and 2 neutrons. Of course the nuclei which remain after alpha decay will not be the same as before. We might represent the emission of an α particle from a nucleus of the americium source as follows:

$$^{241}_{95}\text{Am} \rightarrow\; ^{4}_{2}\text{He} +\; ^{237}_{93}\text{Np}$$

It will be helpful to consider this equation in a little detail. The americium (Am) and neptunium (Np) are sometimes known as parent and daughter nuclei in a decay reaction of this kind. The helium (He) is the alpha particle which is emitted. You will notice that both the top and bottom numbers balance across the equation. The bottom numbers are the numbers of protons in each nucleus. They are also the atomic numbers and they determine the chemistry of the atoms. The top numbers are called the nucleon numbers (number of protons + number of neutrons). You will find that the nucleon numbers are approximately equal to the atomic masses of the elements.

Beta Radiation

A fairly simple experiment will give us a clue to the nature of the beta radiation.

Experiment 49.7. Place the strontium 90 β source a few centimetres from the end window of a G-M tube which is connected to a scaler. Restrict the beam of β radiation by placing two slits cut in fairly thick brass between the source and the detector (Fig. 49.12a). You should get a count rate substantially above background. Now bring a strong magnet up to the β radiation beam (Fig. 49.12b). You will get a large reduction in the count rate. Beta rays are, in fact, readily deflected by a magnetic field. This suggests that they are charged particles and perhaps electrons. Precise measurements on the magnetic deflection of β rays have confirmed this suggestion. It must be emphasised that although β particles are electrons, they originate in the nucleus of an atom and not from among the orbiting electrons.

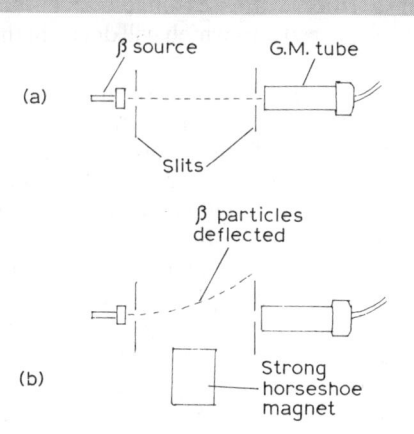

Fig. 49.12. Deflection of β particle by a magnet

Beta decay also changes the parent nucleus, but since the mass of an electron is so small, it is really only the nuclear charge which is affected. Thus:

$$^{90}_{38}\text{Sr} \rightarrow\; ^{90}_{39}\text{Y} +\; ^{0}_{-1}\text{e}$$

This equation tells us that the total number of nucleons is the same in parent and daughter nuclei, but there is one more proton and one fewer neutron in the yttrium than in the strontium.

N.B. α particles are also charged and consequently may be deflected in a magnetic field. However on account of their relatively large mass, a very strong field indeed is needed to make this deflection significant.

Gamma Rays

γ Radiation, the most penetrating of the three types we have discussed, has been shown to be unaffected by a magnetic field. More important still, it has been shown to have wave properties. γ rays are, as was mentioned in chapter 32, part of the electromagnetic spectrum and travel with the same velocity as light.

The emission of a gamma ray by an atomic nucleus changes neither the charge nor mass of it. It is believed that gamma emission is accompanied by a rearrangement of the protons and neutrons in the nucleus.

Radioactive Decay

We mentioned earlier in this chapter that the emission of radiation from a nucleus is 'spontaneous'. At present scientists are completely unable to predict when an individual atom of uranium will emit its alpha particle. It may be in the next second, it may not be for a million years. However if we take a lump of uranium containing a million atoms, we can predict with reasonable accuracy the number of those atoms which will decay in the next minute.

It helps to compare this with throwing dice. If we throw one die we have a poor chance of predicting how it will fall. If, on the other hand we throw a thousand dice, we may predict with fair accuracy that one sixth of them will come up 'one'. To carry this analogy a little further, if we remove all the 'ones', throw all the rest again and keep repeating the process, we could plot a graph of the type shown in Fig. 49.13. It is a graph of the number of dice left against the number of throws. In the case of radioactive decay we might get a similar graph if we plotted the number of atoms not yet decayed in a sample, against time. The time scale, of course, could be anything. With the natural radioactive substances it would be in terms of millions of years. In other cases it might be only fractions of a second. Such short lived atoms would not exist in any quantity in nature. We talk about the half-life of a radioactive substance. This means the time taken for the number of undecayed atoms in a sample to fall to half of its original value. You will see from the graph that the half-life for the dice throwing analogy is between three and four throws.

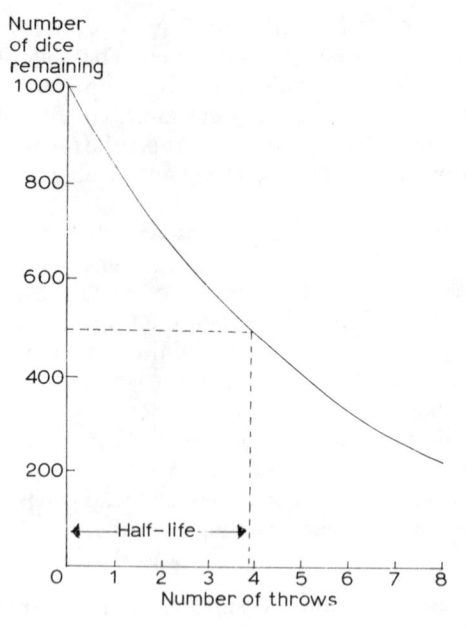

Fig. 49.13

It would not, of course, be an easy matter to determine the number of undecayed atoms in a sample at any instant. However it is found that the number of decays per second is proportional to the number of radioactive atoms present. Hence to determine half-life a graph of the count rate shown by a G.-M. tube, placed in a fixed position near the source, against time, would do. Unfortunately you would need many lifetimes to carry out such an experiment with the long lived sources which are likely to be available to you.

Radioactive Decay Series

The picture we have given in this chapter so far is a rather simplified one, in that a radioactive atom does not generally decay once and then remain stable. In most cases a particular nucleus is part of a radioactive decay series. The first few processes of the decay series starting with uranium 238 are described in the following:

$$^{238}_{92}\text{U} \xrightarrow[4.5 \times 10^9 \text{ years}]{\alpha} {}^{234}_{90}\text{Th} \xrightarrow[24.5 \text{ days}]{\beta}$$

$$^{234}_{91}\text{Pa} \xrightarrow[74 \text{ seconds}]{\beta} {}^{234}_{92}\text{U} \dashrightarrow$$

$$\dashrightarrow {}^{206}_{82}\text{Pb} \text{ (stable)}$$

The end product after many decays most of which are not quoted is the stable substance lead 206. The times quoted are the half-lives for each decay.

Isotopes

The beta source we referred to in the experiments on radioactivity was $^{90}_{38}$Sr. If you refer to the list of elements at the beginning of this book, you will find that strontium does indeed have an atomic number of 38, but in the most common type, there are 38 protons and 50 neutrons. In other words, the nucleon number is 88 and this substance is $^{88}_{38}$Sr. We are thus talking about two substances, $^{90}_{38}$Sr and $^{88}_{38}$Sr. These are said to be two isotopes of strontium. Chemically they are the same because they have the same number of protons (and electrons) per atom. They differ in the number of neutrons present in their nuclei. All the elements can be obtained in more than one isotopic form. Most isotopes are radioactive.

Nuclear Energy

If accurate measurements are made of the masses of the nuclei involved in, for example, alpha decay, it is found that the total mass of the daughter nucleus and the alpha particle is appreciably less than the mass of the parent nucleus. It would seem then that some mass disappears in the reaction. On the other hand, the alpha particle is emitted with considerable kinetic energy. The suggestion here, that mass may be converted to energy, had been made by Einstein in 1905 well before radioactivity was understood. Einstein's prediction was that if quite small amounts of mass could be destroyed, enormous quantities of energy would be released. As a result of this theory, great efforts have been made in modern times to bring about this conversion of mass to energy.

Nuclear Fission

Large quantities of nuclear energy were first obtained during the 1940s. This was achieved in an uncontrolled way with the explosion of nuclear bombs and in a controlled way with the construction of nuclear reactors. These developments followed on discoveries made during the 30s. First the British scientist, Chadwick, succeeded in obtaining free neutrons. Then two Germans, Hahn and Strassmann, discovered in 1938 that if uranium is bombarded with neutrons, a reaction sometimes occurs which is called fission. This consists of the target uranium nucleus splitting into two roughly equal pieces, together with some more free neutrons, (generally 3). A number of different reactions of this type are possible but a typical example is:—

$$^{235}_{92}U + ^{1}_{0}n \rightarrow ^{143}_{56}Ba + ^{90}_{36}Kr + 3^{1}_{0}n$$

In fission reactions a significant amount of mass disappears and so the fission products are produced with large kinetic energy.

You will note that one free neutron here has released three neutrons. If, on average, at least one of these newly released particles goes on to cause the fission of another uranium nucleus, we have the possibility of a continuous chain reaction (Fig. 49.14). The rate at which this reaction takes place depends on how many, on average, of the three neutrons produced go on to cause further fission. If it is barely one, the reaction proceeds in a controlled way. This is what has to be achieved in a power generating reactor. Clearly the construction of reactors is complicated and varied. However, there are certain features common to

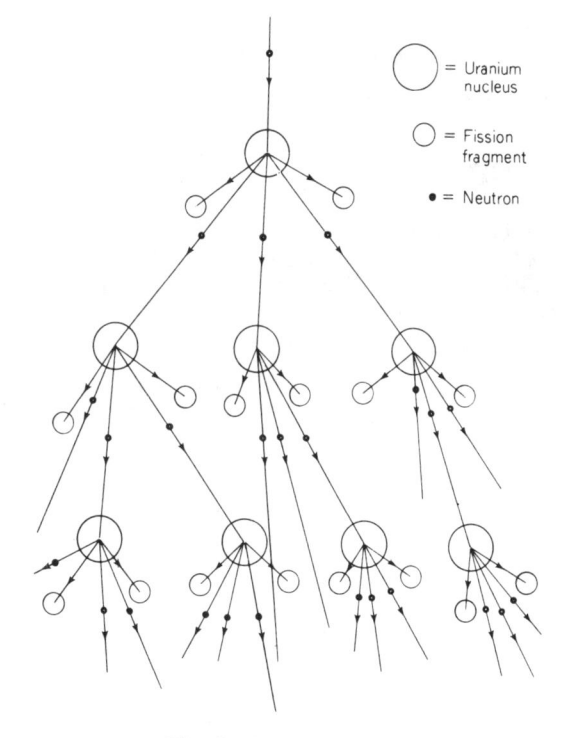

Fig. 49.14. Chain reaction

them all (Fig. 49.15) There is the fuel, usually uranium in the form of rods. There will be a control material. In most British reactors these are rods of cadmium metal whose function is to 'soak up' the spare neutrons. These rods may be adjusted in position so that they take just enough neutrons to keep the reactor barely 'critical'. There may also

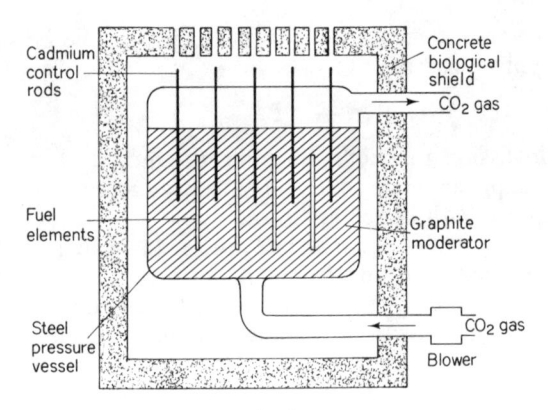

Fig. 49.15. Nuclear reactor

be a 'moderator', often graphite, which slows down the neutrons, since slow neutrons may be better at causing fissions than fast ones. The whole reactor must be surrounded by a thick layer of concrete or steel to contain the vast amounts of radiation produced. The heat energy generated in a reactor is taken away by a coolant which may be carbon dioxide gas (see Chapter 46).

Nuclear Fusion

So far nuclear energy has only been obtained in a controlled way from the fission of heavy nuclei. It is possible also to obtain energy from the fusion of light nuclei, for example, hydrogen. Since hydrogen is very much more abundant then the heavy elements, in the long term fusion reactions are likely to be of crucial importance in meeting the energy needs of man. At present great research efforts are being made to bring about controlled fusion reactions. The problem about them is that they only occur at temperatures of millions of degrees. Such temperatures exist on the surfaces of the sun and other stars, and fusion reactions are the source of solar energy. The hydrogen bomb also depends on fusion reactions and here the very high temperatures are caused by the explosion of a fission bomb. There are several chains of fusion reaction possible. They all amount in the end to the fusion of four hydrogen nuclei to give a helium nucleus.

Exercise 49

1 *Cathode rays may be considered as a stream of electrons.* State evidence to confirm this statement, and indicate how a cathode ray tube can produce a stream of cathode rays and focus them to make a luminous dot on a fluorescent screen at the end of the tube. (*C*)

2 Explain the meaning of **each** of the following terms: *X-rays, alpha particles, beta particles, gamma rays.*

State, indicating the forms of radiation chosen, a possible use for **two** of these forms of radiation.

Explain how X-rays may be produced and describe how it is possible to distinguish between alpha particles and gamma rays. (*C*)

3 Explain the terms: *fission, fusion, isotope.*

Under what circumstances have fission and fusion reactions been used to release energy?

50
Small-scale Preparations and Identifications in Chemistry

Many of the preparations and tests carried out in chemistry can conveniently be performed with small-scale apparatus as used in semi-micro analysis. This results in a saving of time and expense and encourages neater and more careful work.

The requirements for such work include a 150 × 25 mm test-tube with side-arm,

hard-glass test-tubes 125 × 19 mm and 75 × 9 mm (semi-micro),

a holder for the above (tongs are not very suitable),

a dropping or teat-pipette, 3–4 cm^3 capacity (Fig. 50.1(*a*)),

a semi-micro burner (if not available the barrel of a Bunsen can be unscrewed and the gas lit at the jet to give a flame about 2 cm high),

a small spatula made by softening and then flattening the end of a piece of 3 mm glass rod (Fig. 50.1(*b*)),

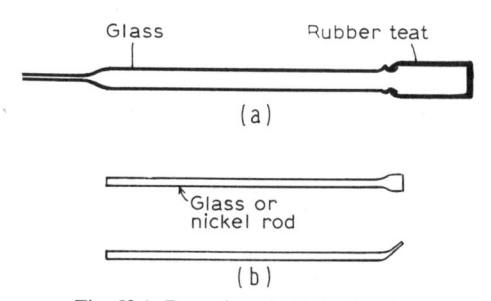

Fig. 50.1. Dropping pipette and spatula

a wash-bottle, preferably of polythene (those which have contained household detergents can easily be adapted),

a large beaker (in which the pipette will be regularly washed and stored when not in use),

a pair of forceps (the type for fractional weights is suitable),

one or two microscope slides.

The amounts of gases or solids which can be prepared are, of course, limited. In the case of gases it is not always feasible to collect them, but many tests can be carried out as the gas issues from the tube. The tube with side-arm will generally be used as the preparation vessel.

The semi-micro tubes are too small for liquids to be poured in or out conveniently and transfer is made by using the pipette. This is held vertically against the palm of the hand and the teat squeezed with the first finger and the thumb to expel some of the air. It is then inserted in the liquid to be transferred and the pressure released. The liquid enters the pipette and is carried to another vessel and squeezed out there. After use, the pipette must be filled and emptied twice, using the distilled water stored in the large beaker. It is left standing full of water in the beaker until required. If it becomes contaminated it must be thoroughly cleaned before use. On the small scale cleanliness is extremely important.

The tubes are heated by holding them just above the flame of the burner (this avoids deposition of soot if the Bunsen jet is being used), but care must be taken to prevent the contents being splashed about if the liquid boils. Addition of a small piece of unglazed pot, or one or two 'anti-bumping' granules, or a long piece of capillary tubing often helps to lessen bumping. Boiling can be better carried out by embedding the bottom of the tube in a small pile of sand heated on a tin lid or asbestos-centred gauze.

Whenever a test-tube is mentioned in the following pages a *small* one, 75 × 9 mm, is intended unless otherwise stated. Test papers about 15 × 6 mm can be cut from the usual strips and held in forceps. The amount of solid which will just cover the flattened end of the spatula will be referred to as one measure.

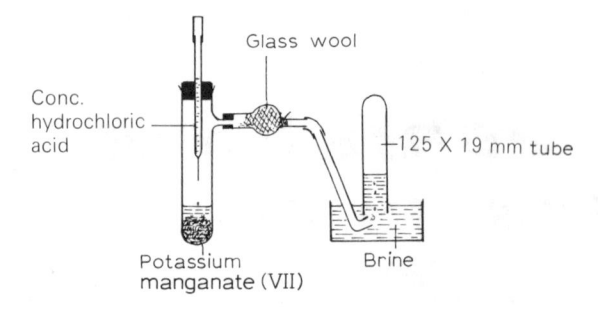

Fig. 50.2. Preparation of Chlorine

Experiment 50.1. Preparation and Properties of Chlorine (see p. 23).

Fig. 50.2 explains the method.* Concentrated hydrochloric acid is dripped on to potassium manganate(VII) and any dust carried over is held back by the glass wool. As the gas is fairly soluble in water it is collected over strong brine. The tubes should be corked *immediately* they are full and replaced. They should be held in position during collection by a clamp fastened to a stand. Test samples as follows, but refer to Chapter 5 for full details. Before carrying out the tests place the generator in the fume cupboard out of the way.
(i) Insert a piece of moistened blue litmus and notice what happens.
(ii) Unravel the end of a wax taper and cut off about one inch of one or two strands. Hold this in the forceps and light. Now place it in one of the tubes. Is there any change in the colour of the flame?
(iii) Add 1 cm³ of a 10% solution of potassium iodide to another tube of gas. Notice the iodine released.

$$2KI + Cl_2 \rightarrow 2KCl + I_2$$
$$\text{or} \quad 2I^- + Cl_2 \rightarrow 2Cl^- + I_2$$

Allotropes of Sulphur (see p. 65). These can be made by similar methods to those on the larger scale, the amounts being reduced to suit the small semi-micro test-tubes. Crystallization should be carried out on a microscope slide or watch glass. When making the monoclinic variety the tube containing the toluene must not be heated directly (toluene is inflammable), but should be held in boiling water until all the sulphur is dissolved. Alternatively, small-scale organic apparatus can be used and refluxing carried out in a 20 or 50 cm³

* In all these preparations the dropping pipette may be replaced by a 25 cm³ tap funnel and the side-arm tube by a small filter flask if desired.

flask. A third method which involves crystallizing molten sulphur is described here.

Experiment 50.2. Preparation of Monoclinic Sulphur.
Stand a crucible full of sulphur or an evaporating basic half-full of sulphur (powder or broken pieces) on a sand bath or asbestos-centred gauze and heat gently until all the sulphur has melted. Have ready a moistened filter paper folded in the usual way in a funnel. Carefully pour the molten sulphur into the filter and let it cool down. When it has nearly set, open the paper and notice the needle-shaped crystals that have formed.

Experiment 50.3. Preparation and Properties of Sulphur Dioxide (see p. 68).
Use the apparatus shown in Fig. 50.3 with sodium hydrogensulphite in the tube and dilute sulphuric acid in the dropping pipette. If the reaction is very slow, warm a little. Test the issuing gas with blue litmus paper and bubble the gas through solutions of (i) potassium manganate(VII) and (ii) potassium dichromate(VI). Notice what happens.

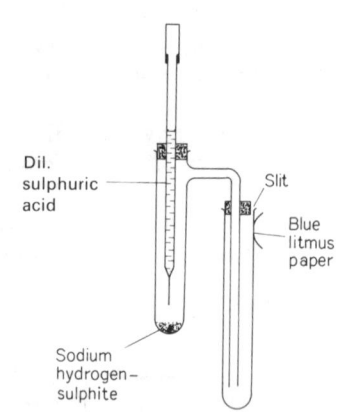

Fig. 50.3. Preparation of Sulphur Dioxide

Experiment 50.4. Some Properties of Nitric Acid.
The preparation of nitric acid using small-scale organic apparatus was described on p. 136. One or two tests can be carried out using about ten drops of the concentrated acid in a semi-micro test-tube.
(i) Add a little copper and notice the brown nitrogen dioxide given off.
(ii) Dilute the acid with an equal volume of water and add a little copper. The gas evolved, nitrogen monoxide, is colourless and only becomes brown on meeting the air outside the tube.
(iii) Add a measure of copper(II) oxide and warm. Here, as in (i) and (ii) a blue-green solution of copper(II) nitrate is formed.

(iv) Add two drops of the acid to a little potassium iodide solution and notice the iodine liberated.

(v) Add five drops of solution of ammonium thiocyanate to the same amount of iron(II) sulphate solution. No colour should develop. Now add two drops of concentrated nitric acid (warm if necessary). A red colour should develop indicating the conversion of the iron(II) ions (Fe^{2+}) to iron (III) (Fe^{3+}) ones. Both (iv) and (v) indicate the oxidizing nature of nitric acid.

Experiment 50.5. Preparation of Nitrogen Dioxide (see p. 138).

Fig. 50.4 clearly explains how nitrogen dioxide can be prepared on the small scale. When the test-tube (125 × 16 mm) is full it can easily be replaced by another and corked. The energy given out in the reaction causes a rise in temperature and if the brown gas is tested when *freshly* collected it should relight a glowing splint. If 50/50 nitric acid is used nitrogen monoxide is produced and should be collected over water.

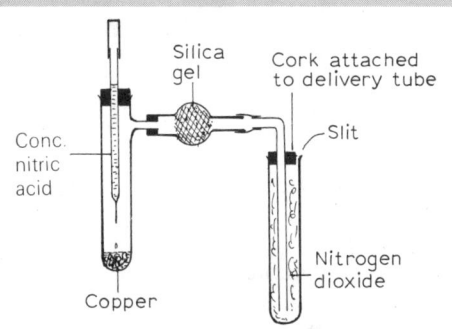

Fig. 50.4. Preparation of Nitrogen Dioxide (in fume cupboard

Experiment 50.6. Preparation of Ammonia (see p. 127).

Ammonium chloride (sal ammoniac) and calcium hydroxide are mixed together and placed in a tube with side-arm as in Fig. 50.5. The asbestos wool soaks up some of the moisture produced and prevents it running back on to the hot part of the tube; the rest is removed by the silica gel. The tube (125 × 19 mm) in which the gas is collected can be considered full when a piece of moist red litmus on the *outside* of the tube turns blue. The tube containing the reactants should be turned before heating so that it is at an angle of about 45° and not vertical. For equation and tests see p. 127 and p. 129.

Experiment 50.7. Preparation of Carbon Monoxide (see p. 145).

The apparatus of Fig. 50.2 can be used for

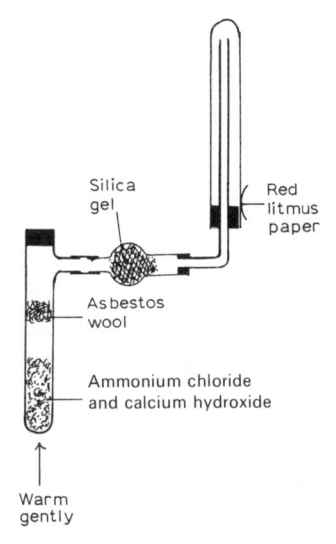

Fig. 50.5. Preparation of Ammonia

preparing carbon monoxide. The pipette contains concentrated sulphuric acid and the test-tube methanoic acid. Soda-lime in the bulb-tube absorbs any carbon dioxide formed and the gas is collected in a normal-size tube over water. If the pipette needs recharging it should be removed along with its cork and the reaction tube temporarily sealed with a bung. As carbon monoxide is so very poisonous this experiment should be carried out in a fume cupboard with a good draught. For tests see p. 146.

An elegant method of preparing carbon monoxide on the small scale by reduction of the dioxide has been devised by J. W. Davis* and is illustrated in Fig. 50.6. The silica gel tube should be heated strongly before the gas is passed through in order to remove any trace of water. Unchanged carbon dioxide is absorbed in soda lime and the monoxide collected over water as in the first experiment.

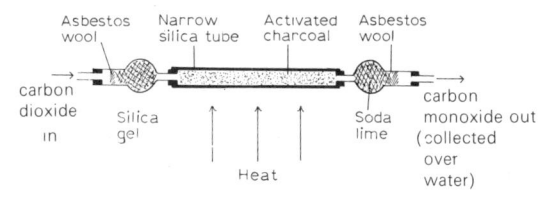

Fig. 50.6. Reduction of Carbon Dioxide by Charcoal

Experiment 50.8. Preparation of Hydrogen Chloride (see p. 21).

A mixture of potassium hydrogensulphate and potassium chloride ground together yields hydro-

---

School Science Review No. 133, p. 415.

gen chloride when heated strongly (see Fig. 43.8). Hydrogensulphates are acidic and can often be used in place of sulphuric acid.

$$KHSO_4(s) + KCl(s) \rightarrow K_2SO_4(s) + HCl(g)$$

For properties and reactions see p. 22.

Experiment 50.9. Preparation of Oxygen (see p. 117).

Oxygen is easily prepared in the cold by the interaction of hydrogen peroxide and potassium manganate(VII) solution. 20-volume peroxide sould be in the tube and the manganate(VII) acidified with dilute sulphuric acid in the pipette.

$$2H_2O_2(aq) \rightarrow 2H_2O(l) + O_2(g)$$

The gas should be collected over water and shown to relight a glowing splint.

Experiment 50.10. Preparation of Hydrogen. Roll about 8 cm of magnesium ribbon round a piece of lead or copper (to act as a sinker) and place it in the bottom of the side-arm test-tube and cover with water (about 5 cm up the tube). When concentrated hydrochloric acid is dropped from the pipette (one drop at a time) hydrogen is rapidly evolved and can be collected in 125 × 19 mm test-tubes over water.

$$Mg(s) + 2HCl(aq) \rightarrow MgCl_2(aq) + H_2(g)$$

Much energy is released in this reaction, both by the dilution of the acid and by its action on the magnesium, and the evolution of hydrogen may be slowed down if necessary by surrounding the reaction tube with a beaker of cold water. The usual properties of hydrogen can be demonstrated as easily with test-tubes of the gas as with jars of it.

Experiment 50.11. Preparation of Carbon Dioxide.

The apparatus used is the same as for hydrogen and oxygen. Carbon dioxide is prepared by placing 2 or 3 small lumps of marble in the tube and covering them with water. Concentrated hydrochloric acid is dropped in from the pipette and the gas collected either over water or by upward displacement of air from 125 × 19 mm test-tubes. These can be used to show that the gas does not support combustion, turns limewater milky, etc.

$$CaCO_3(s) + 2HCl(aq) \rightarrow CaCl_2(aq) + CO_2(g) + H_2O(l)$$

Experiment 50.12. Preparation of Nitrogen (see p. 129).

Pour some 880 ammonia into a 125 × 19 mm test-tube to a depth of 2–3 cm and add wisps of asbestos wool to soak it up. Push them down with a glass rod. Now put in some copper(II) oxide—preferably in wire or granular form—to fill a further 5 cm and clamp in a horizontal position. Fit a Bunsen valve (if you have forgotten what this is refer to p. 267) to the end of a delivery tube and set up the apparatus as in Fig. 50.7 so as to collect the gas over water. Heat the copper(II) oxide with a Bunsen, occasionally transferring the flame to the asbestos. Nitrogen will be produced by oxidation of the ammonia.

$$2NH_3(g) + 3CuO(s) \xrightarrow{\Delta} 3Cu(s) + 3H_2O(g) + N_2(g)$$

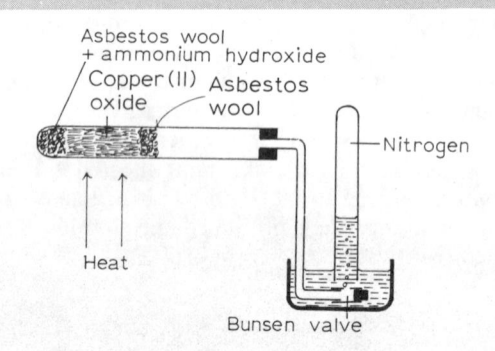

Fig. 50.7. Preparation of Nitrogen

Identification of Gases

The common gases you have prepared are listed in Table 50.1 with the method of identification. Most tests are suitable for use with large or small scale preparations and reference should be made to the appropriate pages of this book.

Identification of Cations present in Compounds

Qualitative analysis is concerned with identifying the various elements or groups present in compounds. Many of the tests for metallic ions are carried out on a solution and involve the production of a precipitate by double decomposition. An alternative method which is of help involves heating some of the substance on a platinum or nichrome wire in a Bunsen flame and noticing the colour produced. For this test the wire is dipped into concentrated hydrochloric acid and held in the flame until there is no noticeable colour. It is then moistened with a little more acid and applied to some of the salt on a watch glass and held just inside the edge of the flame where it is hottest. As a general rule this method gives an *indication* of an element present and should be confirmed by some other test.

Wet and dry tests should be carried out in semi-

Table 50.1 Identification of Gases

Gas		Test	Equation (if any)
Oxygen	O_2	Relights a glowing splint	
Hydrogen	H_2	Burns with a 'pop' when a light is applied	$2H_2 + O_2 \rightarrow 2H_2O$
Carbon monoxide	CO	Burns with a blue flame to give the dioxide	$2CO + O_2 \rightarrow 2CO_2$
Carbon dioxide	CO_2	Turns lime water milky	$Ca(OH)_2 + CO_2 \rightarrow CaCO_3 + H_2O$
Hydrogen chloride	HCl	Acid gas; white fumes with ammonia	$NH_3 + HCl \rightarrow NH_4Cl$
Chlorine	Cl_2	Bleaches damp litmus; pungent smell; green colour	
Sulphur dioxide	SO_2	Acid gas; turns filter paper dipped in solution of potassium chromate green	
Ammonia	NH_3	Alkaline gas; white fumes with hydrogen chloride; yellow-brown colour with Nessler's reagent	$NH_3 + HCl \rightarrow NH_4Cl$
Dinitrogen oxide (nitrous oxide)	N_2O	Relights a glowing splint, but test fails on issuing gas owing to presence of steam	
Nitrogen monoxide (nitric oxide)	NO	Turns brown in the presence of air	$2NO + O_2 \rightarrow 2NO_2$
Nitrogen dioxide	NO_2	Brown gas; relights a glowing splint when freshly collected.	

micro test-tubes, liquids being added or gases removed by means of the pipette. At this stage, all the substances to be identified will be soluble in water or dilute hydrochloric acid so a solution can easily be made. The tests employed are, in the main, ones you have met before, but they are grouped together here for convenience.

Lithium, Sodium and Potassium
Most of the compounds of these elements are soluble in water so there are no *simple* wet tests suitable for identification at this level. However, they can be identified by flame tests and this must suffice. Sodium compounds give an intense persistent yellow colour and potassium ones give a lilac colour. This is often masked by the yellow due to the presence of sodium, but the latter can be cut out by looking at the flame through a piece of cobalt-blue glass. The potassium flame now appears to be crimson. Lithium compounds give the flame a reddish colour.

Copper
(i) The flame test gives flashes of a blue-green colour. You may have noticed that in the reduction of copper(II) oxide by coal gas the flame is often tinged with green and that this happens also if a Bunsen 'lights-back'. The barrel is made of brass and as this gets hot the copper in it colours the flame.

(ii) One measure of copper(II) sulphate should be dissolved in a little water and 3 or 4 drops of aqueous sodium hydroxide added. A light-blue precipitate of copper(II) hydroxide is formed

$$CuSO_4 + 2NaOH \rightarrow Cu(OH)_2 + Na_2SO_4$$
or $\quad Cu_2^+(aq) + 2OH^-(aq) \rightarrow Cu(OH)_2(s)$

and this becomes black on heating.

$$Cu(OH)_2 \xrightarrow{\Delta} CuO + H_2O$$

When this test is repeated with ammonia solution precipitate forms, but addition of a few more drops of ammonia causes this to dissolve leaving a deep bluish-purple solution which is characteristic of copper compounds.

Zinc
(i) If to a little sodium hydroxide solution in a test-tube is added an equal amount of a solution of zinc sulphate a white gelatinous (jelly-like) precipitate of zinc hydroxide forms.

$$ZnSO_4 + 2NaOH \rightarrow Zn(OH)_2 + Na_2SO_4$$
or $\quad Zn^{2+}(aq) + 2OH^-(aq) \rightarrow Zn(OH)_2(s)$

When more sodium hydroxide is added drop-wise with shaking the precipitate *dissolves*, forming sodium zincate (see p. 118).

$$Zn(OH)_2 + 2NaOH \rightarrow Na_2ZnO_2 + 2H_2O$$

Only a few hydroxides dissolve in this way; they are

all amphoteric ones, i.e. soluble in both acid and alkali.
(ii) When a little of the zinc hydroxide precipitate is heated to dryness a *yellow* powder is formed which becomes *white* on cooling. This is zinc oxide.

$$Zn(OH)_2 \rightarrow ZnO + H_2O$$

Aluminium
Addition of sodium or potassium hydroxide solution to an aluminium salt in solution produces a white precipitate of the hydroxide, an action similar to that with a zinc salt.

$$Al_2(SO_4)_3(aq) + 6NaOH(aq) \rightarrow 2Al(OH)_3(s) + 3Na_2SO_4(aq)$$

or $\quad Al^{3+}(aq) + 3OH^-(aq) \rightarrow Al(OH)_3(s)$

Again like zinc hydroxide, this precipitate dissolves when more alkali is added; sodium aluminate is formed.

$$Al(OH)_3 + NaOH \rightarrow NaAlO_2 + H_2O$$

or
$$Al(OH)_3(s) + OH^-(aq) \rightarrow AlO_2^- + 2H_2O(l)$$

However, if the reaction is repeated with ammonia solution the zinc hydroxide first formed dissolves as excess ammonia is added, but the aluminium hydroxide precipitated remains *undissolved*. In this way the two can be distinguished.

Calcium
(i) Calcium compounds colour the flame a brick-red; as with most other metals the colour is not usually a persistent one, but comes in flashes.
(ii) Sodium hydroxide produces a faint white precipitate but only from concentrated solutions of calcium ones.

$$CaCl_2 + 2NaOH \rightarrow Ca(OH)_2 + 2NaCl$$

or $\quad Ca^{2+}(aq) + 2OH^-(aq) \rightarrow Ca(OH)_2(s)$

(iii) Dilute sulphuric acid precipitates white calcium sulphate from concentrated solutions.

$$CaCl_2 + H_2SO_4 \rightarrow CaSO_4 + 2HCl$$

or $\quad Ca^{2+}(aq) + SO_4^{2-}(aq) \rightarrow CaSO_4(s)$

Lead
(i) Addition of sodium hydroxide solution produces a heavy white precipitate of lead hydroxide which quickly settles down. It will dissolve in excess alkali, but not so easily as the zinc hydroxide.

$$Pb(NO_3)_2 + 2NaOH \rightarrow Pb(OH)_2 + 2NaNO_3$$

or $\quad Pb^{2+}(aq) + 2OH^-(aq) \rightarrow Pb(OH)_2(s)$

(ii) Addition of dilute sulphuric acid (or other soluble sulphate) gives a heavy white precipitate of lead sulphate.

$$Pb(NO_3)_2 + H_2SO_4 \rightarrow PbSO_4 + 2HNO_3$$

or $\quad Pb^{2+}(aq) + SO_4^{2-}(aq) \rightarrow PbSO_4(s)$

This precipitate can be produced using a dilute solution of calcium sulphate whereas the latter has no effect on solutions of calcium salts.

(iii) Addition of dilute hydrochloric acid produces a heavy white precipitate of lead chloride; this dissolves on heating, but reappears on cooling.

$$Pb(NO_3)_2 + 2HCl \rightarrow PbCl_2 + 2HNO_3$$

or $\quad Pb^{2+}(aq) + 2Cl^-(aq) \rightarrow PbCl_2(s)$

(iv) Addition of potassium iodide solution produces a heavy yellow precipitate of lead iodide; this is the best test.

$$Pb(NO_3)_2 + 2KI \rightarrow PbI_2 + 2KNO_3$$

or $\quad Pb^{2+}(aq) + 2I^-(aq) \rightarrow PbI_2$

Iron
Here we have to distinguish between iron in the iron(II) state and iron in the iron(III) state.
(i) Sodium hydroxide solution or ammonia solution added to an iron(III) compound in solution produces a red-brown precipitate of iron(III) hydroxide.

$$FeCl_3 + 3NaOH \rightarrow Fe(OH)_3 + 3NaCl$$

or $\quad Fe^{3+}(aq) + 3OH^-(aq) \rightarrow Fe(OH)_3(s)$

(ii) With an iron(II) compound the precipitate is a dirty green colour, impure iron(II) hydroxide.

$$FeSO_4 + 2NaOH \rightarrow Fe(OH)_2 + Na_2SO_4$$

or $\quad Fe^{2+}(aq) + 2OH^-(aq) \rightarrow Fe(OH)_2(s)$

This precipitate gradually goes brown as it oxidizes to the iron(III) state.
(iii) Addition of ammonium thiocyanate (NH_4CNS) solution to an iron(III) salt gives a blood-red colour immediately; with an iron(II) salt there should not be any colour (or only a faint one) until a drop of concentrated nitric acid or hydrogen peroxide is added. Then the red colour appears as oxidation of Fe^{2+} ions to Fe^{3+} ions quickly takes place.

Ammonium

Although the ammonium ion is not a metallic one it does behave in a similar way to sodium and other ions (see p. 130) and carries a positive charge.

Ammonium salts evolve ammonia when warmed with sodium hydroxide solution.

$$NH_4Cl + NaOH \rightarrow NH_3 + NaCl + H_2O$$
or $NH_4^+(s) + OH^-(aq) \rightarrow NH_3(g) + H_2O(l)$

One measure of an ammonium salt should be put in a tube with a pellet of sodium hydroxide and just covered with water and warmed. The issuing gas has a pungent smell and will:
(a) turn a small piece of red litmus held at the mouth of the tube blue,
(b) produce white fumes of ammonium chloride in contact with a small piece of filter paper dipped in concentrated hydrochloric acid,
(c) stain a piece of filter paper dipped in Nessler's reagent a yellow-brown,
(d) turn a spot of copper(II) sulphate solution on a filter paper a deep blue colour.

The various reactions with sodium hydroxide and ammonium hydroxide may be summarized as in the table.

Chloride (see p. 23).

(i) When the dry substance is heated with about 5–6 drops of concentrated sulphuric acid and a filter paper strip dipped in 880 ammonia is held at the mouth of the tube, with fumes of ammonium chloride indicate the evolution of hydrogen chloride.

(ii) If, in the above test, the chloride is mixed with a like amount of manganese (IV) oxide before being heated with the acid, the hydrogen chloride is oxidized to chlorine which can be recognised by its bleaching action on moist litmus paper.

(iii) Addition of silver nitrate solution to a solution containing chloride ions produces a white precipitate of silver chloride which becomes violet in colour on exposure to light.

$$AgNO_3 + NaCl \rightarrow AgCl + NaNO_3$$
or $Ag^+(aq) + Cl^-(aq) \rightarrow AgCl(s)$

The suspension should be divided into two parts and each treated separately.
(a) Addition of dilute ammonia solution dissolves the precipitate.
(b) Addition of dilute nitric acid does not dissolve the precipitate.

Ions	Precipitate	Excess NaOH	Excess NH$_4$OH
Fe^{2+}	Dirty green	Insoluble	Insoluble
Fe^{3+}	Red-brown	Insoluble	Insoluble
Cu^{2+}	Pale blue	Insoluble	Soluble (deep blue)
Ca^{2+}	White (faint)	Insoluble	Insoluble
Pb^{2+}	White (heavy)	Soluble	Insoluble
Zn^{2+}	White (gelatinous)	Soluble	Soluble
Al^{3+}	White (gelatinous)	Soluble	Insoluble

If hydrogen sulphide is passed into the original solution it will serve to distinguish between the metals above which have *white* hydroxides. It will give a black precipitate with lead salts, a white one with zinc salts (insoluble in aqueous sodium hydroxide) and aluminium salts (soluble in aqueous sodium hydroxide) and none at all with calcium salts. (Iron(II) and copper sulphides are also black.)

Identification of Anions present in Compounds

Tests for anions are of two kinds: (i) identification of a precipitate formed on double decomposition (ii) identification of a gas evolved when the substance is heated with either dilute hydrochloric acid or concentrated sulphuric acid. For each test one measure of the substance should be used; for the wet tests it is dissolved in a little water.

Carbonate.

2–3 cm³ of dilute hydrochloric acid should be boiled in a test-tube (see p. 315 for method) and then one measure of the suspected carbonate dropped into the hot liquid. If a carbonate is present effervescence will be noticed. Some of the gas should be withdrawn by means of a pipette and bubbled through 2–3 cm³ of lime water in another tube. A white precipitate of calcium carbonate occurs.

$$Na_2CO_3 + 2HCl \rightarrow 2NaCl + H_2O + CO_2$$
or $CO_3^{2-}(s) + 2H^+(aq) \rightarrow H_2O(l) + CO_2(g)$
and then $Ca(OH)_2(aq) + CO_2(g) \rightarrow CaCO_3(s) + H_2O(l)$

Sulphate (see p. 73).

Addition of 2–3 drops of dilute hydrochloric

acid followed by 2 drops of barium chloride solution gives a white precipitate of barium sulphate if a sulphate is present.

$$BaCl_2 + Na_2SO_4 \rightarrow BaSO_4 + 2NaCl$$
or $$Ba^{2+}(aq) + SO_4^{2-}(aq) \rightarrow BaSO_4(s)$$

If a white precipitate is obtained when the hydrochloric acid is first added then the original solution will contain a lead salt.

$$Pb^{2+}(aq) + 2Cl^-(aq) \rightarrow PbCl_2(s)$$

Sulphite.
(i) If the test given above for a carbonate is carried out with a sulphite, sulphur dioxide is evolved.

$$Na_2SO_3 + 2HCl \xrightarrow{\Delta} 2NaCl + H_2O + SO_2$$
or $$SO_3^{2-}(s) + 2H^+(aq) \rightarrow H_2O(l) + SO_2(g)$$

If some of the gas is removed with the pipette and bubbled through potassium dichromate(VI) (dichromate) solution containing a little dilute sulphuric acid the solution goes green.

(ii) Addition of 3–4 drops of barium chloride solution to a sulphite dissolved in water produces a white precipitate of barium *sulphite*.

$$BaCl_2 + Na_2SO_3 \rightarrow BaSO_3 + 2NaCl$$
or $$Ba^{2+}(aq) + SO_3^{2-}(aq) \rightarrow BaSO_3(s)$$

This precipitate *dissolves* in dilute hydrochloric acid (or nitric acid) and so can be distinguished from barium *sulphate*.

Nitrate (see p. 138).
(i) If a little copper powder or a small copper turning is placed in a test-tube with a little solid nitrate and the mixture covered with concentrated sulphuric acid, brown fumes of nitrogen dioxide are evolved on heating.
(ii) Half a measure of the solid nitrate is put on a watch glass with a crystal of iron(II) sulphate and a little water is added so that they dissolve when stirred with a spatula. Next, with the teat pipette, one drop of concentrated sulphuric acid is placed in the centre of the solution. If the glass is stood on a white paper the brown ring produced can easily be seen.

If you wish to distinguish between the above radicals by a systematic method you may find the table at the bottom of this page helpful.

The carbonate and sulphite can be separately identified as previously described.

Formation of Salts
If a centrifuge is available many salts can be prepared quite quickly on the small-scale. The general method in many cases is illustrated by the following example.

Experiment 50.13. Preparation of Lead Nitrate and Lead Sulphate.
Half-fill a semi-micro test-tube with dilute nitric acid and heat it to boiling by embedding the bottom of it in a pile of sand being heated on a tin lid or asbestos-centred gauze. (It may be advisable to add one or two 'anti-bumping' granules.) Add successive small portions of lead(II) oxide or lead carbonate until some remains undissolved even when boiled. Centrifuge the suspension unless the undissolved solid has settled at the bottom (your teacher will show you how to do this) and transfer some of the liquid to a small crucible using the pipette. Place one or two drops on a microscope slide and see if crystallization occurs on cooling. If it does not, evaporate some of the water from the solution in the crucible (stand it on a sand heap as above) and try again. When crystallization is about to occur, pour the solution on to a watch glass, cover with another similar glass and leave.

$$PbO + 2HNO_3 \rightarrow Pb(NO_3)_2 + H_2O$$
or $$PbCO_3 + 2HNO_3 \rightarrow Pb(NO_3)_2 + H_2O + CO_2$$

If, after centrifuging, some of the solution is transferred to another tube and a few drops of dilute sulphuric acid are added, double decomposition occurs and lead sulphate is precipitated.

Anion	Addition of lead-acetate solution	Effect of boiling suspension	Addition of dilute nitric acid to the suspension
Chloride	White precipitate	Precipitate dissolves	No change
Carbonate or Sulphite	White precipitate	No change	Precipitate dissolves with effervescence
Sulphate	White precipitate	No change	No change
Nitrate	No precipitate	—	—

> $Pb(NO_3)_2 + H_2SO_4 \rightarrow PbSO_4 + 2HNO_3$
> or $Pb^{2+}(aq) + SO_4^{2-}(aq) \rightarrow PbSO_4(s)$
>
> Centrifuge the suspension, remove and discard the liquid. Wash the precipitate by adding a little water and blowing air through the suspension by means of the pipette. Further centrifuging will leave pure lead sulphate. This can be removed with a glass rod to a filter paper and left to dry.

By varying the acid and by using different oxides or carbonates (or sometimes the metal) it is possible to prepare many other salts such as the chlorides, nitrates and sulphates of copper, calcium, iron, zinc, magnesium, etc.

Salts such as sodium chloride, potassium sulphate, etc., which are normally prepared by neutralization can be scaled down to use about 1 cm³ of dilute acid and the alkali gradually added by the pipette until a little indicator paper indicates neutrality. Alternatively, they may be produced by adding the appropriate carbonate to the acid *in the cold* until effervescence ceases and then proceeding as for lead nitrate.

Exercise 50

1 You are provided with the following reagents and any laboratory apparatus you may require:
lime-water, water, litmus solution, platinum wire, hydrochloric acid, granulated zinc.
By means of any two tests, say how you would tell the difference between the following pairs of substances:
(a) dilute sulphuric acid and a dilute solution of sodium chloride,
(b) a gas-jar of hydrogen and a gas-jar of carbon dioxide,
(c) anhydrous copper(II) sulphate and calcium oxide,
(d) sodium hydrogencarbonate and zinc carbonate,
(e) potassium chloride and ammonium chloride.
(SR)

2 Small quantities of the table salt, sugar, baking powder, ground chalk and potassium chlorate(V) are put into separate test tubes. Describe simple tests by which you could find which substance is in each tube.
(WR)

3 By what simple experiments would you show that:
(a) Sodium nitrate contains sodium and a nitrate?
(b) Barium chloride crystals contain a chloride and water of crystallization?
(c) Iron(III) chloride solution contains iron(III) ions?
(L)

4 For each of the following pairs of substances, describe ONE chemical test which, when it is applied to each member of the pair, will distinguish between them. State what you observe when the tests are carried out:
(a) copper(II) oxide and manganese(IV) oxide;
(b) metallic zinc and metallic magnesium;
(c) lead(II) carbonate and calcium carbonate;
(d) a sulphite and a sulphate;
(e) sulphur dioxide and hydrogen chloride.
(O)

5 For each of the following pairs of substances give TWO chemical tests which would enable you to distinguish between the substances:
(a) ammonium chloride and ammonium sulphate,
(b) carbon and copper(II) oxide,
(c) sodium nitrate and sodium carbonate,
(d) lead(II) nitrate and zinc nitrate.
Equations are not required.
(J)

6 Give, without equations, a simple chemical test in each case, to identify with reasonable certainty (a) chlorine, (b) sulphur dioxide, (c) oxygen, (d) gaseous hydrogen chloride.
(S)

7 Explain why:
(a) anhydrous copper(II) sulphate can be used to detect water,
(b) lime water is used to detect carbon dioxide,
(c) lead(II) nitrate (or ettranoate) paper can be used to test for hydrogen sulphide,
(d) starch iodide paper is used to detect chlorine.
How may starch iodide paper be used to determine the polarity of the terminals of a torch battery? (L)

8 Give TWO chemical tests in each case which would enable you to distinguish between:
(a) zinc carbonate *and* barium carbonate;
(b) copper(II) nitrate *and* copper(II) sulphate;
(c) anhydrous sodium carbonate *and* sodium hydrogencarbonate;
(d) red mercury(II) oxide *and* tri-lead tetroxide.
(J)

9 For any TWO of the following pairs of substances give (a) ONE reaction in which their behaviour is different:
(i) the metals sodium and copper,
(ii) sodium carbonate and copper(II) carbonate;
(iii) sodium nitrate and copper(II) nitrate.
For each reaction you give, describe what you would observe and name the products. (C)

10 A white salt A melts when gently heated and gives off steam and a gas B that ignites a glowing splint but does not give brown fumes with nitrogen monoxide.
When a mixture of A with sodium hydroxide solution is heated there is evolved a pungent smelling gas C, which is very soluble in water giving an alkaline solution D.

When a mixture of the salt A, copper, and concentrated sulphuric acid is heated brown fumes D are evolved.

Identify A, B, C, D and E and write equations for the reactions described. Draw the apparatus you would use to prepare D from A. (L)

11 State clearly how you would distinguish between the following pairs of substances, giving ONE test in each case and a description of the result of each test. In each case give ONE reason why the two substances might be mistaken for one another:
  (a) sulphur dioxide and hydrogen chloride;
  (b) carbon monoxide and hydrogen;
  (c) nitrogen and carbon dioxide;
  (d) potassium nitrate and potassium chlorate.
(O)

12 As a result of a flood in a laboratory, the labels were washed off five bottles, each containing a colourless liquid. The labels were found to read: 'dilute hydrochloric acid', 'dilute nitric acid', 'dilute sulphuric acid', 'sodium hydroxide', 'sodium carbonate'.

Describe in detail what you would do to find out how to replace the labels correctly. All the usual laboratory chemical reagents and apparatus are available. Equations are not required. (J)

13 An unlabelled bottle containing a colourless oily liquid is found. The following were the results of tests applied.
  (i) The liquid turns litmus solution red.
  (ii) The liquid added to moist sugar, the mixture swells into a black steamy mass.
  (iii) When the liquid is added to a little water cautiously in a test tube the water becomes very hot.

Say what you conclude from each of these observations. What do you suppose the liquid to be? What further test would you apply to prove your idea?
(MR)

14 A green powder, when heated in a test tube, gives off a colourless, odourless gas, which when it is bubbled through lime water turns it milky and then after a few minutes turns it clear again. Account for these observations, and state and explain what you would expect to happen if the solution was now boiled.

It is found that during the experiment the green powder has turned black. When this new substance is placed in a tube that is open at both ends and heated in a current of dry ammonia, the black powder turns pink, drops of a colourless liquid condense further along the tube and an unreactive gas comes out from the end of the tube.

Suggest what the original green powder was and give reasons for your answer. Sketch and label an apparatus for repeating the last experiment so that the colourless liquid and the unreactive gas can be collected for examination. (L)

15 A series of experiments was carried out to investigate the properties of white powder N (the results are shown in the table at the bottom of the page).
  (i) What type of compound is the original powder N?
  (ii) What type of compound is the residue Q?
  (iii) Name the gases evolved when the white powder S is heated.
  (iv) The yellow-green gas W evolved and the metal X formed are elements. What are W and X?
  (v) Why does the white substance V conduct in the molten state and not in the solid state?
  (vi) What is the nature of the bonding in a compound such as V?
  (vii) Write a balanced equation for the reaction which occurs when the solid S is heated.
  (viii) Write a balanced equation for the reaction between the original powder N and dilute nitric acid.
  (ix) What do you observe when sodium hydroxide solution is added to an aqueous solution of the metal ion contained in N? Write an equation for each reaction that occurs.

EXPERIMENT	RESULT
(a) The powder N was heated.	A colourless gas P was evolved which turned lime water milky. A yellow residue Q remained. The yellow colour of the residue persisted when the residue was cold.
(b) Excess dilute nitric acid was added to the original powder N.	The same colourless gas P as in (a) was evolved. A colourless solution R was obtained.
(c) Half of the solution R was evaporated.	A white solid S remained.
(d) The white solid S was strongly heated in a test tube.	There was a crackling sound. A brown gas T was evolved. When a glowing splint was introduced into the test tube it ignited. A yellow solid U remained.
(e) To the other half of the solution R a solution of sodium chloride was added.	A white precipitate was obtained.
(f) The mixture was filtered and the residue V was washed. The residue V was then placed in water and the mixture heated and allowed to cool slowly.	The precipitate V was soluble in hot water but formed needle-like white crystals on cooling.
(g) The crystals V were heated and they melted at about 500°C. The molten V was electrolyzed.	A yellow-green gas W was evolved and a metal X was deposited on the cathode.

51
Simple Organic Chemistry; Long Chains and Giant Molecules

Organic chemistry may be defined as the chemistry of carbon compounds, excluding such substances as the oxides and sulphides of the element, the carbonates and hydrogencarbonates and the metallic carbides. The number of compounds formed by different elements varies from one to another, but the compounds containing carbon outnumber all the others put together. This is one of the reasons why they are considered separately, but there are more important ones. Originally, it was thought that all substances present in animal or vegetable material (all of them carbon compounds) possessed some 'vital force' which distinguished them from those of the insert and lifeless mineral kingdom; the former were termed 'organic' and the latter inorganic'. However, in 1828, a compound called urea, present in urine, was synthesized from completely inorganic materials and it was shown that no real distinction between the two types existed. In addition, in inorganic chemistry the formula $CuSO_4$ represents one substance only, but there are four different compounds having the formula $C_4H_{10}O$. Amongst the complexity of carbon compounds the 'Law' of Multiple Proportions does not hold; in the three different hydrocarbons $C_{20}H_{38}$, $C_{21}H_{40}$ and $C_{22}H_{42}$ there is no *simple* ratio between the weights of hydrogen uniting with some fixed weight of carbon.

Even so, there are many resemblances. Just as inorganic chemistry has its classes of acids, bases and salts and sub-classes such as oxides, sulphates, etc., so organic chemistry has its groups, alcohols, acids, esters, hydrocarbons, etc. All but a very few of the 104 known elements may be present in inorganic compounds, but comparatively few are found in large amounts in organic ones.

The majority of organic compounds also contain hydrogen, and other elements often present are oxygen, fluorine, chlorine, bromine, iodine, nitrogen, phosphorus and sulphur. You will notice that these are all nonmetallic elements and all these compounds are covalent ones (see Chapter 6). They do not conduct an electric current (but see p. 223). Metals may also be present in certain classes, such as salts, and these may be electrovalent, e.g. sodium acetate.

Experiment 51.1. Testing for Carbon and Hydrogen.
Take one measure of sugar or starch and grind it in a mortar with about ten measures of copper(II) oxide (previously heated to ensure that it is dry and then cooled in a desiccator). Place the mixture in a semi-micro test-tube and heat strongly in the bunsen flame.
(i) Hold a piece of cobalt chloride paper (made by soaking filter paper in cobalt chloride solution and drying in an oven) at the mouth of the tube and notice that it turns from blue to pink, indicating the production of water and therefore the presence of hydrogen in the original compound.

$$CuO + H_2 \rightarrow Cu + H_2O$$

(ii) Squeeze the teat-pipette and insert it in the tube and release the pressure. Now bubble the gas taken in through lime water and notice the milkiness produced. Carbon dioxide has been formed so carbon must be present in the original compound.

$$2CuO(s) + C(s) \rightarrow 2Cu(s) + CO_2(g)$$

Alkanes or Paraffins
Carbon alone of all the elements is able to form long chains in which the atoms are joined to each other by covalent bonds. There may be hundreds or even thousands of such carbon atoms and the chains may be straight as in

$$-\underset{|}{\overset{|}{C}}-\underset{|}{\overset{|}{C}}-\underset{|}{\overset{|}{C}}-\underset{|}{\overset{|}{C}}- \quad \text{or} \quad -\underset{|}{\overset{|}{C}}-\underset{|}{\overset{|}{C}}-\underset{|}{\overset{|}{C}}-\underset{|}{\overset{|}{C}}-\underset{|}{\overset{|}{C}}-\underset{|}{\overset{|}{C}}-$$
butane $\qquad\qquad\qquad$ hexane

or branched as in the diagram overleaf:

```
      |  |  |                    -C-
     -C-C-C-     or         |  |  |  |
      |  |  |              -C-C-C-C-
         -C-                  |  |  |
          |                     -C-
                                 |
  2-methylpropane          2,2-dimethylbutane
```

Each C— represents a C—H bond, the hydrogen atoms not being shown in the diagrams. The formulae for these compounds can be written in full—and more precisely—in another way like this:

$$CH_3.CH_2.CH_2.CH_3$$
butane

$$CH_3.CH(CH_3).CH_3$$
2-methylpropane

$$CH_3.CH_2CH_2.CH_2.CH_2.CH_3$$
hexane

$$CH_3.CH(CH_3).CH(CH_3).CH_3$$
2,2-dimethylbutane

Compounds such as these in which only carbon and hydrogen are present are called hydrocarbons and many of them are very important as fuels. The above are examples of a group called the alkanes (paraffins). The simplest members such as methane, CH_4, ethane, C_4H_6 and propane, C_3H_8 are all gases at room temperature. As the number of carbon atoms (and therefore the density) increases the compounds become liquids and finally waxy solids. The general formula is C_nH_{2n+2}.

The formula of each alkane increases by CH_2 compared with the one before and the molar mass therefore increases by 14. Such a series, where this increase is regular and where the change in properties on passing from one compound to the next is only slight, is called a *homologous series*.

If you look at the formulae above and put all the carbons and hydrogens together you will find that both butane and 2-methylpropane (also called isobutane) have the same formula, C_4H_{10}; Similarly, both hexane and 2,2-dimethylbutane have the formula C_6H_{14}. The only difference in the formulae is in the way in which the carbon atoms are joined together to form either a straight chain or a branched chain. Such compounds as these are called *isomers* and as the number of carbon atoms increases the number of possible isomers becomes immense. The properties of any two isomers in the same group with the same formula do not differ very greatly.

Methane
The first of the alkanes (n = 1) is methane. This is formed sometimes in stagnant ponds by the decomposition of decaying vegetation and is known as marsh gas. It is often present in coal mines and may form an explosive mixture with air. If it is burnt and there is insufficient air present for its complete combustion poisonous carbon monoxide is formed. It is the main constituent of natural gas (p. 149) which is often found associated with oil (itself of animal or vegetable origin) and is produced during the bacterial treatment of sewage.

Experiment 51.2. Preparation of Methane.
The chemicals needed for this preparation are anhydrous sodium ethanoate (acetate) and soda-lime. If the former is not available heat some crystalline sodium ethanoate in a crucible or evaporating dish. It first melts and then solidifies and once again melts. Stir the melt and allow it to cool.

Grind approximately equal masses of anhydrous sodium ethanoate and soda-lime together in a mortar along with a small amount of iron filings (this helps the reaction). Soda-lime (formed by slaking calcium oxide with sodium hydroxide instead of water) reacts in the same way as sodium hydroxide but is easier to handle. Half fill the test-tube shown in Fig. 51.1 and heat gently.

$$CH_3COONa(s) + NaOH(s) \rightarrow CH_4(g) + Na_2CO_3(s)$$

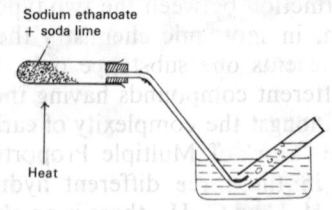

Fig. 51.1. Preparation of Methane

Collect several tubes of the gas and test as instructed.
(a) Apply a light—what kind of flame is produced?
(b) Shake with a little bromine water—is there any change in the colour?
(c) Shake with a little lime water—is there any change here? Repeat with a tube in which the gas has been burnt.
(d) Add a little dilute hydrochloric acid to the cold residue in the test-tube; remove some of the gas liberated and identify it by bubbling it through lime water. What does this tell you about the residue? Add some acid to the original mixture and test in the same way—does this behave similarly?

The alkanes are called *saturated* hydrocarbons as they cannot add on any more elements or groups without first losing hydrogen atoms, carbon always having a valency of four. They are relatively inert, although they do form compounds such as trichloromethane (chloroform), $CHCl_3$, by substitution. They burn easily to form carbon dioxide and water. With methane the reaction is

$$CH_4 + 2O_2 \rightarrow CO_2 + 2H_2O$$

By suitable treatment methane can be:
(i) converted into a mixture of carbon monoxide and hydrogen which is used in the manufacture of both methanol (methyl alcohol) CH_3OH, and ammonia (it is the hydrogen that is required here),
(ii) decomposed to give carbon black for use in printers' ink and for improving the toughness of rubber tyres,
(iii) used to produce ethyne (acetylene), C_2H_2, important for the production of the plastic polymer, polyvinyl chloride (PVC).

Petroleum

Crude petroleum or oil is a mixture of many hydrocarbons, liquid, gaseous and solid. Most of them belong to the alkane series and can be obtained from petroleum by fractional distillation. The oil is heated to vaporize it and it then passes up a large tower where the more volatile components rise to the top and are removed. Less volatile portions are collected from intermediate positions up the tower. The lightest fraction with the lowest boiling point contains gaseous alkanes and is called 'refinery gas'. The next fraction obtained is 'petrol', a mixture of various liquid hydrocarbons, mainly those in the range C_5H_{12} to $C_{10}H_{22}$ with boiling points between 30°C and about 200°C. As the number of carbon atoms increases we get denser fractions with larger molecules such as 'paraffin' or 'kerosine', 'gas oil', 'diesel oil' and 'pitch'. There is most demand for petrol and additional petrol hydrocarbons can be obtained (i) by breaking down (cracking) some of the larger molecules in the oil or (ii) by combining some of the simple molecules of gaseous hydrocarbons such as butane, C_4H_{10}, into larger molecules which are liquids. This latter process is called *polymerization* and will be further considered later in this chapter.

Alkenes or Olefines

If methane, CH_4, is shaken with a little bromine water, the latter remains brown. No combination occurs as there are no spare valency bonds on the carbon to take up additional atoms. If, however, a gas called ethene (ethylene), C_2H_4, is used, the bromine water becomes colourless and an oily liquid is formed. Evidently the bromine atoms have been added on to the ethene molecule; a hydrocarbon such as this which can form additional compounds is said to be *unsaturated*.

Look at the formula for ethene, C_2H_4. In order to satisfy the four valencies of carbon and the one of hydrogen it must be written

$$\begin{array}{c} H \\ \diagdown \\ C=C \\ \diagup \\ H \end{array} \begin{array}{c} H \\ \diagup \\ \\ \diagdown \\ H \end{array} \quad \text{or } CH_2:CH_2$$

where two of the carbon valency bonds are shared between the two carbon atoms. This is a *double bond* and such compounds can add on other atoms as in

$$\begin{array}{c} H \quad H \\ | \quad | \\ Br-C-C-Br- \\ | \quad | \\ H \quad H \end{array} \quad \text{or } CH_2Br.CH_2Br$$

1,2-dibromoethane; the double bond is broken.

Ethene (ethylene)

Ethene is the simplest member of a series of olefines (or alkenes); these have a general formula, C_nH_{2n}, where n can be any number from 2 upwards.

In industry ethene is formed by the 'thermal cracking' of alkanes present in petroleum, e.g.

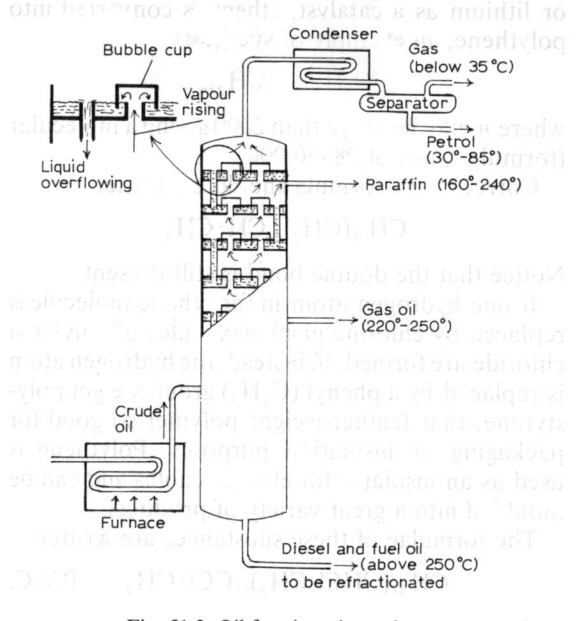

$$\underset{\text{propane}}{C_3H_8} \xrightarrow{\Delta} \underset{\text{methane}}{CH_4} + \underset{\text{ethene}}{C_2H_4}$$

Fig. 51.2. Oil fractionating column

Table 51.1 Some of the Differences between Organic and Inorganic Compounds

Organic	Inorganic
Mostly covalent compounds, but metal salts such as sodium ethanoate are electrovalent.	Mostly electrovalent, but some such as PCl_5 and HCl are covalent.
Usually insoluble in water (except salts and some alcohols and acids).	Many are soluble in water, but a number are not.
Poor electrolytes (except salts).	Many are good electrolytes in solution.
Generally have a low melting point.	Many have a very high melting point.

This can be carried out using the apparatus of Fig. 51.1. In the tube is placed about 3 cm³ of medicinal paraffin or motor oil and this is soaked up by mineral or asbestos wool as previously explained. With the tube horizontal several pieces of porous pot are placed in it and the delivery tube connected. The pot is heated strongly, the liquid is vaporized and then 'cracked' as it passes through the hot zone. A flammable gas containing ethene is collected.

In the laboratory ethene is usually obtained by dehydrating ethanol (ethyl alcohol) by heating it with concentrated sulphuric acid or by passing the vapour over strongly heated aluminium oxide.

$$C_2H_5OH \xrightarrow{H_2SO_4} C_2H_4 + H_2O$$

When ethene is tested in the following ways the results give an indication of some of the properties of these unsaturated hydrocarbons.

(i) It will burn and when lime water is added to the jar after the combustion it turns milky, showing carbon dioxide to have been formed.

(ii) When bromine water is added it is decolorized.

(iii) When it is shaken with a solution of potassium manganate(VII) containing a little sodium hydroxide the latter soon goes brown as manganese(IV) oxide is precipitated. The ethene forms an addition product called ethane-1,2-diol or 'ethylene glycol'; this is used widely as an antifreeze in car radiators and for preventing the icing-up of aeroplane wings. These tests show that alkenes are more reactive than the alkanes.

The uses of ethene include:

(a) the production of 1,2-dibromoethane—a little of this is blended with most petrols to prevent the deposition of lead in the cylinders of car engines when tetraethyl lead (see p. 252) is added to petrol as an anti-knock',

(b) the artificial ripening of fruits such as oranges and bananas which are picked when green,

(c) the formation of polyethene (polythene) by the joining together of many ethene molecules, another example of polymerization.

Polymerization

Ethene and other compounds containing a double bond are able under suitable conditions to react in such a way that the simple molecules join together, end to end, to form long chains which may be straight or branched. These substances are called polymers. They have the same empirical formula as the simple substance (the monomer), but an extremely high molecular or formula mass. Some naturally occurring polymers are rubber and cellulose and artificial ones include nylon, terylene, orlon, bakelite, perspex, polythene, polystyrene and polyvinyl chloride. The number of simple groups in the giant molecules of these and other polymers may run into thousands.

At ordinary temperatures and pressures, with the addition of an organic compound of aluminium or lithium as a catalyst, ethene is converted into polythene, an example of synthesis.

$$C_2H_4 \rightarrow (CH_2)_n$$

where n may be more than 2000 giving a molecular (formula) mass of 28–30 000.

Correctly the formula should be written

$$CH_3.(CH_2)_n.CH:CH_2$$

Notice that the double bond is still present.

If one hydrogen atom in the ethene molecule is replaced by chlorine giant molecules of polyvinyl chloride are formed; if, instead, the hydrogen atom is replaced by a phenyl (C_6H_5) group we get polystyrene, that feather-weight polymer so good for packaging or insulation purposes. Polythene is used as an insulator for electric cables and can be moulded into a great variety of products.

The formulae of these substances are written

$$CH_3.(CHCl.CH_2)_n.CCl:CH_2 \quad \text{P.V.C.}$$

and

$$CH_3.(CH(C_6H_5).CH_2)_n.C(C_6H_5)_n:CH_2$$
Polystyrene

Do NOT try to remember these formulae, they are only given as examples of polymerization. Do note, however, that the general formula in each case is

$$CH_3-\square-CH_2-\square-CH_2-\square-CH_2-$$

and only the part in the box varies, being CHCl for P.V.C. and $CH(C_6H_5)$ for polystyrene.

Alcohols

One important class of organic compounds is the alcohols of which ethane-1,2-diol (ethylene glycol) is a member. This, as we have seen, can be obtained from ethene as can the most important alcohol of all, ethanol (or ethyl alcohol) C_2H_5OH.

In this process ethene, from the petroleum cracking plant, is absorbed in concentrated sulphuric acid at 100°C to form ethyl hydrogensulphate, $C_2H_5HSO_4$.

$$C_2H_4 + H_2SO_4 \xrightarrow{\Delta} C_2H_5HSO_4$$

When this is mixed with water and heated, alcohol distils over.

$$C_2H_5[HSO_4 + H]OH \xrightarrow{\Delta} C_2H_5OH + H_2SO_4$$

The sulphuric acid is regained and concentrated for further use.

Fermentation

Although more and more alcohol (at present about 75% of requirements) for industrial use is being made by the above method, the classical process of fermentation is still used to produce all the alcohol required for beverages such as gin, whisky, sherry, etc.

Cereals, fruits and potatoes contain a carbohydrate called starch. This has a molecular formula of $(C_6H_{10}O_5)_n$ where n may be as high as 50 000, another example of a giant molecule.

When this is ground with water and malt, fermentation takes place converting the starch to a double sugar called maltose, $C_{12}H_{22}O_{11}$. This is brought about by an *enzyme* called diastase present in the malt (and also in saliva which can cause the same change). Enzymes are natural catalysts with complex chemical structures, each effecting some specific conversion.

Maltose, molasses, cane sugar and beet sugar can be further fermented by the addition of yeast. The double sugars are broken down into simple ones

$$C_{12}H_{22}O_{11} + H_2O \xrightarrow{enzyme} 2C_6H_{12}O_6 \text{ (glucose)}$$

and then into alcohol and carbon dioxide

$$C_6H_{12}O_6 \xrightarrow{enzyme} 2C_2H_5OH + 2CO_2$$

Yeast is a very small organism consisting of a single cell only. It contains several different enzymes and these bring about the conversion outlined.

Experiment 51.3. Breaking down the Sucrose Molecule.

Dissolve about 10 g of cane sugar (sucrose) in 100 cm³ of water and add about one gram of fresh brewer's yeast (obtainable from your baker); put in also about 0·2 g of ammonium phosphate and potassium nitrate to act as a kind of 'food' for the yeast. If you have a water bath available set this at 30°C and place the flask in it (put a loose plug of cotton wool in the neck) and leave for two days. If there is no water bath of the thermostatic kind (one where the temperature is kept constant) put the flask on a radiator or just leave it in the room; in the last case the time required will be longer and you will have to judge by the vinous smell when fermentation is over. Alternatively you may connect a delivery tube and bung to the flask and have the end of the tube dipping into some lime water; this should go cloudy as carbon dioxide is given off.

After fermentation is complete, decant the liquid from any solid matter into a distillation flask. (If several pupils have carried out the experiment take a sample of the liquor from each.) Fit the flask with a condenser and distil, collecting the distillate up to about 95°C. This has now to be redistilled in a flask fitted with a fractionating column. This comprises a tube containing glass beads, small pieces of glazed pot, metal rings or even glass wool and its purpose (as explained when dealing with the fractionation of crude petroleum) is to hold back the constituents with the higher boiling points. Distil again and now collect only the portion boiling below 86°C. This may still contain some water, but it can be dried with a lump of calcium oxide and then redistilled. The liquid obtained should burn.

Fermentation is an example of *degradation*, i.e. the breaking down of a carbon compound into one with less carbon atoms. Here, cane sugar with twelve carbon atoms (or starch with many thousands) is converted into alcohol (two carbon atoms) and carbon dioxide.

Ethanol

Ethanol is indispensable to industry and medicine; in order to prevent its use as a beverage (for which high taxes are payable) it is 'denatured', i.e. mixed with some other substances so as to make it unfit to drink. The usual denaturant used is methanol (methyl alcohol), CH_3OH, which is poisonous. Often a dye is also added and we have methylated spirits.

Ethanol has a vinous smell (i.e. like wine) and is soluble in water in all proportions. It has a very low freezing point ($-112°C$.) and so is used in minimum thermometers. It is suitable for use as a fuel, either alone or blended with petrol. On burning the products are carbon dioxide and water, some of the necessary oxygen coming from the alcohol.

Experiment 51.4 The Oxidation of Ethanol.
(i) Mix 2 cm³ of ethanol with 1 cm³ of concentrated sulphuric acid in a 100 mm × 16 mm tube, add a measure of sodium dichromate(VI) and warm. Notice the penetrating odour; it is that of ethanal formed by the mild oxidation of ethanol.

$$CH_3CH_2OH + [O] \longrightarrow CH_3CHO + H_2O$$
from the dichromate (VI) — ethanal

The solution remaining will be green.

(ii) Warm 1 cm³ of ethanol with 2 cm³ of a solution of potassium manganate(VII) acidified with a little dilute sulphuric acid. The manganate(VII) goes colourless and the sharp odour of ethanoic (acetic) acid can be smelt.

$$CH_3.CH_2OH + 2[O] \xrightarrow{\Delta} CH_3COOH + H_2O$$
ethanoic acid

Look at the equations for these two reactions and then answer the question: 'Which is the stronger oxidant, the dichromate(VI) or the manganate(VII)?'

(iii) Mix 2 cm³ of ethanol with 2 cm³ of ethanoic acid and 2-3 drops of concentrated sulphuric acid and warm. Pour into cold water. Notice the very pleasant smell now produced; it is of ethyl ethanoate, an ester.

$$CH_3.COO[H + HO].C_2H_5 \xrightarrow{\Delta} CH_3.COOC_2H_5 + H_2O$$

(iv) Add a small piece of dried sodium to some ethanol and try to identify any gas evolved. Note the rise in temperature of the liquid. Write a possible equation. When all action has ceased evaporate the solution to dryness: the solid left is called sodium ethoxide. Can you suggest a formula for it?

(v) Add some of the white solid, phosphorus pentachloride, PCl_5 to ethanol and notice the vigorous reaction. A similar action occurs with the solid and water as the hydroxide group in the liquids is replaced by a chlorine atom.

$$C_2H_5OH + PCl_5 \longrightarrow C_2H_5CL + POCl_3 + HCl$$
chloroethane

Hydrogen chloride is given off and fumes in the moist air. This reaction is best carried out in the fume cupboard.

Ethanoic Acid (Acetic Acid)

Organic acids resemble inorganic acids in many of their properties, e.g. they all contain hydrogen replaceable by a metal to form a salt. In the main, they conduct electricity *in solution*, but to a much slighter extent than do mineral acids (see p. 223). They are called 'weak' acids because they possess few hydrogen ions by comparison with mineral acids. They all have a common group of atoms in the carboxylic group, —COOH. Ethanoic acid is $CH_3.COOH$ and it is the last hydrogen atom which is replaced in forming salts such as sodium acetate, $CH_3.COONa$.

It is made from ethanol by letting the alcohol trickle over wood shavings containing bacteria called *mycoderma aceti*. Warm air is blown through and the alcohol is oxidized to the acid. The dilute impure solution of acetic acid obtained in this way is *vinegar*.

Industrially, the acid is made by the atmospheric oxidation of alcohol or ethanal (acetaldehyde) in the presence of a manganese catalyst.

Ethanoic acid possesses the usual properties of mineral acids; it will:
(a) dissolve magnesium with the liberation of hydrogen and the formation of magnesium ethanoate,
(b) dissolve washing soda or marble with evolution of carbon dioxide,
(c) react with bases with the production of a salt and water,
(d) react with alcohols to form esters.
This acid used to be called acetic acid, and ethanoates were called acetates.

Esters

Esters are compounds made by the action of an alcohol on an acid, usually an organic acid.

Ethyl ethanoate can be made on the small-scale as in Experiment 52.4 (iii) above and recognized by its pleasant fruity smell. Many esters occur in and give scent to fruit and flowers and are used in flavourings and perfumes. Some are good solvents for lacquers and others are used in the manufacture of rayon and some plastics. When ethanoic acid acts on cellulose $(C_6H_{10}O_5)_n$ in the presence of concentrated sulphuric acid it forms cellulose acetates which are precipitated on adding water. These are used in making lacquers for car bodies and in the manufacture of aeroplane dope, nail varnish and nail polish remover. Some characteristic odours are indicated in Table 51.2 (do not attempt to learn these names).

Table 51.2. Characteristic Odours of some Esters.

Ester	Odour
Amyl ethanoate (acetate)	pear; banana
Octyl chanoate (acetate)	orange
Ethyl butanoate	banana
Amyl valerate	apple
Methyl anthranilate	grape

Esters are similar in one respect to inorganic salts in that they are formed from an acid with a replaceable hydrogen atom and a substance with a hydroxide group (they are, however, covalent compounds).

$$HCl + NaOH \rightarrow NaCl + H_2O$$
$$HCl + C_2H_5OH \rightarrow C_2H_5Cl + H_2O$$
$$CH_3COOH + C_2H_5OH \rightarrow$$
$$CH_3COOC_2H_5 + H_2O$$
ethyl ethanoate

Many of the esters found in the bodies of animals or the seeds of plants such as the coconut palm are fats or oils and are compounds formed from glycerol (propane-1,2,3-triol) (glycerine, an alcohol) and 'fatty' acids such as stearic, palmitic, oleic, valeric or butanoic. Butter is a mixture of such esters, the numbers of carbon atoms varying from 4 to 18. Most animal fats are 'saturated' compounds and are solid at ordinary temperatures whereas vegetable oils are 'unsaturated' and are liquids. (If you have forgotten what these terms mean look back a few pages.) Liquid vegetable oils may be changed into solid fats by heating with hydrogen in the presence of powdered nickel as a catalyst and this is done in the production of margarine and various cooking fats.

Oils, like fats, are complex mixtures of glyceryl esters, the *main* constituents of some being shown in Table 51.3.

When esters are boiled with water they are reconverted into the alcohol and acid from which they were made. If sodium or potassium hydroxide

Table 51.3 Glyceryl Esters in Oils and Fats.

Fat or Oil	Main Constituents	
Soya bean oil	Glyceryl	linoleate
Sunflower oil	Glyceryl	oleate
Cottonseed oil	Glyceryl	
Butter	—	butanoate
Lard	—	palmitate
Beef dripping	—	stearate
Olive oil		
Almond oil	—	oleate
Ground nut oil		
Castor oil	—	ricinoleate
Linseed oil	—	linoleate
Coconut oil	—	oleate, myristate, laurate
Palm oil	—	oleate, palmitate

is used instead of water this *hydrolysis* (another example of degradation) occurs more readily, the sodium or potassium salt of the acid being formed. If the ester is a fat or oil the sodium or potassium salt can be used as a soap.

Just as hydrocarbons may polymerize to form compounds with higher molar masses—polymers—so may some other types of organic compounds including esters. One particularly important long-chain polyester is terylene which has a formula continually repeating the unit.

$$-OC.C_6H_4.COO.CH_2.CH_2.OOC.C_6H_4.CO-$$

Soap

Soap, of a sort, has been known since biblical times and was reported by the Romans as being used by the Huns in central Europe. Although the import of olive oil into southern Europe led to the production of soap in Marseilles by the early middle ages, it still remained a luxury. It was made mainly in rural districts or large houses. All forms of grease were saved during the year and in the soap-making 'season' this grease was boiled in soap-kettles with lye or liquor obtained by mixing the ashes from wood fires with rain water. ('*Alkali*' means 'the ashes' and refers to the liquor obtained by this method, weak potassium carbonate solution, an alkali.) It was said of Queen Elizabeth I that she had a bath every three months whether she needed it or not! In the eighteenth century, having baths became fashionable amongst those who could afford servants to do the fetching and carrying, but it was not until the middle of the last century that soap began to be manufactured on the large scale. In England it was classified as one of the 'objectionable' trades and soap factories were mainly set up outside city limits, e.g. in Stratford on the eastern edge of London.

The essential method by which soap is produced consists of steam heating a fat or oil with sodium hydroxide until the soap is formed along with glycerol. A saturated solution of salt—in which the glycerol is soluble but not the soap—is added and thoroughly mixed. The soap coagulates and rises to the top, the glycerol and brine being drawn off at the bottom; from this the glycerol is recovered by distillation. The soap is washed and dried and mixed at the right stage with dyes and perfumes.

Glyceryl stearate + sodium hydroxide $\xrightarrow{\Delta}$
(fat) (alkali)

sodium stearate + glycerol
(soap) (alcohol)

Experiment 51.5. Soap Making.

Obtain a lump of dripping or lard about the size of a walnut (or use 10 cm³ of cooking oil) and place it in a 100 cm³ flask along with 25 cm³ of dilute sodium hydroxide solution and 25 cm³ of ethanol (this is only to help the solution and can be recovered later). Fit a condenser as in Fig. 51.3 and heat. This process, whereby none of a mixture boils away, is known as *refluxing*; any alcohol which vaporizes is condensed again and returned to the flask. After about 30 minutes—or when the fat has completely dissolved—remove the condenser and fit the flask for distillation. Heat and collect the portion which comes over between 78°–82°C. This is the alcohol. (If it is not desired to collect the alcohol, it can all be driven off after refluxing is complete by running the water out of the condenser and heating for a further 15 minutes.)

Allow the solution in the flask to cool and then pour it into about 100 cm³ of saturated brine. Notice the white precipitate of soap produced. Fit up a filter flask and Buchner Funnel (Fig. 43.9, p. 236) and connect to the filter pump. Pour the soap suspension on to the paper. The glycerol and brine pass through and the crude soap remains.

The soap can be refined by transferring about 10 g of it to a beaker, adding 20 cm³ of boiling distilled water plus 2 drops of methyl violet, pouring again into brine and once more filtering. Add a few drops of some perfume such as oil of lavender, mould into shape and dry.

Soap made from dripping is mainly sodium stearate, from lard, sodium palmitate and from olive oil, sodium oleate. Sodium stearate is a salt of stearic acid, $C_{17}H_{35}COOH$ (octadecanoic acid). Hence soap may be called sodium octadecanoate.

Soaps made using sodium hydroxide are known as 'hard soaps'; if potassium hydroxide is used we get a 'soft soap'.

Proteins

Protein is the name given to those substances which form the main bulk of living tissue and which, along with fats and carbohydrates, make up almost the entire structure of all living creatures. Proteins all contain carbon, hydrogen, oxygen and nitrogen; most of them also have sulphur and possibly phosphorus as one of the combining elements present. They all have very high molar masses (this value may be as much as five million), but can be broken down to simpler substances called aminoacids. Proteins may be regarded as polymers of aminoacids.

The simplest aminoacid is glycine, formed by replacing one of the hydrogen atoms in the CH_3 group of ethanoic (acetic) acid by the NH_2 group; it is also called aminoethanoic acid.

$CH_3.COOH$ $CH_2(NH_2).COOH$
ethanoic   glycine

The simplest protein would be formed by joining two molecules of glycine as shown here:

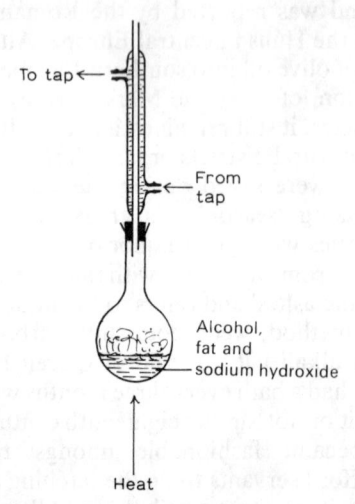

$$\begin{array}{ll} CH_2.COOH & + \quad HNH \\ | & \quad\quad\quad | \\ NH_2 & \quad\quad CH_2.COOH \end{array}$$

Fig. 51.3. Refluxing in preparation of soap

$$\begin{array}{cc} CH_2.CO.NH \\ | & | \\ NH_2 & CH_2.COOH + H_2O \end{array}$$

Examples of complex proteins are albumin (in egg-white), haemoglobin—MM approximately 68 000—(in the blood), keratin (in the hair), caseinogen (in milk), chlorophyll (in green plants), fibroin (in silk), gelatin.

Experiment 51.6. Reactions of Proteins.
Stir some egg-white vigorously with about four times its volume of distilled water and filter through muslin (or use a solution of gelatin).

(*a*) To 5 cm³ of this solution add an equal volume of dilute sodium hydroxide solution and then a few drops of aqueous copper(II) sulphate. A violet colour is produced (this is called the *biuret* test).

(*b*) Boil 5 cm³ of the solution with 2 cm³ of concentrated nitric acid, cool in water and add ·880 ammonia (in a fume cupboard). A yellow-orange colour appears.

(*c*) Heat 5 cm³ of the solution with an equal volume of concentrated hydrochloric acid plus a few pieces of porous pot to prevent 'bumping'. (It is better if the mixture can be heated under reflux—see Fig. 51.3—for half an hour.) Cool and dilute with an equal volume of water. Try tests (*a*) and (*b*) on this solution. You should find that the protein has been broken down into simple aminoacids which do not respond to these tests.

Carbohydrates

Mention of this important class of foods has been made on p. 151 in connexion with respiration; you should read this section again. Carbohydrates all contain carbon, hydrogen and oxygen as the combining elements and the number of atoms of hydrogen in each molecule is always twice as many as the atoms of oxygen. A simple formula would be $C_x(H_2O)_y$. The ones we most often meet are glucose and dextrose which both have the formula $C_6H_{12}O_6$ (they are isomers), sucrose (cane sugar) and maltose—both $C_{12}H_{22}O_{11}$ and cellulose and starch, $(C_6H_{10}O_5)_n$. These last are giant molecules with extremely high molar masses of the order of one million. They are, of course, all covalent compounds and do not conduct electricity when dissolved in water.

The large starch molecules can be broken down into glucose molecules by boiling with dilute hydrochloric acid or by digesting with saliva (you may already have noticed that if you chew bread for a long time it begins to taste sweet). As already explained on p. 329 glucose can be further broken down by enzyme fermentation to ethanol and carbon dioxide.

Exercise 51

1 (*a*) Give the names of substances with the following formulae and say what you understand by the term 'hydrocarbon'; CH_4, C_2H_6, C_2H_4.

(*b*) We talk about the distillation of water and the *fractional* distillation of crude oil: what is the difference between the two processes and why do we not fractionally distil water?

(*c*) What do you understand by the term 'cracking' as applied to the petroleum industry? Why is the process of cracking done and how is it carried out? (*EM*)

2 (*a*) What do you understand by the term 'carbohydrate'? Name two sources of cane sugar.

(*b*) Give the formulae for ethanol (ethyl alcohol) and ethanoic (acetic) acid. Name two sources of starch in the diet.

(*c*) What two products are obtained if a fat is boiled with sodium hydroxide solution? Name (i) a solid fat, (ii) a liquid fat. (*EM*)

3 (*a*) Which of the following phrases explains what is meant by a 'saturated' organic compound? Has been soaked in water—contains only single bonds—cannot be concentrated any further—contains double bonds—is a liquid.

(*b*) Give one use for polyethene (polythene). Complete the following sentence: 'Polyethene is made by joining together .... and this is done by applying great .... (*MR*)

4 How would you prepare a sample of soap in the laboratory? What are the differences between a common household soap and a good toilet soap? What is soft soap? (*WR*)

5 Draw a simple diagram to illustrate the preparation of ethene from ethanol and concentrated sulphuric acid. Write the formula for ethanol and explain the purpose of (i) the sulphuric acid and (ii) the liquid in the wash bottle through which the ethene is passed. At what temperature is the mixture in the flask kept? (*MR*)

6 Explain how you would test a substance to find out if it contained both carbon and hydrogen.

7 What products form when sugar ferments? Explain how this fermentation process is made use of in baking.

Describe how you could prepare a sample of soap in the laboratory from a fat. State what fat you would use. (*SR*)

8 Explain briefly, with a diagram if it will help you,

how crude oil is fractionally distilled. List the main products which are obtained by the first fractionation of the oil, and from any three products you mention go on to say for what purposes they are used.

What are (a) natural gas, (b) coal gas? In each case say briefly how they are obtained, and what gases they are composed of chiefly. (SR)

9 Explain briefly how, using cane sugar, you would carry out the process of fermentation.

Draw a fully labelled diagram of the apparatus you would use to obtain some ethanol (ethyl alcohol) from the fermented liquor.

Wine will 'go sour' if it is left exposed to the air for some time.

Explain (a) what is formed when the wine goes sour, and (b) what it is that causes the wine to turn sour.

10 Describe briefly how you would show that a given organic compound contains (a) carbon and (b) hydrogen.

Use a diagram or diagrams if they help you.

What do you understand by the term 'ester'?

Explain what is meant by a detergent, and say how a detergent acts in the washing of clothes.

Explain briefly how rayon is made. (EM)

11 The following are all changes involving compounds of carbon.

(i) Changing starch into sugar by the action of hydrochloric acid.

(ii) Changing sugar into alcohol by the action of yeast.

(iii) Changing polystyrene into styrene by distillation.

(a) Which one of these changes takes place in the body?

(b) Which one of these changes is part of the process of making wine?

(c) The reverse of one of these changes is an important industrial process. Which one is it? (WR)

12 A solid hydrocarbon $C_{20}H_{42}$ was heated in a tube sufficiently to vaporize it. The vapour of this compound was passed over red hot porous pot and eventually a mixture of gases was collected over water. This mixture was found to decolourize bromine water.

(a) To which family of compounds does the hydrocarbon $C_{20}H_{42}$ belong and what is the general formula for this family?

(b) What evidence is there that the solid hydrocarbon has broken down to simpler molecules?

(c) What name is given to reactions in which larger molecules are broken down into smaller molecules by heating in the absence of air?

(d) Name one gas which could be present in the gas mixture and give its formula.

(e) Name the compound with high molecular mass formed when molecules of the gas named in (d) are linked together chemically.

(f) What is the name given to the process whereby molecules of a substance of low molecular mass are repeatedly linked with themselves to form a new substance with a high relative molecular mass? (L)

13 Some crude oil was heated in a side-arm tube and the distillate passed into a tube surrounded by cold water. Two mixtures, D and E, were collected. D boiled in the temperature range 20°C to 70°C whilst E boiled from 170°C to 220°C.

(a) What is the name given to the process used to separate a liquid into mixtures of different boiling points?

(b) What is the purpose of the beaker of water?

(c) Give (i) two differences, other than boiling point, and (ii) two similarities, between D and E. These differences and similarities may be either physical or chemical. (L)

14 (i) A certain volatile liquid is found to have the following approximate percentage composition by mass: carbon 14·1%, hydrogen 2·35%, chlorine 83·5%. Calculate its simplest formula. Given that its relative molecular mass (molecular weight) is 85, what is its molecular formula? Make a diagram showing the structure of its molecule.

(ii) The hydrocarbon propane, C_3H_8, burns in a good supply of air according to the following equation

$$C_3H_8 + 5O_2 \rightarrow 3CO_2 + 4H_2O.$$

What is the minimum volume of oxygen needed for this reaction to take place, given that the original volume of propane is 56 cm³?

What volume of carbon dioxide would be produced? (Assume that all gas volumes are measured under s.t.p. conditions.)

Calculate also, the mass of water formed in this case.

What changes would occur in the above reaction if the supply of oxygen was restricted?

15 Methane is sometimes liberated from fissures in coal seams during the mining of coal, and if the ventilating system of the mine is not adequate an explosion sometimes occurs. Give an equation that accounts for the formation of carbon monoxide from methane and the air in the mine when this happens. Ethene, C_2H_4, is another gaseous hydrocarbon. Write the equation for its complete combustion in air.

What would be the composition and volume of the gas left if a mixture of 8 cm³ of ethene and 150 cm³ of air were exploded in a closed vessel by a series of sparks. Assume that the gas is completely oxidized, also that all measurements are made at atmospheric pressure and temperature and that the air contains one-fifth of its volume of oxygen.

16 Make a clearly labelled diagram of the apparatus you would use for collecting a few cubic centimetres of water formed when ethanol burns in air, using a spirit lamp. How would you prove (a) that the liquid collected was water, (b) that it had been obtained from the combustion of ethanol, and not from the air?

What other product is formed when ethanol burns freely in air? Describe and explain a chemical test by means of which you could detect this.

Write an equation for the complete combustion of propane in oxygen.

How much oxygen would be needed to burn completely 50 cm³ of propane? (C)

Miscellaneous Questions

1 (a) Name the metallic elements present in: (i) limestone, (ii) washing soda, (iii) steel, (iv) common salt.

(b) What is the meaning of $2NH_4OH$ and how many atoms of hydrogen does the expression contain?

(c) Give the symbols for the four ions present in copper(II) sulphate solution.

(d) Name:
 (i) A gas which is used in the purification of water supplies,
 (ii) a poisonous gas which occurs in the exhaust fumes of a car,
 (iii) a gas which forms about 11% by weight of water.

(e) What is the gram molecular mass of sulphuric acid?

(f) A hydrometer was found to read 1·120 when floated in the acid from a car battery.
 (i) What does this figure represent?
 (ii) What does it indicate about the condition of the battery?

(g) Arrange the following temperatures in rising order of hotness: $10°C$, $10K$, $10°F$.

(h) What special properties have (i) the material making the filament, (ii) the gas in the bulb, of an electric lamp?

(i) Show by a diagram the least height a mirror must be for a man standing upright to see himself from head to feet, the mirror being vertical. (SR)

2 (a) State the law of constant composition.

(b) Describe or explain the meaning of the following terms: *alkali; reduction; sublimation; diffusion; catalyst.*

(c) Complete the following equations:

$$ZnCO_3 + 2HNO_3 =$$
$$Mg + H_2SO_4 =$$
$$NaOH + NaHCO_3 =$$
$$ZnSO_4 + 2NaOH =$$

(d) Give the *names* of (i) **two** gases which burn readily in air, (ii) **one** gas which is at least twice as dense as air, (iii) the **two** most abundant elements in the earth's crust, (iv) **two** gases present in air besides oxygen, nitrogen, and water vapour, (v) **two** common alloys of which copper is an important constituent. (OC)

(a) Give the name and formula in each case of (i) an acid salt, (ii) a dibasic acid, (iii) an acidic oxide, (iv) an alkali, (v) an oxidizing agent (other than oxygen).

(b) Calculate the percentage of sulphur by mass in the compound ion(III) sulphate $[Fe_2(SO_4)_3]$.

(c) A compound containing hydrogen, carbon and oxygen has the following percentage composition by mass: $H = 4·35$, $C = 26·09$, $O = 69·56$. Calculate the empirical (i.e. simplest) formula for the compound. (J)

4 (a) State (i) the law of multiple proportions, (ii) the law of constant composition.

(b) Define or explain the meaning of the following terms: *molecule; atomic mass; precipitation; salt; cathode.*

(c) Write down equations for the reactions between the following substances and give the **names** of the products:
 (i) calcium oxide and nitric acid;
 (ii) calcium carbonate and hydrochloric acid;
 (iii) carbon dioxide and carbon.

(d) Write down (i) the **name** of **one** salt present in permanently hard water, (ii) the **names** of the two chief constituents of producer gas, (iii) the atomicity of the nitrogen molecule, (iv) the **name** of **one** metal the hydroxide of which is insoluble in water, but soluble in both acids and alkalis. (OC)

5 Describe what you would observe, and say what substances are formed, when each of any **four** of the following compounds are heated separately in ignition tubes:
(a) ammonium chloride,   (d) lead(II) carbonate,
(b) sodium carbonate crystals,   (e) zinc nitrate,
(c) potassium chlorate(V).   (f) sodium nitrate. (C)

6 Explain briefly why:
(a) Limestone becomes lighter when strongly heated,
(b) Iron becomes heavier when it rusts,
(c) Moist chlorine bleaches,
(d) Chlorine is given off when a mixture of common salt, manganese(IV) oxide and sulphuric acid is heated,
(e) Silverware tarnishes in ordinary air. (L)

7 You are provided with ammonium chloride, concentrated sulphuric acid, manganese(IV) oxide and sodium hydroxide, but no other chemicals except water. Using ordinary laboratory apparatus, but not a battery, outline methods by which you could prepare **three** different gases. (J)

8 (a) What is the minimum mass of zinc needed to prepare 10 g of crystalline zinc sulphate $(ZnSO_4.7H_2O)$?

(b) On complete combustion of the gas propene (C_3H_6), carbon dioxide and water are formed according to the equation

$$2C_3H_6 + 9O_2 = 6CO_2 + 6H_2O.$$

335

What is the minimum volume of oxygen required to burn completely 20 cm³ of propene, and what is the volume of carbon dioxide produced, all volumes being measured at the same temperature and pressure?

(c) Sketch and label the apparatus you would use to purify water by distillation. (OC)

9 Explain why:
(a) lime water is used to test for carbon dioxide;
(b) anhydrous cobalt chloride can be used to detect the presence of water;
(c) a solution of chlorine in water becomes turbid when hydrogen sulphide is passed into it;
(d) drops of water form on the outside of a beaker containing cold water that is heated by a bunsen flame;
(e) washing soda can be used to soften water that exhibits permanent hardness. (L)

10 Describe what you would observe, and say what reactions would occur in **four** of the following cases:
(a) calcium metal is dropped into cold water;
(b) drops of water fall upon a lump of calcium oxide;
(c) a lump of 'fused' calcium chloride is left exposed to the air for a day or two;
(d) a large crystal of sodium carbonate is left out in the air for several days;
(e) a solution of sodium carbonate is mixed with calcium chloride solution. (C)

11 Explain **four** of the statements below:
(a) Sugar may be obtained in the following manner:
   (i) crushing sugar beet (or sugar cane),
   (ii) stirring the crushed material with water,
   (iii) straining off the liquid,
   (iv) heating this liquid for a time,
   (v) allowing the liquid to cool.
(b) When drops of water fall on a lump of quicklime, the lime becomes hot and soon crumbles.
(c) Limestone hills often contain caves.
(d) Concentrated sulphuric acid may be used for drying hydrogen chloride but not ammonia; quicklime may be used for drying ammonia but not hydrogen chloride.
(e) Iron pyrites (FeS_2) is often used in preparing one of the gases from which sulphuric acid is manufactured.

12 (a) Write equations for the following reactions, and name the products:
   (i) The action of dilute hydrochloric acid on iron.
   (ii) The action of sodium on water.
   (iii) The reaction between hydrogen sulphide and copper(II) sulphate solution.
   (iv) The action of dilute nitric acid on copper oxide.
(b) Calculate the percentage by mass of iron in crystalline iron(II) sulphate ($FeSO_4.7H_2O$).
(c) When the gas methane (CH_4) is burnt, carbon dioxide and water are formed according to the equation

$$CH_4 + 2O_2 = CO_2 + 2H_2O.$$

What is the minimum volume of oxygen required to burn completely 10 cm³ of methane, and what is the volume of carbon dioxide produced (all volumes being measured at the same temperature and pressure)?

(d) Sketch the apparatus you would use to purify water by distillation. (OC)

13 Explain as fully as you can **four** of the following:
(a) Calcium chloride crystals on heating and then cooling form a shapeless mass, but the solid remaining, on exposure to air, turns slowly to a colourless liquid.
(b) A little lime is often added to hard water to soften it, but too much makes the water hard again.
(c) Moist sulphur dioxide bleaches straw, but on exposure to air and sunlight the straw turns yellow again.
(d) In one process for manufacturing sulphuric acid, sulphur dioxide and oxygen are passed over hot platinum, spread as a fine powder over asbestos.
(e) Limestone is part of the 'charge' fed in at the top of the blast furnace when smelting iron. (C)
[NO diagrams are required.]

14 Give the chemical name of each substance underlined below. No other information is required.
(a) White crystals which yield oxygen on heating and leave a residue. When added to silver nitrate solution, this residue gives a white precipitate, insoluble in dilute nitric acid.
(b) A white powder which sublimes on heating alone, and which yields a pungent alkaline gas when mixed and warmed with damp slaked lime.
(c) A colourless gas with a characteristic smell which gives a dark brown precipitate when it is bubbled through lead nitrate solution.
(d) Pale green crystals which, on heating, first give steam, then white, pungent, acidic fumes, leaving a red powder. (S)

15 (a) State (i) the law of Avogadro, (ii) the law of multiple proportions.
(b) Define or explain the meaning of the following terms: sublimation; molecule; diffusion; precipitation; oxidation.
(c) Give the chemical names and formulae of the following substances: quicklime; washing-soda; sal-ammoniac; marble.
(d) Arrange the following elements in the order of their relative abundance in the earth's crust, putting the most abundant element first: aluminium; potassium; oxygen; iron. (OC)

16 Account for the formation of:
(a) an inflammable gas when zinc dust is treated with sodium hydroxide solution;
(b) a pungent smelling gas when ammonium chloride is treated with sodium hydroxide solution;
(c) a crystalline precipitate when carbon dioxide is passed into a cold, fairly concentrated solution of sodium hydroxide;
(d) a reddish-brown precipitate when sodium hydroxide solution is added to iron(III) chloride solution. (L)

17 (a) Define *molecular mass* and *molecular formula*.
(b) Sodium carbonate crystals ($Na_2CO_3.10H_2O$) on exposure to the air slowly give off water of crystallization, forming the monohydrate ($Na_2CO_3.H_2O$). What mass of the finely powdered crystals will, if left long enough, lose 72·9 g of water?

(c) Ammonia is manufactured by making nitrogen combine with hydrogen according to the equation

$$N_2 + 3H_2 = 2NH_3.$$

What volume of hydrogen, measured at s.t.p., would be used if 4 g of nitrogen entered into combination? Write out the law which is illustrated by this reaction. (C)

18 Describe in outline (a) how hydrogen is manufactured, starting from coke and water; (b) how nitrogen is obtained from the air; (c) how hydrogen and nitrogen are combined to give ammonia. (Diagrams and technical details of the plants involved are **not** required.)

Some ammonia is absorbed in 100 cm³ of molar hydrochloric acid. It is found that 20 cm³ of molar sodium hydroxide are needed to neutralize the excess acid. How many grams of ammonia were absorbed? (C)

19 Describe fully how you would determine the relative density of a given liquid; state briefly how an increase of temperature would affect the relative density of (i) ice cold water, (ii) water at 15°C.

A stone having a volume of 25 cm³ has a mass of 120 g in air. What would be its mass when totally immersed in (a) water, (b) sea water of density 1·02 g/cm³? What is the relative density of the stone? (O)

20 State the principle of the conservation of energy. Trace the energy changes which take place in **three** of the following cases:

(a) a pendulum bob is pulled to one side and then allowed to swing;

(b) an athlete jumps over a bar in a high jump;

(c) a steam train on a level track starts from rest and is later brought to rest by the application of its brakes;

(d) an electric bulb connected to the electricity mains is used to light a room. (O)

21 (a) Explain the use of wire gauze in a miner's lamp, or a car radiator heater.

(b) A glass beaker containing some ether is placed on a wet bench. Describe and explain what happens when air is blown through the ether.

(c) Explain how an eclipse of the sun occurs, showing on a diagram the regions of total and partial eclipse. (L)

22 Answer any *two* of the following:

(a) Define *specific latent heat of fusion* and describe an experiment to determine its value for ice.

(b) Describe a clinical thermometer. Point out, and give reasons for, particular features in its construction.

(c) Define coefficient of linear expansion for a solid. 'When expansion or contraction of a metal is resisted, a considerable force is exerted.' Describe one instance in which this fact is made use of, and one in which it has to be allowed for. (L)

23 (a) State two ways in which boiling is different from evaporation and describe an experiment to show that a liquid loses heat during evaporation.

(b) What is meant by the statement that heat is a form of energy?

Under what conditions does a body possess (i) potential energy, (ii) kinetic energy? In each of these cases, state the formula by which the amount of energy may be calculated.

Describe a simple experiment to show that mechanical energy may be transformed into heat energy. (L)

24 Explain the scientific principles involved in each of the following statements:

(i) water pipes may burst during very cold weather;

(ii) snow continues to lie on the ground after a heavy fall, even though the air is quite warm;

(iii) a flask containing boiling water is sealed, and the water inside continues to boil even when cold water is poured over the outside of the flask;

(iv) the pipe connected to the top of a domestic hot-water storage tank feels hot but the pipe connected to the bottom of the tank feels cold.

(O)

25 Explain the scientific principles involved in each of the following:

(i) warming the neck of a glass bottle to release the glass stopper that has jammed;

(ii) the apparent bending of a stick that is partly immersed in water;

(iii) using an ammeter for measuring larger currents than those shown on its scale, by linking a suitable resistance across its terminals. (O)

26 State the laws of reflection and the laws of refraction of light.

Explain by ray diagrams (a) the appearance of a straight rule when placed slantwise into water so that part only is submerged, (b) the production of a virtual image by a concave mirror. (J)

27 (a) State a formula for the heating effect of a current in a wire.

A 2 kW electric fire is used on a 240 V mains circuit. What current does it take? If it is used on an average, eight hours a day, find the weekly running cost. (Cost of electricity is 2p a unit.)

(b) Given a U-shaped piece of soft iron and a coil of insulated copper wire, draw a diagram to show how you could make an electromagnet having a N and S pole.

Show on your diagram (i) the direction of the current, (ii) the method of winding, (iii) the resulting polarity. (L)

28 Describe a simple experiment to demonstrate (a) the magnetic effect, and (b) the chemical effect of a current, stating in each instance the effect you would expect to obtain.

Explain with the aid of a diagram, the construction of any piece of apparatus which depends on the magnetic effect. (L)

(a) Which of the following have (i) molecular structures, (ii) ionic structures: ethanol, carbon dioxide, sodium sulphate, propane, ammonia, magnesium oxide? Illustrate your answer by structural formulae.

(b) Give **two** tests by means of which you could distinguish between the following liquids: (i) a solution of ethanol in water, (ii) a dilute solution of sulphuric acid.

(c) How, starting with lead(II) nitrate (lead nitrate) crystals, would you prepare (i) lead(II) oxide (lead monoxide), (ii) lead(II) sulphate (lead sulphate)?
(C)

30 In an experiment to determine the number of g-atoms of silver precipitated by 1 g-atom of copper, 0·32 g of copper powder was added to 10 cm³ of a warm solution of silver nitrate. The temperature of the silver nitrate increased, the solution boiled and a precipitate of silver was formed in the test-tube. This precipitate was washed, first with water and then with acetone before the test-tube was stood in a beaker of boiling water. Upon weighing the precipitate of silver weighed 1·08 g.

(a) What fraction of a g-atom of (i) copper, and (ii) silver, were involved in the reaction? (1 g-atom of copper has a mass of 64 g, and 1 g-atom of silver has a mass of 108 g.)

(b) How many g-atoms of silver would be precipitated by one g-atom of copper?

(c) Write the ionic equation for the reaction.

(d) Give, and explain, the colour change which the solution undergoes during the reaction.

(e) Why did the solution boil?

(f) Why was the precipitate of silver washed with acetone and then stood in a beaker of boiling water?

(g) Give the name for this type of chemical reaction.

(h) Name a suitable metal and solution which could be used to determine, in a similar manner, the number of g-atoms of lead formed from 1 g-atom of a metal.

(i) Name one metal which could *not* be used in (h) and give a reason for your answer. (L)

31 (a) State **three** ways in which chemical changes differ from physical changes.

(b) What type of change occurs in the following examples? In **each** case give **one** reason for your choice

(i) The effect observed upon removing the stopper from a bottle of effervescent ('fizzy') drink.

(ii) Striking a match.

(iii) Dissolving sugar in tea.

(iv) Adding chalk to dilute hydrochloric acid.

(v) Obtaining pure water from sea water.

(c) How would you separate ammonium chloride from sodium chloride given a finely powdered mixture of each? (C)

32 Explain the following observations

(a) dilute sulphuric acid has little action with marble, but it reacts violently with copper carbonate,

(b) concentrated sulphuric acid turns sugar (i.e. $C_{12}H_{22}O_{11}$) black,

(c) when concentrated sulphuric acid is heated with sulphur, a poisonous gas with a pungent odour is produced,

(d) copper(II) sulphate crystals turn white when heated,

(e) steam is produced when water is added to quicklime,

(f) drops of colourless liquid condense at the mouth of a test-tube in which sodium hydrogencarbonate is being heated. (C)

33 The following table shows some properties of six elements. (The letters used are not the symbols for the elements.)

Element	Relative Atomic mass	Melting Point °C	Boiling Point °C	Mass of 1 cm³ at 0°C
A	200	−39	357	13·6 g
B	56	1539	2887	7·90 g
C	27	659	1997	2·70 g
D	32	113	444	2·07 g
E	4	−269	−268	0·00017 g
F	32	119	444	1·96 g

(a) In answering this question a letter may be used once, more than once, or not at all.
Write down the letters corresponding to

(i) one element which is a solid at room temperature,

(ii) an element which, for a given mass, has a fixed volume but no fixed shape, at room temperature,

(iii) the element whose molecules are most widely spaced at 0°C,

(iv) *two* polymorphic forms of the same element,

(v) an element which would change its physical state if cooled from room temperature to −50°C.

(b) (i) What is the volume of 1 g-atom of element C?

(ii) How many atoms are there in 2 g of element E? (1 g-atom of any element contains 6 × 10²³ atoms.) (L)

34 (a) Under what conditions and with what results does water react with (i) sodium, and (ii) carbon?

(b) The apparatus in the diagram below was used to discover the number of g-atoms of hydrogen combining with one g-atom of oxygen.

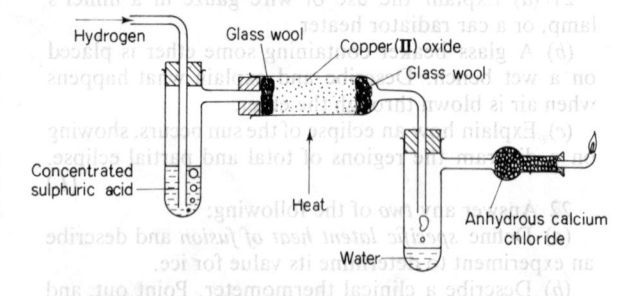

(i) Why was the hydrogen passed through concentrated sulphuric acid?

(ii) Give the colour change, the equation and the name of the type of reaction taking place as the hydrogen passed over the hot copper(II) oxide.

(iii) Why was the glass wool placed in the reaction tube?

(iv) What precaution should have been taken before igniting the excess hydrogen?

(v) What was the purpose of the anhydrous calcium chloride?

(c) In the above experiment 3·24 g of copper(II) oxide

gave 2·60 g of copper, and 0·72 g of water was collected. (1 g-atom of hydrogen has a mass of 1 g, and 1 g-atom of oxygen has a mass of 16 g.)

How many grams, and g-atoms, of
(i) oxygen had been removed from the copper(II) oxide,
(ii) hydrogen had combined with the oxygen to form the water?

How many g-atoms of hydrogen had therefore combined with one g-atom of oxygen? (L)

Answers to Numerical Questions

Atomic Structure etc. p. 2
1. 11
2. 68, 39, 45, 21, 32, 270, 44, 42, 40

Measuring Length etc. p. 10
2. Glass volume = 8 cm^3, Density = 2·5 gm./cm^3
   Water volume = 0·25 m^3, relative density = 1
4. Relative density = 4
5. Density = 0·89 g/cm^3
6. Density = 0·84 g/cm^3
7. (a) 3·5 cm^2, (b) 165·9 g
8. 1·3 g/cm^3

Time, Velocity and Acceleration p. 16
1. 200 m/s
2. (a) 96 m, (b) 5$\frac{1}{3}$ m/s
3. 54 m
4. 1·3 m/s^2, 40 m/s
5. 30·1 m/s, 6·3 s

Forces and Motion p. 20
1. 2780 N
2. 50 m/s
3. 0·22 m/s^2, 52 s, 11·6 m/s, 2320 N

Molecular Properties p. 37
3. 330 cm^3

Pressure in Liquids and Gases p. 44
1. Greatest pressure = 15·7 kN/m^2
2. Pressure = 392 N/m^2 Thrust = 7·9 N
4. 7812 N
6. 129 kN/m^2
7. 10^5 N/m^2

Chemical Arithmetic p. 48
1. H 1·59%, N 22·22%, O 76·19%
2. K 8·2%, Al 5·8%, S 13·5%, O 26·9%, H$_2$O 45·6%
3. Na 58·97%, N 35·90%, H 5·13%
4. Mg 11·82%, Cl 34·98%, H$_2$O 53·20%
5. Fe 20·14%, S 11·51%, O 23·02%, H$_2$O 45·33%
6. K 24·68%, Mn 34·80%, O 40·52%
7. Mg 34·29%, Si 20·00%, O 45·71%
8. Na 17·04%, S 47·40%, O 35·56%
9. H 0·143 g, S 2·286 g, O 4·571 g
10. C 2·446 g, H 0·306 g, O 1·630 g, Cl 3·618 g
11. Na 4·423 g, H 0·192 g, S 6·154 g, O 9·231 g
12. N 1·05 g, H 0·30 g, Cl 2·65 g
13. N 1·217 g, H 0·434 g, S 2·783 g, O 5·566 g
14. C 6·546 g, H 1·091 g, O 4·363 g
15. POCl$_3$
16. Fe$_3$O$_4$
17. CaH$_2$C$_2$O$_6$ [Ca(HCO$_3$)$_2$]
18. Li$_2$S$_2$O$_3$
19. FeSO$_4$.7H$_2$O
20. C$_6$H$_{14}$O$_6$
21. MgCl$_2$.6H$_2$O
22. 5
23. C$_{12}$H$_{22}$O$_{11}$
24. C$_5$H$_7$
25. CH$_2$O; C$_6$H$_{12}$O$_6$
26. (a) (i) CH$_2$O, (ii) C$_2$H$_4$O$_2$, (b) 2·8 dm^3, 30·75 g

Equations etc. p. 51
1. 10 g
2. 1200 kg
3. 4·91 g, 10·24 g
4. 8·5 g, 3·5 g
5. 44·4 g, 40·5 g
6. 4·0%
7. 54·3 g, 78·6 g, 26·4 g
8. 2·88 g
9. 7·6%
10. 2·47 g
13. 10, 62·9 kg
14. (a) 0·50 g, (b) 66·67%
15. 0·1 mole, 1·62 g
16. (a) 10·2 g, 6·2 g, (b) 1·12 dm^3
18. (a) 4·03 g, (b) 45·3%, (c) 77·2%

339

19 21·9 g HCl, 72 g NaHSO$_4$
20 (i) 1·74 g, (ii) CuCl$_2$, (iii) 1·12 dm^3, (iv) 36·1%
21 (a) 5·4 g, (b) 44%
22 (a) Na 32·4%, S 22·5%, O 45·1%, (b) CH$_4$, 22 g
23 (a) 257·6 g, (b) XSO$_4$.5H$_2$O
24 (a) 11·6 g, (b) 5 g

Archimedes' Principle etc.   p. 56
1 (a) 5000 kg/m^3, (b) 4·9 kN/m^2
  (c) 980 N/m^2, (d) 3·92 N
2 (a) 1·84 N, (b) 19 × 10^{-5}m^3, (c) 4200 kg/m^3
3 764 Kg
4 0·92 g/cm^3
5 6·25 cm
6 4·375 cm^3
7 Paraffin = 0·8, solid = 6·0
8 0·5 m
9 Volume = 8cm^3, Mass = 7·2 g, Weight = 0·072N, Apparent weight = 0·428N
10 Mass of cork = 7·5 g, Mass of water = 7·5 g, Mass of salt solution = 7·5 g, Volume = 6·25 cm^3.

Statics   p. 64
1 90 g
3 63 N
4 735 N; 980 N
5 43 N

Sulphur   p. 73
6 10 kg
13 326·5 g

Machines Work, Energy and Power   p. 84
1 (i) 588 J, (ii) 600 J, (iii) 12 J
2 M.A. = 2·9, V.R. = 4, Efficiency = 72%
3 23%
4 (a) 78·4 kJ, (b) 325 W, (c) 1·62 kW
5 (a) 90%, (b) 198 W
6 5·5 W
7 1960 J of potential energy
9 170 N, 56%

Expansion of Solids and Liquids   p. 90
4 0·000012°C
5 0·000012°C
6 48 cm

Expansion of Gases   p. 96
1 830 cm^3
2 70 cm mercury, 63·6 cm mercury
3 3 atmospheres extra
4 12·0 dm^3
5 1·319 g
6 10 strokes
7 1·8 m^3

Heat and Mechanical Energy   p. 105
3 0·23°C

Quantities of Heat   p. 108
1 (i) 24 J, (ii) 7200 J, (iii) 18°C
2 1890 J/kg°C

3 2·8p
4 28 kJ/g

Change of State   p. 114
1 2·31 MJ/kg
4 315 kJ/kg, 6·2°C
5 4·3°C
6 873°C

Air   p. 122
6 66·6 g
10 90% N$_2$, 10% O$_2$

Some Simple Chemical Laws   p. 126
2 Yes
4 8:1
6 PbO
7 Yes, X$_2$O$_3$
8 X$_3$N$_2$
10 X$_2$Y$_3$

Nitric Acid   p. 139
15 150 cm^3

More Laws   p. 157
1 200, 100; 1, 2; MCl, MCl$_2$
2 9·75%
3 NO
4 17
5 (a) 51·22%, (b) CH$_3$, C$_2$H$_6$, (c) 9·6 dm^3
6 (a) 7·5 dm^3, (b) 3300 g CO$_2$
7 (a) 448 cm^3 (b) 480 cm^3 (c) 0·04 g, (d) 249 cm^3, 269 cm^3 (e) 0·022 g
8 (b) C$_2$H$_4$O, C$_4$H$_8$O$_2$, (c) 1·12 l, (d) 16 g
9 8·42 g, 2·47 dm^3, chlorate(V)
10 (a) 1·12 dm^3, (b) 100 cm^3, (c) 25·5%
11 (a) (i) 3·24 g, (ii) 112 cm^3, (b) (ii) 12·1, (iii) 24·2, MO, 95·2
12 (a) 672 cm^3
14 (a) 51·2%, (b) (i) 0·049, (ii) 56, (iii) 127
16 (b) 1·12 dm^3, (c) 20·9%, (CaSO$_4$)$_2$. H$_2$O

Reflection   p. 170
6 7·5 cm in front, 0·5 cm high
7 (a) 3·75 cm, (b) 7·5 cm

Refraction   p. 175
1 28°, 3·3 cm

Lenses   p. 182
2 40 cm, 5
3 8 cm, 2, virtual
4 60 cm, 20 cm square
5 36 cm, 9 cm square
8 50 mm

Waves   p. 191
2 1·2 MHz
3 1·5 s
4 2 × 10^8 m/s

Sound   p. 199
2 (a) 330 m/s, (b) 790 m

4 1·03 m
5 111·4 cm
6 610 m
7 1600 m/s, 4 m
8 (b) 1300 Hz

Solution p. 206
  5 1 g
10 (a) 47·5, (b) 29·8, (c) 17·7 g, (d) 56·5°
12 (a) 90 g
14 (b) (i) 20 g, (ii) none, (iii) 5 g

Electric Currents p. 215
  1 2·5 uA, 1·5 uA, 1·0 uA
  2 4 Ω
  3 (a) 8 Ω, (b) $1\frac{7}{8}$ Ω
  4 (a) 12 Ω, (b) 0·9 A
  5 0·22 Ω resistor in parallel
  6 300 Ω
  7 (a) 0·0167 Ω resistor in parallel,
     (b) 1195 Ω in series

Heating Effect of Electric Currents p. 220
  1 3·6 MJ
  2 1·5 Ω
  3 3 A 308 s
  4 60 W, 0·30 p
  5 53%, 1·5 p
  7 2·7 Ω
  8 480 W, 240 W

Electrical Energy in Chemical Changes p. 233
  9 16 min. 53 s

Standard Solutions p. 244
  1 0·04M, 4·24 g

2 1·158 M
3 5·18 g
4 0·945 g
5 30·09 g
6 (a) 30 cm³, (b) 10 cm³, (c) 63, 0·125M
7 (a) (i) 9·8 g, (ii) 5·6 g, (b) (i) 0·071M
   (ii) 0·125M, (c) 0·075M
8 24

Periodic Table p. 248
  1 (i) (a) 12·3 m³, (b) 0·136 m³
  2 (e) 22·4, 11·2; 1, 2

Metals p. 268
  9 9520 kg
11 MO
20 25 g
26 17·8 g

States of Matter p. 302
  2 (c) (i) 8·2 cm³, (ii) 200 s; (d) (i) 12 cm³,
    (ii) 60 cm³; (e) 20

Miscellaneous p. 335
  1 (e) 98
  3 (b) 28%, (c) H_2CO_2
  8 (a) 2·28 g, (b) 90 cm³, 60 cm³
12 (b) 20·1%, (c) 20 cm³, 10 cm³
17 (b) 128·7 g, (c) 9·6 dm³
18 1·36 g
19 (a) 95 g, (b) 94·5 g, (c) 4·8
27 (a) 8·3 A, 18·8 p
30 (a) $\frac{1}{200}, \frac{1}{100}$ (b) 2
33 (c) (i) 0·64, 0·04,
    (ii) 0·08, 0·08; 2

Useful Data

Standard Temperature = 0°C or 273 K
Standard Pressure = 760 mm of mercury
1 atmosphere (760 mm of mercury) = 101 kN/m²
Acceleration due to gravity at the earth's surface = 9·8 m/s²
Specific latent heat of fusion of ice = 336 kJ/kg
Specific latent heat of vaporization of water = 2·27 MJ/kg
Velocity of light in a vacuum = 3 × 10⁸ m/s
Velocity of sound in air at 0°C = 330 m/s

$$\text{Resistance (ohm)} = \frac{\text{potential difference (volt)}}{\text{current (ampere)}}$$

Power (watt) = potential difference (volt) × current (ampere)
Charge (coulomb) = current (ampere) × time (second)
1 g of Hydrogen occupies 11·12 dm³ (litres) at s.t.p. = 11·92 dm³ (litres) at r.t.p.

2·016 g of Hydrogen occupy 22·4 dm³ (litres) at s.t.p. = 24 dm³ (litres) at r.t.p.
Atomic mass standard, Carbon = 12
Molar volume = 22·4 dm³ (litres) at s.t.p. = 24 dm³ (litres) at r.t.p.
General Gas Equation

$$\frac{P_1 V_1}{T_1} = \frac{P_2 V_2}{T_2} \text{ (Kelvin temperatures)}$$

Some Common Chemical Formulae

Ammonia	NH_3		Lime water (calcium hydroxide)	$Ca(OH)_2$
Ammonium hydroxide	NH_4OH		Methane	CH_4
Benzene	C_6H_6		Methanoic acid (formic acid)	$HCOOH$
Copper(II) sulphate crystals	$CuSO_4.5H_2O$		Nitric acid	HNO_3
Calcium hydrogencarbonate	$Ca(HCO_3)_2$		Potassium chlorate(V) (chlorate)	$KClO_3$
Calcium sulphate	$CaSO_4$		Potassium manganate(VII)	
Chalk, limestone, marble	$CaCO_3$		(permanganate)	$KMnO_4$
Common salt (sodium chloride)	$NaCl$		Sand, silica	SiO_2
Ethane	C_2H_6		Silver nitrate	$AgNO_3$
Ethanoic acid (acetic acid)	CH_3COOH		Sodium hydrogencarbonate	$NaHCO_3$
Ethyne (acetylene)	C_2H_2		Sodium carbonate	Na_2CO_3
Ethanol (alcohol)	C_2H_5OH		Sodium hydroxide (caustic soda)	$NaOH$
Ethene (ethylene)	C_2H_4		Sulphuric acid	H_2SO_4
Ethanedioic acid (oxalic acid)	$(COOH)_2$		Washing soda crystals	$Na_2CO_3.10H_2O$
Glucose	$C_6H_{12}O_6$		Water	H_2O
Hydrogen chloride, hydrochloric acid	HCl			

Approximate Relative Atomic Masses

(for use in calculations)

Element	Symbol	Relative Atomic Mass	Element	Symbol	Relative Atomic Mass
Aluminium	Al	27	Mercury	Hg	200·5
Bromine	Br	80	Nitrogen	N	14
Calcium	Ca	40	Oxygen	O	16
Carbon	C	12	Phosphorus	P	31
Chlorine	Cl	35·5	Potassium	K	39
Copper	Cu	63·5	Silver	Ag	108
Hydrogen	H	1	Sodium	Na	23
Iron	Fe	56	Sulphur	S	32
Lead	Pb	207	Tin	Sn	118·5
Lithium	Li	7	Titanium	Ti	48
Magnesium	Mg	24	Zinc	Zn	65·5
Manganese	Mn	55			

International Relative Atomic Masses

Name	Symbol	Atomic No.	Relative atomic mass	Name	Symbol	Atomic No.	Relative atomic mass
Actinium	Ac	89	[227]	Mendelevium	Md	101	[256]
Aluminium	Al	13	26·98	Mercury	Hg	80	200·61
Americium	Am	95	[243]	Molybdenum	Mo	42	95·95
Antimony	Sb	51	121·76	Neodymium	Nd	60	144·27
Argon	Ar	18	39·94	Neon	Ne	10	20·18
Arsenic	As	33	74·91	Neptunium	Np	93	[237]
Astatine	At	85	[210]	Nickel	Ni	28	58·71
Barium	Ba	56	137·36	Niobium	Nb	41	92·91
Berkelium	Bk	97	[245]	Nitrogen	N	7	14·00
Beryllium	Be	4	9·01	Nobelium	No	102	[253]
Bismuth	Bi	83	209·00	Osmium	Os	76	190·2
Boron	B	5	10·82	Oxygen	O	8	16
Bromine	Br	35	79·91	Palladium	Pd	46	106·4
Cadmium	Cd	48	112·41	Phosphorus	P	15	30·97
Caesium	Cs	55	132·91	Platinum	Pt	78	195·09
Calcium	Ca	20	40·08	Plutonium	Pu	94	[242]
Californium	Cf	98	[246]	Polonium	Po	84	[210]
Carbon	C	6	12	Potassium	K	19	39·10
Cerium	Ce	58	140·13	Praseodymium	Pr	59	140·92
Chlorine	Cl	17	35·45	Promethium	Pm	61	[145]
Chromium	Cr	24	52·01	Protactinium	Pa	91	[231]
Cobalt	Co	27	58·94	Radium	Ra	88	[226·05]
Copper	Cu	29	63·54	Radon	Rn	86	[222]
Curium	Cm	96	[243]	Rhenium	Re	75	186·22
Dysprosium	Dy	66	162·51	Rhodium	Rh	45	102·91
Einsteinium	Es	99	[254]	Rubidium	Kb	37	85·48
Erbium	Er	68	167·27	Ruthenium	Ru	44	101·1
Europium	Eu	63	152·0	Samarium	Sm	62	150·35
Fermium	Fm	100	[253]	Scandium	Sc	21	44·96
Fluorine	F	9	19·00	Selenium	Se	34	78·96
Francium	Fr	87	[223]	Silicon	Si	14	28·09
Gadolinium	Gd	64	157·26	Silver	Ag	47	107·88
Gallium	Ga	31	69·72	Sodium	Na	11	22·99
Germanium	Ge	32	72·60	Strontium	Sr	38	87·63
Gold	Au	79	197·0	Sulphur	S	16	32·06
Hafnium	Hf	72	178·50	Tantalum	Ta	73	180·95
Helium	He	2	4·00	Technetium	Tc	43	[99]
Holmium	Ho	67	164·94	Tellurium	Te	52	127·61
Hydrogen	H	1	1·008	Terbium	Tb	65	158·93
Indium	In	49	114·82	Thallium	Tl	81	204·39
Iodine	I	53	126·91	Thorium	Th	90	232·05
Iridium	Ir	77	192·2	Thulium	Tm	69	168·94
Iron	Fe	26	55·85	Tin	Sn	50	118·70
Krypton	Kr	36	83·80	Titanium	Ti	22	47·90
Kurchatovium	Ku	104	—	Tungsten	W	74	183·86
Lanthanum	La	57	138·92	Uranium	U	92	238·07
Lawrencium	Lw	103	[257]	Vanadium	V	23	50·95
Lead	Pb	82	207·21	Xenon	Xe	54	131·30
Lithium	Li	3	6·94	Ytterbium	Yb	70	173·04
Lutetium	Lu	71	174·99	Yttrium	Y	39	88·92
Magnesium	Mg	12	24·32	Zinc	Zn	30	65·38
Manganese	Mn	25	54·94	Zirconium	Zr	40	91·22

The figures in brackets refer to the most stable or the most abundant isotope of the element concerned.

Logarithms

	0	1	2	3	4	5	6	7	8	9	1	2	3	4	5	6	7	8	9
																	Differences		
10	0000	0043	0086	0128	0170	0212	0253	0294	0334	0374	4	8	12	17	21	25	29	33	37
11	0414	0453	0492	0531	0569	0607	0645	0682	0719	0755	4	8	11	15	19	23	26	30	34
12	0792	0828	0864	0899	0934	0969	1004	1038	1072	1106	3	7	10	14	17	21	24	28	31
13	1139	1173	1206	1239	1271	1303	1335	1367	1399	1430	3	6	10	13	16	19	23	26	29
14	1461	1492	1523	1553	1584	1614	1644	1673	1703	1732	3	6	9	12	15	18	21	24	27
15	1761	1790	1818	1847	1875	1903	1931	1959	1987	2014	3	6	8	11	14	17	20	22	25
16	2041	2068	2095	2122	2148	2175	2201	2227	2253	2279	3	5	8	11	13	16	18	21	24
17	2304	2330	2355	2380	2405	2430	2455	2480	2504	2529	2	5	7	10	12	15	17	20	22
18	2553	2577	2601	2625	2648	2672	2695	2718	2742	2765	2	5	7	9	12	14	16	19	21
19	2788	2810	2833	2856	2878	2900	2923	2945	2967	2989	2	4	7	9	11	13	16	18	20
20	3010	3032	3054	3075	3096	3118	3139	3160	3181	3201	2	4	6	8	11	13	15	17	19
21	3222	3243	3263	3284	3304	3324	3345	3365	3385	3404	2	4	6	8	10	12	14	16	18
22	3424	3444	3464	3483	3502	3522	3541	3560	3579	3598	2	4	6	8	10	12	14	15	17
23	3617	3636	3655	3674	3692	3711	3729	3747	3766	3784	2	4	6	7	9	11	13	15	17
24	3802	3820	3838	3856	3874	3892	3909	3927	3945	3962	2	4	5	7	9	11	12	14	16
25	3979	3997	4014	4031	4048	4065	4082	4099	4116	4133	2	3	5	7	9	10	12	14	15
26	4150	4166	4183	4200	4216	4232	4249	4265	4281	4298	2	3	5	7	8	10	11	13	15
27	4314	4330	4346	4362	4378	4393	4409	4425	4440	4456	2	3	5	6	8	9	11	13	14
28	4472	4487	4502	4518	4533	4548	4564	4579	4594	4609	2	3	5	6	8	9	11	12	14
29	4624	4639	4654	4669	4683	4698	4713	4728	4742	4757	1	3	4	6	7	9	10	12	13
30	4771	4786	4800	4814	4829	4843	4857	4871	4886	4900	1	3	4	6	7	9	10	11	13
31	4914	4928	4942	4955	4969	4983	4997	5011	5024	5038	1	3	4	6	7	8	10	11	12
32	5051	5065	5079	5092	5105	5119	5132	5145	5159	5172	1	3	4	5	7	8	9	11	12
33	5185	5198	5211	5224	5237	5250	5263	5276	5289	5302	1	3	4	5	6	8	9	10	12
34	5315	5328	5340	5353	5366	5378	5391	5403	5416	5428	1	3	4	5	6	8	9	10	11
35	5441	5453	5465	5478	5490	5502	5514	5527	5539	5551	1	2	4	5	6	7	9	10	11
36	5563	5575	5587	5599	5611	5623	5635	5647	5658	5670	1	2	4	5	6	7	8	10	11
37	5682	5694	5705	5717	5729	5740	5752	5763	5775	5786	1	2	3	5	6	7	8	9	10
38	5798	5809	5821	5832	5843	5855	5866	5877	5888	5899	1	2	3	5	6	7	8	9	10
39	5911	5922	5933	5944	5955	5966	5977	5988	5999	6010	1	2	3	4	5	7	8	9	10
40	6021	6031	6042	6053	6064	6075	6085	6096	6107	6117	1	2	3	4	5	6	8	9	10
41	6128	6138	6149	6160	6170	6180	6191	6201	6212	6222	1	2	3	4	5	6	7	8	9
42	6232	6243	6253	6263	6274	6284	6294	6304	6314	6325	1	2	3	4	5	6	7	8	9
43	6335	6345	6355	6365	6375	6385	6395	6405	6415	6425	1	2	3	4	5	6	7	8	9
44	6435	6444	6454	6464	6474	6484	6493	6503	6513	6522	1	2	3	4	5	6	7	8	9
45	6532	6542	6551	6561	6571	6580	6590	6599	6609	6618	1	2	3	4	5	6	7	8	9
46	6628	6637	6646	6656	6665	6675	6684	6693	6702	6712	1	2	3	4	5	6	7	7	8
47	6721	6730	6739	6749	6758	6767	6776	6785	6794	6803	1	2	3	4	5	5	6	7	8
48	6812	6821	6830	6839	6848	6857	6866	6875	6884	6893	1	2	3	4	4	5	6	7	8
49	6902	6911	6920	6928	6937	6946	6955	6964	6972	6981	1	2	3	4	4	5	6	7	8
50	6990	6998	7007	7016	7024	7033	7042	7050	7059	7067	1	2	3	3	4	5	6	7	8
51	7076	7084	7093	7101	7110	7118	7126	7135	7143	7152	1	2	3	3	4	5	6	7	8
52	7160	7168	7177	7185	7193	7202	7210	7218	7226	7235	1	2	2	3	4	5	6	7	7
53	7243	7251	7259	7267	7275	7284	7292	7300	7308	7316	1	2	2	3	4	5	6	6	7
54	7324	7332	7340	7348	7356	7364	7372	7380	7388	7396	1	2	2	3	4	5	6	6	7
	0	1	2	3	4	5	6	7	8	9	1	2	3	4	5	6	7	8	9

	0	1	2	3	4	5	6	7	8	9	1	2	3	4	5	6	7	8	9
																Differences			
55	7404	7412	7419	7427	7435	7443	7451	7459	7466	7474	1	2	2	3	4	5	5	6	7
56	7482	7490	7497	7505	7513	7520	7528	7536	7543	7551	1	2	2	3	4	5	5	6	7
57	7559	7566	7574	7582	7589	7597	7604	7612	7619	7627	1	2	2	3	4	5	5	6	7
58	7634	7642	7649	7657	7664	7672	7679	7686	7694	7701	1	1	2	3	4	4	5	6	7
59	7709	7716	7723	7731	7738	7745	7752	7760	7767	7774	1	1	2	3	4	4	5	6	7
60	7782	7789	7796	7803	7810	7818	7825	7832	7839	7846	1	1	2	3	4	4	5	6	6
61	7853	7860	7868	7875	7882	7889	7896	7903	7910	7917	1	1	2	3	4	4	5	6	6
62	7924	7931	7938	7945	7952	7959	7966	7973	7980	7987	1	1	2	3	3	4	5	6	6
63	7993	8000	8007	8014	8021	8028	8035	8041	8048	8055	1	1	2	3	3	4	5	5	6
64	8062	8069	8075	8082	8089	8096	8102	8109	8116	8122	1	1	2	3	3	4	5	5	6
65	8129	8136	8142	8149	8156	8162	8169	8176	8182	8189	1	1	2	3	3	4	5	5	6
66	8195	8202	8209	8215	8222	8228	8235	8241	8248	8254	1	1	2	3	3	4	5	5	6
67	8261	8267	8274	8280	8287	8293	8299	8306	8312	8319	1	1	2	3	3	4	5	5	6
68	8325	8331	8338	8344	8351	8357	8363	8370	8376	8382	1	1	2	3	3	4	4	5	6
69	8388	8395	8401	8407	8414	8420	8426	8432	8439	8445	1	1	2	2	3	4	4	5	6
70	8451	8457	8463	8470	8476	8482	8488	8494	8500	8506	1	1	2	2	3	4	4	5	6
71	8513	8519	8525	8531	8537	8543	8549	8555	8561	8567	1	1	2	2	3	4	4	5	5
72	8573	8579	8585	8591	8597	8603	8609	8615	8621	8627	1	1	2	2	3	4	4	5	5
73	8633	8639	8645	8651	8657	8663	8669	8675	8681	8686	1	1	2	2	3	4	4	5	5
74	8692	8698	8704	8710	8716	8722	8727	8733	8739	8745	1	1	2	2	3	4	4	5	5
75	8751	8756	8762	8768	8774	8779	8785	8791	8797	8802	1	1	2	2	3	3	4	5	5
76	8808	8814	8820	8825	8831	8837	8842	8848	8854	8859	1	1	2	2	3	3	4	5	5
77	8865	8871	8876	8882	8887	8893	8899	8904	8910	8915	1	1	2	2	3	3	4	4	5
78	8921	8927	8932	8938	8943	8949	8954	8960	8965	8971	1	1	2	2	3	3	4	4	5
79	8976	8982	8987	8993	8998	9004	9009	9015	9020	9025	1	1	2	2	3	3	4	4	5
80	9031	9036	9042	9047	9053	9058	9063	9069	9074	9079	1	1	2	2	3	3	4	4	5
81	9085	9090	9096	9101	9106	9112	9117	9122	9128	9133	1	1	2	2	3	3	4	4	5
82	9138	9143	9149	9154	9159	9165	9170	9175	9180	9186	1	1	2	2	3	3	4	4	5
83	9191	9196	9201	9206	9212	9217	9222	9227	9232	9238	1	1	2	2	3	3	4	4	5
84	9243	9248	9253	9258	9263	9269	9274	9279	9284	9289	1	1	2	2	3	3	4	4	5
85	9294	9299	9304	9309	9315	9320	9325	9330	9335	9340	1	1	2	2	3	3	4	4	5
86	9345	9350	9355	9360	9365	9370	9375	9380	9385	9390	1	1	2	2	3	3	4	4	5
87	9395	9400	9405	9410	9415	9420	9425	9430	9435	9440	0	1	1	2	2	3	3	4	4
88	9445	9450	9455	9460	9465	9469	9474	9479	9484	9489	0	1	1	2	2	3	3	4	4
89	9494	9499	9504	9509	9513	9518	9523	9528	9533	9538	0	1	1	2	2	3	3	4	4
90	9542	9547	9552	9557	9562	9566	9571	9576	9581	9586	0	1	1	2	2	3	3	4	4
91	9590	9595	9600	9605	9609	9614	9619	9624	9628	9633	0	1	1	2	2	3	3	4	4
92	9638	9643	9647	9652	9657	9661	9666	9671	9675	9680	0	1	1	2	2	3	3	4	4
93	9685	9689	9694	9699	9703	9708	9713	9717	9722	9727	0	1	1	2	2	3	3	4	4
94	9731	9736	9741	9745	9750	9754	9759	9763	9768	9773	0	1	1	2	2	3	3	4	4
95	9777	9782	9786	9791	9795	9800	9805	9809	9814	9818	0	1	1	2	2	3	3	4	4
96	9823	9827	9832	9836	9841	9845	9850	9854	9859	9863	0	1	1	2	2	3	3	4	4
97	9868	9872	9877	9881	9886	9890	9894	9899	9903	9908	0	1	1	2	2	3	3	4	4
98	9912	9917	9921	9926	9930	9934	9939	9943	9948	9952	0	1	1	2	2	3	3	4	4
99	9956	9961	9965	9969	9974	9978	9983	9987	9991	9996	0	1	1	2	2	3	3	3	4
	0	1	2	3	4	5	6	7	8	9	1	2	3	4	5	6	7	8	9

Logs

Antilogarithms

	0	1	2	3	4	5	6	7	8	9	1	2	3	4	5	6	7	8	9
·00	1000	1002	1005	1007	1009	1012	1014	1016	1019	1021	0	0	1	1	1	1	2	2	2
·01	1023	1026	1028	1030	1033	1035	1038	1040	1042	1045	0	0	1	1	1	1	2	2	2
·02	1047	1050	1052	1054	1057	1059	1062	1064	1067	1069	0	0	1	1	1	1	2	2	2
·03	1072	1074	1076	1079	1081	1084	1086	1089	1091	1094	0	0	1	1	1	1	2	2	2
·04	1096	1099	1102	1104	1107	1109	1112	1114	1117	1119	0	1	1	1	1	2	2	2	2
·05	1122	1125	1127	1130	1132	1135	1138	1140	1143	1146	0	1	1	1	1	2	2	2	2
·06	1148	1151	1153	1156	1159	1161	1164	1167	1169	1172	0	1	1	1	1	2	2	2	2
·07	1175	1178	1180	1183	1186	1189	1191	1194	1197	1199	0	1	1	1	1	2	2	2	2
·08	1202	1205	1208	1211	1213	1216	1219	1222	1225	1227	0	1	1	1	1	2	2	2	3
·09	1230	1233	1236	1239	1242	1245	1247	1250	1253	1256	0	1	1	1	1	2	2	2	3
·10	1259	1262	1265	1268	1271	1274	1276	1279	1282	1285	0	1	1	1	1	2	2	2	3
·11	1288	1291	1294	1297	1300	1303	1306	1309	1312	1315	0	1	1	1	2	2	2	2	3
·12	1318	1321	1324	1327	1330	1334	1337	1340	1343	1346	0	1	1	1	2	2	2	2	3
·13	1349	1352	1355	1358	1361	1365	1368	1371	1374	1377	0	1	1	1	2	2	2	3	3
·14	1380	1384	1387	1390	1393	1396	1400	1403	1406	1409	0	1	1	1	2	2	2	3	3
·15	1413	1416	1419	1422	1426	1429	1432	1435	1439	1442	0	1	1	1	2	2	2	3	3
·16	1445	1449	1452	1455	1459	1462	1466	1469	1472	1476	0	1	1	1	2	2	2	3	3
·17	1479	1483	1486	1489	1493	1496	1500	1503	1507	1510	0	1	1	1	2	2	2	3	3
·18	1514	1517	1521	1524	1528	1531	1535	1538	1542	1545	0	1	1	1	2	2	2	3	3
·19	1549	1552	1556	1560	1563	1567	1570	1574	1578	1581	0	1	1	1	2	2	3	3	3
·20	1585	1589	1592	1596	1600	1603	1607	1611	1614	1618	0	1	1	1	2	2	3	3	3
·21	1622	1626	1629	1633	1637	1641	1644	1648	1652	1656	0	1	1	2	2	2	3	3	3
·22	1660	1663	1667	1671	1675	1679	1683	1687	1690	1694	0	1	1	2	2	2	3	3	3
·23	1698	1702	1706	1710	1714	1718	1722	1726	1730	1734	0	1	1	2	2	2	3	3	4
·24	1738	1742	1746	1750	1754	1758	1762	1766	1770	1774	0	1	1	2	2	2	3	3	4
·25	1778	1782	1786	1791	1795	1799	1803	1807	1811	1816	0	1	1	2	2	2	3	3	4
·26	1820	1824	1828	1832	1837	1841	1845	1849	1854	1858	0	1	1	2	2	3	3	3	4
·27	1862	1866	1871	1875	1879	1884	1888	1892	1897	1901	0	1	1	2	2	3	3	3	4
·28	1905	1910	1914	1919	1923	1928	1932	1936	1941	1945	0	1	1	2	2	3	3	4	4
·29	1950	1954	1959	1963	1968	1972	1977	1982	1986	1991	0	1	1	2	2	3	3	4	4
·30	1995	2000	2004	2009	2014	2018	2023	2028	2032	2037	0	1	1	2	2	3	3	4	4
·31	2042	2046	2051	2056	2061	2065	2070	2075	2080	2084	0	1	1	2	2	3	3	4	4
·32	2089	2094	2099	2104	2109	2113	2118	2123	2128	2133	0	1	1	2	2	3	3	4	4
·33	2138	2143	2148	2153	2158	2163	2168	2173	2178	2183	0	1	1	2	2	3	3	4	4
·34	2188	2193	2198	2203	2208	2213	2218	2223	2228	2234	1	1	2	2	3	3	4	4	5
·35	2239	2244	2249	2254	2259	2265	2270	2275	2280	2286	1	1	2	2	3	3	4	4	5
·36	2291	2296	2301	2307	2312	2317	2323	2328	2333	2339	1	1	2	2	3	3	4	4	5
·37	2344	2350	2355	2360	2366	2371	2377	2382	2388	2393	1	1	2	2	3	3	4	4	5
·38	2399	2404	2410	2415	2421	2427	2432	2438	2443	2449	1	1	2	2	3	3	4	4	5
·39	2455	2460	2466	2472	2477	2483	2489	2495	2500	2506	1	1	2	2	3	3	4	5	5
·40	2512	2518	2523	2529	2535	2541	2547	2553	2559	2564	1	1	2	2	3	4	4	5	5
·41	2570	2576	2582	2588	2594	2600	2606	2612	2618	2624	1	1	2	2	3	4	4	5	5
·42	2630	2636	2642	2649	2655	2661	2667	2673	2679	2685	1	1	2	2	3	4	4	5	6
·43	2692	2698	2704	2710	2716	2723	2729	2735	2742	2748	1	1	2	3	3	4	4	5	6
·44	2754	2761	2767	2773	2780	2786	2793	2799	2805	2812	1	1	2	3	3	4	5	5	6
·45	2818	2825	2831	2838	2844	2851	2858	2864	2871	2877	1	1	2	3	3	4	5	5	6
·46	2884	2891	2897	2904	2911	2917	2924	2931	2938	2944	1	1	2	3	3	4	5	5	6
·47	2951	2958	2965	2972	2979	2985	2992	2999	3006	3013	1	1	2	3	3	4	5	5	6
·48	3020	3027	3034	3041	3048	3055	3062	3069	3076	3083	1	1	2	3	4	4	5	6	6
·49	3090	3097	3105	3112	3119	3126	3133	3141	3148	3155	1	1	2	3	4	4	5	6	6
	0	1	2	3	4	5	6	7	8	9	1	2	3	4	5	6	7	8	9

	0	1	2	3	4	5	6	7	8	9	1	2	3	4	5	6	7	8	9
														Differences					
·50	3162	3170	3177	3184	3192	3199	3206	3214	3221	3228	1	1	2	3	4	4	5	6	7
·51	3236	3243	3251	3258	3266	3273	3281	3289	3296	3304	1	2	2	3	4	5	5	6	7
·52	3311	3319	3327	3334	3342	3350	3357	3365	3373	3381	1	2	2	3	4	5	5	6	7
·53	3388	3396	3404	3412	3420	3428	3436	3443	3451	3459	1	2	2	3	4	5	6	6	7
·54	3467	3475	3483	3491	3499	3508	3516	3524	3532	3540	1	2	2	3	4	5	6	6	7
·55	3548	3556	3565	3573	3581	3589	3597	3606	3614	3622	1	2	2	3	4	5	6	7	7
·56	3631	3639	3648	3656	3664	3673	3681	3690	3698	3707	1	2	3	3	4	5	6	7	8
·57	3715	3724	3733	3741	3750	3758	3767	3776	3784	3793	1	2	3	3	4	5	6	7	8
·58	3802	3811	3819	3828	3837	3846	3855	3864	3873	3882	1	2	3	4	4	5	6	7	8
·59	3890	3899	3908	3917	3926	3936	3945	3954	3963	3972	1	2	3	4	5	5	6	7	8
·60	3981	3990	3999	4009	4018	4027	4036	4046	4055	4064	1	2	3	4	5	6	6	7	8
·61	4074	4083	4093	4102	4111	4121	4130	4140	4150	4159	1	2	3	4	5	6	7	8	9
·62	4169	4178	4188	4198	4207	4217	4227	4236	4246	4256	1	2	3	4	5	6	7	8	9
·63	4266	4276	4285	4295	4305	4315	4325	4335	4345	4355	1	2	3	4	5	6	7	8	9
·64	4365	4375	4385	4395	4406	4416	4426	4436	4446	4457	1	2	3	4	5	6	7	8	9
·65	4467	4477	4487	4498	4508	4519	4529	4539	4550	4560	1	2	3	4	5	6	7	8	9
·66	4571	4581	4592	4603	4613	4624	4634	4645	4656	4667	1	2	3	4	5	6	7	8	10
·67	4677	4688	4699	4710	4721	4732	4742	4753	4764	4775	1	2	3	4	5	7	8	9	10
·68	4786	4797	4808	4819	4831	4842	4853	4864	4875	4887	1	2	3	4	6	7	8	9	10
·69	4898	4909	4920	4932	4943	4955	4966	4977	4989	5000	1	2	3	5	6	7	8	9	10
·70	5012	5023	5035	5047	5058	5070	5082	5093	5105	5117	1	2	4	5	6	7	8	9	11
·71	5129	5140	5152	5164	5176	5188	5200	5212	5224	5236	1	2	4	5	6	7	8	10	11
·72	5248	5260	5272	5284	5297	5309	5321	5333	5346	5358	1	2	4	5	6	7	9	10	11
·73	5370	5383	5395	5408	5420	5433	5445	5458	5470	5483	1	3	4	5	6	8	9	10	11
·74	5495	5508	5521	5534	5546	5559	5572	5585	5598	5610	1	3	4	5	6	8	9	10	12
·75	5623	5636	5649	5662	5675	5689	5702	5715	5728	5741	1	3	4	5	7	8	9	10	12
·76	5754	5768	5781	5794	5808	5821	5834	5848	5861	5875	1	3	4	5	7	8	9	11	12
·77	5888	5902	5916	5929	5943	5957	5970	5984	5998	6012	1	3	4	5	7	8	10	11	12
·78	6026	6039	6053	6067	6081	6095	6109	6124	6138	6152	1	3	4	6	7	8	10	11	13
·79	6166	6180	6194	6209	6223	6237	6252	6266	6281	6295	1	3	4	6	7	9	10	11	13
·80	6310	6324	6339	6353	6368	6383	6397	6412	6427	6442	1	3	4	6	7	9	10	12	13
·81	6457	6471	6486	6501	6516	6531	6546	6561	6577	6592	2	3	5	6	8	9	11	12	14
·82	6607	6622	6637	6653	6668	6683	6699	6714	6730	6745	2	3	5	6	8	9	11	12	14
·83	6761	6776	6792	6808	6823	6839	6855	6871	6887	6902	2	3	5	6	8	9	11	13	14
·84	6918	6934	6950	6966	6982	6998	7015	7031	7047	7063	2	3	5	6	8	10	11	13	15
·85	7079	7096	7112	7129	7145	7161	7178	7194	7211	7228	2	3	5	7	8	10	12	13	15
·86	7244	7261	7278	7295	7311	7328	7345	7362	7379	7396	2	3	5	7	8	10	12	13	15
·87	7413	7430	7447	7464	7482	7499	7516	7534	7551	7568	2	3	5	7	9	10	12	14	16
·88	7586	7603	7621	7638	7656	7674	7691	7709	7727	7745	2	4	5	7	9	11	12	14	16
·89	7762	7780	7798	7816	7834	7852	7870	7889	7907	7925	2	4	5	7	9	11	13	14	16
·90	7943	7962	7980	7998	8017	8035	8054	8072	8091	8110	2	4	6	7	9	11	13	15	17
·91	8128	8147	8166	8185	8204	8222	8241	8260	8279	8299	2	4	6	8	9	11	13	15	17
·92	8318	8337	8356	8375	8395	8414	8433	8453	8472	8492	2	4	6	8	10	12	14	15	17
·93	8511	8531	8551	8570	8590	8610	8630	8650	8670	8690	2	4	6	8	10	12	14	16	18
·94	8710	8730	8750	8770	8790	8810	8831	8851	8872	8892	2	4	6	8	10	12	14	16	18
·95	8913	8933	8954	8974	8995	9016	9036	9057	9078	9099	2	4	6	8	10	12	14	17	19
·96	9120	9141	9162	9183	9204	9226	9247	9268	9290	9311	2	4	6	8	11	13	15	17	19
·97	9333	9354	9376	9397	9419	9441	9462	9484	9506	9528	2	4	7	9	11	13	15	17	20
·98	9550	9572	9594	9616	9638	9661	9683	9705	9727	9750	2	4	7	9	11	13	16	18	20
·99	9772	9795	9817	9840	9863	9886	9908	9931	9954	9977	2	5	7	9	11	14	16	18	20
	0	1	2	3	4	5	6	7	8	9	1	2	3	4	5	6	7	8	9

Anti-Logs

Index

Absolute zero of temperature, 92
Abundance of elements, 250
Acceleration, 12–16, 18; due to gravity, 15–16
Accommodation (sight), 181
Accumulator, 215, 236–7
Acetaldehyde, 330
Acetic acid, 330
Acetylene, 259, 327
Acidic anhydrides, 69, 132
    oxides, 118, 248
Acidity, 243
Acids, (see also individual acids), 248
Acoustics, 195
Actinons, 246
Activated charcoal, 144
Activity (metals), 1, 232–3, 250–1
Adsorption, 143–4
Aerosols, 26
Air, chapter 22; composition, 115; resistance, 15
Alcohol, 150, 329–30, 331; combustion, 150
Alkali metals, 245
    earths, 245
Alkalinity, 243
Alkalis, 118, 130, 143, 248, 259–61, 331
Alkanes, 325–6
Alkenes, 327–8
Alloys, 257, 258
Alpha particles, 311, 312, 313
Alternating current, 285
Aluminium, 252; abundance, 250; anodizing, 120, 232
Aluminium compounds:
    hydroxide, 261
    oxide, 252
    sulphate, 264
Americium, 311
Amino acids, 232
Ammeters, 213–4, 277; moving iron, 277
Ammonal, 131
Ammonia, chapter 21, 24, 111, 204, 262, 298, 317;
    covalent compound, 36;
    identification, 129, 319, 321; manufacture, 129;
    oxidation, 119, 128, 135;
    preparation, 127, 317;
    properties, 128–9; in water softening, 205
    soda process, 262
Ammoniacal liquor, 131, 148
Ammonium compounds:
    chloride, 21, 131, 133, 262, 317
    hydroxide, 130, 321

ion, 130, 322
nitrate, 131, 133, 134, 138, 139
salts, 130
solubility, 203
sulphate, 131, 139, 148
Thiocyanate, 320
Ampère, 209
Amphoteric oxides, 118, 248, 261, 320
Amplitude (sound) 193; (oscillation), 11
Anaesthetic, 133
Aneroid barometer, 43
Anhydrite, 69, 131, 266
Animal charcoal, 143–4
Anions, identification, 322
Anodizing, 120, 232
Anthracite, 147
Anti-chlor, 26, 69
Anti-freeze, 90, 329
Apparent depth, 172–3
Aquadag, 142
Aqua regia, 137
Archimedes' principle, chapter 11; experiment to verify, 53–4
Argon 115, 218
Argonons, 115, 150, 245, 297
Armature, 277
Astatine, 26, 246
Atmospheric pressure, 42, 95
Atomic energy, 152
    mass, chapter 9
    number, 29, 245, 311
    structure, chapters 1 and 6; 244
    volume, 247–8
Atomicity, 155
Avogadro, Amadeo, 154
    constant, 45, 155, 255, 247
Avogadro's Hypothesis, chapter 27

Bacteria, 139
Baking powder, 263
    soda, 263
Balance wheel, 89
Balanced reactions, 302
Balloons, 55
Barium chloride, 73
Barometers, 43
Bases, 118
Basic anhydrides, 129
    oxides, 118, 248
Basicity, 70
Battery, chapters 38 and 39, 214–5
Bauxite, 252
Beats, 197
Becquerel, Henri, 309

Benzene, 144, 148
Bessemer process, 254
Beta particles, 311–2
Bicarbonates (see hydrogen-carbonates)
Bimetal strips, 87
Binoculars, 174
Bismuth chloride, 302
Biuret test, 333
Bisulphates (see hydrogen-sulphates)
Bisulphites (see hydrogen-sulphites)
Black iron oxide (see iron (II) (III) oxide)
Blast furnace, 253, 256
Bleaching, 24, 69
    powder, 25, 265
Blue vitriol (see copper (II) sulphate)
    water gas, 149
Board of Trade unit, 216
Body temperature, 87
Bohr, Neil, 245, 246
Boiling, 111
    point, 112, 114
Bonding, 145, 247, 297, 298
Bond strength, 298
Bone black, 143
Bordeaux mixture, 267
Boyle's law, 93
Brakes, hydraulic, 40
Brass, 257
Breathing, 151
Brewing industry, 206
Brine, 259–60, 262, 265; electrolysis of, 259–60
British thermal unit, 147
Bromides, 26
Bromine, 26, 68, 144
Brownian motion, 35
Brown-ring test, 138, 322
Buchner funnel, 266, 332
Bunsen flame, 150–1
    valve, 22, 267, 318
Buoyancy, 54
Butane, 147, 326

Cadmium,
Calcite, 261
Calcium, 250–1, 258, 320; reaction with water, 258
Calcium compounds:
    carbide, 259
    carbonate (see also limestone and chalk), 152, 203–6, 259, 261–2, 263
    chloride, 115, 128, 262, 265
    hydrogencarbonate, 152, 263; in hard water, 203–6
    hydrogensulphite, 78

hydroxide, 127, 258–9; in water softening, 204
    ions, 203–5
    oxide, 128, 258–9
    permutit, 205
    phosphate, 143, 265
    stearate, 203
    sulphate, 266, 320; in hard water, 203–5
Calorie, 108
Calorific value, 147, 148, 149
Calorimeter, bomb, 108
Camera, 179
Capillary rise, 37
Carbohydrates, 151, 333
Carbon (see also charcoal, coke, etc.), chapter 26; mineral, 141; oxidation, 119, 137; presence in compounds, 325
    structure, 142; vegetable, 142
    cycle, 151–2
Carbon compounds:
    dioxide, 72, 116, 253, 262, 263–4; absorption, 259; identification, 319, in air, 115, 152, 206, 255; in blood, 151; in Solvay process, 262; preparation, 318; reduction, 146, 253
    disulphide, 62, 133
    monoxide, chapter 26, 72, 132, 149, 253, 317; preparation, 145, 317; properties, 146
    tetrachloride (see tetra-chloromethane)
Carbonates, 261–4; detection, 321; preparation, 118; solubility, 203
Carbonic acid, 203, 206, 261
Carboxyhaemoglobin, 147
Carburetted water gas, 149
Cast-iron, 253
Catalysis, 69, 117, 129, 135, 149, 151, 318, 329
Catalyst, effect on reaction rate, 301
Cathode ray oscilloscope, 308
    rays, 306
Cations, identification, 318, 321
Cavendish, Henry, 115
Caves, 206
Cells, chapter 39, 235–6; fuel, 237
Cellulose, 138, 151, 329, 333
Celsius scale, 86
Cement, 253, 261, 266
Centre of Gravity, 59
Chadwick, James, 313
Chain reaction, 313

Chalk, 151, 152, 203, 243, 259, 261
Change of state, chapter 21
Charcoal, 143–5, 148
Chemical cells, chapter 39
  energy, 152; converted to eletrical energy, 235
Chlorides, 264–6; detection, 23, 321; preparation, 22, 25, 266; solubility, 203
Chlorine, chapter 5, 31, 119, 265, 316, 319; bleaching by, 24 identification, 319 preparation, 23, 316; properties, 23, uses, 26
Chlorophyll, 151, 332
Chromium plating, 232
Clocks, 11, 88–9
Coal, 147, 148
  gas, 148
Cobalt, 60, 310–11
Coke, 143, 147, 148–9
  fire, 146
Colour, chapter 32
Coloured lights, 184
Combustion, chapter 26; of alcohol 150; of coal gas, 149; of coke, 146
Commutator, 281
Compass, magnetic 271; plotting, 273
Concave mirrors, 165–9
Concentration, effect on reaction rate 299, 300
Condenser lens system, 180
Conduction, of electricity, chapter 36; of heat, 98–100
Conductivity, of solids, 221–2; of solutions, 203, 222–30
Conductors, electrical, 290
  thermal, 99
Conservation of energy, Law of, 76
  of mass, Law of 123
Constant Composition, Law of, 124
Contact process, 69, 301, 302
Convection, in air, 97; in water, 89, 96–8
Convector fire, 98
Converging mirrors, 165–9
Convex mirrors, 166–9
Coolant, 230, 252, 314
Cooling curve, 109
Copper, 257–8, 319; action of nitric acid, 132, 137; action of sulphuric acid, 68, 72, 267
  wire, 211–2
Copper compounds;
  chloride, 125–6
  chromate, 267
  hydroxide, 118, 130, 319
  ions, 226; detection, 129, 321
  nitrate, 124, 138
  oxide, 260, 261, 266, 319; reduction, 124
  sulphate, 68, 78, 227, 231, 266–7
Coulomb, 209
Covalency, 30

Covalent bonding, 297
  compounds, 30, 222, 248, 296, 297, 325, 330
Cracking, 147, 149, 327
Critical angle, 173
Crowbar, 83
Cryolite, 252
Crystallization, fractional, 202
Curie, Pierre and Marie, 309
Current measurement (see also ammeters), 226

Dalton, John, 154
Dams, 41
Davy, Humphrey, 100, 251
  lamp, 100
DDT, 26
Declination, angle of, 275
Decrepitation, 132
Definite Proportions, Law of, 124
Degradation, 329
Dehydration, 71
Deliquescence, 262, 265
Density, 7–9; of air, 9; of liquids, 54–5; relative, 8–9
  bottle, 9
Depolarizer, 235
Detergents, 205
Dettol, 26
Dewar, James, 101
Diamond, 141, 248, 297
Diatomic gases, 155
  molecules, 155
Dibromoethane, 327, 328
Diesel fuel, 147, 327,
Diffusion, 35, 295–6
Dinitrogen oxide, 133, 138, 319
  tetroxide, 132
Diode, 305
Dip, angle of, 275
Discharging action of points, 292–3
Dispersion, 183
Displacement of metals, 156–7
Dissociation, 131
Distillation, 200, 204; fractional, 148, 327
Distribution of electrical energy, 285–6
Diverging mirrors, 166–9
Domain theory, 274
Down's cell, 251
Dry battery, 236
  cleaning fluids, 26
Duralumin, 252, 257
Dutch metal, 25
Dynamo, 283–4

Earth, 219–20
Eclipses, 159, 160, 162
Eddy currents, 287
Efficiency, 76, 78
Efflorescence, 262
Einstein, 313
Electric bell, 277
  currents, chapter 36
  heaters, 217–8
  lighting, 218
  motors, 280
  wiring, domestic, 218–20

Electrical energy, chapter 38, 166–7; distribution, 285–6; production, 223, 234, 285; safety, 219–20
Electrochemical series, 1, 138, 145, 232, 251; and electrolysis, 227
  equivalent, 224
Electrodes, 143, 226, 227, 229–31
Electrolysis, chapter 38; factors affecting, 227–8; industrial applications, 231–2; in extraction of metals, 231, 251, 252; mechanism, 236; of copper(II) sulphate solution; 231; of sodium chloride solution, 230; of sodium hydroxide solution, 229; of sodium sulphate, solution, 230; of sulphuric acid, 229; of water, 223; products, 227
Electromagnet, 277
Electromagnetic induction, chapter 46; experiment to show, 282; laws of, 283
Electromagnetism, chapter 45,
Electromotive force, 214–5
Electron, chapters 1, 6 and 49, 244; charge on, 304; in oxidation and reduction, 120, 138, 250
  gun, 306, 308
Electrophorus, 291–2
Electroplating, 232, 267
Electroscope, charging, 290–1; gold leaf, 289, 292
Electrostatic charge, 289; distribution, 130, 131, 292; testing, 290
  forces, 289
  induction, 290–1
Electrostatics, chapter 47
Electrotyping, 232
Electrovalency, 30, 222
Electrovalent compounds, 30, 222
Elements, chapter 41; combination of, chapter 6
Empirical formula, 47
Endothermic reactions, 131, 134, 149, 152, 298
Energy, 76, 152, 244, 297; chemical, 76, 216; conservation of, 76; electrical, chapter 38, 152; heat, 147, 152, 216; Kinetic, 76, 313; light, 152; mechanical, 152; nuclear, 152, 313, 314; potential, 77; solar, 152, 264; sound, 152; sources of, 63 levels, 29, 244 values for fuels, 147–8
Enlarger, photographic, 180
Enzymes, 329
Equations, chapter 10; ionic, 50
Equilibrium, 301; stability of, 60; three forces, 61–2; two forces, 57
Esters, 330–2

Ethanoic acid, 330
Ethanol, 329–30
Ethene, 327–8
Ethylene (see ethene)
  glycol, 328, 329
Ethyl ethanoate, 330
  hydrogensulphate, 329
Ethyne, 327
Evaporation, 111, 200, 203, 206; cooling by, 111
Ewing, James, 274
Exothermic reactions, 134, 147, 149, 152, 298, 300
Expansion, chapter 16; of gases, chapter 17
Extraction of metals, 231
Eye, 181

Faraday, 225
  Michael, 282
Faraday's laws, of electrolysis, 224, 226; of electromagnetic induction, 283
Fats, 331
Fermentation, 329, 333
Ferric compounds (see iron(III) compounds)
Ferromagnetism, chapter 44
Ferromanganese, 254
Ferrous compounds (see iron(II) compounds)
Fertilizers, 130, 131, 149
Filters, colour, 138, 139
Filtration, 201
Fire extinguishers, 263–4
Fission, 152, 313
Fixation of nitrogen, 139
Flame, 147, 150–1
Fleming's left-hand rule, 279, 307
Floating, 54
Fluorine, 26
Focal length of curved mirrors, 166
Focus of curved mirrors, 166
Force, components, 60; parallelogram of, 62; relation to acceleration, 17; resolving, 63; triangle of, 62; units of, 5
Formic acid (see methanoic acid)
Formula, determination of, 47–8
  mass, 247
Formulae, 31–2
Fountain experiment, 22, 128
Fractional crystallization, 202
  distillation, 148, 327, 332
Fraunhofer, Joseph von, 184
Free fall, 15
Freezing point, 109
Freon, 27, 111
Frequency, of oscillation, 11; of sound, 198
Friction, 18
Fuel cells, 237
Fuels, chapter 26, 147–50; combustion, 149; definition, 147
Fulcrum, 83
Fur, 204
Fuse, electric, 219
Fusion, 314

349

Galileo, 161, 175
Galvanometer, moving coil, 175, 213, 279–80
Gamma rays, 186, 312
Gas carbon, 143, 250
equation, 95
laws, chapter 17
pressure, measurement of, 41
reforming, 149
Gases, structure, 296
Gaseous fuels, 147–9
Gay Lussac's law of combining volumes, 153–4
Geiger Muller counter, 310–11
Generation of electrical energy, 285–6
Generators, 285; electrostatic, 293–4
Giant structure, 248, 297, 329
Gilbert, William, 271, 275, 289
Glycerol, 331
Glyceryl esters, 331
stearate, 332
Glycine, 332
Goitre, 27
Gold leaf electroscope, 289, 292
Gram-atom, 226, 242, 247
formula, 46
ion, 247
molecular volume (see molar volume)
molecular weight (see molar mass)
molecule (see mole)
Graphite, 141, 297, 314
Graphs, 3–4, 6, 13, 201–2
Gravity, 15; centre of, 60
Green vitriol (see iron(II) sulphate)
Grid system, 286
Grove, William, 238
Gunpowder, 267
Gypsum, 266

Haber process, 129, 301
Haematite, 253
Haemoglobin, 332
Half-life, 312
Halogens, chapter 5, 26, 245
Hamilton's mill, 293
Hard water, 203–6, 261, 266; advantages, 205; softening methods, 204–5
Health salts, 263
Heat, chapter 20; radiant, 100–101, 186; transfer of, chapter 18
energy, 147, 148, 152, 221, 247
of neutralization, 221
Heating effect of electric current, chapter 37
Helium, 152, 311, 314
Helmholtz coils, 307
Hoffman voltameter, 223
Homogeneous reaction, 300
Homologous series, 326
Hooker cell, 259
Hot water system, 98
Humidity, 112
Hydraulic press, 40
Hydrobromic acid, 68
Hydrocarbons, 325–8

Hydrochloric acid, chapter 5, 265, 298; pH, 243; preparation, 21; properties, 21
Hydrogen, 115, 265, 314, 325; affinity for chlorine, 24; atomicity, 155; calorific value, 148; identification, 319; mass, 45; nascent, 120; reductant, 120
Hydrogen compounds:
chloride, chapter 5, 265, 319; identification, 24, 319; preparation, 21, 265, 318; proof of formula, 155; properties, 22
ions, 205; concentration of, 243
peroxide, 117, 120, 267, 301; in preparation of oxygen, 117
sulphide, synthesis, 67
Hydrogencarbonates, 261–4; solubility, 203
Hydrogensulphates, 70, 266
Hydrogensulphites, 67
Hydrolysis, 331
Hydrometer, 55
Hydroxides, 258–60; precipitation, 130; solubility, 203
Hydroxonium ions, 226
Hydroxyl ions, 205, 226, 260
Hygrometer, 112
Hypermetropia, 182
Hypo (see sodium thiosulphate)
Hypochlorous acid, 24
Hysometer, 86

Ignition temperature, 147, 300
Images, graphical treatment, 178; in curved mirrors, 167–9; in lenses, 177–8; in pinhole cameras, 160; in plane mirrors, 164
Immersion heater, 106–7, 113
Inclined plane, 70; acceleration on, 16
Induced currents, chapter 46
Induction, electrostatic, 290–1 coil, 288
Infra-red, 100, 186
Ink, 144, 267
Insulators, 290
Internal combustion engine, 105 resistance of cell, 214–5
Inversion, lateral, 166, 174
Invisible radiations, 185–6
Iodine, 26, 120, 138, 265
Ion exchange, 205
pair, 247
Ionic bonding, 297
compounds, 30, 248, 297
equations, 50, 226–7, 297
reactions, speed of, 298
Ions, chapter 6, 226, 243, 248, 250, 259, 309
Ionization, 226, 310
Iron, 252–6; action on steam, 255, 302; catalyst, 301; oxidation, 255; rusting, 255–6

Iron(II) compounds:
chloride, 25, 265
hydroxide, 320
ions, 256, 320, 321
sulphate, 132, 138; decomposition, 260; preparation, 71, 267
Iron(III) compounds:
chloride, 25, 265
hydroxide, 130, 320
ions, 320, 321
oxide, 119, 255, 260
Iron(II)(III) oxide, 255, 260
Isomers, 333
Isotopes, 313

Jeweller's rouge, 260
Joule, James, 103, 217
Joule, 75, 147, 261
Joule's Law, 217

Keepers (magnet), 273
Kellner-Solvay process, 259
Kelvin, Lord, 92
scale of temperature, 92
Kilowatt, 216
-hour, 216
Kinetic energy, 76
Krypton, 115

Laminations, 288
Lamp black, 143
Lanthanons, 246
Lard, 331
Lavoisier, Antoine, 115, 123
Latent heat, chapter 21; of fusion, 113; of vaporization, 110
Lateral inversion, 166, 174
Laws:
Avogadro's, 154
Boyle's, 93–4
Charles, 91–2
Combining volumes, 154
Conservation of energy, 76
Conservation of mass, 123
Constant composition, 124
Definite proportions, 124
Faraday's, 224, 225, 283
Gay Lussac's, 154
Lenz's, 283
Multiple proportions, 125, 325
Newton's, 18
of Reflection, 163
of Refraction, 171
Snell's, 171
L-D process, 254
Lead, 256–7, 299;
poisoning, 205, 257
Lead compounds:
acetate (see ethanoate)
carbonate, 205, 257, 261
chloride, 261, 320
ethanoate, 67
hydroxide, 257, 261, 320
iodide, 320
ions, detection, 320, 321
nitrate, 132, 261, 267; preparation, 205, 250, 322
(II) oxide, 257, 261
(II)(IV) oxide, 261
sulphate, 205, 267, 261; preparation, 322

sulphide, 256, 321
tetra-ethyl, 252, 257, 328
Leclanché cell, 131, 236
Length, measurement of, 5–6
Lenses, chapter 31; focus of, 176
Lenz's law, 283
Leslie cube, 101
Levers, 83
Light, coloured, 184; invisible, 161; speed of, 161; straight line travel of, 159; white, 183
energy, 152
rays, chapter 28
water, 264
Lighting-back, 150
Limelight, 258
Limestone, 203, 243, 253, 261, 262, 263; 'burning' 152, 259; uses, 261
Limewater (see calcium hydroxide)
Lines of force, due to coil, 278; due to conductor, 278; due to magnet, 272
Liquids, structure, 296.
Litharge, 261
Lithium, 319; solubility of compounds, 203
Local action, 235
Lodestone, 271
Longsight, 181–2
Loudness (sound), 193
Loudspeaker, moving coil, 281
Low-temperature physics, 92
Lubricants, 142
Lucricating oils, 148

Machines, chapter 15
Magnalium, 252, 257
Magnesium, 251
Magnesium compounds:
chloride, 265
nitride, 121
oxide, 118
Magnetic field, due to current, 278; due to earth, 275; effect on electron beams, 306–7; effect on beta particles, 311
induction, 273
lines of force, 272
materials, 274–5
poles, 271
Magnetism, terrestrial, 275; theories of, 274
Magnets, chapter 44; demagnetizing, 272; making, 272
Magnification, of curved mirrors, 167; of lenses, 177
Magnifying glass, 179
Manganese dioxide (see manganese(IV) oxide)
Manganese(IV) oxide, as a catalyst, 117, 301; in preparation of chlorine, 23, in preparation of oxygen, 117
Manometer, 39
Marble, 261, 262
Mass, 5; conservation of, 123;

converted into energy, 123; standard of, 6 number, 29
Massicot, 261
Maxwell's corkscrew rule, 278
Mayow, John, 115
Mechanical advantage, 79
Megajoule, 147
Mendeleev, Dmitri, 245, 246
Mercury, 259
  barometer, 43
Melting point, 109; effect of impurities, 114; effect of pressure, 113
Metallic bonding, 297
Metals, 246, 297; extraction, chapter 42, 231; properties, 248, 250; refining, 231; specific heat capacity, 107
Meteorology, 43
Methane, 147, 148, 149, 326–7
Methanoic acid, 72, 147, 148
Methanol, 327, 330
Methylated spirit, 148, 330
Meyer, Lothar, 245
Micrometer screw gauge, 5
Microphone, 308
Mineral acids, 21
Mirrors, curved, 165–9; plane, 163–5
Molar mass, 155, 239
  solution, 238
  volume, 155
Molarity, 238–41
Mole, 45, 238, 241–2, 247
Molecular formula, 47, 154
  mass, chapter 9; relation to vapour density, 155
  motion, 295–6
  properties of matter, chapter 7
  theory of magnetism, 274
Molecules, 30, 34–7
Moments, principle of, 61
Mortar, 259
Motor car batteries, 237
  car cooling system, 98
  car engine, 104
  effect, 279
Moving coil galvanometer, 213, 279–80
  coil loudspeaker, 281
  iron ammeter, 277
Multiple Proportions, Law of, 125, 325
Musical instruments, 196, 198
Myopia, 181

Naphtha, 149
Naphthalene, cooling curve, 109
Nascent hydrogen, 120
  oxygen, 120
Natural gas (see methane)
Nelson cell, 259, 260
Neon, 115
Neptunium, 311
Nessler's reagent, 121, 129
Neutral points (magnetism), 272–3
Neutralization, 221, 239, 243, 298
Neutron, chapters 1 and 6, 45, 244, 312–4
Newton, Isaac, 18, 183
Newton, 38
Newton's laws of motion, 18
Nitrates, 267; decomposition, 138, identification, 138, 322; preparation, 118, solubility, 203
Nitric acid, chapter 25, 132, 145; decomposition, 137; manufacture, 135; preparation, 136; properties, 131, 316; uses, 138
  oxide (see nitrogen oxide)
Nitrides, 121
Nitrocellulose, 138
Nitrogen, 120–1; atmospheric, 115–7; manufacture, 117; preparation, 121, 129; 318; uses, 121
  cycle, 139–40
  mixture, 121
Nitrogen compounds:
  dioxide, 132, 135–8, 317, 319, 322
  oxide, 132, 135, 138–9, 267, 317
  tetroxide (see dinitrogen tetroxide)
Nitrous oxide (see dinitrogen oxide)
Noble gasses (see Argonons)
Normal salts, 71
Non-conductors, 99, 221
Non-metals, 248
Nuclear bomb, 313
  charge, 312
  energy, 152, 313, 314
  fission, 152, 313
  fusion, 314
  power stations, 285
  reactions, 99, 311–3
  reactors, 252, 313
Nucleon number, 313
Nucleus, chapters 1 and 49, 244

Oersted, Hans Christian, 277
Ohm, Georg, 210
Ohm, 210
Ohm's law, 209–11
Oil, in gas making, 149, 152; of vitriol, 267
Oildag, 142
Oils and esters, 331
Olefines (see alkenes)
Oleum, 70
Open-hearth process, 254
Oscilloscope, 308
Oxalic acid, 72, 146
Oxidation, 72, 119–20, 128, 132, 134, 137, 138, 144, 250, 254, 257, 267; as a loss of electrons, 120, 138 number, 31
Oxides, 258–61; classification, 118; preparation, 117
Oxygen, chapter 22, 133, 151; atmospheric, 115–6; identification, 133, 134, 319; manufacture, 117; preparation, 117, 318; properties, 117–8 mixture, 117

Oxyhaemoglobin, 147
Ozone, 297

Paint drying, 120
Paraboloidal mirror, 167
Paraffins (see Alkanes)
Parallax errors, 165, 172
Parallel forces, 58
  resistances, 212
Particle size, 298
Pascal's vases, 40
Passivity, 136, 232
Peat, 147
Pendulum, simple, 11, 88
Penicillin, 65
Percentage composition, 46
Periodic Table, chapter 41, 2
Periodicity, 247–8
Periscope, mirror, 165; prism, 174
Permanent hardness, 204–5, 266; magnets, 273
Permutit, 205
Perrin tube, 306
Petrol, 147
  engine, 105
Petroleum, 149, 327
pH, 243, 262, 263
Phosphorus, oxidation, 137
Photographic enlarger, 180
Photography, 179–180
Photosynthesis, 151–2
Phototransistor, 185
Pig iron, 253, 254
Pigments, 185
Pinhole camera, 160
Pitch, 148; (of screw) 6; (of sound) 193
Plane mirrors, 163–5
Plaster of Pairs, 266
Plastics, 26
Platinum, 69
Plimsoll line, 55
Points, discharging action of, 292
Polarization, 235
Polymerization, 328–9, 331
Polystyrene, 328
Polythene, 328
Polyvinyl chloride, 26, 328
Potassium, 319
Potassium compounds:
  carbonate, 143, 262
  chlorate, 117, 301
  chloride, 266, 318
  chromate, 120, 267
  dichromate, 120, 267
  ferricyanide, 256
  hydrogensulphate, 266, 318
  hydroxide, 116
  iodide, 120, 138
  nitrate, 138, 267
  permanganate, 117, 120
  solubility, 203
Potential difference, 208–10
  energy, 77
Pot-holing, 206
Power, 78; electrical, 216; units of 78 stations, 285
Precipitation, 50
Presbyopia, 181
Press, hydraulic, 40
Pressure, chapter 8; effect on volume of gas, 93–4; in liquids, 38–41;

measurement of, 39; units of, 38
Priestley, Joseph, 133
Primary cells, 236
Principle of moments, 61
Prismatic binoculars, 174
Prisms, 174, 176, 183–4
Producer gas, 147, 148, 149
Proportions, Law of definite, 124; Law of multiple, 125
Proteins, 139–40, 332–3
Proton, chapters 1 and 6, 45, 244, 311–3
Pulleys, 78–81; Weston Differential, 81
Purification of crystals, 202
P.V.C. 76, 328
Pyrites, 70
Pyrophoric lead, 299

Qualitative analysis, 46, 318–22
Quartz, 257
Quicklime (see calcium oxide)

Radiation, 100, 185; types of 310–12
Radioactive changes, 152, 312; decay series, 312
Radioactivity, chapter 49, 313
Radio waves, 186
Radium, 309
Railway lines, expansion of, 88
Rain, 203
Rare gases (see Argonons)
Raster, 308
Rayon, 259, 260, 263
Reaction rates, 298–301
Real images, 168
Recorder, 199
Red lead (see lead(II)(IV) oxide)
Reduction, 68, 119–20, 129, 133, 134, 144–5, 146, 250, 252, 253; as a gain of electrons, 120, 250; of gas volumes to standard conditions, 95
Refining of metals, 232
Reflection, chapter 29; laws of, 163; total internal, 174; of waves, 188
Refraction, chapter 30; laws of 171; of waves, 188
Refrigeration, 26, 111
Relative density, 9–10
  vapour density, 45
Relays, 277
Resistance, 210; measurement of, 211
Resistances, 212–3
Resistivity, 212
Respiration, 151
Resultant, 63
Retardation, 12
Retina, 181
Reversible reactions, 302
Ring main, 219
Ripple tank, 187
Rivets, 88
Rock salt, 21, 264, 265
Römer, 162
Rubber, artificial, 26
Rusting, 255–6; 261
Rutherford, Ernest, 31

351

Safety lamp, Davy, 100
Salts, preparation, 70, 322
Saponin, 264
Saturated compounds, 326, 331
 solutions, 201
Scalar quantities, 64
Scale, 204
Scheele, Carl Wilhelm, 25
Screw, 82
 Jack, 82
Sea water, 264
Secondary cells, 236
 coils, 286, 287
Series resistances, 212
Shadows, 159–60
Short sight, 181
Shunt, 213–4
Silica, 118; in slag, 253
 gel, 116, 128, 132
Silver nitrate, in chloride tests, 23, 321
 plating, 232
Simple cell, 235
 pendulum, 11–12, 88
Slaked lime (see calcium hydroxide)
Slide projector, 180
 viewer, 179
Slip rings, 283, 284
Small-scale preparations, chapter 50
Snell's law, 171
Soap, 203, 259, 263, 265, 331
Sodamide, 252
Sodium, 230, 251–2, 265
Sodium compounds:
 aluminate, 320
 carbonate, 262–3, 264, 265; in water softening, 205; uses 263
 chloride, 21, 167, 264–5; electrolysis of, 230, 259–60
 cyanide, 252
 ethanoate, 330, 331
 hydrogencarbonate, 71, 243, 263–4
 hydrogensulphate, 21, 71, 136, 266
 hydrogensulphite, 68
 hydroxide, 130, 252, 259–61, 265; electrolysis, 229, 259–60
 hypochlorite, 265
 lauryl sulphate, 206
 monoxide, 258
 nitrate, 133, 136, 138, 267
 nitrite, 121
 oxalate, 146
 permutit, 205
 peroxide, 252, 258
 stearate, 203, 331–2
 solubility of, 203
 sulphate, 71, 230, 266
 sulphite, 68
 thiosulphate, 26, 300
 zincate, 118, 261, 319
Soft water, 203–6
Softening water, methods of, 204–6

Solar energy, 152, 264
Solder, 257,
Solenoid, 276; magnetic field of, 276
Solubility, chapter 35; determination, 201; in non-acqueous solvents, 200; of common compounds, 203; retrograde, 201 curve, 201–2
Solute, 200
Solution, chapter 35; effect on melting and boiling point, 200; molar, 238; saturated, 201; standard, chapter 40
Solvay-process, 262
Solvent, 200; non-aqueous, 200
Sonometer, 197
Sound, chapter 34; musical, 196; needs a medium, 192; reflection of, 195; sources of, 192–3; velocity of, 194–5
 energy, 152
 waves, 192–3
Specific heat capacity, 103; measurement of, 106–7
Spectacles, 181–2
Spectrum, chapter 32, pure 184
Speed, 12; of reactions, 298–301
Speleology, 206
Split ring, 280, 284
Spontaneous combustion, 300
Spring, spiral, 6–7
 balance, 6–7
Stability of equilibrium, 59
Stainless steel, 254, 257
Stalactites, 206
Stalagmites, 206
Standard solutions, chapter 40 temperature and pressure, 95
Starch, 151, 329, 333
Static electricity, chapter, 47
Statics, chapter 12
Steam engines, 104
 turbine, 285
Steel, 253–4, 257
Storage heaters, 218
Stringed instruments, 196
Strontium–90, 311, 313
Structure, chapter 16, 295–7
Sublimation, 133, 265
Sucking-back, precautions against, 128
Sugar, 71, 145, 329, 333; oxidation, 120; production, 151
 charcoal, 71, 145
Sulphates, 70, 260, 266–7; detection, 73, 321; solubility, 203
Sulphides, 67
Sulphites, 67–8; detection, 321
Sulphur, chapter 13; allotropes, 65–6, 317;

extraction, 65; structure, 66.
Sulphur compounds:
 dioxide, chapter 14, 256–7, 260;
 identification, 68, 321;
 preparation, 68, 316;
 uses, 69
 trioxide, 69, 260
Sulphuric acid, chapter 14; decomposition, 72; manufacture, 69; preparation, 69; properties, 70–3; uses, 69
Sulphurous acid, 67, 260
Sun, radiation from, 185
Surface tension, 367
Suspension, 200

Tar, 148
Tartaric acid, 263
Teflon, 27
Telescope, reflecting, 168–9
Television tube, 308
Temperature, effect on reaction rate, 300–1; effect on volume of a gas, 91; Kelvin scale, 92; measurement of, 85; scales, 86, 92
Temporary hardness, 204–5
Terrestrial magnetism, 275
Terylene, 331
Tetrachloromethane, 203
Tetra-ethyl lead, 252, 257, 328
Tetrahydrothiophene, 149
Therm, 147
Thermal dissociation, 131, 132
Thermionic effect, 305
Thermometers, calibration of, 86; clinical, 87; maximum and minimum, 87
Thermopile, 101
Thermostats, 87
Thompson, Benjamin, 103
Thorium, 309
Time, chapter 3
 base, 308
Tin, 257
Toluene, 66
Torricelli, Evangelista, 43
Town's gas, 149
Transformer, 286–7
Transition metals, 246
 temperature, 66
Triangle of forces, 61–2
Tri-iron tetroxide (see iron (II)(III) oxide
Tri-lead tetroxide (see lead(III)(IV) oxide)
Turbine, steam, 104, 285
Type metal, 257

Ultra-violet, 100, 186
Unit of electricity, 216
Universal indicator and pH, 243
Unsaturated compounds, 327, 331
Upthrust, 54
Uranium, 285, 309, 313
Urea, 325

Vacuum flask, 101
Valency, chapter 6; variable, 32, 245
 electrons, 244–5
Van de Graaff generator, 293
Vanadium (V) oxide, 70, 302
Vapour density, 45, 155
Vectors, 64
Velocity, 12, of light, 161–2; of sounds, 194–5; of waves, 190; -ratio, 79; -time graphs, 13–14
Vernier, 4
 calipers, 5
Vinegar, 330
Virtual images, 168
Vision, 181; defects, of, 181
Volt, 208
Voltameter, 223
Voltmeter, 213–4, 308
Volumes of gases corrected to s.t.p., 95; in gaseous reactions, 153
Volumetric composition of stream, 156
Vulcanizing, 65

Washing powders, 205
 soda (see sodium carbonate)
Water, a covalent compound, 30; freezing of, 89; hard, 203–6; soft, 203–6; softening of, 204–6; unusual expansion of, 89–90
 of crystallization, 46
 gas, 147, 148, 149
 softener, 205,
 supply, 41
 vapour in air, 115
 waves, chapter 34
Watt, 78, 216
Waves, chapter 33
Wavelength, of light, 186; of sound, 195
Weber, Ernst, 274
Weight, 7; measurement of, 6–7
Weston differential pulley, 68
Wet and dry bulb thermometer, 112
Wind instruments, 199
Windlass, 81
Wood, 147, 148
 -tar, 143
Work, 75–6; units of, 73

Xenon, 115
X-rays, 186, 309
Xylene, 66

Zeolites, 205
Zinc, 256; action of sulphuric acid on, 72; uses, 256
Zinc compounds:
 hydroxide, 118, 261, 319
 ions, identification, 319, 321
 oxide, 118, 256, 260, 261, 320; uses, 260
 sulphate, 267
 sulphide, 67
Zones of combustion, 150